PROPERTIES OF MATERIALS

Properties of Materials

Anisotropy, Symmetry, Structure

ROBERT E. NEWNHAM

Pennsylvania State University

UNIVERSITY PRESS

Great Clarendon Street, Oxford, OX2 6DP,
United Kingdom

Oxford University Press is a department of the University of Oxford.
It furthers the University's objective of excellence in research, scholarship,
and education by publishing worldwide. Oxford is a registered trade mark of
Oxford University Press in the UK and in certain other countries

© Oxford University Press 2005

The moral rights of the authors have been asserted

First published 2005

All rights reserved. No part of this publication may be reproduced, stored in
a retrieval system, or transmitted, in any form or by any means, without the
prior permission in writing of Oxford University Press, or as expressly permitted
by law, by licence or under terms agreed with the appropriate reprographics
rights organization. Enquiries concerning reproduction outside the scope of the
above should be sent to the Rights Department, Oxford University Press, at the
address above

You must not circulate this work in any other form
and you must impose this same condition on any acquirer

Published in the United States of America by Oxford University Press
198 Madison Avenue, New York, NY 10016, United States of America

British Library Cataloguing in Publication Data

Data available

Library of Congress Cataloging in Publication Data

Data available

ISBN 978-0-19-852076-4

Preface

This book is about anisotropy and structure–property relationships. Tensors and matrices provide the mathematical framework, and symmetry is helpful in determining which coefficients are absent, and which must be equal, but they say nothing about the sizes of the property coefficients. Magnitudes depend more on atomistic arguments. I have tried to point out some of the crystallochemical parameters (such as bond lengths, coordination numbers, and electronic structure) that correlate with property coefficients. These relationships provide a qualitative understanding of the molecular mechanisms which underlie the choice of materials for various engineering applications.

The book contains 32 chapters and about 370 pages. It covers a wide range of topics and is suitable as an introduction to the physical properties of materials. The use of tensors and matrices provides a common theme which ties the topics together. I have taught the course at the advanced undergraduate level and to beginning graduate students. Instructors can select topics for a one semester course or use the entire book for a two-semester course.

The only prerequisites for the course are college-level physics and chemistry. Such courses are commonly offered during the first year or two at American universities. No special training in tensor or matrix algebra is required, but knowledge of basic crystallography is helpful.

In teaching the course and in writing the book, I had the following questions in mind.

How do physical properties depend on direction?

How are these properties described mathematically, and what do the geometric representations look like?

How do matrix representations differ from the tensor descriptions?

How do polar tensor properties differ from axial tensor properties? And what determines the tensor rank?

How are the property coefficients measured and how are they influenced by measurement conditions such as temperature, frequency, pressure, and external fields?

How does symmetry influence the physical properties?

Which coefficients are zero by symmetry, which are equal, and how many measurements are required to specify the property?

How is anisotropy related to crystal structure and texture?

Are there chains or layers in the structure that correlate with the directional properties?

How do the magnitudes of the tensor components depend on bond lengths, bond strengths, and chemical composition?

What are the most important engineering applications?

What combination of properties are involved in the "figure of merit"?

How might future improvements be made in the properties?

The goal of this process—a process some call "Molecular Engineering"—is the optimization of materials and devices through an understanding of structure–property relationships.

Many, many colleagues and friends helped me prepare this book: The faculty, staff, and students at the Pennsylvania State University and the Hong Kong Polytechnic University were especially helpful. The sustained support of the Office of Naval Research is also gratefully acknowledged.

I dedicate this book to my wife Patricia, the mother of our two wonderful children, and my lifelong companion in the House of Love and Good Suppers.

Contents

1 Introduction   1
    1.1 Outline   1
    1.2 Structure–property relationships   3
    1.3 Symmetry of physical properties   4
    1.4 Atomistic arguments: Density   5

2 Transformations   9
    2.1 Why transformations?   9
    2.2 Axis transformations   9
    2.3 Orthogonality conditions   10
    2.4 General rotation (Eulerian angles)   12

3 Symmetry   14
    3.1 Symmetry operations   14
    3.2 Symmetry elements and stereographic projections   15
    3.3 Point groups and their stereograms   17
    3.4 Crystallographic nomenclature   20
    3.5 Point group populations   20

4 Transformation operators for symmetry elements   23
    4.1 Transformation operators for the crystallographic symmetry elements   23
    4.2 Transformation operations for the thirty-two crystal classes   25
    4.3 Standard settings   26
    4.4 Curie group symmetries   26

5 Tensors and physical properties   30
    5.1 Physical properties   30
    5.2 Polar tensors and tensor properties   31
    5.3 Axial tensor properties   32
    5.4 Geometric representations   33
    5.5 Neumann's Principle   34
    5.6 Analytical form of Neumann's Principle   34

6 Thermodynamic relationships   37
    6.1 Linear systems   37
    6.2 Coupled interactions: Maxwell relations   38
    6.3 Measurement conditions   40

7 Specific heat and entropy — 43
- 7.1 Heat capacity of solids — 43
- 7.2 Lattice vibrations — 46
- 7.3 Entropy and the magnetocaloric effect — 48

8 Pyroelectricity — 50
- 8.1 Pyroelectric and electrocaloric tensors — 50
- 8.2 Symmetry limitations — 50
- 8.3 Polar axes — 52
- 8.4 Geometric representation — 53
- 8.5 Pyroelectric measurements — 54
- 8.6 Primary and secondary pyroelectric effects — 54
- 8.7 Pyroelectric materials — 55
- 8.8 Temperature dependence — 55
- 8.9 Applications — 57

9 Dielectric constant — 58
- 9.1 Origins of the dielectric constant — 58
- 9.2 Dielectric tensor — 60
- 9.3 Effect of symmetry — 62
- 9.4 Experimental methods — 63
- 9.5 Geometric representation — 67
- 9.6 Polycrystalline dielectrics — 69
- 9.7 Structure–property relationships — 69

10 Stress and strain — 72
- 10.1 Mechanical stress — 72
- 10.2 Stress transformations — 74
- 10.3 Strain tensor — 75
- 10.4 Matrix transformation for strain — 77

11 Thermal expansion — 79
- 11.1 Effect of symmetry — 79
- 11.2 Thermal expansion measurements — 81
- 11.3 Structure–property relations — 82
- 11.4 Temperature dependence — 85

12 Piezoelectricity — 87
- 12.1 Tensor and matrix formulations — 87
- 12.2 Matrix transformations and Neumann's Law — 89
- 12.3 Piezoelectric symmetry groups — 91
- 12.4 Experimental techniques — 93
- 12.5 Structure–property relations — 94
- 12.6 Hydrostatic piezoelectric effect — 97
- 12.7 Piezoelectric ceramics — 99
- 12.8 Practical piezoelectrics: Quartz crystals — 100

13 Elasticity — 103
- 13.1 Tensor and matrix coefficients — 103

13.2	Tensor and matrix transformations	105
13.3	Stiffness–compliance relations	106
13.4	Effect of symmetry	107
13.5	Engineering coefficients and measurement methods	109
13.6	Anisotropy and structure–property relations	110
13.7	Compressibility	113
13.8	Polycrystalline averages	114
13.9	Temperature coefficients	116
13.10	Quartz crystal resonators	118

14 Magnetic phenomena 122
14.1	Basic ideas and units	122
14.2	Magnetic structures and time reversal	124
14.3	Magnetic point groups	125
14.4	Magnetic axial vectors	130
14.5	Saturation magnetization and pyromagnetism	131
14.6	Magnetic susceptibility and permeability	134
14.7	Diamagnetic and paramagnetic crystals	135
14.8	Susceptibility measurements	137
14.9	Magnetoelectricity	138
14.10	Piezomagnetism	142
14.11	Summary	146

15 Nonlinear phenomena 147
15.1	Nonlinear dielectric properties	147
15.2	Nonlinear elastic properties	148
15.3	Electrostriction	151
15.4	Magnetostriction	153
15.5	Modeling magnetostriction	154
15.6	Magnetostrictive actuators	159
15.7	Electromagnetostriction and pseudopiezoelectricity	160

16 Ferroic crystals 162
16.1	Free energy formulation	162
16.2	Ferroelasticity	165
16.3	Ferromagnetism	168
16.4	Magnetic anisotropy	170
16.5	Ferroelectricity	174
16.6	Secondary ferroics: Ferrobielectricity and ferrobimagnetism	177
16.7	Secondary ferroics: Ferrobielasticity and ferroelastoelectricity	179
16.8	Secondary ferroics: Ferromagnetoelectrics and ferromagnetoelastics	182
16.9	Order parameters	183

17 Electrical resistivity 188
| 17.1 | Tensor and matrix relations | 188 |

	17.2	Resistivity measurements	189
	17.3	Electrode metals	191
	17.4	Anisotropic conductors	193
	17.5	Semiconductors and insulators	194
	17.6	Band gap and mobility	196
	17.7	Nonlinear behavior: Varistors and thermistors	199
	17.8	Quasicrystals	202
18	**Thermal conductivity**		**203**
	18.1	Tensor nature and experiments	203
	18.2	Structure–property relationships	206
	18.3	Temperature dependence	208
	18.4	Field dependence	210
19	**Diffusion and ionic conductivity**		**211**
	19.1	Definition and tensor formulation	211
	19.2	Structure–property relationships	212
	19.3	Ionic conductivity	217
	19.4	Superionic conductors	219
	19.5	Cross-coupled diffusion	220
20	**Galvanomagnetic and thermomagnetic phenomena**		**223**
	20.1	Galvanomagnetic effects	224
	20.2	Hall Effect and magnetoresistance	226
	20.3	Underlying physics	227
	20.4	Galvanomagnetic effects in magnetic materials	229
	20.5	Thermomagnetic effects	232
21	**Thermoelectricity**		**234**
	21.1	Seebeck Effect	234
	21.2	Peltier Effect	235
	21.3	Thomson Effect	235
	21.4	Kelvin Relations and absolute thermopower	236
	21.5	Practical thermoelectric materials	238
	21.6	Tensor relationships	239
	21.7	Magnetic field dependence	240
22	**Piezoresistance**		**243**
	22.1	Tensor description	243
	22.2	Matrix form	244
	22.3	Longitudinal and transverse gages	245
	22.4	Structure–property relations	247
23	**Acoustic waves I**		**249**
	23.1	The Christoffel Equation	249
	23.2	Acoustic waves in hexagonal crystals	252
	23.3	Matrix representation	255

	23.4	Isotropic solids and pure mode directions	256
	23.5	Phase velocity and group velocity	258

24	**Acoustic waves II**	261
	24.1 Acoustic impedance	261
	24.2 Ultrasonic attenuation	262
	24.3 Physical origins of attenuation	264
	24.4 Surface acoustic waves	265
	24.5 Elastic waves in piezoelectric media	266
	24.6 Nonlinear acoustics	270

25	**Crystal optics**	274
	25.1 Electromagnetic waves	274
	25.2 Optical indicatrix and refractive index measurements	276
	25.3 Wave normals and ray directions	278
	25.4 Structure–property relationships	280
	25.5 Birefringence and crystal structure	282

26	**Dispersion and absorption**	286
	26.1 Dispersion	286
	26.2 Absorption, color, and dichroism	288
	26.3 Reflectivity and luster	291
	26.4 Thermo-optic effect	292

27	**Photoelasticity and acousto-optics**	294
	27.1 Basic concepts	294
	27.2 Photoelasticity	295
	27.3 Static photoelastic measurements	296
	27.4 Acousto-optics	298
	27.5 Anisotropic media	300
	27.6 Material issues	300

28	**Electro-optic phenomena**	302
	28.1 Linear electro-optic effect	303
	28.2 Pockels Effect in KDP and ADP	304
	28.3 Linear electro-optic coefficients	308
	28.4 Quadratic electro-optic effect	309

29	**Nonlinear optics**	313
	29.1 Structure–property relations	313
	29.2 Tensor formulation and frequency conversion	315
	29.3 Second harmonic generation	316
	29.4 Phase matching	318
	29.5 Third harmonic generation	322

30	**Optical activity and enantiomorphism**	325
	30.1 Molecular origins	325

	30.2	Tensor description	327
	30.3	Effect of symmetry	329
	30.4	Relationship to enantiomorphism	331
	30.5	Liquids and liquid crystals	333
	30.6	Dispersion and circular dichroism	337
	30.7	Electrogyration, piezogyration, and thermogyration	340

31 Magneto-optics 342

	31.1	The Faraday Effect	342
	31.2	Tensor nature	343
	31.3	Faraday Effect in microwave magnetics	345
	31.4	Magneto-optic recording media	346
	31.5	Magnetic circular dichroism	348
	31.6	Nonlinear magneto-optic effects	350
	31.7	Magnetoelectric optical phenomena	351

32 Chemical anisotropy 354

	32.1	Crystal morphology	354
	32.2	Growth velocity	356
	32.3	Crystal growth and crystal structure	358
	32.4	Surface structures and surface transformations	360
	32.5	Etch figures and symmetry relations	361
	32.6	Micromachining of quartz and silicon	363
	32.7	Tensor description	366

Further Reading 369

Index 375

Introduction

The physical and chemical properties of crystals and textured materials often depend on direction. An understanding of anisotropy requires a mathematical description together with atomistic arguments to quantify the property coefficients in various directions.

Tensors and matrices are the mathematics of choice and the atomistic arguments are partly based on symmetry and partly on the basic physics and chemistry of materials. These are subjects of this book: tensors, matrices, symmetry, and structure–property relationships.

1.1	Outline	1
1.2	Structure–property relationships	3
1.3	Symmetry of physical properties	4
1.4	Atomistic arguments: Density	5

1.1 Outline

We begin with transformations and tensors and then apply the ideas to the various symmetry elements found in crystals and textured polycrystalline materials. This brings in the 32 crystal classes and the 7 Curie groups. After working out the tensor and matrix operations used to describe symmetry elements, we then apply Neumann's Law and the Curie Principle of Symmetry Superposition to various classes of physical properties.

The first group of properties is the standard topics of classical crystal physics: pyroelectricity, permittivity, piezoelectricity, elasticity, specific heat, and thermal expansion. These are the linear relationships between mechanical, electrical, and thermal variables as laid out in the Heckmann Diagram (Fig. 1.1). These standard properties are all polar tensors ranging in rank from zero to four.

Axial tensor properties appear when magnetic phenomena are introduced. Magnetic susceptibility, the relationship between magnetization and magnetic

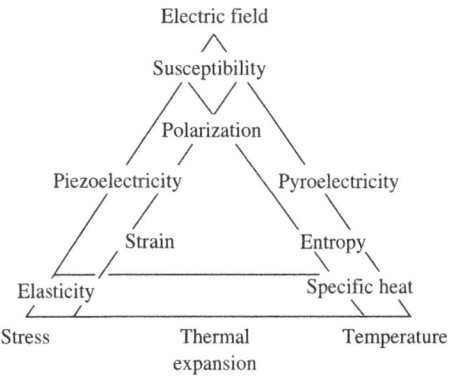

Fig. 1.1 The Heckmann Diagram relating mechanical, electrical, and thermal variables.

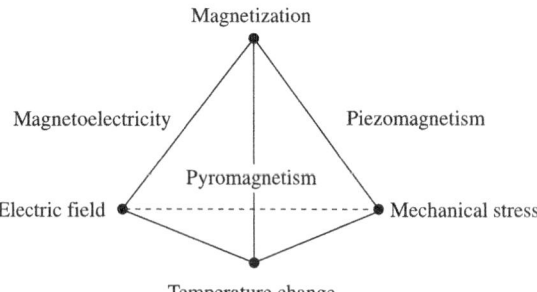

Fig. 1.2 Magnetic properties are coupled to thermal, electrical, and mechanical effects through axial tensor relationships.

Table 1.1 Polar and axial tensor properties of rank 0, 1, 2, 3, and 4. The rank of a tensor is nothing mysterious. It is simply the number of directions involved in the measurement of the property. A few examples are included

Rank	Polar	Axial
Zero	Specific heat	Rotatory power
First	Pyroelectricity	Pyromagnetism
Second	Thermal expansion	Magnetoelectricity
Third	Piezoelectricity	Piezomagnetism
Fourth	Elastic compliance	Piezogyrotropy

field, is a polar second rank tensor, but the linear relationships between magnetization and thermal, electrical, and mechanical variables are all axial tensors. As shown in Fig. 1.2, magnetization can be added to the Heckmann Diagram converting it into a tetrahedron of linear relationships. Pyromagnetism, magnetoelectricity, and piezomagnetism are the linear relationships between magnetization and temperature change, electric field, and mechanical stress.

Examples of tensors of rank zero through four are given in Table 1.1. In this book we will also treat many of the nonlinear relationships such as magnetostriction, electrostriction, and higher order elastic constants.

The third group of properties is transport properties that relate flow to a gradient. Three common types of transport properties relate to the movement of charge, heat, and matter. Electrical conductivity, thermal conductivity, and diffusion are all polar second rank tensor properties. In addition, there are cross-coupled phenomena such as thermoelectricity, thermal diffusion, and electrolysis in which two types of gradients are involved. All these properties are also influenced by magnetic fields and mechanical stress leading to additional cross-coupled effects such as piezoresistivity and the Hall Effect. The transport properties analogous to the Heckmann Diagram are shown in Fig. 1.3.

A fourth family of directional properties involves hysteresis and the movement of domain walls. These materials can be classified as primary and secondary ferroics according to the types of fields and forces required to move the walls. The primary ferroics include such well-known phenomena as ferromagnetism, ferroelectricity, and ferroelasticity (Fig. 1.4). Less well known are the secondary ferroic effects in materials like quartz which show both ferrobielasiticity and the ferroelastoelectric effect.

The final portions of the book concern wave phenomena. Crystal optics has long been an important part of crystal physics with roots in classical optical

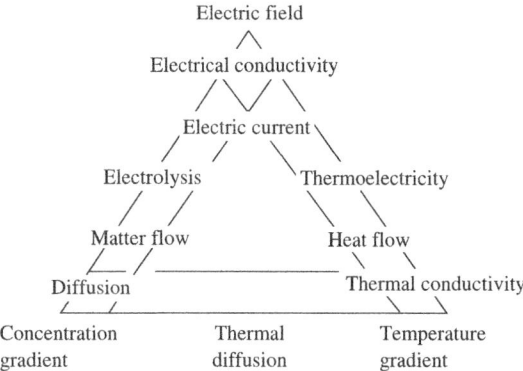

Fig. 1.3 Cross-coupled transport phenomena.

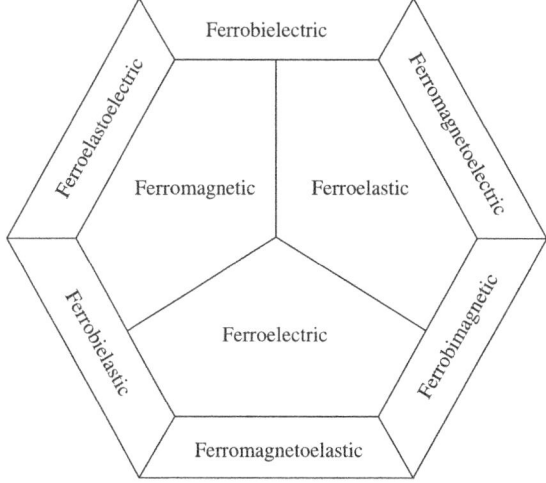

Fig. 1.4 Nine types of primary and secondary ferroics involving domain wall motion.

mineralogy. In modern times this has become an important component of the information age through the applications of nonlinear optics, magneto-optics and electro-optics. Linear and nonlinear ultrasonic phenomena are an important part of physical acoustics which are analogous to the optical effects.

Crystalline and noncrystalline media with handedness exhibit several types of gyrotropy in which the plane of polarization rotates as the wave passes through the medium. This leads to an interesting group of tensor properties known as optical activity, acoustical activity, and the Faraday Effect.

The last chapter of the book deals with chemical anisotropy, with emphasis on crystal morphology and etching phenomena.

1.2 Structure–property relationships

This book is about crystal physics and crystal chemistry, and their applications to engineering problems. When faced with the task of identifying useful materials, the materials scientist uses atomic radii, chemical bond strengths, anisotropic atomic groupings, electronic band structure, and symmetry arguments as criteria in the material selection process.

Crystal physics is mainly concerned with the relationships between symmetry and the directional properties of crystals. Symmetry and its relationships to physical properties is reviewed in the next section. In general, symmetry arguments are useful in determining which property coefficients are absent and which are equal, but not in estimating the relative sizes of the coefficients. Magnitudes depend more on the atomistic arguments based on crystal chemistry and solid-state physics. Using examples drawn from engineering technology, I have tried to point out the crystallochemical parameters most important to the understanding of a molecular mechanism, and to the choice of new materials.

For most scientists and engineers, an important goal is to develop fundamental understanding while at the same time remaining alert for possible applications. In solving solid-state problems it is helpful to ask, what atoms are involved and what are their electron configurations? What types of chemical bonds are formed? What is the symmetry of the crystal? How are the atoms arranged in the crystal structure? Are there chains or layers that give rise to anisotropy? Do these arrangements promote certain mechanisms for electronic or atomic motions or distortions? How do these mechanisms give rise to the observed properties? Which properties are important in engineering applications?

These are the questions I had in mind when I wrote the book.

1.3 Symmetry of physical properties

In determining the effect of symmetry on physical properties, there are four symmetries to be considered: (1) the symmetry of the material, (2) the symmetry of the external forces, (3) the symmetry of the resulting change or displacement, and (4) the symmetry of the physical property relating displacement to external force. Here we are using the terms *force* and *displacement* in the general sense to include electric, magnetic, and thermal quantities as well as mechanical effects.

All materials—whether crystalline or not—show some kind of symmetry. Single crystals have symmetry belonging to one of the 32 crystal classes (Chapter 3). Ferromagnetic, ferrimagnetic, and antiferromagnetic crystals exhibit long-range magnetic order. Additional symmetry groups involving time reversal operators are used to describe magnetic structures (Chapter 14).

Amorphous materials, glasses, and liquids have spherical symmetry, $\infty\infty m$, one of the seven Curie groups (Chapter 4). Liquid crystals and liquids made of the enantiomorphic molecules exhibit somewhat lower symmetry (Chapter 30). Ceramics, metals, rocks, and other polycrystalline solids made up of grains with random orientation have spherical symmetry. The properties of platy or fibrous materials with aligned crystallites conform to cylindrical symmetry, ∞/mm. A rectangular plank cut from a large tree has orthorhombic symmetry because of the different properties associated with the longitudinal, tangential, and radial directions. Thus wood has nine independent elastic constants just as orthorhombic crystals do.

The physical forces of importance in materials science are mechanical stress, electric field, magnetic field, and temperature. Tensile stresses possess cylindrical symmetry (∞/mm) while shear stresses have orthorhombic symmetry (mmm). Both are centric because a balance of forces is required to prevent translational or rotational motion. Electric fields can be represented by a vector with polar cylindrical symmetry, ∞m. Moving electric charges produce magnetic fields so that a current loop can be used as the symmetry representation. Magnetic fields have axial cylindrical symmetry, point group ∞/mm'. Temperature is a scalar quantity with spherical symmetry, $\infty\infty m$.

When a polycrystalline solid is hot-pressed, it adopts the cylindrical symmetry of the compressive stress. Poled ferroelectric ceramics are cooled through the Curie temperature in the presence of an electric field to influence the domain structure. The resulting symmetry is ∞m with the ∞-fold rotation axis parallel to the applied field. Magnetically-poled ceramics have symmetry ∞/mm'. When subjected to two forces, a ceramic retains the symmetry elements common to both. An electrically-poled, hot-pressed ceramic has symmetry ∞m when field and stress are parallel and $mm2$ when perpendicular. According to Curie's Principle, "when certain causes lead to certain effects, the symmetry elements of the causes should be observed in these effects."

The same principle applies to the symmetry of any change or displacement: a crystal under an external influence will exhibit only those symmetry elements that are common to the crystal without the influence and the influence without the crystal. As an illustration, when a tensile stress is applied along the [111] direction of a cubic crystal belonging to $m3m$, the symmetry of the strained crystal is $3m$, the highest group common to $m3m$ and cylindrical symmetry about [111]. This principle is true regardless of the size of the force or its effect on the material. The displacement may be permanent (plastic flow), semi-permanent (domain change), or reversible (elastic). In the latter case, the symmetry of the crystal reverts to its original class when the force is removed, but when the external forces produces permanent changes, the crystal retains the symmetry of the displacement after the force is removed.

Symmetry arguments like these are used to determine the number of independent property coefficients required for each material. Using Neumann's Principle (Chapter 5), it will be shown that quartz crystals (point group 32) are not pyroelectric (Chapter 8), but have two independent dielectric constants (Chapter 9), two independent piezoelectric constants (Chapter 12), and six independent elastic constants (Chapter 13). In contrast, polycrystalline quartz (point group $\infty\infty m$) is neither pyroelectric nor piezoelectric, and has only one dielectric constant and only two independent elastic constants.

1.4 Atomistic arguments: Density

The relationship between density and crystal structure illustrates how the crystal chemistry can sometimes be used to predict the magnitude of a physical property.

The density of a crystal is intimately related to chemical composition and crystal structure through the relation

$$\rho = \frac{MZ}{N_0 V}.$$

The density ρ (g/cm^3) is determined by the molecular weight M, the number of molecules per unit cell Z, Avogadro's number N_0, and the unit cell volume V, measured in cm^3. For a triclinic crystal, V is $abc(1 - \cos^2\alpha - \cos^2\beta - \cos^2\gamma)^{1/2}$, where a, b, c are the cell dimensions and α, β, γ are the interaxial

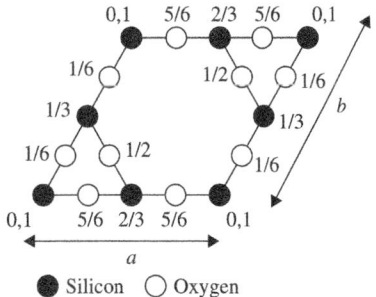

Fig. 1.5 The unit cell of β-quartz viewed along c. The heights of the oxygen and silicon atoms are expressed in cell fractions.

angles. Unit cell volumes for higher symmetry crystals are easily deduced from the triclinic formula.

A sample density calculation from quartz is carried out for β-quartz (Fig. 1.5). The drawing outlines the structure as viewed along the c-axis. β-quartz (SiO_2) belongs to the hexagonal crystal system and has a molecular weight of 60.09 g/mole. Avogadro's number N_0 is 6.023×10^{23} molecules/mole. The unit cell dimensions are $a = b = 4.913$ Å, $c = 5.405$ Å. Interaxial angles for the hexagonal system are $\alpha = \beta = 90°$, $\gamma = 120°$. From this information, the unit cell volume is

$$V = a^2 c(\sqrt{3}/2) = 113 \text{ Å}^3 = 1.13 \times 10^{-22} \text{ cm}^3.$$

Examining the unit cell drawing, it can be seen that there are eight silicons at the corners of the unit cell, and four more on the side faces. Since each corner belongs to eight cells and each face to two cells, the number of silicons in the cell is $8/8 + 4/2 = 3$. The number of oxygens can be counted in a similar way. There are eight O on side faces and two more inside the hexagonal cell. This adds up to $8/2 + 2 = 6$ oxygens. Thus there are three formula units of SiO_2 per cell: $Z = 3$. Collecting all numbers, the density

$$\rho = MZ/N_0 V = 2.65 \text{ g/cm}^3.$$

The most important factor affecting density is molecular weight because atomic weights vary over a wider range than do atomic volumes. Among difluorides, densities increase with cation atomic weight: BeF_2 1.99, MgF_2 3.14, CaF_2 3.18, SrF_2 4.24, BaF_2 4.89, PbF_2 8.24. The densities of MgF_2 and CaF_2 are nearly the same because the fluorite structure is more open than the rutile arrangement.

A similar trend is observed in alkali halides when density is plotted as a function of molecular weight (Fig. 1.6). If all the atoms were similar in size, density would be linearly proportional to molecular weight. Obviously this is not the case. Going from LiF to CsI there is an overall increase in density but the relationship is far from linear. In fact, the fluorides are denser than the chlorides because of the large increase in ionic radius going from F^- to Cl^-. This gives a dip in the density vs. molecular weight curve that is also observed in some oxides, semiconductors, and metals.

As a rule, crystals with close-packed atoms have larger densities than those that do not. Thus the densities of corundum, spinel, chrysoberyl, and olivine

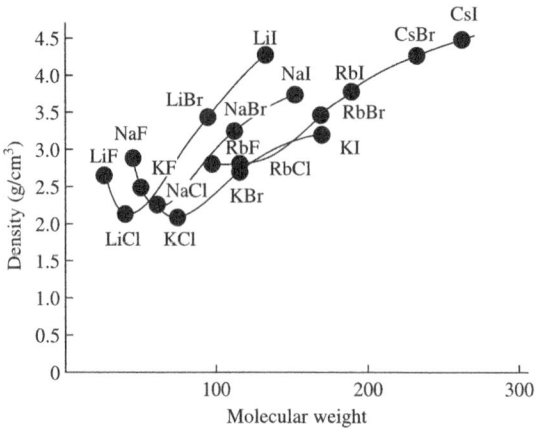

Fig. 1.6 The density of alkali halides with the rocksalt structure plotted as a function of molecular weight. Density usually increases with weight but size changes cause anomalies.

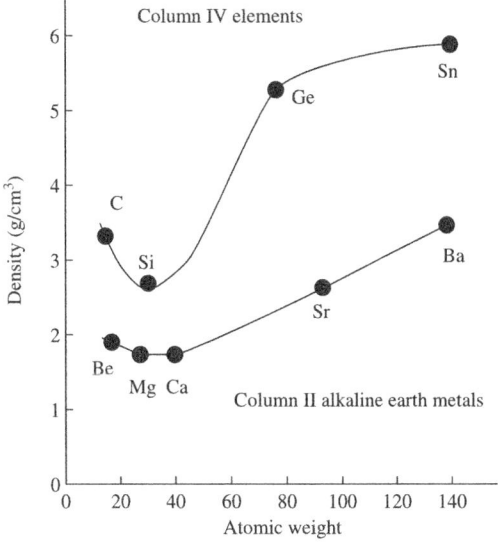

Fig. 1.7 Density vs. atomic weight curves for column IV and column II elements. Note that the densities of column IV elements with four bonding electrons per atom are much higher than the densities of the alkaline earth metals.

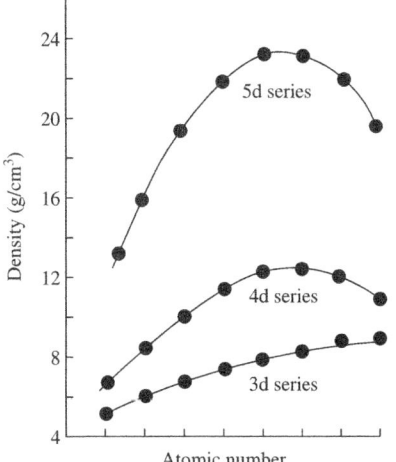

Fig. 1.8 Transition metal elements are very dense and have high melting points.

lie in the range 3.3–4.0, considerably larger than those of quartz, beryl, and feldspar that are near 2.7 g/cm^3. The differences in densities of the minerals are useful in separating them by flotation methods.

Among polymorphs, high-pressure phases are denser than low-pressure phases because of the importance of the PV term in the free energy when P is large. Kyanite, the stable high-pressure form of Al$_2$SiO$_5$, has a density of 3.63, compared to 3.15 for andalusite. For the same reason, low-temperature polymorphs are often denser than high-temperature forms. When T is very small, the PV term dominates the TS term in the free energy function. As an example, quartz ($\rho = 2.65$) is denser than its high-temperature polymorph cristobalite ($\rho = 2.32$).

The density vs. molecular weight curves for semiconductors and metals resemble those of the alkali halides. Silicon has a lower density than diamond, and calcium is less dense than beryllium and magnesium (Fig. 1.7). Among metals the highest densities are found in the 5d transition metal series (Fig. 1.8).

Note the peaks near the middle of the 4d and 5d series where the d electrons participate strongly in interatomic bonding. The d electrons in 3d transition metals are buried further inside the atoms and participate less in bonding, but give rise to impressive magnetic behavior.

Density is a scalar quantity that does not depend on direction. In the next chapter, we begin the discussion of anisotropic properties.

Problem 1.1

Sanidine feldspar, $KAlSi_3O_8$, is monoclinic with lattice parameters $a = 8.56$, $b = 13.03$, $c = 7.17$ Å, $\beta = 116°$. There are four molecular units in the unit cell. Calculate the density of this common rock-forming mineral.

Problem 1.2

a. Silver and palladium both crystallize with a face-centered cubic crystal structure. Make a sketch of the structure and count the number of atoms per unit cell. The cubic cell edges are 3.890 for palladium and 4.086 Å for silver.
b. Compute the densities of these two metals.
c. Ag–Pd alloys are often used as electrodes. As pointed out in Chapter 17, they form a complete solid solution with Pd atoms substituting randomly for Ag. The measured density of a certain Ag–Pd alloy is 11.00 g/cm^3. Estimate the composition of this $Ag_{1-x}Pd_x$ alloy in atomic% Pd and weight% Pd. There are several ways of doing this. One method presumes Vegard's Law, and another is based on Retger's Rule. *Vegard's Law* states that the lattice parameter of a solid solution changes linearly with composition, and *Retger's Rule* states that the unit cell volume changes linearly with composition. How much difference does it make for the estimated composition of the Ag–Pd alloy?

Transformations

2

Many physical properties depend on direction and the resulting anisotropy is best described with the use of tensors. Tensors are classified according to how they transform from one coordinate system to another. Therefore, we begin by describing transformations.

2.1	Why transformations?	9
2.2	Axis transformations	9
2.3	Orthogonality conditions	10
2.4	General rotation (Eulerian angles)	12

2.1 Why transformations?

There are several reasons why we want to do this: (1) transformations help us define tensors, (2) these tensors can be used to describe physical properties, (3) the effects of symmetry on physical properties can be determined by how the tensor transforms under a symmetry operation, (4) the magnitude of a property in any arbitrary direction can be evaluated by transforming the tensor, (5) using these numbers, we can draw a geometric representation of the property, and (6) the transformation procedure provides a way of averaging the properties over direction. This is useful when relating the properties of polycrystalline materials to those of the single crystal.

Mathematically, there is nothing fancy about these transformations. We are simply converting one set of orthogonal axes (Z_1, Z_2, Z_3) into another (Z'_1, Z'_2, Z'_3). The two sets of axes are related to one another by nine direction cosines: a_{11}, a_{12}, a_{13}, a_{21}, a_{22}, a_{23}, a_{31}, a_{32}, and a_{33}. Collectively all nine can be written as a_{ij} where $i, j = 1, 2, 3$.

The axes and direction cosines are illustrated in Fig. 2.1. It is important not to confuse the subscripts of the direction cosines. As defined in the drawing, a_{12} is the cosine of the angle between Z'_1 and Z_2, whereas a_{21} is the cosine of the angle between Z'_2 and Z_1. The first subscript always refers to the "new" or transformed axis. The second subscript is the "old" or original axis. The original or starting axes is usually a right-handed set, but it need not be. The transformed "new" axes may be either right- or left-handed, depending on the nature of the transformation. This will become clearer when we look at some transformations representing various symmetry operations. In any case, both the old and new axes are orthogonal.

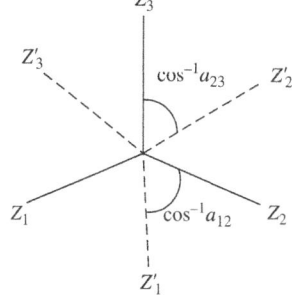

Fig. 2.1 Direction cosines relate "old" (unprimed) and "new" (primed) axial systems.

2.2 Axis transformations

We now proceed to write out these transformations in mathematical form. The "new" coordinate system Z'_1, Z'_2, Z'_3 is related to the "old" axes by the equations

$$Z'_1 = a_{11}Z_1 + a_{12}Z_2 + a_{13}Z_3$$
$$Z'_2 = a_{21}Z_1 + a_{22}Z_2 + a_{23}Z_3 \qquad (2\text{-}1)$$
$$Z'_3 = a_{31}Z_1 + a_{32}Z_2 + a_{33}Z_3.$$

These equations can also be written in matrix form

$$\begin{pmatrix} Z'_1 \\ Z'_2 \\ Z'_3 \end{pmatrix} = \begin{pmatrix} a_{11} & a_{12} & a_{13} \\ a_{21} & a_{22} & a_{23} \\ a_{31} & a_{32} & a_{33} \end{pmatrix} \begin{pmatrix} Z_1 \\ Z_2 \\ Z_3 \end{pmatrix}$$

or in shortened form as

$$\begin{array}{ccc} 3 \times 1 & 3 \times 3 & 3 \times 1 \\ (Z') & = (a) & (Z). \end{array}$$

In tensor form the same equations are

$$Z'_i = \sum_j a_{ij} Z_j = a_{ij} Z_j,$$

where $i, j = 1, 2, 3$. Here we adopt the Einstein convention in which repeated subscripts imply summation.

In dealing with physical property tensors, sometimes we transform from "old" to "new", as we have just done, and sometimes we transform in the reverse direction from "new" to "old":

$$Z_1 = a_{11}Z'_1 + a_{21}Z'_2 + a_{31}Z'_3$$
$$Z_2 = a_{12}Z'_1 + a_{22}Z'_2 + a_{32}Z'_3$$
$$Z_3 = a_{13}Z'_1 + a_{23}Z'_2 + a_{33}Z'_3.$$

Note that rows and columns have interchanged with respect to Equation 2-1. In matrix form,

$$\begin{pmatrix} Z_1 \\ Z_2 \\ Z_3 \end{pmatrix} = \begin{pmatrix} a_{11} & a_{21} & a_{31} \\ a_{12} & a_{22} & a_{32} \\ a_{13} & a_{23} & a_{33} \end{pmatrix} \begin{pmatrix} Z'_1 \\ Z'_2 \\ Z'_3 \end{pmatrix}$$

$$\begin{array}{ccc} 3 \times 1 & 3 \times 3 & 3 \times 1 \\ (Z) & = (a)_t & (Z'). \end{array}$$

Matrix $(a)_t$ is the transpose of (a) with rows and columns interchanged. In tensor form the new to old transformation is

$$Z_i = \sum_j a_{ji} Z'_j = a_{ji} Z'_j,$$

where $i, j = 1, 2, 3$.

2.3 Orthogonality conditions

In writing out these equations for a complicated transformation there is always the possibility of numerical error. Therefore, it is helpful to check the orthogonality conditions that arise because of transforming between two orthogonal systems. In an orthogonal system all the angles are right angles. This places restrictions on the direction cosine matrix.

$$(a) = \begin{pmatrix} a_{11} & a_{12} & a_{13} \\ a_{21} & a_{22} & a_{23} \\ a_{31} & a_{32} & a_{33} \end{pmatrix}.$$

The restrictions are such that for any row or column, the sum of the squares is one. Therefore there will be nine equations of the following form:

$$a_{11}^2 + a_{12}^2 + a_{13}^2 = 1$$

and

$$a_{11}^2 + a_{21}^2 + a_{31}^2 = 1.$$

For adjacent rows or adjacent columns, the products sum to zero:

$$a_{11}a_{12} + a_{21}a_{22} + a_{31}a_{32} = 0$$

$$a_{11}a_{21} + a_{12}a_{22} + a_{13}a_{23} = 0.$$

Taken all together, the orthogonality conditions can be represented by the expressions

$$a_{ik}a_{jk} = a_{ki}a_{kj} = \delta_{ij},$$

where $i, j, k = 1, 2, 3$, and δ_{ij} is the Kronecker δ. $\delta_{ij} = 1$ if $i = j$ and $\delta_{ij} = 0$ if $i \neq j$.

Because of the orthogonality conditions not all the a_{ij} values are independent, and this provides a valuable check on the numerical values.

As an example, the direction cosine matrix for a clockwise rotation of $\phi°$ about axis Z_1 is

$$(a) = \begin{pmatrix} 1 & 0 & 0 \\ 0 & \cos\phi & -\sin\phi \\ 0 & \sin\phi & \cos\phi \end{pmatrix}.$$

When dealing with a transformation such as this, it is helpful to keep track of the determinant $|a|$. Its value will be ± 1. No other values are permitted. If the transformation does not involve a handedness change, $|a| = +1$. For the rotation of $\phi°$ about Z_1,

$$|a| = \cos^2\phi + \sin^2\phi = 1.$$

Therefore the original right-handed axes remain right-handed in the new axial system. For mirror planes and other symmetry operations involving a change in handedness, $|a| = -1$. As explained later, certain physical properties such as optical activity are associated with handedness.

Problem 2.1

    a. Work out the direction cosines for a counterclockwise rotation of 30° about Z_3. Check the orthogonality conditions.
    b. How do the values change if the 30° rotation is clockwise rather than counterclockwise?
    c. Suppose the rotation was carried out about Z_3' rather than Z_3. How do the (a) matrices compare?

2.4 General rotation (Eulerian angles)

Three angles (ϕ, θ, ψ) are required in specifying the mutual orientation of two sets of orthogonal axes. Two angles (θ, ϕ) are needed to specify the orientation of a single direction in space. A third angle ψ fixes the other two axes of an orthogonal set of axes. This can be pictured as three consecutive rotation operations. The standard way of deriving the direction cosine matrix for a general rotation is as follows.

I. *First rotation*: a counterclockwise rotation of $\phi°$ about Z_3.

$$(a)_\text{I} = \begin{pmatrix} \cos\phi & \sin\phi & 0 \\ -\sin\phi & \cos\phi & 0 \\ 0 & 0 & 1 \end{pmatrix}.$$

II. *Second rotation*: a counterclockwise rotation of $\theta°$ about Z_1'

$$(a)_\text{II} = \begin{pmatrix} 1 & 0 & 0 \\ 0 & \cos\theta & \sin\theta \\ 0 & -\sin\theta & \cos\theta \end{pmatrix}.$$

III. *Third rotation*: a counterclockwise rotation of $\psi°$ about Z_3''

$$(a)_\text{III} = \begin{pmatrix} \cos\psi & \sin\psi & 0 \\ -\sin\psi & \cos\psi & 0 \\ 0 & 0 & 1 \end{pmatrix}.$$

The three rotations are illustrated in Fig. 2.2.

The general rotation is the product of the three individual rotations:

$$(a) = (a)_\text{III}(a)_\text{II}(a)_\text{I}$$

$$(a) = \begin{pmatrix} \cos\psi & \sin\psi & 0 \\ -\sin\psi & \cos\psi & 0 \\ 0 & 0 & 1 \end{pmatrix} \begin{pmatrix} 1 & 0 & 0 \\ 0 & \cos\theta & \sin\theta \\ 0 & -\sin\theta & \cos\theta \end{pmatrix} \begin{pmatrix} \cos\phi & \sin\phi & 0 \\ -\sin\phi & \cos\phi & 0 \\ 0 & 0 & 1 \end{pmatrix}$$

$$(a) = \begin{pmatrix} \begin{pmatrix} \cos\phi\cos\psi \\ -\cos\theta\sin\phi\sin\psi \end{pmatrix} & \begin{pmatrix} \cos\psi\sin\phi \\ +\cos\theta\cos\phi\sin\psi \end{pmatrix} & \sin\theta\sin\psi \\ \begin{pmatrix} -\cos\theta\cos\psi\sin\phi \\ -\cos\phi\sin\psi \end{pmatrix} & \begin{pmatrix} \cos\theta\cos\phi\cos\psi \\ -\sin\phi\sin\psi \end{pmatrix} & \cos\psi\sin\theta \\ \sin\theta\sin\phi & -\cos\phi\sin\theta & \cos\theta \end{pmatrix}.$$

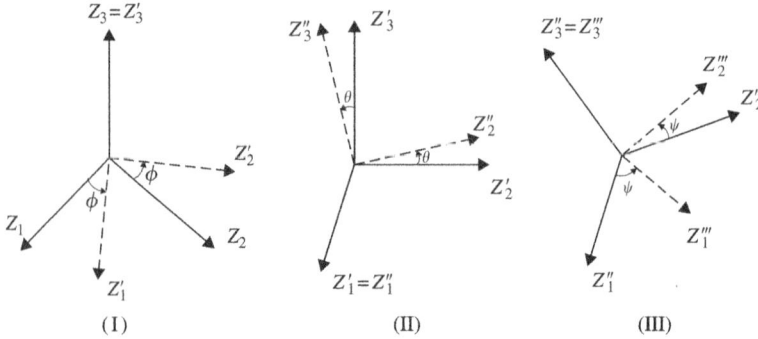

Fig. 2.2 Three angular rotations required to specify a general rotation of orthogonal axes.

Problem 2.2
Like all transformations, the general Eulerian rotation must satisfy the orthogonality conditions. Check this result and determine the value of the determinant $|a|$.

The general rotation is not used very often in common crystal physics, but in advanced product development—as in the quartz crystal industry—doubly- and triply-rotated cuts are used to eliminate unwanted temperature and stress effects. In these cases three orientation angles are used to specify the optimum orientation. Usually, however, we will be dealing with only one rotation.

Symmetry will be discussed in the next chapter, and transformation matrices for various symmetry operations will be given in Chapter 4. This will lead to Neumann's Law and the effect of symmetry on anisotropic physical properties. Symmetry greatly simplifies many of the property matrices and reduces the number of measurements required to describe anisotropy.

Problem 2.3
For cubic crystals, the $[100] = Z_1$, $[010] = Z_2$, and $[001] = Z_3$ directions constitute a set of orthogonal right-handed axes. The $[11\bar{2}] = Z'_1$, $[\bar{1}10] = Z'_2$, and $[111] = Z'_3$ directions are also perpendicular to one another.

   a. Write out the set of direction cosines relating the new axes to the old.
   b. Describe this transformation by a set of Eulerian angles ϕ, θ, and ψ.

3 Symmetry

3.1 Symmetry operations 14
3.2 Symmetry elements and stereographic projections 15
3.3 Point groups and their stereograms 17
3.4 Crystallographic nomenclature 20
3.5 Point group populations 20

All single crystals possess translational symmetry, and most possess other symmetry elements as well. In this chapter we describe the 32 crystallographic point groups used for single crystals. The seven Curie groups used for textured polycrystalline materials are enumerated in the next chapter.

3.1 Symmetry operations

We live in a three-dimensional world which means that there are basically four kinds of geometric symmetry operations relating one part of this world to another. The four primary types of symmetry are translation, rotation, reflection, and inversion. As pictured in Fig. 3.1, these symmetry operators operate on a point with coordinates Z_1, Z_2, Z_3 and carry it to a new position. By definition, all crystals have a unit cell that is repeated many times in space, a point Z_1, Z_2, Z_3 is repeated over and over again as one unit cell is translated to the next.

A mirror plane perpendicular to one of the principal axes is a two-dimensional symmetry element that reverses the sign of one coordinate. Rotation axes are

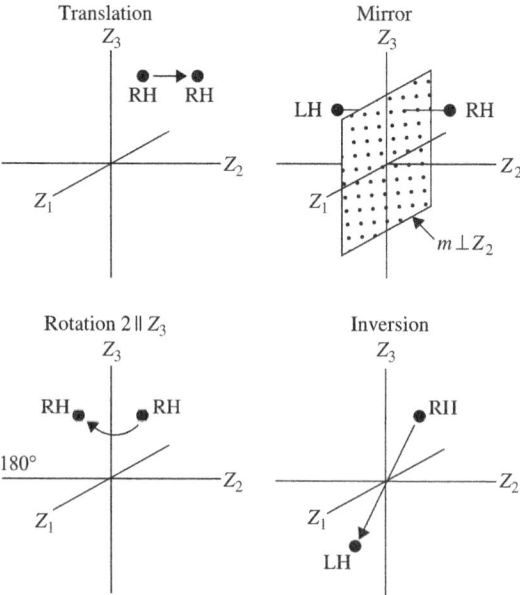

Fig. 3.1 Four types of symmetry elements. Mirror and inversion transformation are accompanied by a handedness change; rotations and translations are not.

one-dimensional symmetry elements that change two coordinates, while an inversion center is a zero-dimensional point that changes all three coordinates.

In developing an understanding of the macroscopic properties of crystals, we recognize that the scale of physical property measurements is much larger than the unit cell dimensions. It is for this reason that we are not concerned about translational symmetry and work with the 32 point group symmetries rather than the 230 space groups. This greatly simplifies the structure–property relationships in crystal physics.

3.2 Symmetry elements and stereographic projections

Aside from the identity operator 1, there are only four types of rotational symmetry consistent with the translation symmetry common to all crystals. Fig. 3.2 shows why. Parallelograms, equilateral triangles, squares, and hexagons will pack together to fill space but, pentagons* and other polygons will not. This means that only 2-, 3-, 4-, and 6-fold symmetry axes are found in crystals. This is the starting point for generating the 32 crystal classes. When taken in combination with mirror planes and inversion centers, these four types of rotation axes are capable of forming 32 self-consistent three-dimensional symmetry patterns around a point. These are the so-called 32 crystal classes or crystallographic point groups.

To visualize these crystal classes, we make use of stereographic projections (Fig. 3.3). Stereographic projections are an easy way of representing three-dimensional objects in two dimensions. The projection begins with an object located at the center of a sphere. Any point of interest is then projected to the surface of the sphere by drawing a straight line from the center of the sphere through the point in question, and on to the surface of the sphere. If the point ends in the northern hemisphere, then connecting it to the South Pole completes the operation. Where this line intersects the equatorial plane is the point on the stereographic projection. If the point on the sphere is in the southern hemisphere, then it is connected to the North Pole, again noting the intersection with the equatorial plane. Closed and open circles identify points originating

* Some quasicrystalline alloys have fivefold symmetry but do not possess the translational periodicity of single crystals (see Section 17.8).

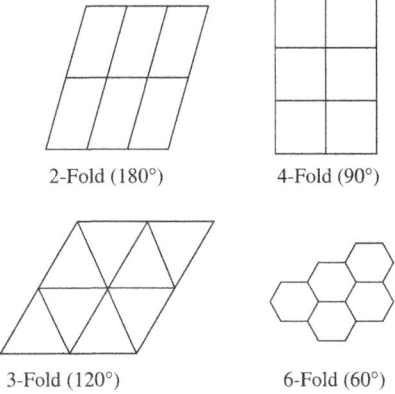

Fig. 3.2 Planar figures with 2-, 3-, 4-, or 6-fold symmetry generate geometric patterns that fill space.

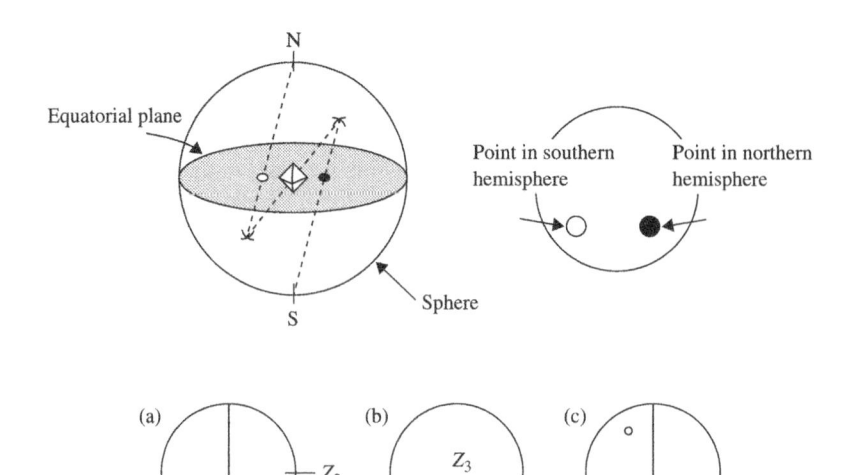

Fig. 3.3 The stereographic projection used to visualize three-dimensional objects in two dimensions.

Fig. 3.4 Stereograms for (a) a mirror (*m*) perpendicular to Z_2, (b) a mirror (*m*) perpendicular to Z_3, and (c) an inversion center ($\bar{1}$), sometimes called a center of symmetry.

from the northern and southern hemispheres, respectively. In general, then, the stereographic projection is simply the equatorial plane with its collection of open and closed circles.

To illustrate the use of the stereographic projection, we draw stereograms for several of the symmetry elements that occur in crystals. The usual orientation for the stereographic projection is to place Z_3 along the N–S axis and Z_1 and Z_2 in the equatorial plane. Consider first a mirror plane perpendicular to Z_2, as in the monoclinic crystal system. The mirror plane shown in Fig. 3.4(a) is drawn as a straight line. The solid circle to the right of the mirror plane is an arbitrarily chosen point in the northern hemisphere. To generate the symmetry-related point to the left of the mirror plane, a perpendicular is drawn from the reference point to the mirror plane, and then extended an equal distance on the other side. The second point is also in the northern hemisphere, as shown in the diagram.

The stereogram looks rather different for a mirror plane perpendicular to Z_3 (Fig. 3.4(b)). In this case the mirror plane is the equatorial plane, and is denoted by a heavy line drawn around the equator. Dropping a perpendicular from a point in the northern hemisphere to the equatorial plane, and bringing it out on equal distance on the other side, generates a symmetry-related point in the southern hemisphere. In the stereogram this appears as an open circle directly beneath the solid circle.

As a third example, consider a center of symmetry or inversion center. The stereogram shown in Fig. 3.4(c) has a center of symmetry at the center. Beginning with a reference point in the northern hemisphere, a line is drawn to the center and brought out an equal distance on the other side. This gives an equivalent point in the southern hemisphere.

Problem 3.1
Draw the stereograms for a twofold rotation axis along Z_1 and then along Z_3. Compare these stereograms with those of the mirror planes and the inversion center.

3.3 Point groups and their stereograms

Point group symmetry controls anisotropy and therefore it is an important concept. A point group is a self-consistent set of symmetry elements operating around a point. By "self-consistent" we mean that the symmetry elements return the starting point to its original position in a finite number of steps. Since handedness also changes for certain symmetry operations such as mirror planes or inversion centers, it is also important the reference point return to its original position with the same handedness.

Some point groups have only one symmetry element and are rather easy to visualize. Ten of the 32 crystal classes fall into this category: $1, \bar{1}, 2, m, 3, \bar{3}, 4, \bar{4}, 6$, and $\bar{6}$. Stereograms are grouped into six crystal systems in Figs. 3.5–3.11.

In the triclinic system there are only two point groups, 1 and $\bar{1}$. Other than translational symmetry, there is only the inversion symmetry of point group $\bar{1}$.

Twofold symmetry is characteristic of the monoclinic system that has three point groups 2, m, and $2/m$. By convention, Z_2 is chosen as the twofold symmetry axis, the b crystallographic axis. The mirror plane m is equivalent to a twofold inversion axis ($\bar{2}$) in which 180° rotation is accompanied by inversion. Point group $2/m$ has two independent symmetry elements, a twofold axis and a mirror plane perpendicular to it.

Orthorhombic crystal classes (Fig. 3.7) possess three orthogonal twofold axes or twofold inversion axes ($\bar{2} = m$). Again there are three point groups in this system: 222, $mm2$, and mmm. Comparing Fig. 3.7, Fig. 3.6, and Fig. 3.5 it is immediately apparent which point groups contain inversion symmetry, and which do not. The two equivalent points for point group $\bar{1}$ are present

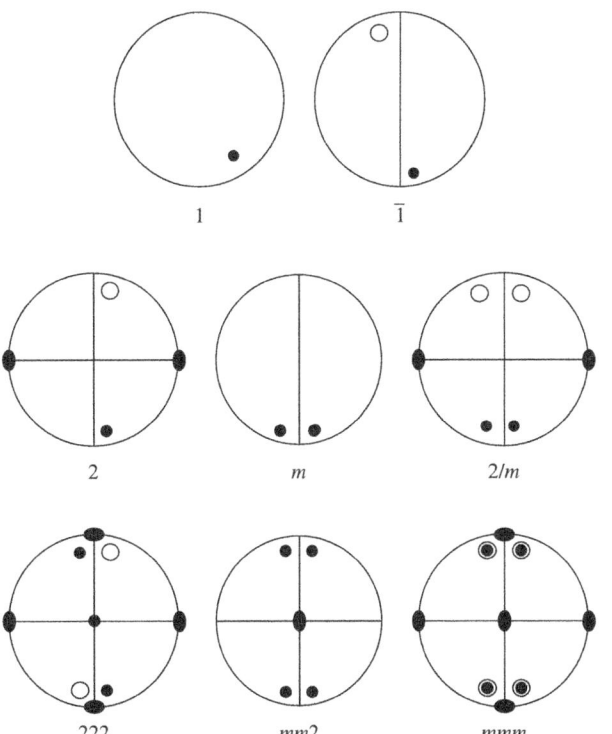

Fig. 3.5 Point groups 1 and $\bar{1}$ in the triclinic system.

Fig. 3.6 Three monoclinic point groups with a twofold axis, a mirror plane, or both.

Fig. 3.7 Orthorhombic point groups 222, $mm2$, and mmm.

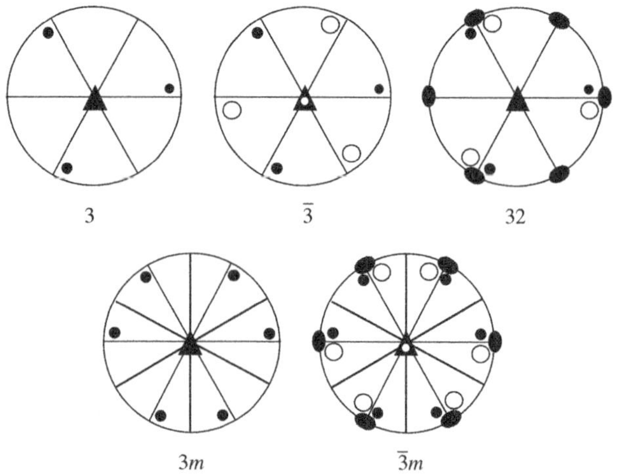

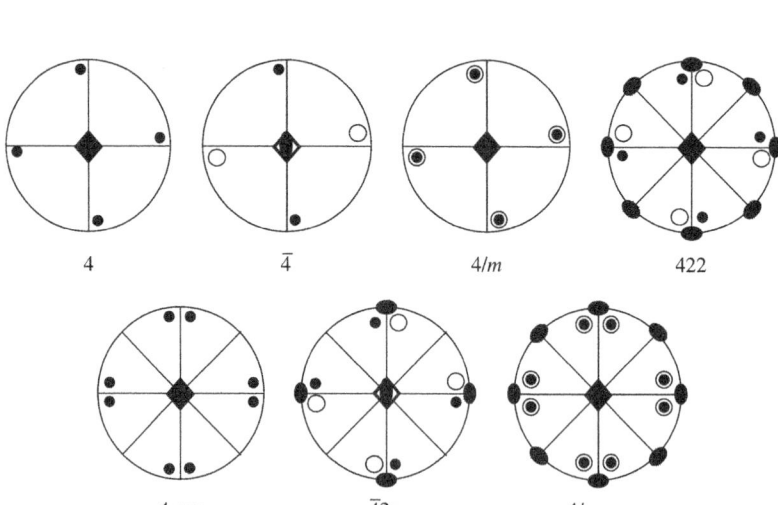

Fig. 3.8 Crystal classes in the trigonal system.

Fig. 3.9 Seven tetragonal point groups.

in point groups $2/m$ and mmm but not in the other five groups 1, 2, m, 222, and $mm2$. Inversion symmetry eliminates odd-rank physical properties such as pyroelectricity and piezoelectricity.

Continuing to the trigonal, tetragonal, and hexagonal crystal systems, Z_3 is always chosen as the axis of high symmetry. There are five trigonal classes with a threefold rotation or inversion axis along Z_3 (Fig. 3.8): 3, $\bar{3}$, $3m$, 32, and $\bar{3}m$. The secondary symmetry axis, typified by the twofold axis in point group 32, is chosen along Z_1. These choices for orientations are only for convenience. There is no reason why Z_3 must be the threefold symmetry axis, but it causes less confusion if everyone follows the same convention.

Tetragonal crystals possess a single fourfold symmetry axis along Z_3. There are seven crystal classes 4, $\bar{4}$, $4/m$, $\bar{4}2m$, $4mm$, 422, and $4/mmm$ (Fig. 3.9). It is worth noting that there are often more symmetry elements than appear in the symbol. Point group $4/mmm$ possesses one 4, five m, four 2, and $\bar{1}$.

Hexagonal crystal classes are analogous to the trigonal case with seven members: 6, $\bar{6}$, $6/m$, $\bar{6}m2$, $6mm$, 622, and $6/mmm$ (Fig. 3.10). The question often arises, which symmetry elements are needed to generate the group? For the point groups like $\bar{6}$ with only one symmetry element, the answer is obvious. For those with two elements like $6/m$, both are needed. For those with three or four like $6mm$ and $6/mmm$ often the last element is redundant and is not needed to generate the group. So why are unnecessary symmetry elements included in the symbol? Partly it is custom, but they also serve as a reminder that they are there. In the case of $6mm$, for example, the second mirror plane is crystallographically distinct from the first one.

The remaining five crystal classes belong to the cubic crystal system (Fig. 3.11). As a minimum, all the cubic classes contain four threefold axes along the body-diagonal $\langle 111 \rangle$ directions of cube. In the five classes, twofold, fourfold, or fourfold inversion axes lie along the $\langle 100 \rangle$ cube axes corresponding to the Z_1, Z_2, and Z_3 property axes. In three of the classes the $\langle 110 \rangle$ face diagonals are also symmetry directions.

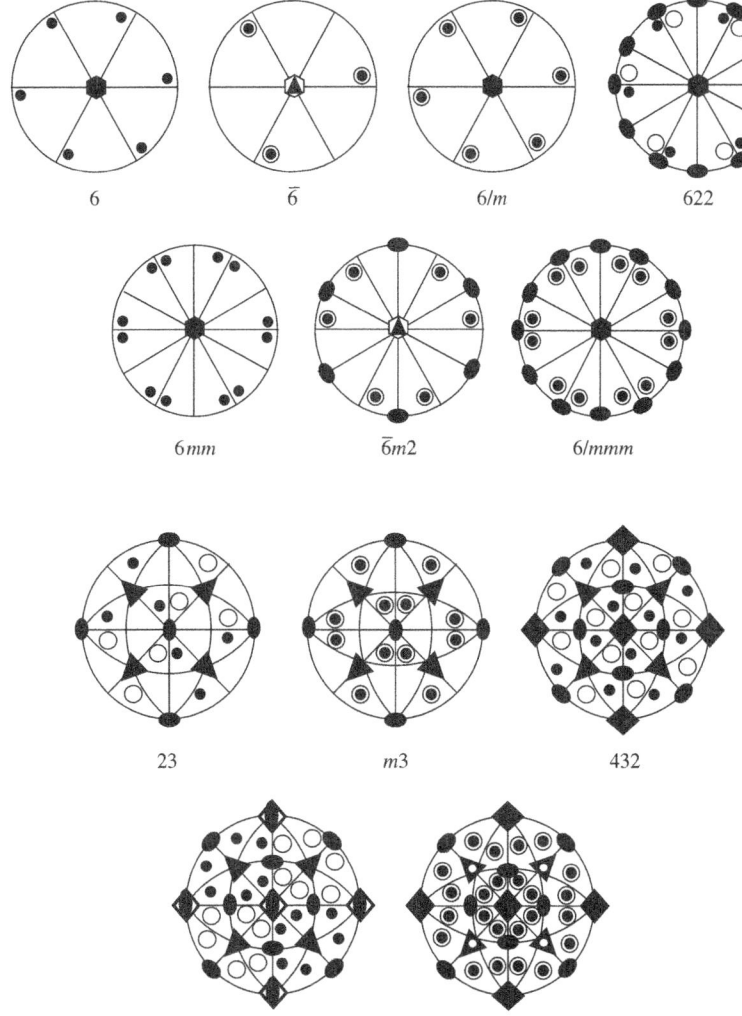

Fig. 3.10 Seven hexagonal crystal classes.

Fig. 3.11 Cubic crystal classes.

Problem 3.2
Fivefold symmetry is observed in quasicrystals and certain composite materials. Draw the stereographic projections of point groups 5, $\bar{5}$, $5m$, 52, $5/m$, and $\bar{5}m$.

3.4 Crystallographic nomenclature

The symbols used for the 32 crystal class are those recommended by the International Union of Crystallography, and sometimes referred to as the Herman–Mauguin symbols. The earlier Schoenflies symbols are still used by some chemists and spectroscopists. Equivalences between the two sets are listed in Table 3.1 along with representative compounds.

3.5 Point group populations

Not all symmetries are equally abundant. A survey of 127,000 inorganic and 156,000 organic compounds shows that most crystals have low symmetry. Table 3.2 lists the percentages for the 32 crystal classes. A total of 94% of organic crystals and 72% of inorganic crystals are orthorhombic, monoclinic, or triclinic. Point group $2/m$ is by far the most common in both classes of

Table 3.1 Examples of the 32 crystal classes

$1 = C_1$	Kaolinite	$Al_2Si_2O_5(OH)_4$
$\bar{1} = C_i$	Copper sulfate	$CuSO_4 \cdot 5H_2O$
$2 = C_2$	Sucrose	$C_{12}H_{12}O_{11}$
$m = C_S$	Potassium nitrite	KNO_2
$2/m = C_{2h}$	Orthoclase	$KAlSi_3O_8$
$222 = D_2$	Iodic acid	HIO_3
$mm2 = C_{2V}$	Sodium nitrite	$NaNO_2$
$mmm = D_{2h}$	Forsterite	Mg_2SiO_4
$3 = C_3$	Nickel tellurate	Ni_3TeO_6
$\bar{3} = C_{3i}$	Ilmenite	$FeTiO_3$
$32 = D_3$	Low-quartz	SiO_2
$3m = C_{3V}$	Lithium niobate	$LiNbO_3$
$\bar{3}m = D_{3d}$	Corundum	Al_2O_3
$4 = C_4$	Iodosuccinimide	$C_4H_4INO_2$
$\bar{4} = S_4$	Boron phosphate	BPO_4
$4/m = C_{4h}$	Scheelite	$CaWO_4$
$422 = D_4$	Nickel sulfate	$NiSO_4 \cdot 6H_2O$
$4mm = C_{4V}$	Barium titanate	$BaTiO_3$
$\bar{4}2m = D_{2d}$	Potassium dihydrogen phosphate	KH_2PO_4
$4/mmm = D_{4h}$	Rutile	TiO_2
$6 = C_6$	Nepheline	$NaAlSiO_4$
$\bar{6} = C_{3h}$	Lead germanate	$Pb_5Ge_3O_{11}$
$6/m = C_{6h}$	Apatite	$Ca_5(PO_4)_3F$
$622 = D_6$	High-quartz	SiO_2
$6mm = C_{6V}$	Zincite	ZnO
$\bar{6}m2 = D_{3h}$	Benitoite	$BaTiSi_3O_9$
$6/mmm = D_{6h}$	Beryl	$Be_3Al_2Si_6O_{18}$
$23 = T$	Sodium chlorate	$NaClO_3$
$m3 = T_h$	Pyrite	FeS_2
$432 = O$	Manganese	β-Mn
$\bar{4}3m = T_d$	Zincblende	ZnS
$m3m = O_h$	Rocksalt	$NaCl$

Table 3.2 Population statistics for the 32 crystallographic point groups gathered from more than 280,000 chemical compounds. Inorganic (I) and organic (O) have somewhat different percentages (data collected by G. Johnson)

	I	O		I	O
1	0.67%	1.24%	422	0.40%	0.48%
$\bar{1}$	13.87	19.18	4mm	0.30	0.09
2	2.21	6.70	$\bar{4}2m$	0.82	0.34
m	1.30	1.46	4/mmm	4.53	0.69
2/m	34.63	44.81	6	0.41	0.22
222	3.56	10.13	$\bar{6}$	0.07	0.01
mm2	3.32	3.31	6/m	0.82	0.17
mmm	12.07	7.84	622	0.24	0.05
3	0.36	0.32	6mm	0.45	0.03
$\bar{3}$	1.21	0.58	$\bar{6}m2$	0.41	0.02
32	0.54	0.22	6/mmm	2.82	0.05
3m	0.74	0.22	23	0.44	0.09
$\bar{3}m$	3.18	0.25	m3	0.84	0.15
4	0.19	0.25	432	0.13	0.01
$\bar{4}$	0.25	0.18	$\bar{4}3m$	1.42	0.11
4/m	1.17	0.67	m3m	6.66	0.12

materials. Within each system, centrosymmetric crystals are more common than noncentrosymmetric classes. Among organic crystals, 75% are centric, and among inorganic it is even higher at 82%. This means that odd-rank tensor properties such as piezoelectricity and pyroelectricity are absent in most crystals. Only about 22% are potentially piezoelectric and 12% potentially pyroelectric.

Regarding optical properties, more than 95% of crystals show double refraction, 87% are biaxial, 8% uniaxial, and 5% optically isotropic. About 15% of all crystals are enantiomorphic and potentially optically active.

There are some interesting differences between inorganics and organics. Higher symmetry crystals (trigonal, tetragonal, hexagonal, and cubic) are more abundant among metals, oxides, and halides with relatively simple chemical formulae. As a result, cubic crystals in point group $m3m$ are very common. Rocksalt, spinel, garnet, perovskite, diamond, and fluorite are all in $m3m$, as are face-centered cubic metals, body-centered cubic metals, and several of the important intermetallic Laves phases. Many of these structures are based on close-packing principles.

Close packing is also important in organic crystals, but it is the molecules that are close-packed, rather than individual atoms or ions. Most organic molecules have complex shapes that do not conform to rotational symmetry higher than twofold axes.

But if complex molecular shapes and complex chemistry favor low symmetry, why is it that centrosymmetric crystals are three times more common than noncentrosymmetric crystals? It is unclear why most crystals possess a center of symmetry. What physical principle favors inversion symmetry? Is it that atoms and molecules reside in potential wells where the interatomic forces are balanced? Inversion symmetry leads to equal forces in opposing directions favoring stability.

Another interesting difference has to do with enantiomorphism. In biochemistry, inversion symmetry and mirror planes of all types are practically nonexistent, because most living systems, from people on down, have a handedness. A dominant handedness is present at the molecular level as well, often in the form of helices, and in crystals as screw axes. This is quite different from minerals like quartz where the right- and left-handed forms are found in equal abundance.

The main theme of this book is anisotropy, and Table 3.2 makes it clear why anisotropy is important. Most crystals have anisotropic physical properties because of their low symmetry. It is important to know how to describe this anisotropy and explain its causes, and make use of the variation in properties with direction.

Problem 3.3
Which type of handedness dominates at the molecular level in living systems? Give some examples.

Transformation operators for symmetry elements

4

Since the physical property tensors depend on symmetry, mathematical methods for determining the influence of symmetry are needed. More specifically, transformation matrices are required for the symmetry elements that generate the crystal class. As explained earlier, some classes require only one symmetry element while others may require two or three. None require more than three. This chapter also includes a brief discussion of the seven Curie groups used for textured solids and liquids.

4.1	Transformation operators for the crystallographic symmetry elements	23
4.2	Transformation operations for the thirty-two crystal classes	25
4.3	Standard settings	26
4.4	Curie group symmetries	26

4.1 Transformation operators for the crystallographic symmetry elements

As explained in Chapter 2, transformations from one coordinate system to another can be specified by a set of nine direction cosines, a_{ij} where $i, j = 1, 2, 3$. The first index, i, refers to the "new" or transformed axis, the second index j to the "old" or reference axis.

There are four types of symmetry elements that require discussion: rotation axes, mirror planes, inversion centers, and inversion axes in which rotation is accompanied by inversion. All rotations are assumed to be in the counterclockwise direction. Other symmetry elements such as rotoreflection axes are not needed.

The fourteen symmetry elements in Table 4.1 generate the 32 crystal classes. No proof is offered here but this statement can be verified geometrically using the stereographic projections in Chapter 3.

There are two final points concerning these transformation matrices. First, keep in mind that they must obey the orthogonality conditions described in Chapter 2. This provides a useful way of avoiding mistakes.

The second point concerns handedness. Symmetry elements involving reflection or inversion reverses the handedness of the coordinate system. This of course includes inversion axes such as $\bar{1}, \bar{2} = m, \bar{3}, \bar{4}$, and $\bar{6}$. The handedness change can be verified by showing that the determinant of the transformation matrix is -1. For ordinary rotation axes such as 2, 3, 4, and 6, there is no change in handedness, and the determinant is $+1$.

Problem 4.1
In cubic crystals, there is sometimes a twofold axis parallel to $\langle 110 \rangle$ axes. Draw the stereographic projection and work out the transformation matrix for this symmetry element.

Table 4.1 Transformation matrices for selected crystallographic symmetry operations

Identity operator	1	$\begin{pmatrix} 1 & 0 & 0 \\ 0 & 1 & 0 \\ 0 & 0 & 1 \end{pmatrix}$
Inversion center	$\bar{1}$	$\begin{pmatrix} -1 & 0 & 0 \\ 0 & -1 & 0 \\ 0 & 0 & -1 \end{pmatrix}$
Twofold rotation (2) parallel to Z_1	$2 \parallel Z_1$	$\begin{pmatrix} 1 & 0 & 0 \\ 0 & -1 & 0 \\ 0 & 0 & -1 \end{pmatrix}$
Twofold rotation (2) parallel to Z_2	$2 \parallel Z_2$	$\begin{pmatrix} -1 & 0 & 0 \\ 0 & 1 & 0 \\ 0 & 0 & -1 \end{pmatrix}$
Mirror (m) perpendicular to Z_1	$m \perp Z_1$	$\begin{pmatrix} -1 & 0 & 0 \\ 0 & 1 & 0 \\ 0 & 0 & 1 \end{pmatrix}$
Mirror (m) perpendicular to Z_2	$m \perp Z_2$	$\begin{pmatrix} 1 & 0 & 0 \\ 0 & -1 & 0 \\ 0 & 0 & 1 \end{pmatrix}$
Mirror (m) perpendicular to Z_3	$m \perp Z_3$	$\begin{pmatrix} 1 & 0 & 0 \\ 0 & 1 & 0 \\ 0 & 0 & -1 \end{pmatrix}$
Threefold rotation (3) parallel to Z_3	$3 \parallel Z_3$	$\begin{pmatrix} -1/2 & \sqrt{3}/2 & 0 \\ -\sqrt{3}/2 & -1/2 & 0 \\ 0 & 0 & 1 \end{pmatrix}$
Threefold rotation (3) parallel to [111]	$3 \parallel [111]$	$\begin{pmatrix} 0 & 1 & 0 \\ 0 & 0 & 1 \\ 1 & 0 & 0 \end{pmatrix}$
Threefold inversion axis ($\bar{3}$) parallel to Z_3	$\bar{3} \parallel Z_3$	$\begin{pmatrix} 1/2 & -\sqrt{3}/2 & 0 \\ \sqrt{3}/2 & 1/2 & 0 \\ 0 & 0 & -1 \end{pmatrix}$
Threefold inversion axis ($\bar{3}$) parallel to [111] in cubic crystals	$\bar{3} \parallel [111]$	$\begin{pmatrix} 0 & -1 & 0 \\ 0 & 0 & -1 \\ -1 & 0 & 0 \end{pmatrix}$
Fourfold rotation (4) parallel to Z_3	$4 \parallel Z_3$	$\begin{pmatrix} 0 & 1 & 0 \\ -1 & 0 & 0 \\ 0 & 0 & 1 \end{pmatrix}$
Fourfold inversion axis ($\bar{4}$) parallel to Z_3	$\bar{4} \parallel Z_3$	$\begin{pmatrix} 0 & -1 & 0 \\ 1 & 0 & 0 \\ 0 & 0 & -1 \end{pmatrix}$
Sixfold rotation (6) parallel to Z_3	$6 \parallel Z_3$	$\begin{pmatrix} 1/2 & \sqrt{3}/2 & 0 \\ -\sqrt{3}/2 & 1/2 & 0 \\ 0 & 0 & 1 \end{pmatrix}$
Sixfold inversion axis ($\bar{6}$) parallel to Z_3	$\bar{6} \parallel Z_3$	$\begin{pmatrix} -1/2 & -\sqrt{3}/2 & 0 \\ \sqrt{3}/2 & -1/2 & 0 \\ 0 & 0 & -1 \end{pmatrix}$

Table 4.2 Minimum symmetry elements for the 32 crystal classes

Crystal class	Symmetry elements
1	1
$\bar{1}$	$\bar{1}$
2	$2 \parallel Z_2$
m	$m \perp Z_2$
$2/m$	$2 \parallel Z_2, m \perp Z_2$
222	$2 \parallel Z_1, 2 \parallel Z_2$
$mm2$	$m \perp Z_1, m \perp Z_2$
mmm	$m \perp Z_1, m \perp Z_2, m \perp Z_3$
3	$3 \parallel Z_3$
$\bar{3}$	$\bar{3} \parallel Z_3$
32	$3 \parallel Z_3, 2 \parallel Z_1$
$3m$	$3 \parallel Z_3, m \perp Z_1$
$\bar{3}m$	$\bar{3} \parallel Z_3, m \perp Z_1$
4	$4 \parallel Z_3$
$\bar{4}$	$\bar{4} \parallel Z_3$
$4/m$	$4 \parallel Z_3, m \perp Z_3$
422	$4 \parallel Z_3, 2 \parallel Z_1$
$4mm$	$4 \parallel Z_3, m \perp Z_1$
$\bar{4}2m$	$\bar{4} \parallel Z_3, 2 \parallel Z_1$
$4/mmm$	$4 \parallel Z_3, m \perp Z_3, m \perp Z_1$
6	$6 \parallel Z_3$
$\bar{6}$	$\bar{6} \parallel Z_3$
$6/m$	$6 \parallel Z_3, m \perp Z_3$
622	$6 \parallel Z_3, 2 \parallel Z_1$
$6mm$	$6 \parallel Z_3, m \perp Z_1$
$\bar{6}m2$	$\bar{6} \parallel Z_3, m \perp Z_1$
$6/mmm$	$6 \parallel Z_3, m \perp Z_3, m \perp Z_1$
23	$2 \parallel Z_1, 3 \parallel [111]$
$m3$	$m \perp Z_1, 3 \parallel [111]$
432	$4 \parallel Z_3, 3 \parallel [111]$
$\bar{4}3m$	$\bar{4} \parallel Z_3, 3 \parallel [111]$
$m3m$	$m \perp Z_1, 3 \parallel [111], m \perp [110]$

4.2 Transformation operations for the thirty-two crystal classes

In Table 4.2 we list the minimum symmetry requirements for each of the 32 crystal classes. These are the transformation operations needed to develop the physical property matrices for single crystals. These techniques will be described in the next chapter.

As explained later, in simplifying property matrices these symmetry operations are applied in consecutive steps. For example, for point group $\bar{4}2m$, first the property tensor is simplified using the $\bar{4}$ symmetry element and then it is further simplified using the twofold axis. It is unnecessary to use the m operator.

Problem 4.2

a. For any point group, the product of any two operators generates another symmetry element. Show that for point group $2/m$, the product of the twofold axis and the mirror plane generates a center of symmetry.

b. For cubic class 432, what symmetry element is generated by the product of the first two symmetry elements?

Problem 4.3
Some symmetry elements are automatically included within other symmetry elements. What elements are included in 6 and $\bar{6}$?

4.3 Standard settings

The question often arises how are the physical property axes (Z_1, Z_2, Z_3) related to the crystallographic axes (a, b, c)? It is obvious that the two coordinate systems will not always coincide since property axes are always orthogonal, whereas crystallographic axes are not. In selecting the initial measurement axes we recommend the IEEE standard settings:

In the *triclinic system*, Z_3 is chosen parallel to c, and Z_2 is normal to the ac plane (010). The Z_1 axis lies in the (010) plane and is perpendicular to Z_2 and Z_3 forming a right-handed coordinate system.

For *monoclinic crystals*, Z_2 is chosen parallel to the b [010] axis. It is parallel to the twofold symmetry axis or perpendicular to the mirror plane, as is customary in crystallography. Axis Z_3 is chosen along c, and Z_1 is perpendicular to Z_2 and Z_3, again forming a right-handed system.

In the *orthorhombic system* it is customary to select a unit cell in which $c < a < b$. The Z_1, Z_2, and Z_3 property axes are taken as a, b, and c, respectively.

Trigonal crystals have threefold symmetry along [001], the c crystallographic axis. For physical properties this is the Z_3 direction. The Z_1 axis is parallel to [100], a crystallographic direction, with Z_2 completing the right-handed orthogonal set.

For the *tetragonal system*, Z_3 and c lie along the fourfold axis, Z_1 and Z_2 are parallel to a and b, respectively.

Among *hexagonal crystals*, Z_3 is again selected along c, the sixfold symmetry axis. As with trigonal crystals, Z_1 is parallel to a, and Z_2 is mutually perpendicular to Z_1 and Z_3.

Cubic crystals have Z_1, Z_2, and Z_3 parallel to a, b, and c, respectively.

For crystallographic groups possessing more than one symmetry axis these settings will turn out to be the principal axes (Section 9.4). For groups with only one symmetry axis only one of the principal axes is fixed. For the two triclinic groups, none of the principal axes will correspond to crystallographic axes. There are further complications in certain point group such as $\bar{4}2m$, $\bar{6}2m$, but the generating symmetry elements in Table 4.2 and the stereographic projections in Chapter 3 make the labeling clear. When dealing with polar axes, the positive end of the axis is normally chosen to make the principal piezoelectric coefficient a positive number. It should be remembered that not all authors follow the IEEE conventions.

4.4 Curie group symmetries

In many practical applications engineers work with amorphous or polycrystalline materials rather than single crystals. Such materials are often processed

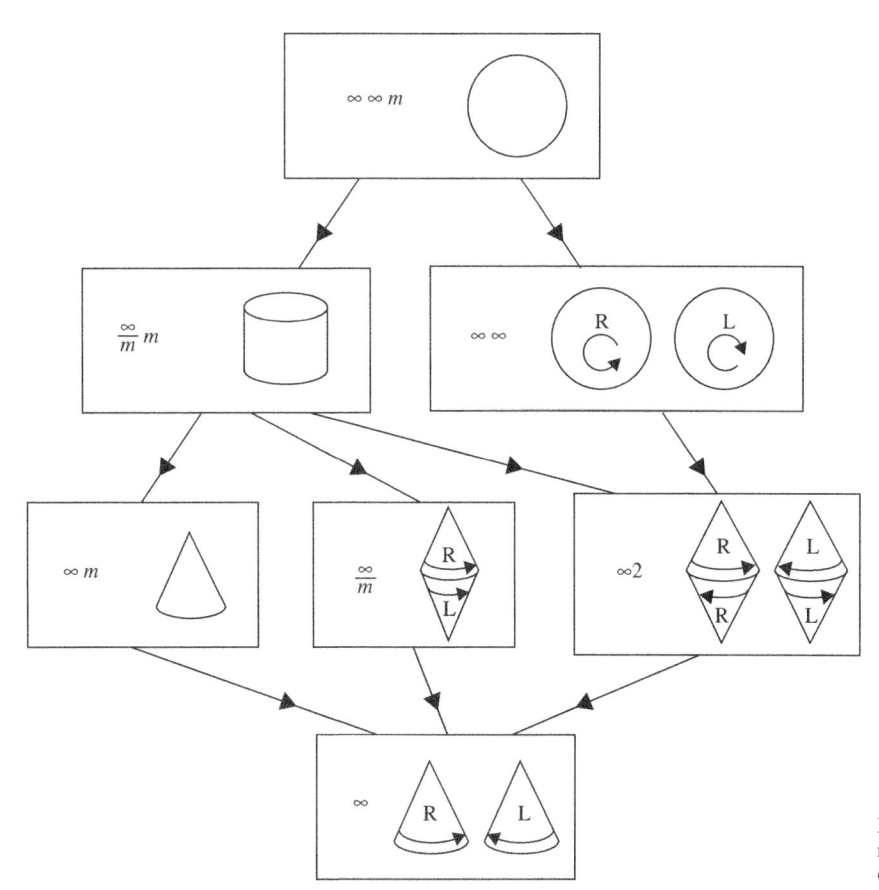

Fig. 4.1 The seven Curie groups can be represented geometrically as spheres, cylinders, and cones, with or without handedness.

in such a way that their properties are anisotropic. The symmetry of these textured materials can often be described by Curie group symmetries (Fig. 4.1). The Curie groups are sometimes called the limiting groups or the continuous groups.

The Curie groups all have a common symmetry element represented by an ∞-fold rotation axis. The transformation matrix for an ∞-fold axis parallel to Z_3 is

$$\infty \parallel Z_3 \begin{pmatrix} \cos\theta & \sin\theta & 0 \\ -\sin\theta & \cos\theta & 0 \\ 0 & 0 & 1 \end{pmatrix},$$

which holds true for all values of θ. Required symmetry elements for the seven Curie groups are listed in Table 4.3.

In materials science and engineering there are many examples of textured materials belonging to the Curie groups. Polycrystalline metals and ceramics with randomly oriented grains possess spherical symmetry, $\infty\infty m$. Organic substances with randomly oriented right-handed crystallites, like dextrose, belong to point group $\infty\infty$. Polycrystalline levose with its left-handed molecules also belongs to $\infty\infty$, but a racemic mixture of dextrose and levose grains restores a statistical mirror plane, shifting the symmetry up to $\infty\infty m$.

Table 4.3 Minimum symmetry operations for textured materials in limiting groups

Curie group	Symmetry operators
∞	$\infty \parallel Z_3$
∞m	$\infty \parallel Z_3, m \perp Z_1$
$\infty 2$	$\infty \parallel Z_3, 2 \parallel Z_1$
∞/m	$\infty \parallel Z_3, m \perp Z_3$
∞/mm	$\infty \parallel Z_3, m \perp Z_3, m \perp Z_1$
$\infty\infty$	$\infty \parallel Z_3, \infty \parallel Z_1$
$\infty\infty m$	$\infty \parallel Z_3, \infty \parallel Z_1, m \perp Z_1$

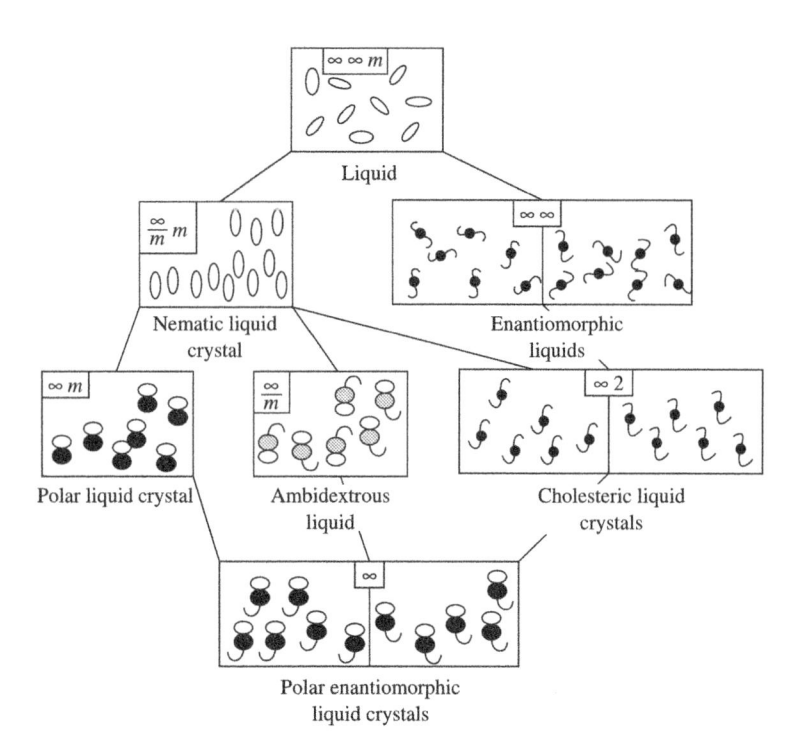

Fig. 4.2 Liquids and liquid crystals exhibit a number of different Curie group symmetries.

Preferred orientation is often observed after uniaxial hot-pressing polycrystalline materials. The resultant symmetry is cylindrical, point group ∞/mm, with the ∞-fold axis in the direction of the applied force. If such a procedure is carried out on a right- or left-handed medium, the symmetry is $\infty 2$.

Poled polycrystalline ferroelectrics belong to ∞m, and magnetized polycrystalline ferromagnets are in ∞/m. A poled right- or left-handed polycrystalline substance belongs to the lowest Curie group, ∞.

Processed polycrystalline materials may also adopt crystallographic symmetry. Ferroelectric polymers such as polyvinylidene fluoride, are often stretched in one direction to align the molecules, and then electrically poled in a perpendicular direction. This results in an orthorhombic texture, point group $mm2$.

Liquids and liquid crystals may also exhibit Curie group symmetries. An ordinary liquid such as water or a molten salt, with randomly oriented molecular structure, has spherical symmetry $\infty\infty m$. When sugar is dissolved in water it induces optical activity causing the plane of polarization of light to be rotated. This changes the symmetry of liquid to $\infty\infty$. In a nematic liquid crystal the molecules align in a parallel fashion to give cylindrical symmetry ∞/mm, with a preferred direction. Cholesteric liquid crystals adopt a helical patterning conforming to point group $\infty 2$, while ferroelectric liquid crystals possess a polar axis. They belong to ∞m or ∞, depending on whether or not the molecules have a handedness. Liquid crystals are discussed further in Section 30.5.

A schematic illustration of limiting group liquids is shown in Fig. 4.2.

Problem 4.4

    a. Molecules sometimes have symmetry elements that are not allowed in crystals. Molecular symmetry is not constrained by the need for translational periodicity. Some simple molecules have Curie group symmetry. Examples include He, N_2, CO_2, and CO. Sketch the molecules and show the location of the symmetry elements.

b. Other molecules like H_2O and C_6H_6 belong to crystallographic point groups. Sketch the molecular structures of water and benzene and assign them to the appropriate crystal class.

c. Boron hydride, $B_{12}H_{12}$, forms pentagonal dodecahedra. What are symmetry elements of this molecule? Fivefold symmetry is also observed in quasicrystals and buckyballs.

5 Tensors and physical properties

5.1 Physical properties 30
5.2 Polar tensors and tensor properties 31
5.3 Axial tensor properties 32
5.4 Geometric representations 33
5.5 Neumann's Principle 34
5.6 Analytical form of Neumann's Principle 34

In this chapter we introduce the tensor description of physical properties along with Neumann's Principle relating symmetry to physical properties.

5.1 Physical properties

As pointed out in the introduction, many different types of anisotropic properties are described in this book, but all have one thing in common: a physical property is a relationship between two measured quantities. Four examples are illustrated in Fig. 5.1.

Elasticity is one of the standard equilibrium properties treated in crystal physics courses. The elastic compliance coefficients relate mechanical strain, the dependent variable, to mechanical stress, the independent variable. For small stresses and strains, the relationship is linear, but higher order elastic constants are needed to describe the departures from Hooke's Law.

Thermal conductivity is typical of the many transport properties in which a gradient leads to flow. Here the dependent variable is heat flow and the independent variable is a temperature gradient. Again the relationship is linear for small temperature gradients.

Hysteretic materials such as ferromagnetic iron exhibit more complex physical properties involving domain wall motion. In this case magnetization is the dependent variable responsive to an applied magnetic field. The resulting magnetic susceptibility depends on the past history of the material. If the sample is initially unmagnetized, the magnetization will often involve only

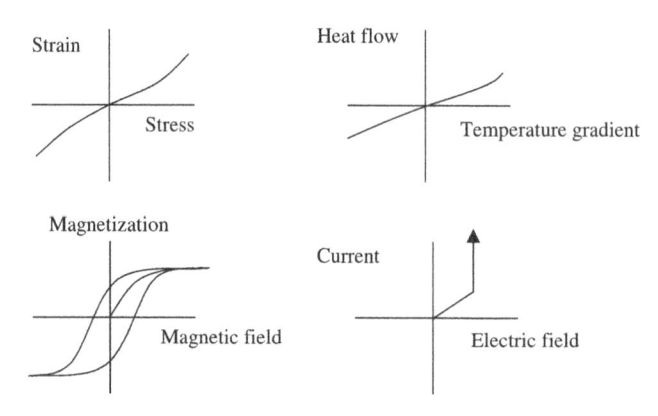

Fig. 5.1 Four types of physical properties. Elasticity is a typical equilibrium property relating stress and strain. Thermal conductivity is representative of the transport properties. Ferromagnetism is hysteretic in nature while electric breakdown is an irreversible property in which the material is permanently altered.

reversible domain wall motion for small magnetic fields. In this case the susceptibility is anhysteretic, but for large fields the wall motion is only partly reversible leading to hysteresis.

The fourth class of properties leads to permanent changes involving irreversible processes. Under very high electric fields, dielectric materials undergo an electric breakdown process with catastrophic current flow. Under small fields Ohm's Law governs the relationship between current density and electric field with a well-defined resistivity, but high fields lead to chemical, thermal, and mechanical changes that permanently alter the sample. Irreversible processes are sometimes anisotropic but they will not be discussed in this book.

5.2 Polar tensors and tensor properties

Measured quantities such as stress and strain can be represented by tensors, and so can physical properties like elastic compliance that relate these measurements. This is why tensors are so useful in describing anisotropy.

All tensors are defined by the way in which they transform from one coordinate system to another. As explained in Chapter 2, all these transformations involve a set of direction cosines a_{ij}, where $i, j = 1, 2, 3$.

In this book we deal mainly with two kinds of tensors: polar tensors and axial tensors. Axial tensors change sign when the handedness changes, whereas polar tensors do not. Their transformation laws are slightly different.

For a polar tensor, the general transformation law for a tensor of rank N is

$$T'_{ijk...} = a_{il} a_{jm} a_{kn} \ldots T_{lmn...},$$

where $T'_{ijk...}$ is the tensor component in the new axial system, $T_{lmn}\ldots$ is a tensor component in the old system, and $a_{il}, a_{jm}, a_{kn} \ldots$ are the direction cosines relating the two coordinate systems. In this expression, each tensor component has N subscripts and there are N direction cosines involved in the product $a_{il}\, a_{jm}\, a_{kn} \ldots$. The tensor rank N has a very simple meaning. It is simply the number of directions involved in measuring the property. As an example, the thermal conductivity k relates the heat flow h to the temperature gradient dT/dZ:

$$h = -k \frac{dT}{dZ}.$$

There are two directions involved in measuring k: the direction in which we set up the temperature gradient, and the direction that the heat flow is measured. In general the two directions will not be the same. In tensor form this equation becomes

$$h_i = -k_{ij} \frac{dT}{dZ_j}.$$

The minus sign in these two expressions remind us that heat always flows down the temperature gradient from hot to cold. Here there are three tensors: h_i and dT/dZ_j are first rank polar tensor *quantities* that transform as

$$h'_i = a_{ij} h_j$$

and
$$\frac{dT'}{dZ_i} = a_{ij}\frac{dT}{dZ_j}$$

while the thermal conductivity, that depends on both measurement directions, is a second rank tensor *property*.

$$k'_{ij} = a_{il}a_{jm}k_{lm}.$$

There are two important points to remember here. First, repeated subscripts always imply summation so there will be nine terms on the right side of the last equation. Second, h_i and dT/dZ_j are *not* properties of the material. We are at liberty to choose these experimental conditions in any way we wish, but the thermal conductivity is a property that belongs to the material. It therefore depends on the symmetry of the material, whereas the heat flow and temperature gradient do not.

The tensor rank of other physical properties is determined in a similar way. Pyroelectricity describes a relationship between thermal and electrical variables: a change in temperature ΔT creates a change in the electric polarization P. Polarization is a vector (= first rank tensor) and temperature is a scalar (= zero rank tensor). Therefore the pyroelectric coefficient, defined by $P_i = p_i \Delta T$, is a first rank tensor property.

Four directions are involved in the measurement of elastic constants. There are two directions for mechanical force and two for mechanical strain. Stress is force per unit area, and one direction is needed for the force, and another for the normal to the face on which the force acts. Strain is change in length per unit length, and directions are needed for both the reference line and the direction of the change. Therefore two subscripts are needed for stress X_{ij} and two for strain x_{ij}. Elastic compliance, that relates the two through Hooke's Law, will therefore require four subscripts:

$$x_{ij} = s_{ijkl}X_{kl}.$$

The elastic constants, s_{ijkl}, are represented by a fourth rank tensor.

5.3 Axial tensor properties

Later in the book we will deal with several properties that change sign when the axial system changes from right-handed to left-handed. Properties such as pyromagnetism, optical activity, and the Hall Effect are axial tensors that depend on the handedness. Axial tensors transform in the following manner:

$$T'_{ijk...} = |a|a_{il}a_{jm}a_{kn}\ldots T_{lmn...}$$

which is almost identical to that of a polar tensor. The difference is $|a|$, the determinant of the direction cosine matrix. As explained previously, $|a| = \pm 1$, depending on whether or not the handedness of the axial system changes during the transformation. For symmetry operations involving mirror planes or inversion centers, $|a| = -1$ and the sign of the tensor coefficient changes. No change occurs for rotation axes.

Magnetoelectricity is a good example of an axial tensor property. The magnetoelectric coefficients relate a change in magnetization (a first rank axial tensor) to a change in electric field (a first rank polar tensor). Since two directions are involved in the measurement, magnetoelectricity is a second rank axial tensor property.

5.4 Geometric representations

Tensor properties involve the product of direction cosines, as listed in Table 5.1. A second rank polar tensor will include terms like $\cos^2 \phi$, for example, and can therefore be represented by a quadric surface.

Representative geometries are shown in Fig. 5.2. Scalar properties such as density and specific heat are independent of sample orientation and therefore the property can be visualized as a sphere. A vector property like pyroelectricity will have its maximum value along the polar axis and then fall to zero for directions perpendicular to the polar axis. The pyroelectric coefficients will change sign for opposing directions creating a negative lobe. Other odd-rank polar tensors will also show positive and negative lobes.

Even rank tensor properties will occasionally have positive and negative lobes as well. As discussed later, some physical properties such as permittivity and elasticity are constrained to have positive principal coefficients, while others such as thermal expansion, may have both positive and negative values. The illustrations in Fig. 5.2 are typical for permittivity and elasticity. Numerous examples will be presented later.

Table 5.1 Transformation laws for polar tensors of various ranks. The rank of a tensor denotes the number of different directions that must be specified in carrying out the measurement of a physical property

Tensor rank	Transformation law	Geometric representation
0	$T' = T$	Sphere
1	$T'_i = a_{ij}T_j$	Vector
2	$T'_{ij} = a_{ik}a_{jl}T_{kl}$	Quadric
3	$T'_{ijk} = a_{il}a_{jm}a_{kn}T_{lmn}$	Cubic
4	$T'_{ijkl} = a_{im}a_{jn}a_{ko}a_{lp}T_{mnop}$	Quartic

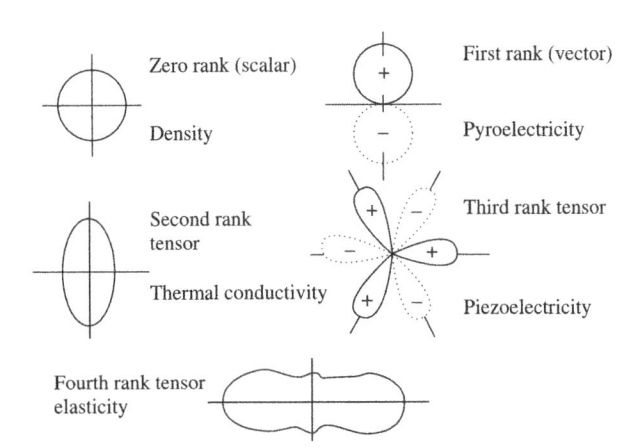

Fig. 5.2 Typical geometric surfaces of physical properties plotted as a function of measurement directions.

5.5 Neumann's Principle

The most important concept in Crystal Physics is Neumann's Principle that states: *"The symmetry of any physical property of a crystal must include the symmetry elements of the point group of the crystal."* The proof of Neumann's Principle is common sense. What it says is that measurements made in symmetry-related directions will give the same property coefficients.

Sodium chloride is a cubic crystal belonging to point group $m3m$. The [100] and [010] directions are equivalent fourfold symmetry axes (Fig. 5.3). Since these directions are physically the same, it makes sense that measurements of permittivity, elasticity, or any other physical property will be the same in these two directions. This means that when the magnitudes of the property are plotted as a function of direction, the resulting figure will show fourfold symmetry when viewed along the [100] or [010] directions. In other words the symmetry of the physical property will include the symmetry elements of the point group.

But the reverse is not true, for the symmetry of the physical property may be much higher than that of the point group. This becomes obvious when we visualize a scalar property such as specific heat. Here the geometric representation is a sphere (symmetry group $\infty\infty m$) that includes the symmetry of sodium chloride (point group $m3m$) but not vice versa.

The argument just applied to [100] directions in NaCl, holds for other directions as well. In NaCl, the [110] and [$\bar{1}$10] directions are symmetry-related twofold axes, and therefore the properties will be the same when measurements are carried out in a similar way along [110] and [$\bar{1}$10].

What about the properties along [100] and [110]? Will they sometimes be the same? For scalar properties the answer is, of course, yes. For higher rank tensor properties, it will depend upon the point group symmetry *and* the tensor rank. In cubic crystals, second rank tensors like permittivity and resistivity, measurements along [100] and [110] will be the identical, but not for fourth rank tensor properties like elastic compliance. The reasons will become clearer after applying Neumann's Principle to a number of different situations.

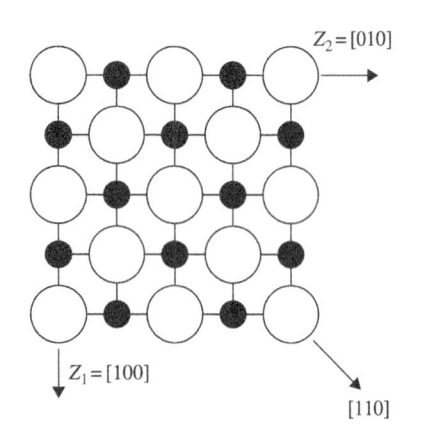

Fig. 5.3 Crystal structure of NaCl showing the equivalence of [100] and [010] directions.

5.6 Analytical form of Neumann's Principle

In expressing Neumann's Principle mathematically, we begin with the definition of a tensor

$$T'_{ijk...} = a_{il}a_{jm}a_{kn}\ldots T_{lmn...}.$$

The direction cosine matrix for any symmetry operation is expressed through the (a) matrix. These (a) coefficients are then substituted into the above equation to transform the tensor coefficients under the action of the symmetry operator. If the crystal possesses this symmetry element, the property coefficient must be left unchanged.

Mathematically this means

$$T'_{ijk...} = T_{ijk...}.$$

As an example, consider a monoclinic crystal belonging to point group m. There is only one symmetry element, a mirror plane perpendicular to

$Z_2 = [010]$, the b crystallographic axis. The direction cosine matrix for $m \perp Z_2$ is

$$(a) = \begin{pmatrix} 1 & 0 & 0 \\ 0 & -1 & 0 \\ 0 & 0 & 1 \end{pmatrix} = \begin{pmatrix} a_{11} & a_{12} & a_{13} \\ a_{21} & a_{22} & a_{23} \\ a_{31} & a_{32} & a_{33} \end{pmatrix}.$$

We now apply Neumann's Principle to a third rank tensor property, beginning with tensor coefficient T'_{111},

$$T'_{111} = a_{11}^3 T_{111} + a_{11}^2 a_{12} T_{112} + \cdots + a_{13}^3 T_{333}.$$

For the mirror plane, all terms go to zero except the first term:

$$T'_{111} = T_{111}.$$

By Neumann's Law this coefficient remains unchanged. This means coefficient T_{111} is unaffected by the mirror plane.

The situation is different for T_{222}.

$$T'_{222} = a_{22}^3 T_{222} + a_{22}^2 a_{21} T_{221} + \cdots.$$

Again all terms disappear except the first, and $T'_{222} = -T_{222}$. In this case Neumann's Principle says $-T_{222} = T_{222}$ which is only possible if $T_{222} = 0$. Therefore this property coefficient must disappear for crystals belonging to point group m.

This is why Neumann's Principle is useful to an experimentalist. It greatly simplifies the description of physical properties by eliminating some coefficients and equalizing others.

To illustrate, consider two of the standard single crystal materials available to scientists and engineers: quartz (SiO_2) and corundum (Al_2O_3). Both belong to the trigonal crystal system but have different point group symmetries. Quartz is in point group 32 while corundum is somewhat higher in $\bar{3}m$.

Symmetry arguments based on Neumann's Principle tell us that neither crystal is pyroelectric since first rank polar tensors disappear for both point groups. In regard to permittivity, resistivity, thermal expansion, and other second rank tensor properties, there will be three nonzero coefficients, but only two measurements are required because two of the three coefficients are equal.

For piezoelectricity and other third rank polar tensors, quartz and corundum are very different. Quartz is an outstanding piezoelectric material while corundum is totally useless. The center of symmetry in point group $\bar{3}m$ causes all piezoelectric coefficients disappear. Quartz, on the other hand, has five nonzero coefficients, two of which are independent. A large number of piezoelectric resonators and timing devices make use of these two coefficients.

Elasticity is a fourth rank tensor property so there are many different directions to consider. For a triclinic crystal with no mirror planes or rotation axes, there would be 18 independent elastic constants, but the higher symmetry of quartz and corundum reduce the required number of experiments considerably. Only six independent elastic coefficients are present in point group 32 and $\bar{3}m$.

Comparable simplifications are found for other polar and axial tensor properties. These ideas will be discussed throughout the book.

Problem 5.1
The cubic crystal structure of rocksalt is pictured in Figs. 5.3 and 13.4. The [100] direction is equivalent to [010], [001], [$\bar{1}$00], [0$\bar{1}$0], and [00$\bar{1}$], a total of six directions. How many directions are related to [110] by symmetry? What about [111] and [210]?

Problem 5.2
There are ten properties listed in Table 1.1. Each is represented by a different tensor. Write out the tensor transformation for each property. For pyroelectricity, a first rank polar tensor, there are three coefficients. How many for the other properties?

Thermodynamic relationships

6

In the next few chapters we shall discuss tensors of rank zero to four which relate the intensive variables in the outer triangle of the Heckmann Diagram (Fig. 1.1) to the extensive variables in the inner triangle. Effects such as pyroelectricity, permittivity, pyroelectricity, and elasticity are the standard topics in crystal physics that allow us to discuss tensors of rank one through four. First, however, it is useful to introduce the thermodynamic relationships between physical properties and consider the importance of measurement conditions.

6.1	Linear systems	37
6.2	Coupled interactions: Maxwell relations	38
6.3	Measurement conditions	40

6.1 Linear systems

Before discussing all the cross-coupled relationships, we first define the coupling within the three individual systems. In a thermal system, the basic relationship is between change in entropy δS [J/m^3] and change in temperature δT [K]:

$$\delta S = C \delta T,$$

where C is the specific heat per unit volume [J/m^3 K] and T is the absolute temperature. S, T, and C are all scalar quantities.

In a dielectric system the electric displacement D_i [C/m^2] changes under the influence of the electric field E_i [V/m]. Both are vectors and therefore the electric permittivity, ε_{ij}, requires two-directional subscripts. Occasionally the dielectric stiffness, β_{ij}, is required as well.

$$D_i = \varepsilon_{ij} E_j$$
$$E_i = \beta_{ij} D_j.$$

Some authors use polarization P rather than electric displacement D. The three variables are interrelated through the constitutive relation

$$D_i = P_i + \varepsilon_0 E_i = \varepsilon_{ij} E_j.$$

The third linear system in the Heckmann Diagram is mechanical, relating strain x_{ij} to stress X_{kl} [N/m^2] through the fourth rank elastic compliance coefficients s_{ijkl} [m^2/N].

$$x_{ij} = s_{ijkl} X_{kl}.$$

Alternatively, Hooke's Law can be expressed in terms of the elastic stiffness coefficients c_{ijkl} [N/m^2].

$$X_{ij} = c_{ijkl} x_{kl}.$$

6.2 Coupled interactions: Maxwell relations

When cross coupling occurs between thermal, electrical, and mechanical variables, the Gibbs free energy $G(T,X,E)$ is used to derive relationships between the property coefficients. Temperature T, stress X, and electric field E are the independent variables in most experiments.

From the exact differential

$$dG = -SdT - x_{ij}dX_{ij} - D_m dE_m \quad [\text{J/m}^3]$$

we obtain the relations:

$$S = -\left(\frac{\partial G}{\partial T}\right)_{X,E}$$

$$D_m = -\left(\frac{\partial G}{\partial E_m}\right)_{T,X}$$

$$x_{ij} = -\left(\frac{\partial G}{\partial X_{ij}}\right)_{T,E}.$$

Physical constants are defined and interrelated by taking the second derivatives. The order of differentiation can be reversed leading to equivalent coefficients for the direct and converse piezoelectric effects, between the pyroelectric coefficient and the electrocaloric effect, and between thermal expansion coefficients α_{ij} and the piezocaloric effect.

Specific heat

$$\frac{C^{X,E}}{T} = -\left(\frac{\partial^2 G}{\partial T^2}\right)_{X,E} = \left(\frac{\partial S}{\partial T}\right)_{X,E}$$

Permittivity

$$\varepsilon_{nm}^{X,T} = -\left(\frac{\partial^2 G}{\partial E_n \partial E_m}\right)_{X,T} = \left(\frac{\partial D_n}{\partial E_m}\right)_{X,T}$$

Elastic compliance

$$s_{ijkl}^{E,T} = -\left(\frac{\partial^2 G}{\partial X_{ij} \partial X_{kl}}\right)_{E,T} = \left(\frac{\partial x_{ij}}{\partial X_{kl}}\right)_{E,T}$$

Direct and converse piezoelectric effects

$$d_{nij}^T = -\left(\frac{\partial^2 G}{\partial X_{ij} \partial E_n}\right)_T = \left(\frac{\partial D_n}{\partial X_{ij}}\right)_{E,T} = \left(\frac{\partial x_{ij}}{\partial E_n}\right)_{X,T}$$

Pyroelectricity and the electrocaloric effect

$$p_n^X = -\left(\frac{\partial^2 G}{\partial T \partial E_n}\right)_X = \left(\frac{\partial D_n}{\partial T}\right)_{X,E} = \left(\frac{\partial S}{\partial E_n}\right)_{X,T}$$

Thermal expansion and the piezocaloric effect

$$\alpha_{ij}^E = -\left(\frac{\partial^2 G}{\partial X_{ij} \partial T}\right)_E = \left(\frac{\partial x_{ij}}{\partial T}\right)_{X,E} = \left(\frac{\partial S}{\partial X_{ij}}\right)_{E,T}.$$

The general expressions for a coupled system are obtained by differentiating the free energy function G.

$$\delta S = \left(C^{X,E}/T\right)\delta T + \alpha_{ij}^E X_{ij} + p_m^X E_m$$

$$x_{ij} = \alpha_{ij}^E \delta T + s_{ijkl}^{E,T} X_{kl} + d_{mij}^T E_m$$

$$D_n = p_n^X \delta T + d_{nkl}^T X_{kl} + \varepsilon_{nm}^{X,T} E_m$$

Nine linear relationships, requiring six sets of property coefficients, are described by these three equations.

The examples just given point out the importance of thermodynamics to the physical properties of crystals. The equivalence of the electrocaloric effect and the pyroelectric effect together with the other Maxwell relations greatly simplifies the subject.

Linear magnetic phenomena can be introduced in a similar way. For a thermomagnetic system the Gibbs free energy is controlled by changes in the temperature T and magnetic field components H_i [A/m]:

$$dG = -S dT - I_i dH_i.$$

The magnetization I_i [Wb/m^2] and entropy S are partial derivatives of the Gibbs potential:

$$I_i = -\left(\frac{\partial G}{\partial H_i}\right)_T, \quad S = -\left(\frac{\partial G}{\partial T}\right)_H.$$

Further differentiation gives:

$$\frac{\partial^2 G}{\partial H_i \partial T} = -\left(\frac{\partial I_i}{\partial T}\right)_H = \frac{\partial^2 G}{\partial T \partial H_i} = -\left(\frac{\partial S}{\partial H_i}\right)_T.$$

This shows the equivalence of the pyromagnetic and magnetocaloric coefficients. Analogous Maxwell relations can be derived for the linear magnetoelectric and piezomagnetic phenomena discussed in Chapter 14.

The direct and converse magnetoelectric coefficients are related through the partial derivatives

$$\left(\frac{\partial I_i}{\partial E_j}\right)_H = \left(\frac{\partial P_i}{\partial H_j}\right)_E.$$

For a piezomagnetic system, the direct and converse effects are

$$\left(\frac{\partial I_i}{\partial X_{jk}}\right)_H = \left(\frac{\partial x_{jk}}{\partial H_i}\right)_X.$$

Higher order effects are handled in a similar way. Electrostriction and magnetostriction are good examples. Strain is proportional to the square of the electric field in electrostrictive materials. Expanding the Gibbs free energy into second order effects we find for an electromechanical system,

$$\frac{\partial^3 G}{\partial X \partial E^2} = \frac{\partial^2 x}{\partial E^2} = M$$

$$= \frac{\partial^3 G}{\partial E \partial X \partial E} = \frac{\partial^2 G}{\partial X \partial E} = \frac{\partial d}{\partial E}$$

$$= \frac{\partial^3 G}{\partial E \partial E \partial X} = \frac{\partial^2 D}{\partial E \partial X} = \frac{\partial \varepsilon}{\partial X}.$$

The electrostriction coefficient M is equal to the field dependence of the piezoelectric coefficient $\partial d/\partial E$ and to the stress dependence of the electric permittivity $\partial \varepsilon/\partial X$. This means that the electrostrictive coefficients can be measured in three different ways.

Magnetostriction, the dependence of strain on the square of the magnetic field, can also be measured in three different ways. It is equivalent to the dependence of the piezomagnetic coefficient on magnetic field and to the stress dependence of the magnetic susceptibility. The tensor form of these higher order effects are presented in Chapter 15.

6.3 Measurement conditions

Fig. 7.1 shows the specific heat measured under two different boundary conditions: at constant pressure p, and constant volume V. Over much of the temperature range C^P and C^V are nearly equal but the two curves begin to diverge above room temperature. Typically they differ by a few percent. Other boundary conditions become important for electrical and magnetic measurements, and for ultrasonic and optical experiments carried out at higher frequencies.

Experimentalists often find it difficult to maintain ideal measurement conditions. *Isothermal* (T constant) experiments are carried out slowly to keep the sample in equilibrium with its surroundings at all times. *Adiabatic* (S constant) measurements are conducted in such a way that heat does not flow in or out of the specimen. This can be done by thermally isolating the sample from the environment or by making measurements faster than the times needed for heat transfer. These adiabatic experiments are referred to as *dynamic*, in contrast with the *static* isothermal tests.

Speed is also important in meeting mechanical boundary constraints. In a *mechanically free* or *unclamped* (constant stress X) test, the sample is allowed to slowly deform. Deformation takes time because strain travels with the speed of sound. By careful control of the mounting scheme it is possible to carry out such tests under static conditions. *Mechanically clamped* (constant strain x) experiments are difficult at low frequencies because they require that the crystal be surrounded by a medium of infinite stiffness. Normally they are carried out at high frequencies in which the deformations are too slow to follow the external fields or forces.

Magnetic and electric boundary conditions are sometimes important as well. *Electrically free* (E constant) conditions are met by keeping the surface at a constant potential. Short-circuiting the sample with a metal coating accomplishes this result. Embedding the specimen within a high permeability matrix ensures a *magnetically free* (H constant) environment. *Electrically clamped* (P constant) or *magnetically clamped* (I constant) boundary conditions are not easy to satisfy since some polarization and magnetization mechanisms are capable of following very high frequencies. Working under open-circuit conditions is often the best that can be done. Mobile domain walls make significant contributions to the polarization and magnetization in ferroelectric and ferromagnetic substances. The nucleation and growth process involved in wall motion requires time so that this process is effectively clamped at high frequencies.

Measurement conditions sometimes make a big difference, and sometimes not. The differences are determined thermodynamically. As an example, consider the electrocaloric effect described in Section 7.3. The heat capacities under electrically free and electrically clamped conditions are evaluated as follows. Changes in electric displacement D_i and entropy S arise from changes in temperature T and electric field E_j:

$$dD_i = \left(\frac{\partial D_i}{\partial E_j}\right)_T dE_j + \left(\frac{\partial D_i}{\partial T}\right)_{E_j} dT$$

$$dS = \left(\frac{\partial S}{\partial E_j}\right)_T dE_j + \left(\frac{\partial S}{\partial T}\right)_{E_j} dT.$$

Eliminating dE_j between these two equations and setting $dD_i = 0$ gives

$$dS = \left(\frac{\partial S}{\partial T}\right)_E dT - \left(\frac{\partial D_i}{\partial T}\right)_E \left(\frac{\partial S}{\partial E_j}\right)_T \left(\frac{\partial E_i}{\partial D_j}\right)_T dT$$

at constant D. Dividing through by dT and using the Maxwell relation $(\partial S/\partial E_j)_T = (\partial D_j/\partial T)_E$ leads to

$$\left(\frac{\partial S}{\partial T}\right)_E - \left(\frac{\partial S}{\partial T}\right)_D = \left(\frac{\partial D_i}{\partial T}\right)_E \left(\frac{\partial D_j}{\partial T}\right)_E \left(\frac{\partial E_i}{\partial D_j}\right)_T.$$

Multiplying through by temperature and converting to property coefficients gives the differences between electrically free and electrically clamped

Table 6.1 Difference between heat capacities (C), pyroelectric coefficients (p), thermal expansion coefficients (α), dielectric permittivities (ε), piezoelectric constants (d), and elastic compliances (s), measured under different boundary conditions

---

Mechanically clamped (C^x) and free (C^X) specific heat measured at constant field.
$C^X - C^x = T\alpha_{ij}\alpha_{kl}c_{ijkl}$

Electrically clamped (C^D) and free (C^E) specific heat measured at constant stress.
$C^D - C^E = Tp_i p_j \beta_{ij}^T$

Pyroelectric and electrocaloric coefficients measured at constant stress (p_i^X) and constant strain (p_i^x).
$p_i^X - p_i^x = \alpha_{jk}^E c_{jklm}^{E,T} d_{ilm}^T$

Mechanically clamped (ε_{ij}^x) and free (ε_{ij}^X) dielectric permittivities measured under isothermal conditions.
$\varepsilon_{ij}^X - \varepsilon_{ij}^x = d_{ikl}d_{jmn}c_{klmn}^E$

Isothermal (ε_{ij}^T) and adiabatic (ε_{ij}^S) permittivities measured at constant stress.
$\varepsilon_{ij}^T - \varepsilon_{ij}^S = p_i p_j T/C^E$

Electrically clamped (α_{ij}^D) and free (α_{ij}^E) thermal expansion and piezocaloric coefficients.
$\alpha_{ij}^E - \alpha_{ij}^D = d_{ijk}^T \beta_{kl}^{X,T} p_l^X$

Isothermal (d_{ijk}^T) and adiabatic (d_{ijk}^S) direct and converse piezoelectric coefficients.
$d_{ijk}^T - d_{ijk}^S = Tp_i^x \alpha_{jk}^E/C^{x,E}$

Isothermal (s_{ijkl}^T) and adiabatic (s_{ijkl}^S) elastic compliances measured at constant field.
$s_{ijkl}^T - s_{ijkl}^S = \alpha_{ij}\alpha_{kl}T/C^x$

Electrically clamped (s_{ijkl}^D) and free (s_{ijkl}^E) elastic compliances measured at constant temperature.
$s_{ijkl}^E - s_{ijkl}^D = d_{mij}d_{nkl}\beta_{mn}^x$

specific heat.

$$C^E - C^D = T p_i p_j \beta_{ij}^T.$$

In this expression, p_i and p_j are components of the pyroelectric effect and β_{ij}^T is a component of the isothermal dielectric stiffness.

Differences in other measured coefficients are similar in form (Table 6.1). In most oxides and other good insulators, the differences are seldom larger than 1%. Pyroelectricity is the one exception. Secondary pyroelectricity arising from piezoelectricity and thermal expansion is sometimes larger than the primary pyroelectric effect. Coefficients can also be quite different in ferroic crystals especially near phase transitions.

Specific heat and entropy

7

Before beginning the discussion of directional properties, we pause to consider specific heat, an important scalar property of solids which helps illustrate the important thermodynamic relationships between measured properties. Heat capacity, compressibility, and volume expansivity are interrelated through the laws of thermodynamics. Based on these ideas, similar relationships are established for other electrical, thermal, mechanical, and magnetic properties. Several atomistic concepts are introduced to help understand the structure–property relationships involved in specific heat measurements.

7.1	Heat capacity of solids	43
7.2	Lattice vibrations	46
7.3	Entropy and the magnetocaloric effect	48

7.1 Heat capacity of solids

The heat capacity or specific heat is the amount of heat required to raise the temperature of a solid by 1 K. It is usually measured in units of J/kg K. Theorists prefer to work in J/mole K, and older scientists sometimes use calories rather than joules. One calorie is 4.186 J. For solids and liquids, the specific heat is normally measured at a constant pressure:

$$C_P = \left(\frac{\Delta Q}{\Delta T}\right)_P,$$

where ΔQ is the heat added to increase the temperature by ΔT. Measurements on gases are usually carried out at constant volume:

$$C_V = \left(\frac{\Delta Q}{\Delta T}\right)_V.$$

Electrical methods are generally employed in measuring specific heat. A heating coil is wrapped around the sample and the resulting change in temperature is measured with a thermocouple. If a current I flows through a wire of resistance R, the heat generated by the wire in a time Δt is given by

$$\Delta Q = I^2 R \Delta t.$$

To eliminate heat loss to the surroundings, especially at low temperatures where C_P is small, the sample is suspended in a vacuum by very thin thread. The temperature change ΔT is measured as a function of time. The molar heat capacity at constant pressure, is given by

$$C_P = \frac{I^2 R \Delta t}{n \Delta T},$$

where n is the number of moles in the sample. The specific heat at constant volume, C_V, is much more difficult to measure and is usually obtained from the thermodynamic relation

$$C_P - C_V = TV\beta^2/K$$

in which V is the molar volume, β the volume expansivity and K the isothermal compressibility. Expansivity will be discussed later in Section 11.2, in connection with linear thermal expansion. Compressibility is described in Section 13.7 with elasticity.

The quantity $A = V\beta^2/KC_P^2$ is found to be nearly temperature independent. For copper, it is 1.55×10^{-5} moles/cal from 100 to 1200 K. Therefore, C_V can be evaluated from the so-called *Nernst–Lindemann* equation:

$$C_P - C_V = AC_P^2 T.$$

C_V is obtained for other materials by measuring $C_P(T)$ and evaluating A at room temperature. It is unnecessary to measure V, β, and K as a function of temperature.

Fig. 7.1 shows the specific heat curves for copper and rocksalt. From this it is apparent that metals and nonmetals have similar temperature dependence, though they differ in detail. For both materials C_P and C_V go to zero as T approaches zero, but the temperature dependence is different. For NaCl and other insulators, C is proportional to T^3 at very low temperatures. Lattice vibrations control the specific heat in these materials. In metals like copper the free electrons make a contribution to specific heat that results in a linear dependence of C on T at low temperature.

Above 10 K the lattice vibrations dominate in both metals and insulators. In most solids the atoms vibrate about their equilibrium positions but there are no rotational degrees of freedom. Classically, each vibrational degree of freedom involves both potential and kinetic energy giving a thermal energy of kT per degree of vibrational freedom. Since there are three independent directions, the average vibrational energy is $3kT$. For N atoms, the internal energy $U = 3NkT$ and the specific heat is

$$C_V = \left(\frac{\partial U}{\partial T}\right)_V = 3Nk = 3R,$$

where k is Boltzmann's Constant and R is the Universal Gas Constant.

An element like Cu contains $N_0 = 6.023 \times 10^{23}$ atoms/mole, giving $C_V = 3N_0 k = 6$ cal/K/mole $= 25$ J/K/mole. Thus the classical result depends only on the number of atoms, and is independent of the atomic species, chemical bonding, and crystal structure. This is the so-called *Law of Dulong and Petit*. Near room temperature and above, measured values agree well with classical theory. Some typical values for solid elements are Al 6.0, Cu 5.7, Ag 5.8, and Pb 6.2 cal/K/mole.

Diatomic solids contain $2N_0$ atoms/mole, doubling the specific heat. The specific heat of sodium chloride at room temperature is 12.0 cal/K/mole. For a mole of triatomic CaF_2, the observed value is 17.1, in accordance with the number of atoms per mole.

At lower temperatures, the specific heat approaches zero at 0 K. Debye developed a theory that describes the temperature dependence of C_V with an expression governed by a characteristic temperature θ_D, the Debye temperature. Debye treated the vibrating solid as a continuous medium rather than a collection of discrete atoms, but disregarded all vibrations with wavelengths shorter than near-neighbor bond lengths.

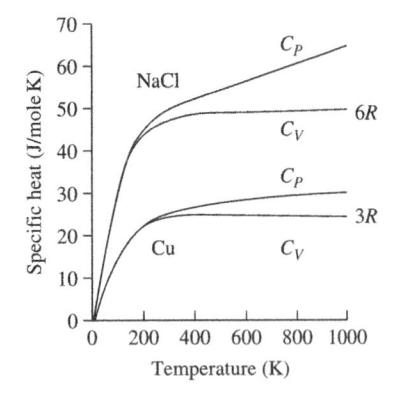

Fig. 7.1 Temperature variation of C_P and C_V in sodium chloride and copper.

In the liquid helium range at very low temperatures, Debye theory predicts a specific heat

$$C_V \cong 3nR\left[\frac{4}{5}\pi^4\left(\frac{T}{\theta_D}\right)^3 + \cdots\right] = 1944n\left(\frac{T}{\theta_D}\right)^3\left[\frac{J}{\text{mole K}}\right],$$

where n is the number of atoms per molecule, R is the Gas Constant, 8.31 J/mole K, and θ_D is the empirical Debye constant. At high temperature, the Debye theory predicts that $C_V \to 3nR$ in agreement with the experimental Law of Dulong and Petit. In between the high- and low-temperature regimes, the specific heat values are expressed as series

$$C_V \cong 3nR\left[1 - \frac{1}{20}\left(\frac{\theta_D}{T}\right)^2 + \frac{1}{560}\left(\frac{\theta_D}{T}\right)^4 + \cdots\right]\left[\frac{J}{\text{mole K}}\right]$$

and fitted to the experimental values with the Debye constant θ_D.

For substances with small θ_D, C_V increases rapidly with temperature, leveling off at the classical value well below room temperature. The characteristic temperature is controlled by the elastic constants and chemical bonding (Table 7.1). Soft metals have low Debye temperatures: Pb 70 K, Cd 168 K. The θ_D values for ionic crystals are somewhat higher (NaCl 280 K, CaF$_2$ 475 K) but much less than diamond (1860 K).

The Debye temperature is largest for solids with high melting points. Lindemann has shown that for many solids θ_D is roughly proportional to the melting point:

$$\theta_D \text{ (K)} \cong \frac{200}{V^{1/3}}\left(\frac{T_m}{M}\right)^{1/2},$$

where V is the molar volume in cm^3/mole, M the molar mass in grams/mole, and T_m the melting point in Kelvin.

Anisotropic crystals show unusual behavior at low temperatures, where the specific heat of most simple structures is proportional to T^3. Layer structures such as arsenic and antimony follow a T^2 dependence while selenium and other chain structures show a linear relation between specific heat and temperature.

Benzol and other molecular crystals have peculiar specific curves because chemical bonding within a molecule is much stronger than between molecules. As a result, the intermolecular vibrations are much easier to excite than the

Table 7.1 Debye temperature θ_D and root-mean-square vibration amplitudes at room temperature

Element, structure, and interatomic distance (Å)			θ_D (K)		$\sqrt{U^2}$ (Å)
Al	FCC	2.86	395		0.101
Au	FCC	2.88	175		0.084
Cu	FCC	2.56	314		0.084
Li	BCC	3.04	316		0.209
Mg	HCP	3.20, 3.21	320	$\parallel c$	0.125
	$c/a = 1.624$			$\perp c$	0.130
Pb	FCC	3.50	70		0.206
Zn	HCP	2.67, 2.91	250	$\parallel c$	0.153
	$c/a = 1.856$			$\perp c$	0.091
C	D	1.54	1860		0.050
Si	D	2.34	550		0.075

internal vibrations within a single molecule, and the material behaves as if it had two characteristic temperatures.

The specific heats of some molecular solids are unusually large because of rotational motions. The onset of molecular rotations is accompanied by large changes in C_V.

In metals such as copper there is a contribution to the specific heat from conduction, but only electrons with energies close to the Fermi Level can be raised to higher energy states. Using Fermi–Dirac statistics it can be shown that the electronic contribution to the specific heat is

$$C_V \cong \frac{Nk\pi^2}{2} \frac{kT}{E_F},$$

where N is the number of conduction electrons per mole, k is Boltzmann's Constant, and E_F is the Fermi energy. For copper, even at high temperature near the melting point, the electronic contribution is less than 0.8 J/mole K. Therefore the specific heat of metals differs little from other inorganic solids. It is only at very low temperatures where the linear dependence dominates the T^3 dependence of lattice vibrations.

Debye temperatures for a number of semiconductor crystals with the zincblende structure are shown in Fig. 7.2. Note the decrease in θ_D with increasing lattice parameter. Short strong bonds give large Debye temperatures.

Problem 7.1
Experimental values of C_P (in cal/mole K) measured as a function of temperature (in K) for water:

T	C_P	T	C_P	T	C_P
0	0.00	100	3.75	280	18.03
40	1.57	150	5.16	300	18.00
60	2.39	200	6.57	340	18.03
80	3.09	273	8.64	373	18.09

a. Compare these values with the Dulong and Petit prediction and explain the structure–property relationships.
b. Calculate C_V at 273 K from C_P using the measured values for expansivity ($\beta = 158 \times 10^{-6}$/K) and compressibility (12.7×10^{-11} m^2/N).

7.2 Lattice vibrations

Since lattice vibrations control many of the thermal properties of solid it is important to have a visual picture of these atomic motions.

Rigid models made of plastic balls and metal rods are often used to represent crystal structures, but the atoms in a real solid are in ceaseless motion, oscillating rapidly about the equilibrium sites. As Dame Kathleen Lonsdale once wrote, "a crystal is like a class of children arranged for drill, but standing at ease, so that while the class as a whole has regularity both in time and space, but each individual child is a little fidgety."

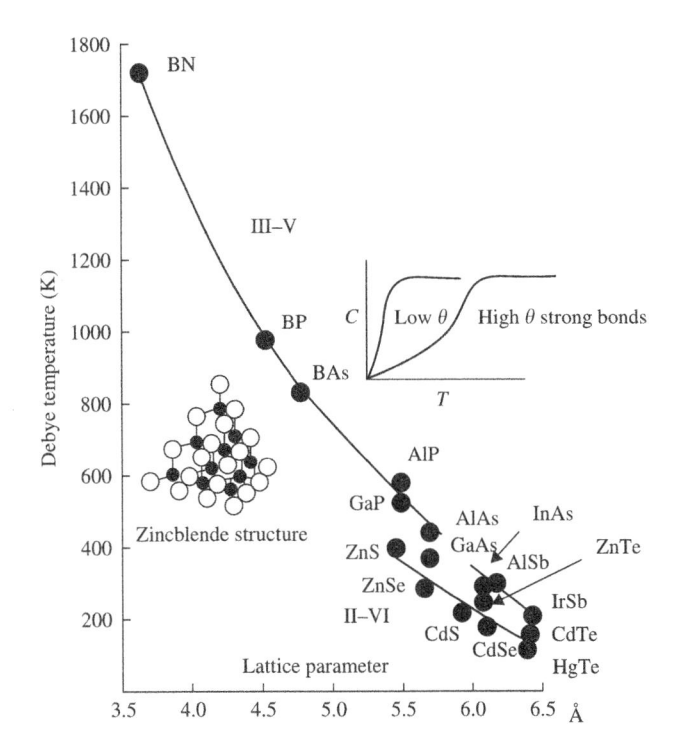

Fig. 7.2 Crystal structures with long, weak bonds have higher specific heat at low temperature because the atoms vibrate very easily.

In a crystal this fidgetiness usually amounts to a few percent of the interatomic distance. For the elements listed in Table 7.1, the vibration amplitudes near room temperature are 3–7% of the nearest neighbor distances. As might be expected, the vibrations are largest for soft materials such as lithium and lead, and smallest for diamond.

Vibration amplitudes increase with temperature, but not very rapidly. The root-mean-square vibration amplitude of aluminum increases from 0.057 Å at 10 K, to 0.152 Å at 600 K. Motion continues to low temperatures because of the presence of zero point energy.

Atoms do not vibrate with equal amplitude in all directions. Magnesium and zinc crystallize in the hexagonal close-packed structure with twelve neighbors around each atom, six in the same (001) layer, three in the layer above, and three below. The packing in magnesium is almost ideal with nearly equal interatomic distances, and the thermal vibrations are nearly isotropic (Table 7.1). In zinc, the c/a ratio exceeds the ideal value (1.633) so that the Zn–Zn distances within a close-packed plane are shorter than those between atoms in adjacent layers. As a result atoms can vibrate more easily along c than perpendicular to c where atomic motion is more restricted.

For diatomic compounds the lighter of the two atoms generally vibrates with the larger amplitude. Thermal vibration amplitudes estimated from X-ray diffraction intensities for LiH show that H^- vibrates more than Li^+ because it is lighter. Measurements on LiF, NaF, NaCl, KCl, and CaF_2 verify this result.

In organic molecular crystals where C, N, and O all have about the same atomic weight, the atoms near the perimeter of the molecule usually undergo larger vibrations than those near the center where the bonding is stronger.

7.3 Entropy and the magnetocaloric effect

For a reversible process an increase in temperature produces a change in entropy:
$$dQ = T\,dS = C_V\,dT,$$
where S is the entropy in J/mole. Entropy is a measure of the disorder of the system, but it is difficult to describe because there are so many ways to introduce disorder into a system (Fig. 7.3). For a solid consisting of atoms, molecules, and conduction electrons, a number of different types of motions, excitations, and defects can occur:

a. Vibrations of atoms and molecules about their lattice positions.
b. Molecular rotations.
c. Spin disorder of conduction electrons.
d. Translational motions of charge carriers.
e. Disordered magnetic and electric dipoles.
f. Vacancies, interstitials, and other defects.
g. Atomic order–disorder.
h. Electronic excitations.

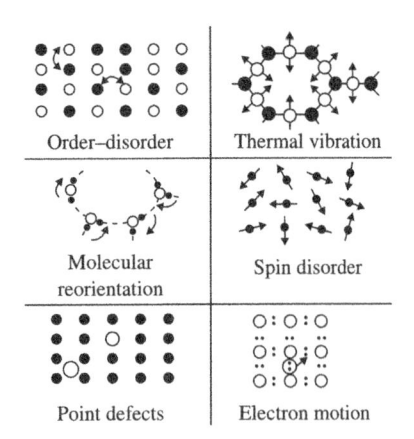

Fig. 7.3 Some of the origins of entropy. Intermetallic compounds frequently show local disorder, while molecular reorientation is common in ice and other hydrogen-bonded systems. Thermal vibrations are present in all solids. Paramagnetic and paraelectric materials possess randomly oriented dipoles, metals have mobile electrons, and point defects are prevalent in many silver salts. In its simplest formulation ($S = k \ln W$) is proportional to W, the number of ways to disorder.

In terms of their contribution to specific heat, none are as important as lattice vibrations, but several interesting interactions take place under external fields and forces.

The *magnetocaloric effect* is a coupling between thermal and magnetic properties that has long been used to cool cryogenic systems to temperatures approaching absolute zero. The process is known as *adiabatic demagnetization* and takes advantage of the entropy stored in paramagnetic salts with randomly oriented magnetic dipoles. To produce temperatures below 1 K, a small paramagnetic crystal is suspended in a cryostat on tiny threads. The vessel contains gaseous helium at low pressure and is immersed in a bath of liquid helium cooled to 1 K by pumping away He vapor. A magnetic field is then switched on, partially aligning the magnetic dipoles. This is an isothermal process because the He gas in the cryostat maintains thermal contact with the liquid He bath. After equilibrating, the gaseous He is pumped out of the cryostat, thermally isolating the paramagnetic crystal and its aligned spins from the environment. The magnetic field is then decreased to zero causing the sample to cool to a very low temperature as the magnetic dipoles return to a disordered state. Heat is removed from the lattice vibrations by the magnetic spin system. At these low temperatures well below 1 K, any residual He gas surrounding the sample condenses, and the sample can be maintained at low temperature for an appreciable period of time.

The magnetocaloric effect just described makes use of one of the Maxwell relations governing reversible changes in thermodynamic systems (Chapter 6). For the adiabatic demagnetization experiment, the change in entropy with magnetic field under isothermal conditions is equal to the change in magnetization with temperature under adiabatic conditions.

In equation form,

$$\left(\frac{\partial S}{\partial H}\right)_T = \left(\frac{\partial I}{\partial T}\right)_H,$$

where S is entropy, H magnetic field, I the induced magnetization, and T temperature. For a dilute paramagnetic salt such as $KCr(SO_4)_2 \cdot 12H_2O$, the unpaired spins of the chromium ions maintain their random orientations to very low temperature. As a result the Curie Law behavior characteristic of a paramagnetic solid is maintained below 1 K. Therefore $(\partial I/\partial T)_H$ is negative and so is $(\partial S/\partial H)_T$. The extraction of heat from the lattice by the chromium spins is a small effect but it is significant at very low temperatures because the lattice vibration energy has almost disappeared.

Effects similar to the magnetocaloric effect occur in electrically- and mechanically-aligned systems. Crystals like KTaO$_3$ are the electric analogs to the paramagnetic salts used in adiabatic demagnetization. The dielectric constants of paraelectric solids follow a Curie Law in which the electric dipoles become very easy to align at low temperature. The *electrocaloric effect* is governed by the Maxwell relation

$$\left(\frac{\partial S}{\partial E}\right)_T = \left(\frac{\partial P}{\partial T}\right)_E,$$

where E is electric field and P the field-induced polarization. Adiabatic depolarization leads to cooling when carried out in a two-step process in which an electric field polarizes the crystal under isothermal conditions, and then the crystal is depolarized adiabatically to lower the temperature.

The field-induced magnetocaloric and electrocaloric effects just described are not quite the same as the magnetocaloric and electrocaloric effects described in later chapters. Pyroelectric crystals possess an electrocaloric effect even in the absence of an applied electric field. The effect is associated with the temperature dependence of the spontaneous polarization P_S. In this case the Maxwell relation is

$$\left(\frac{\partial S}{\partial E}\right)_T = \left(\frac{\partial P_S}{\partial T}\right)_{E=0}.$$

As demonstrated later (Chapter 8), P_S is observed in only ten of the 32 crystal classes. Field-induced electrocaloric effects have no symmetry restrictions.

Pyromagnetism is the magnetic counterpart to pyroelectricity. The temperature dependence of the spontaneous magnetization (I_S) in ferromagnetic and ferrimagnetic materials gives rise to a spontaneous magnetocaloric effect:

$$\left(\frac{\partial S}{\partial H}\right)_T = \left(\frac{\partial I_S}{\partial T}\right)_{H=0}.$$

The spontaneous magnetocaloric effect is found in only 31 of 90 magnetic point groups (Chapter 14).

A coupling between entropy and mechanical stress leads to a *piezocaloric effect* related to thermal expansion:

$$\left(\frac{\partial S}{\partial X}\right)_T = \left(\frac{\partial x}{\partial T}\right)_X,$$

where X is mechanical stress and x is strain. The piezocaloric effect is especially large in rubber and other elastomers with large thermal expansion coefficients. When rubber is in an unstretched state, the molecules are coiled in an amorphous form, but when it is stretched, they align in a semicrystalline state that causes a large decrease in entropy.

8 Pyroelectricity

8.1 Pyroelectric and
    electrocaloric tensors 50
8.2 Symmetry limitations 50
8.3 Polar axes 52
8.4 Geometric
    representation 53
8.5 Pyroelectric
    measurements 54
8.6 Primary and secondary
    pyroelectric effects 54
8.7 Pyroelectric materials 55
8.8 Temperature
    dependence 55
8.9 Applications 57

As the name implies, pyroelectricity is a first rank tensor property relating a change polarization P to a change in temperature δT. The defining relation can also be written in terms of the electric displacement D since no field is applied:

$$P_i = D_i = p_i \delta T \ [\text{C/m}^2]$$

8.1 Pyroelectric and electrocaloric tensors

Pyroelectricity is a first rank polar tensor because of the way it transforms. Being polar vectors, P_i and D_i transform as

$$D'_i = a_{ij} D_j$$

whereas the temperature change transforms as a zero rank tensor, or a scalar:

$$\delta T' = \delta T.$$

Transforming the defining relation for pyroelectricity we get

$$D'_i = a_{ij} D_j = a_{ij} p_j \delta T = a_{ij} p_j \delta T' = p'_i \delta T'.$$

Both the independent variable δT and the dependent variable D_i have now been transformed to the new coordinate system. The property relating D'_i to $\delta T'$ is the transformed pyroelectric coefficient

$$p'_i = a_{ij} p_j.$$

Thus the pyroelectric coefficient is a polar first rank tensor property.

In Sections 6.1 and 7.3 it was shown that the electrocaloric effect and the pyroelectric effect are governed by the same set of coefficients p_i. The change in entropy per unit volume caused by an electric field is

$$\delta S = p_i E_i \ [\text{J/m}^3].$$

The pyroelectric (=electrocaloric coefficient) coefficient is usually expressed in units of $\mu\text{C/m}^2$ K and can be either positive or negative in sign depending on whether the spontaneous (built-in) polarization is increasing or decreasing with temperature.

8.2 Symmetry limitations

Pyroelectricity disappears in all centrosymmetric materials. The proof follows. For a first rank tensor there are, in general, three nonzero coefficients p_1, p_2, and p_3 representing the values of the pyroelectric coefficient along property axes Z_1, Z_2, and Z_3, respectively. The principal axes are perpendicular to each other and are chosen in accordance with the IEEE convention (Section 4.3).

The tensor coefficients are often written in matrix form:

$$\begin{matrix} 3\times 1 & 3\times 3 & 3\times 1 \\ (p') & = & (a) & (p). \end{matrix}$$

By Neumann's Principle $(p') = (p)$ if the direction cosine matrix (a) is a symmetry element of the material. For a center of symmetry operation we have

$$(p') = \begin{pmatrix} p'_1 \\ p'_2 \\ p'_3 \end{pmatrix} = \begin{pmatrix} -1 & 0 & 0 \\ 0 & -1 & 0 \\ 0 & 0 & -1 \end{pmatrix} \begin{pmatrix} p_1 \\ p_2 \\ p_3 \end{pmatrix} = \begin{pmatrix} -p_1 \\ -p_2 \\ -p_3 \end{pmatrix} = -(p).$$

By Neumann's Principle

$$\begin{pmatrix} -p_1 \\ -p_2 \\ -p_3 \end{pmatrix} = \begin{pmatrix} p_1 \\ p_2 \\ p_3 \end{pmatrix},$$

which can only be satisfied if all three pyroelectric coefficients are zero: $p_1 = p_2 = p_3 = 0$. Therefore, pyroelectricity disappears in all point groups containing inversion symmetry (10 of the 32 crystal classes and 3 of the 7 Curie groups).

The pyroelectric effect sometimes disappears in other point groups as well. Quartz is an important piezoelectric crystal that does not exhibit pyroelectricity. The point group is 32 with a threefold axis along Z_3 and a twofold axis along Z_1. Under the threefold transformation,

$$\begin{pmatrix} p'_1 \\ p'_2 \\ p'_3 \end{pmatrix} = \begin{pmatrix} -1/2 & \sqrt{3}/2 & 0 \\ -\sqrt{3}/2 & -1/2 & 0 \\ 0 & 0 & 1 \end{pmatrix} \begin{pmatrix} p_1 \\ p_2 \\ p_3 \end{pmatrix} = \begin{pmatrix} -p_1/2 + \sqrt{3}p_2/2 \\ -\sqrt{3}p_1/2 - p_2/2 \\ p_3 \end{pmatrix} = \begin{pmatrix} p_1 \\ p_2 \\ p_3 \end{pmatrix}$$

the last equality being Neumann's Principle. This can be satisfied only if $p_1 = p_2 = 0$. Next we operate with the second symmetry element, the twofold axis along Z_1, and equate it to the already reduced pyroelectric matrix.

$$\begin{pmatrix} p'_1 \\ p'_2 \\ p'_3 \end{pmatrix} = \begin{pmatrix} 1 & 0 & 0 \\ 0 & -1 & 0 \\ 0 & 0 & -1 \end{pmatrix} \begin{pmatrix} p_1 \\ p_2 \\ p_3 \end{pmatrix} = \begin{pmatrix} p_1 \\ -p_2 \\ -p_3 \end{pmatrix} = \begin{pmatrix} 0 \\ 0 \\ p_3 \end{pmatrix}.$$

The result is that $p_1 = p_2 = p_3 = 0$.

Two final examples will conclude the discussion. Pyroelectricity was first discovered in the mineral tourmaline, point group $3m$. There are two independent symmetry elements, a threefold axis along Z_3 and a mirror plane perpendicular to Z_1. Following the procedure for quartz, the threefold axis causes p_1 and p_2 to disappear. The mirror plane then gives the result.

$$\begin{pmatrix} p'_1 \\ p'_2 \\ p'_3 \end{pmatrix} = \begin{pmatrix} -1 & 0 & 0 \\ 0 & 1 & 0 \\ 0 & 0 & 1 \end{pmatrix} \begin{pmatrix} p_1 \\ p_2 \\ p_3 \end{pmatrix} = \begin{pmatrix} -p_1 \\ p_2 \\ p_3 \end{pmatrix} = \begin{pmatrix} 0 \\ 0 \\ p_3 \end{pmatrix}.$$

Therefore tourmaline is pyroelectric long the Z_3 axis with one nonzero coefficient p_3.

A similar result is found for poled ferroelectric ceramics. Here the grains are randomly oriented but a strong DC field aligns the domains within each grain. Pyroelectricity and piezoelectricity are observed after the poling field is removed. The symmetry of the poled ceramic is ∞m with the ∞-fold axis along Z_3, the direction of poling field. There are an infinite number of mirror planes parallel to this axis.

The appropriate operators for the symmetry operations are given in Section 4.3. For the infinite-fold axis the transformation is

$$\begin{pmatrix} p'_1 \\ p'_2 \\ p'_3 \end{pmatrix} = \begin{pmatrix} \cos\theta & -\sin\theta & 0 \\ \sin\theta & \cos\theta & 0 \\ 0 & 0 & 1 \end{pmatrix} \begin{pmatrix} p_1 \\ p_2 \\ p_3 \end{pmatrix}$$

$$= \begin{pmatrix} p_1 \cos\theta - p_2 \sin\theta \\ p_1 \sin\theta + p_2 \cos\theta \\ p_3 \end{pmatrix} = \begin{pmatrix} p_1 \\ p_2 \\ p_3 \end{pmatrix}.$$

The equality must hold for all θ values. This is only possible if $p_1 = p_2 = 0$. For a mirror plane perpendicular to Z_1,

$$\begin{pmatrix} p'_1 \\ p'_2 \\ p'_3 \end{pmatrix} = \begin{pmatrix} -1 & 0 & 0 \\ 0 & 1 & 0 \\ 0 & 0 & 1 \end{pmatrix} \begin{pmatrix} p_1 \\ p_2 \\ p_3 \end{pmatrix} = \begin{pmatrix} -p_1 \\ p_2 \\ p_3 \end{pmatrix} = \begin{pmatrix} 0 \\ 0 \\ p_3 \end{pmatrix}.$$

As with tourmaline there is no effect on p_3. The same holds true for any mirror plane parallel to Z_3.

8.3 Polar axes

Pyroelectric matrices for the 32 crystal classes and seven Curie groups are listed in Table 8.1. For triclinic crystals in point group 1, three measurements are necessary to specify the pyroelectric effect. For monoclinic point group m, two measurements, p_1 and p_3, are required. All other pyroelectric classes require only one measurement. The direction of the polar axis is fixed by symmetry in these classes, and cannot change with temperature or pressure unless the symmetry changes. The magnitude of p_3 will change but not its orientation.

It is easy to visualize the polar axis directions by examining stereographic projections. Stereograms for the three monoclinic point groups ($2, m, 2/m$) are shown in Fig. 8.1. By drawing vectors from the origin to each equivalent point, the vectors can be summed to give the polar axis direction. For point group 2, the polar axis lies along the twofold symmetry axis, while for point group m the polar axis lies in the plane of the mirror. The four vectors for point group $2/m$ sum to zero, confirming the absence of polar symmetry and pyroelectricity in this point group.

The two ends of a polar axis are *not* related by any symmetry element of the crystal. Symmetry elements that destroy polarity include a center of symmetry or a mirror plane or twofold axis perpendicular to the axis. Pyroelectricity always occurs along a polar axis but not all polar axes exhibit pyroelectricity.

Table 8.1 Matrices for the ten pyroelectric crystal classes and two pyroelectric Curie groups (all others are zero)

Point group 1	$\begin{pmatrix} p_1 \\ p_2 \\ p_3 \end{pmatrix}$
Point group 2	$\begin{pmatrix} 0 \\ p_2 \\ 0 \end{pmatrix}$
Point group m	$\begin{pmatrix} p_1 \\ 0 \\ p_3 \end{pmatrix}$
Point groups $mm2$, 3, $3m$, 4, $4mm$, 6, $6mm$, ∞, ∞m	$\begin{pmatrix} 0 \\ 0 \\ p_3 \end{pmatrix}$

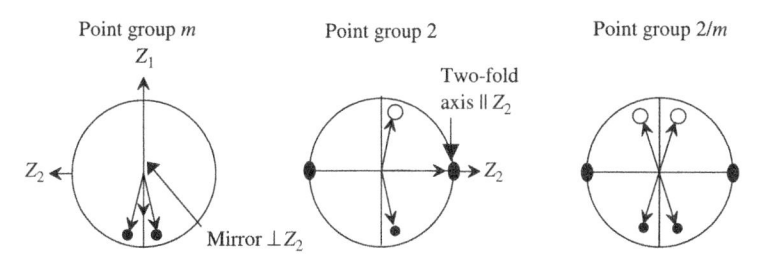

Fig. 8.1 Vector summations in the monoclinic point groups m, 2, and $2/m$. By drawing a vector from the origin at the center of the stereographic projection to each equivalent point, the direction of the polar axes can easily be visualized.

The Z_1 axis of quartz, for example, is a polar axis but pyroelectric charges do not occur in this direction when the temperature is changed. This is because the pyroelectric charges in the three symmetry-related polar axes in point group 32 sum to zero. Pyroelectricity occurs only along *unique* polar axes unrelated to other polar axes.

8.4 Geometric representation

By plotting the pyroelectric coefficient as a function of measurement direction, one obtains a geometric view of the pyroelectric effect. For a crystal plate cut perpendicular to an arbitrarily chosen direction Z_1' (Fig. 8.2), the pyroelectric coefficient is given by

$$p_1' = a_{ij}p_j = a_{11}p_1 + a_{12}p_2 + a_{13}p_3.$$

Converting to spherical coordinates,

$$p_1' = p_1 \sin\theta \cos\phi + p_2 \sin\theta \sin\phi + p_3 \cos\theta.$$

Pyroelectricity was first discovered in the mineral tourmaline that belongs to point group $3m$. The pyroelectric coefficients of tourmaline are $p_1 = p_2 = 0$, $p_3 = 4$ μC/m^2 K. The geometric representation of the pyroelectric effect in tourmaline consists of two spheres, one positive and one negative, centered

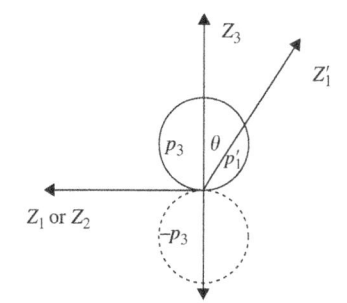

Fig. 8.2 A geometric representation of the pyroelectric representation of the pyroelectric effect in crystals like tourmaline.

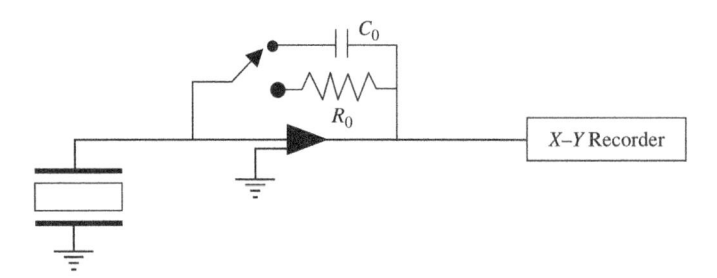

Fig. 8.3 Two methods of measuring the pyroelectric coefficient.

at $Z_3 = \pm 2 \ \mu\text{C/m}^2$ K. The spheres are obtained by plotting the equation $p'_1 = 4\cos\theta$ for all ϕ and θ values.

8.5 Pyroelectric measurements

An electroded sample is heated in a sample chamber and its change in polarization monitored with the measurement circuit shown in Fig. 8.3. The coefficient p is obtained by measuring the voltage V across a reference capacitor C_0 or by measuring the pyroelectric current through a reference resistor R_0.

Pyroelectric voltage method:
$$p = \frac{C_0}{A}\frac{dV}{dT} \quad [\text{C/m}^2 \text{ K}].$$

Pyroelectric current method:
$$p = \frac{i}{A \, dT/dt} \quad [\text{C/m}^2 \text{ K}],$$

where A is the electrode area, T temperature, t time, V voltage, and i current.

8.6 Primary and secondary pyroelectric effects

Pyroelectric coefficients are sometimes measured under different boundary conditions. When measured at constant strain, thermal expansion effects do not contribute to the polarization, and this so-called "primary" pyroelectric effect originates from internal rearrangements in structure. This can be pictured as cations moving relative to anions with no change in unit cell dimensions. But this is not the way pyroelectric coefficients are normally measured.

Most pyroelectric experiments are carried out at constant stress rather than constant strain. This means that the unit cell dimensions can change through thermal expansion, and since all pyroelectric materials are also piezoelectric, there will be a contribution to the pyroelectric effect. This is the so-called "secondary" pyroelectric effect given by the thermodynamic relation.

$$p_i^X - p_i^x = \alpha_{jk}^E c_{jklm}^{E,X} d_{ilm}^X.$$

In this expression α, c, d are thermal expansion, elastic stiffness, and piezoelectric coefficients, respectively. The unclamped pyroelectric coefficient p_i^X measured at constant stress is equal to the sum of the pyroelectric effect caused

Table 8.2 Pyroelectric coefficients measured at room temperature for ferroelectric and nonferroelectric materials. The so-called "primary" pyroelectric coefficient is measured at constant strain and represents the rearrangement of polarization charges inside the unit cell. The "secondary" contribution comes from change in unit cell dimensions with temperature. Together they give the experimental pyroelectric coefficient that is normally measured at constant stress (the units are $\mu C/m^2\ K$)

	Experimental value (p^X)	Secondary effect ($p^X - p^x$)	Primary effect (p_n^x)
Ferroelectrics			
Poled ceramic (∞m)			
$BaTiO_3$	−200	+60	−260
$PbZr_{0.95}Ti_{0.05}O_3$	−268	+37.7	−305.7
Crystal			
$LiNbO_3\ (3m)$	−83	+12.8	−95.8
$LiTaO_3\ (3m)$	−176	−1	−175
$Pb_5Ge_3O_{11}\ (3)$	−95	+15.5	−110.5
$Ba_2NaNb_5O_{15}\ (mm2)$	−100	+41.7	−141.7
$Sr_{0.5}Ba_{0.5}Nb_2O_6\ (4mm)$	−550	−48	−502
$(CH_2CF_2)_n\ (mm2)$	−27	−13	−14
TGS (2)	−270	−330	+60
Nonferroelectrics			
Crystal			
CdSe (6mm)	−3.5	−0.56	−2.94
CdS (6mm)	−4.0	−1.0	−3.0
ZnO (6mm)	−9.4	−2.5	−6.9
Tourmaline (3m)	+4.0	−3.52	−0.48
$Li_2SO_4 \cdot 2H_2O\ (2)$	+86.3	+26.1	+60.2

by the atomic rearrangements inside the unit cell (p_i^x) and the piezoelectric contribution due to thermal expansion.

8.7 Pyroelectric materials

Table 8.2 lists the measured pyroelectric coefficients for several ferroelectric and nonferroelectric crystals together with a few poled polycrystalline materials. Note that the coefficient is negative in most ferroelectrics because the spontaneous polarization decreases as temperature is raised. The polarization goes to zero at T_c, the Curie temperature, where the symmetry changes to a nonpolar point group. Note that, the secondary contribution to the pyroelectric coefficients is generally smaller than the primary effect, especially in ferroelectrics where the atoms involved in the phase transformation are undergoing substantial movements.

8.8 Temperature dependence

Pyroelectric coefficients approach zero at very low temperatures where thermal motion decreases dramatically. This is illustrated for zinc oxide in Fig. 8.4 showing how the pyroelectric coefficient correlates well with specific heat since both are caused by thermal motions. ZnO has the hexagonal wurtzite structure with polar axis along [001] direction. All the ZnO_4 tetrahedra point

Fig. 8.4 (a) Crystal structure of zinc oxide showing the polar orientation of the zinc–oxygen bonds along the hexagonal c-axis. ZnO has the wurtzite structure and belongs to polar group 6mm. (b) The pyroelectric coefficients of nonferroelectric crystals are generally small and correlate well with specific heat.

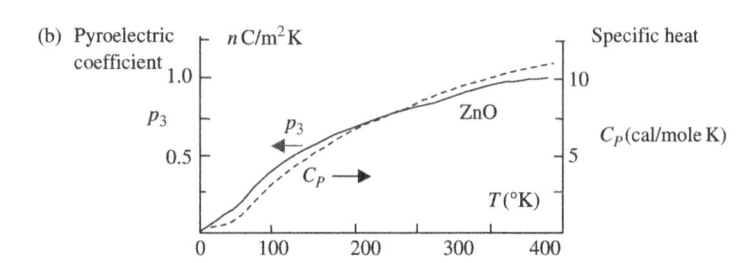

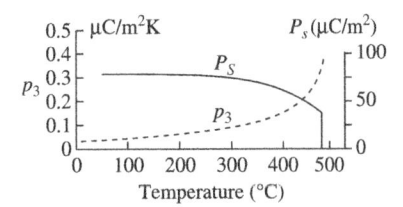

Fig. 8.5 In poled ferroelectric crystals the pyroelectric effect is the temperature derivative of the spontaneous polarization. PbTiO$_3$ polarizes along the tetragonal c-axis. Both P_s and the pyroelectric coefficient p_3 are zero above T_c where the symmetry changes to cubic at 490°C.

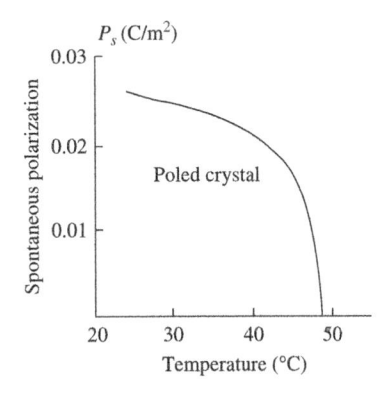

Fig. 8.6 Triglycine sulfate crystals exhibit a substantial pyroelectric effect below the Curie temperature of 47°C. The crystals are monoclinic with the maximum pyroelectric coefficient along the [010] twofold symmetry axis.

along the hexagonal c-axis causing the electric dipoles associated with the cations and anions to add rather than cancel.

At high temperatures the pyroelectric effect increases as thermal motion becomes more pronounced. This is especially true for ferroelectric crystals and poled ceramics in which the spontaneous polarization decreases rapidly near the Curie temperature. Lead titanate (PbTiO$_3$) and triglycine sulfate (TGS = (NH$_2$CH$_2$COOH)$_3$ · H$_2$SO$_4$) are typical ferroelectrics with first- and second-order phase transitions, respectively. Both are widely used in pyroelectric devices.

The spontaneous polarization and pyroelectric coefficients of PbTiO$_3$ and TGS are shown in Figs 8.5 and 8.6, respectively. PbTiO$_3$ has a first-order phase transformation at 490°C at which the spontaneous polarization drops abruptly to zero. The pyroelectric coefficient, which is the temperature derivative of the spontaneous polarization, also disappears at T_c. Above T_c, PbTiO$_3$ has the cubic perovskite structure, point group $m3m$. Below T_c it develops a large spontaneous polarization along the [001] direction as the symmetry changes to polar point group $4mm$. As shown in Fig. 8.5 the pyroelectric coefficient p_3 increases to very large values near T_c.

Similar behavior is observed for ferroelectric triglycine sulfate (Fig. 8.6). In TGS the spontaneous polarization decreases smoothly to zero at its Curie temperature of 47°C. The pyroelectric coefficient increases slowly near room temperature, then more rapidly as it approaches T_c. Large triglycine sulfate crystals are grown easily from water solution and are often doped with alanine to stabilize the polarization state. The ferroelectric effect, spontaneous polarization, and pyroelectric behavior are caused by changes in hydrogen bonding. Protons linking two of the molecular groups undergo an order–disorder phase change at the Curie temperature. Hydrogen bonds parallel to the [010] twofold symmetry axis freeze into either an up- or down-position in a double potential well. This creates dipole moments and spontaneous polarization within each ferroelectric domain. Above T_c the protons oscillate back and forth between the two potential wells effectively destroying the polarization. The point group is $2/m$ above T_c and 2 below T_c.

In a few crystals the pyroelectric coefficient changes sign with temperature. For lithium sulfate p passes through zero mean 120 K, and for barium nitrate, reversal takes place at 160 K. This means that the spontaneous polarization passes through a maximum (or minimum) at the crossover point.

8.9 Applications

Pyroelectric detectors are widely used as burglar alarms, fire detectors, and infrared imaging systems. The material figure of merit for many of these applications is

$$\frac{p}{C_V\sqrt{K}\tan\delta},$$

where p is the pyroelectric coefficient, C_V is the volume specific heat, K is the dielectric constant, and $\tan\delta$ the electrical dissipation factor. Many of the best ferroelectric materials are unsuitable because of their large dielectric constants. It is also important to keep the specific heat and thermal mass small to allow a rapid rise in temperature.

Certain glass composites can be converted to pyroelectric sensors by appropriate heat treatment. Polar glass-ceramics of $Li_2Si_2O_5$ and $Ba_2TiSi_2O_8$ (fresnoite) are made as glasses at high temperature and then annealed in a temperature gradient to produce aligned polar crystallites. The pyroelectric and piezoelectric properties are similar to those of the silicate minerals tourmaline and quartz. The aligned crystallites nucleate on the surface of the glass and then grow into the glass with polar symmetry. The symmetry group is ∞m, the same as a poled ferroelectric, but no poling is required, and there is no phase transition.

Problem 8.1
There are three orthorhombic point groups 222, *mm*2, and *mmm*.

a. Draw the stereographic projections showing the symmetry elements and symmetry-equivalent points. Determine the polar axis directions.
b. Write out the transformation matrices for the symmetry elements required to generate each of the orthorhombic groups.
c. Using Neumann's Principle, determine the pyroelectric matrices for orthorhombic crystals.

Problem 8.2
As shown in Table 8.2, the electrocaloric effect is related to the pyroelectric effect through the Maxwell relation.

$$\frac{\partial S}{\partial E} = \frac{\partial P}{\partial T}.$$

When an electrical field is applied to a pyroelectric crystal, its temperature changes. Using the data shown in Fig. 8.4, estimate the temperature change in ZnO under a field of 10^5 V/m. How large is the electrocaloric effect at room temperature? What is the temperature change?

9 Dielectric constant

9.1 Origins of the dielectric constant 58
9.2 Dielectric tensor 60
9.3 Effect of symmetry 62
9.4 Experimental methods 63
9.5 Geometric representation 67
9.6 Polycrystalline dielectrics 69
9.7 Structure–property relationships 69

The dielectric constant K is a measure of a material's ability to store electric charge. In scalar form the defining relations are as follows:

$$D = \varepsilon E,$$

where D is the electric displacement measured in C/m^2, ε is the electric permittivity in F/m, and E is electric field in V/m. The dielectric constant K is the relative permittivity:

$$K = \varepsilon/\varepsilon_0,$$

where $\varepsilon_0 = 8.85 \times 10^{-12}$ F/m is the permittivity of free space. The electric displacement D is equal to the sum of the charges stored on the electrode plus those originating from the polarization, P [C/m^2]

$$D = \varepsilon_0 E + P.$$

In this chapter we discuss the tensor nature of the dielectric constant, how it is represented geometrically, and some typical structure–property relationships.

9.1 Origins of the dielectric constant

Dielectric constants range over about four orders of magnitude in insulator materials. Because of their low density, gases have dielectric constants only slightly larger than one. At one atmosphere, the dielectric constant of air is 1.0006. Most common ceramics and polymers have dielectric constants in the range between 2 and 10. Polyethylene is 2.3 and silica glass is 3.8. These are low-density dielectrics with substantially covalent bonding. More ionic materials like NaCl and Al_2O_3 have slightly higher K values in the 6–10 range. High K materials like water ($K \sim 80$) and $BaTiO_3$ ($K \sim 1000$) have special polarization mechanisms involving rotating dipoles or ferroelectric phase transformations.

A schematic view of the principal types of polarization mechanisms is illustrated in Fig. 9.1. The *electronic* component of polarization arising from field-induced changes in the electron cloud around each atom is found in all matter. The *ionic* contribution is also common and is associated with the relative motions of cations and anions in an electric field. *Orientational* polarizability arises from the rotation of molecular dipoles in the field. These motions are common in organic substances. Many materials also contain *mobile charge carriers* in the form of ions or electrons that can migrate under applied fields. All four types of polarization are capable of creating anisotropy as well.

By measuring the dielectric constant as a function of frequency, one can separate the various polarization components. Each polarization mechanism has a limiting characteristic frequency. Electrons have very small mass and are therefore able to follow high frequency fields up through the optical range (Fig. 9.2). Ions are a thousand times heavier but continue to follow fields up to the infrared range. Molecules—especially those in liquids and solids—are heavier yet and are severely impeded by their surroundings. Most *rotational* effects, like those in water, are limited to microwave frequencies. *Space charge* effects are often in the kilohertz range or even lower. Fig. 9.2 shows a typical frequency spectrum of a dielectric containing all four types of polarization. Dielectric constants are complex numbers with real (K') and imaginary (K'') components.

In simple inorganic solids, the two most important mechanisms are the electronic and ionic polarizabilities. The dielectric constants of twelve alkali halide crystals are compared in Fig. 9.3. The electronic contribution n^2 is obtained from the optical refractive index n. Subtracting n^2 from the low frequency dielectric constant gives the ionic part. The two parts are nearly equal except for the lithium salts where the ionic polarization dominates. This is probably due to the small size of the Li^+ ion that oscillates more easily than the larger Na^+, K^+, and Rb^+ ions.

Thermal measurements also shed light on the various types of polarization. Lead zirconate titanate (PZT = $PbZr_{1-x}Ti_xO_3$) is a ferroelectric ceramic used in piezoelectric and dielectric applications. Domain walls make an important contribution to dielectric properties of ferroelectric materials. Fig. 9.4 compares the measured dielectric constants of a typical "hard" and "soft" PZT. Soft PZTs have mobile domain walls that make a very large contribution to permittivity at room temperature but freeze out at low temperatures. The effect is much smaller in hard PZT where the defect structure makes it difficult to move domain walls. The intrinsic part of the dielectric constant coming from electronic and ionic motions within each domain can be ascertained from thermodynamic modeling. The chemical compositions of hard and soft PZTs differ only slightly.

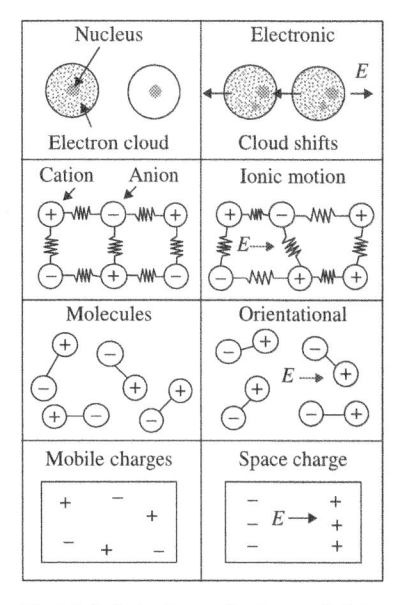

Fig. 9.1 Polarization mechanisms underlying electric permittivity.

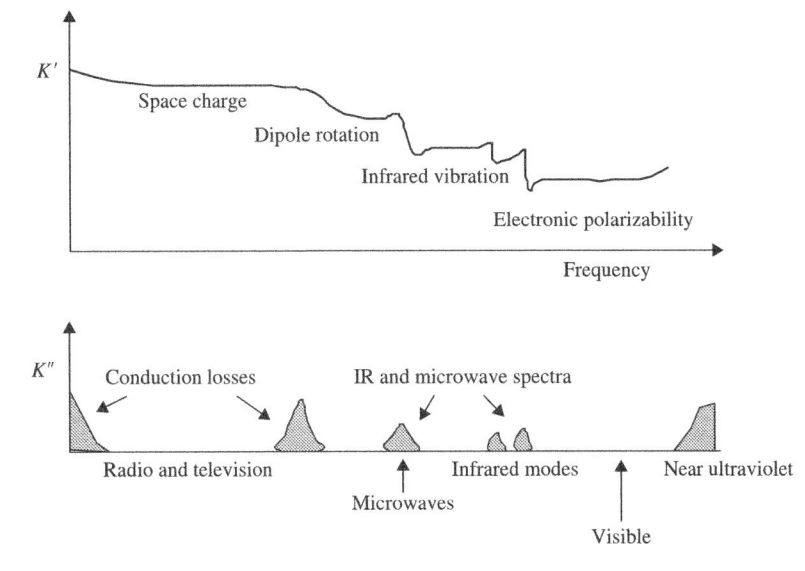

Fig. 9.2 Dielectric spectrum of a complex solid contain several polarization mechanisms.

60 *Dielectric constant*

Fig. 9.3 Electronic and ionic contributions to the dielectric constants of alkali halide crystals.

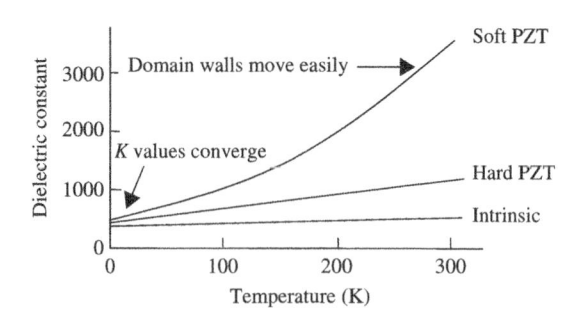

Fig. 9.4 Temperature dependence of the dielectric constant of unpoled PZT. Ferroelectric domain wall contributions to the polarization freeze out of low temperature (Cross).

Hard PZTs are doped with acceptor ions such as Fe^{3+} while soft PZTs are doped with Nb^{5+} or other donor ions. The dopants create defects that either impede or promote domain wall motion.

9.2 Dielectric tensor

Both electric displacement and electric field are vectors, or first rank tensors. Therefore there are two directions involved in measuring the electric permittivity, making it a second rank tensor:

$$D_i = \varepsilon_{ij} E_j.$$

Transforming these tensors between coordinate systems we obtain

$$D'_i = a_{ij} D_j = a_{ij} \varepsilon_{jk} E_k = a_{ij} \varepsilon_{jk} a_{lk} E'_l$$
$$D'_i = \varepsilon'_{il} E'_l.$$

Therefore the permittivity and the dielectric constant transform as a polar second rank tensor.

$$\varepsilon'_{il} = a_{ij}a_{lk}\varepsilon_{jk}$$

$$K'_{il} = \frac{\varepsilon'_{il}}{\varepsilon_0} = a_{ij}a_{lk}K_{jk}.$$

This is the way K transforms in tensor notation. In matrix notation the transformation looks like this

$$\begin{array}{cccccccc} 3\times 1 & & 3\times 3 & 3\times 1 & & 3\times 3 & 3\times 3 & 3\times 1 \\ (D') & = & (a) & (D) & = & (a) & (\varepsilon) & (E) \\ & & 3\times 3 & 3\times 3 & 3\times 3 & 3\times 1 & & 3\times 3 & 3\times 1 \\ & = & (a) & (\varepsilon) & (a)^{-1} & (E') & = & (\varepsilon') & (E') \end{array},$$

where $(a)^{-1}$ is the reciprocal direction cosine matrix. It is equal to $(a)_t$, the transpose of the (a) matrix. In matrix form the dielectric constant transforms as

$$(K') = (a)(K)(a)_t.$$

Based on energy arguments, the (K) matrix must be symmetric. When an electric field is applied to a dielectric, the change in energy is

$$dU = E_i\, dD_i = \varepsilon_{ij}E_i\, dE_j.$$

Expanding,

$$dU = \varepsilon_{11}E_1 dE_1 + \varepsilon_{12}E_1 dE_2 + \varepsilon_{21}E_2 dE_1 + \cdots$$

and taking differentials with respect to the field components E_1 and E_2,

$$\frac{\partial U}{\partial E_1} = \varepsilon_{11}E_1 + \varepsilon_{21}E_2 + \varepsilon_{31}E_3$$

$$\frac{\partial U}{\partial E_2} = \varepsilon_{12}E_1 + \varepsilon_{22}E_2 + \varepsilon_{32}E_3.$$

The second derivatives give the permittivity coefficients.

$$\frac{\partial^2 U}{\partial E_1 \partial E_2} = \varepsilon_{21}$$

$$\frac{\partial^2 U}{\partial E_2 \partial E_1} = \varepsilon_{12}.$$

Since the stored energy is the same regardless of which field component is applied first, $\varepsilon_{21} = \varepsilon_{12}$. The permittivity and dielectric constant matrices are therefore symmetric:

$$\varepsilon_{ij} = \varepsilon_{ji} \quad \text{and} \quad K_{ij} = K_{ji}.$$

Energy arguments also place certain restrictions on the signs and magnitudes of the dielectric constant. The stored electric energy per unit volume must be a positive number.

$$\int_0^E \varepsilon_{ij} E_i \, dE_j = \frac{1}{2} \varepsilon_{ij} E_i E_j > 0$$

$$= \frac{1}{2} \varepsilon_{11} E_1^2 + \frac{1}{2} \varepsilon_{22} E_2^2 + \frac{1}{2} \varepsilon_{33} E_3^2 + \varepsilon_{12} E_1 E_2 + \varepsilon_{13} E_1 E_3$$

$$+ \varepsilon_{23} E_2 E_3 > 0.$$

No matter what combination of fields is applied, the energy is greater than zero. If $E_1 \neq 0 = E_2 = E_3$, then $\frac{1}{2} \varepsilon_{11} E_1^2 > 0$. Therefore $\varepsilon_{11} > 0$ and so is K_{11}. By the same argument K_{22} and K_{33} must also be positive numbers.

Off diagonal components such as K_{12} are not necessarily positive, but there are restrictions which apply. Referring to the energy expression, suppose $E_1 = E_2 \neq 0 = E_3$, then $K_{11} + K_{22} > -2K_{12}$. And if $E_1 = -E_2 \neq 0 = E_3$, then $K_{11} + K_{22} > 2K_{12}$. Combining these two inequalities,

$$K_{11} + K_{22} > |2K_{12}|.$$

And if $K_{11} = K_{22}$ by symmetry, then $K_{11} > |K_{12}|$. Thus the off-diagonal components are generally smaller than the diagonal values.

9.3 Effect of symmetry

We illustrate by deriving the dielectric constant matrix for cubic crystals. The minimum symmetry found in cubic crystals corresponds to point group 23 with twofold axes along $\langle 100 \rangle$ directions and threefold axes along $\langle 111 \rangle$ axes. The two independent symmetry elements needed to generate symmetry group 23 are shown in Fig. 9.5.

To carry out the simplification process for cubic crystals we first apply the twofold symmetry operation.

$$(K') = (a)(K)(a)_t$$

$$= \begin{pmatrix} 1 & 0 & 0 \\ 0 & -1 & 0 \\ 0 & 0 & -1 \end{pmatrix} \begin{pmatrix} K_{11} & K_{12} & K_{13} \\ K_{21} & K_{22} & K_{23} \\ K_{31} & K_{32} & K_{33} \end{pmatrix} \begin{pmatrix} 1 & 0 & 0 \\ 0 & -1 & 0 \\ 0 & 0 & -1 \end{pmatrix}$$

$$= \begin{pmatrix} K_{11} & -K_{12} & -K_{13} \\ -K_{21} & K_{22} & K_{23} \\ -K_{31} & K_{32} & K_{33} \end{pmatrix} = \begin{pmatrix} K_{11} & K_{12} & K_{13} \\ K_{21} & K_{22} & K_{23} \\ K_{31} & K_{32} & K_{33} \end{pmatrix}.$$

By Neumann's Principle, the transformed and untransformed matrices must be equal. Therefore $K_{12} = K_{13} = K_{21} = K_{31} = 0$.

Next we take the simplified (K) matrix and apply the second element, a threefold rotation about [111].

$$(K') = \begin{pmatrix} 0 & 1 & 0 \\ 0 & 0 & 1 \\ 1 & 0 & 0 \end{pmatrix} \begin{pmatrix} K_{11} & 0 & 0 \\ 0 & K_{22} & K_{23} \\ 0 & K_{32} & K_{33} \end{pmatrix} \begin{pmatrix} 0 & 0 & 1 \\ 1 & 0 & 0 \\ 0 & 1 & 0 \end{pmatrix}$$

$$= \begin{pmatrix} K_{22} & K_{23} & 0 \\ K_{32} & K_{33} & 0 \\ 0 & 0 & K_{11} \end{pmatrix} = \begin{pmatrix} K_{11} & 0 & 0 \\ 0 & K_{22} & K_{23} \\ 0 & K_{32} & K_{33} \end{pmatrix}.$$

$$(a) = \begin{pmatrix} +1 & 0 & 0 \\ 0 & -1 & 0 \\ 0 & 0 & -1 \end{pmatrix} \qquad (a) = \begin{pmatrix} 0 & +1 & 0 \\ 0 & 0 & +1 \\ +1 & 0 & 0 \end{pmatrix}$$

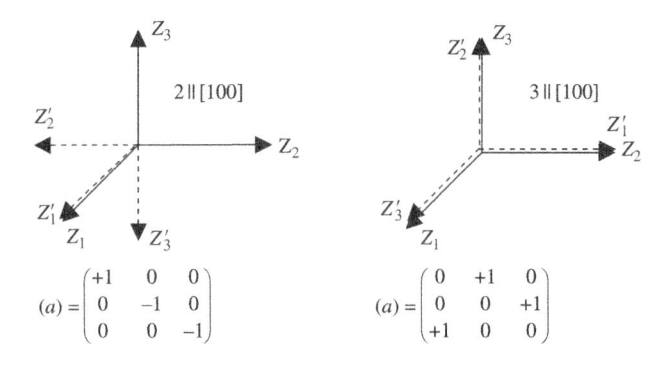

Fig. 9.5 Two symmetry elements found in all cubic crystals.

Again applying Neumann's Principle, the result is $K_{11} = K_{22} = K_{33}$, and $K_{23} = K_{32} = 0$. Therefore the final result is that the dielectric constant for cubic crystals is a scalar.

$$(K) = \begin{pmatrix} K_{11} & 0 & 0 \\ 0 & K_{11} & 0 \\ 0 & 0 & K_{11} \end{pmatrix}$$

The same symmetry-reduction procedure can be done easier and faster by the direct inspection method. For point group 23, the twofold axis along Z_1 carries $Z_1 \to Z_1$, $Z_2 \to -Z_2$, and $Z_3 \to -Z_3$, or in shorthand notation: $1 \to 1$, $2 \to -2$, $3 \to -3$. Next apply these operations to the subscripts of the dielectric constant tensor, and then apply Neumann's Principle.

$$\begin{aligned}
K_{11} &\quad 11 \to 11 = 11 \\
K_{12} &\quad 12 \to -12 = 12 \quad \therefore K_{12} = 0 \\
K_{13} &\quad 13 \to -13 = 13 \quad \therefore K_{13} = 0 \\
&\vdots
\end{aligned}$$

Next apply the same procedure to the second symmetry element, the threefold rotation about the body diagonal $[111]$: $1 \to 2, 2 \to 3, 3 \to 1$. Then

$$\begin{aligned}
K_{11} &\quad 11 \to 22 = 11 \quad \therefore K_{11} = K_{22} \\
&\vdots
\end{aligned}$$

The direct inspection method can be used for symmetry operations involving most symmetry elements, but not for three- or sixfold axes where the angles differ from 90° or 180°. The tensor or matrix techniques are necessary for nonorthogonal transformations.

Applying Neumann's Principle to the 32 crystal classes and seven Curie groups leads to the dielectric matrices in Table 9.1.

9.4 Experimental methods

The dielectric constant is anything but constant. It depends on frequency, electric field, temperature, pressure, and many other variables, and it is a complex

Table 9.1 Dielectric constant matrices for the various symmetry groups. Depending on the symmetry, between one and six measurements are required to specify the dielectric properties

Triclinic crystals
  Classes 1 and $\bar{1}$
  Six coefficients
$$\begin{pmatrix} K_{11} & K_{12} & K_{13} \\ K_{12} & K_{22} & K_{23} \\ K_{13} & K_{23} & K_{33} \end{pmatrix}$$

Monoclinic crystals
  Classes 2, m, and $2/m$
  Four coefficients
$$\begin{pmatrix} K_{11} & 0 & K_{13} \\ 0 & K_{22} & 0 \\ K_{13} & 0 & K_{33} \end{pmatrix}$$

Orthorhombic crystals
  Classes 222, $mm2$, and mmm
  Three coefficients
$$\begin{pmatrix} K_{11} & 0 & 0 \\ 0 & K_{22} & 0 \\ 0 & 0 & K_{33} \end{pmatrix}$$

Uniaxial crystals
  Classes 3, $\bar{3}$, 32, $3m$, $\bar{3}m$, 4, $\bar{4}$, $4/m$, 422,
  $4mm$, $\bar{4}2m$, $4/mmm$, 6, $\bar{6}$, $6/m$, 622,
  $6mm$, $\bar{6}m2$, and $6/mmm$
  Curie groups ∞, ∞m, ∞/m, $\infty 2$, and ∞/mm
  Two coefficients
$$\begin{pmatrix} K_{11} & 0 & 0 \\ 0 & K_{11} & 0 \\ 0 & 0 & K_{33} \end{pmatrix}$$

Cubic crystals
  Classes 23, $m3$, 432, $\bar{4}3m$ and $m3m$
  Curie groups $\infty\infty$ and $\infty\infty m$
  One coefficient
$$\begin{pmatrix} K_{11} & 0 & 0 \\ 0 & K_{11} & 0 \\ 0 & 0 & K_{11} \end{pmatrix}$$

quantity $K = K' - iK''$. The loss factor $\tan\delta = K''/K'$ can be very appreciable near resonant frequencies or in conducting solids. Measurement techniques depend on the frequency range (Fig. 9.6).

For anisotropic solids, measurements must be carried out in two or more directions. To illustrate, we consider the most general case, a triclinic crystal. There are no symmetry directions in triclinic crystals and therefore six measurements in different directions are required. From these six experiments we obtain the three principal dielectric constants and the three angles relating the principal axes to the measurement axes. All six quantities must be remeasured at different temperatures and frequencies because the principal axes are not fixed by symmetry. Moreover, the principal axes for K', the real part of dielectric constant, are not the same as those for the imaginary part, K''.

We begin the process by identifying the measurement axes Z'_i. The choice is arbitrary but must be fixed relative to the triclinic crystallographic axes (Fig. 9.7). Since the property axes are orthogonal and triclinic axes are not, the two coordinate systems cannot coincide. For convenience, we selected the [100] direction for Z'_1, and Z'_2 to lie between [100] and [010] in the (001) plane. In keeping with customary crystallographic notation, the triclinic interaxial angles are α, β, and γ.

In practice the triclinic axes are generally located using X-ray diffraction, although crystal morphology often provides a useful short cut. It is helpful to have a set of back-reflection Laue photographs and a crystal goniometer to orient the triclinic crystal along the measurement axes Z'_i. Using a saw, a cube-shape crystal is cut from the triclinic crystal, and six different measurement directions are identified (Fig. 9.8). For convenience, three directions are chosen along Z'_1, Z'_2, Z'_3. Three plates cut in these orientations are labeled measurement directions I, II, and III, respectively. The normal to the first plate, Z'_1, has direction cosines $a_{11} = 1$, $a_{12} = 0$, $a_{13} = 0$ relative to the measurement

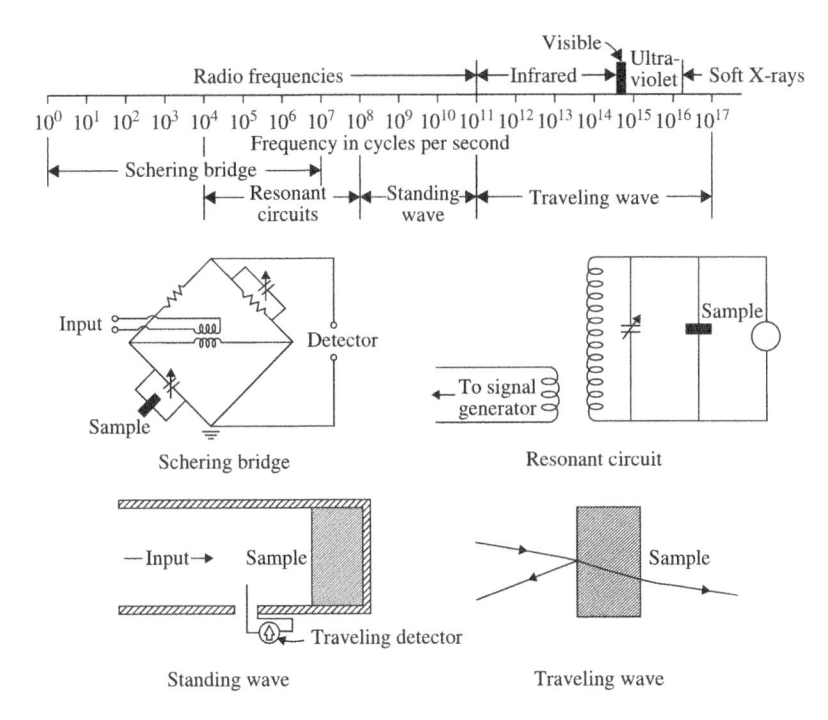

Fig. 9.6 Frequency ranges and measuring techniques (von Hippel).

coordinate system Z'_i. For plate II cut perpendicular to Z'_2 the direction cosines for the field direction are $a_{11} = 0$, $a_{12} = 1$, $a_{13} = 0$. And for plate III, $a_{11} = 0$, $a_{12} = 0$, $a_{13} = 1$. Three more measurements are needed for triclinic crystals. As shown in Fig. 9.8, plates IV, V, and VI are cut perpendicular to three body diagonal directions of the cube. The normal to these plates form angles of 54.7° or 125.3° with respect to the Z'_i coordinate system. The six plates and their direction cosines are illustrated in Fig. 9.8.

After cutting and polishing, the six plates are electroded on their major faces, and the capacitance is measured to give six values of the dielectric constant: K^{I}, K^{II}, K^{III}, K^{IV}, K^{V}, and K^{VI}. In shaping the samples it is important to keep the plate thickness small compared to the lateral dimensions. This ensures that the electric field will be parallel to the surface normal, remembering that in anisotropic solids the electric displacement is parallel to the applied electric field only in principal axis directions. Having the equipotential surface electrodes close together defines the electric field direction.

The dielectric constant matrix (K') for measurement coordinate system are determined from the six experimental values $K^{\mathrm{I}} - K^{\mathrm{VI}}$. K^{I}, the measured value for the plate cut perpendicular to Z'_1, is equal to K'_{11}. And for the other five plates the measured values are given by

$$K^M = a_{1i} a_{1j} K'_{ij} \qquad M = \text{I–VI}.$$

Using the direction cosines for each plate,

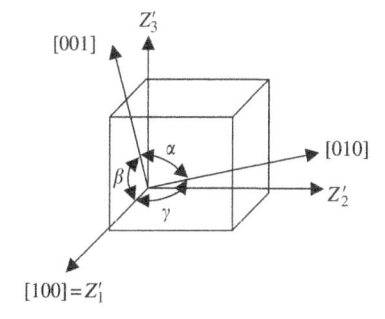

$$K^{\mathrm{I}} = K'_{11}$$
$$K^{\mathrm{II}} = K'_{22}$$
$$K^{\mathrm{III}} = K'_{33}$$

Fig. 9.7 Orientation of the measurement axes Z'_1, Z'_2, Z'_3 relative to the triclinic axes [100], [010], [001].

Dielectric constant

Fig. 9.8 Six plates, labeled I–VI, are cut from a cube whose axes are Z'_1, Z'_2, Z'_3. Dielectric constants K^I, K^{II}, K^{III}, K^{IV}, K^V, and K^{VI} are measured for the six orientations with direction cosines listed below. The six K values are then combined to give the three principal dielectric constants and the three angles required to specify the orientation of the principal axes.

I: (1,0,0) II: (0,1,0) III: (0,0,1)

IV: $(1/\sqrt{3}, 1/\sqrt{3}, 1/\sqrt{3})$ V: $(1/\sqrt{3}, 1/\sqrt{3}, -1/\sqrt{3})$ VI: $(-1/\sqrt{3}, 1/\sqrt{3}, 1/\sqrt{3})$

$$K^{IV} = \frac{1}{3}(K'_{11} + K'_{22} + K'_{33} + 2K'_{12} + 2K'_{13} + 2K'_{23})$$

$$K^{V} = \frac{1}{3}(K'_{11} + K'_{22} + K'_{33} + 2K'_{12} - 2K'_{13} - 2K'_{23})$$

$$K^{VI} = \frac{1}{3}(K'_{11} + K'_{22} + K'_{33} - 2K'_{12} - 2K'_{13} + 2K'_{23}).$$

Solving for the off-diagonal components,

$$K'_{12} = \frac{3}{4}(K^{IV} + K^V) - \frac{1}{2}(K^I + K^{II} + K^{III})$$

$$K'_{13} = \frac{3}{4}(K^V + K^{VI}) + \frac{1}{2}(K^I + K^{II} + K^{III})$$

$$K'_{23} = \frac{3}{4}(K^{IV} + K^{VI}) - \frac{1}{2}(K^I + K^{II} + K^{III}).$$

All six components of K'_{ij} are now known.

The next step is to convert to principal axes Z_1, Z_2, Z_3. In so doing, we need to find the three principal dielectric constants K_{11}, K_{22}, K_{33}, and the nine direction cosines relating the principal axes to the measurement axes (Fig. 9.9).

Beginning with the matrix transformation from the measurement axes to the principal axes, $(K') = (a)(K)(a)_t$. Post-multiplying by the direction cosine matrix (a), we obtain

$$(K')(a) = (a)(K),$$

which is written out as

$$\begin{pmatrix} K'_{11} & K'_{12} & K'_{13} \\ K'_{12} & K'_{22} & K'_{23} \\ K'_{13} & K'_{23} & K'_{33} \end{pmatrix} \begin{pmatrix} a_{11} & a_{12} & a_{13} \\ a_{21} & a_{22} & a_{23} \\ a_{31} & a_{32} & a_{33} \end{pmatrix} = \begin{pmatrix} a_{11} & a_{12} & a_{13} \\ a_{21} & a_{22} & a_{23} \\ a_{31} & a_{32} & a_{33} \end{pmatrix} \begin{pmatrix} K_{11} & 0 & 0 \\ 0 & K_{22} & 0 \\ 0 & 0 & K_{33} \end{pmatrix}.$$

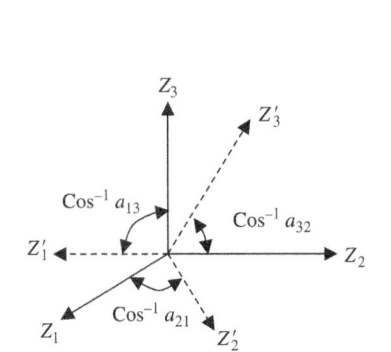

Fig. 9.9 Direction cosines relating the measurement axes Z'_i to the principal axes Z_i.

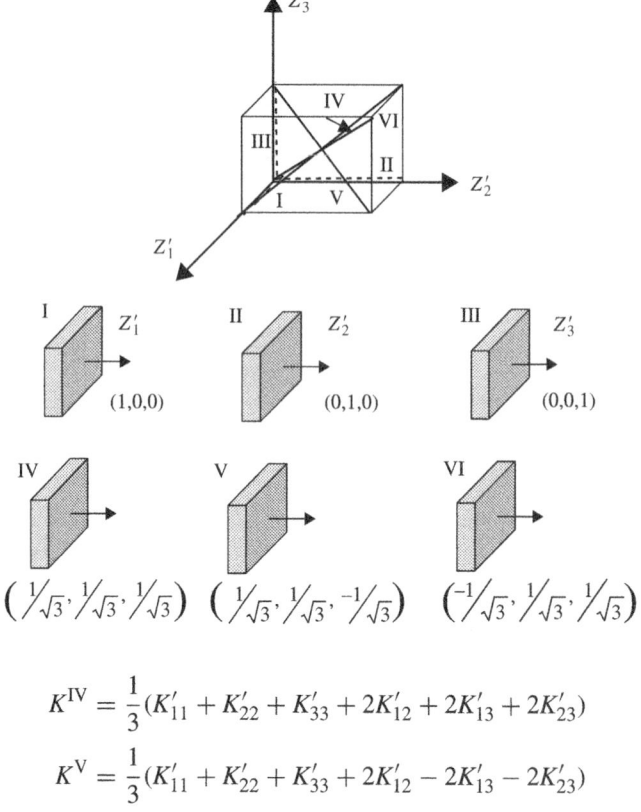

This leads to nine equations in K'_{ij}, a_{ij}, and K_{ii}

$$K'_{11}a_{11} + K'_{12}a_{21} + K'_{13}a_{31} = a_{11}K_{11}$$
$$K'_{11}a_{12} + K'_{12}a_{22} + K'_{13}a_{32} = a_{12}K_{22}$$
$$\vdots \qquad \vdots \qquad \vdots$$
$$\vdots \qquad \vdots \qquad \vdots$$
$$K'_{13}a_{13} + K'_{23}a_{23} + K'_{33}a_{33} = a_{33}K_{33}.$$

In solving for the nine a_{ij} values and the three K_{ij} values, we must also make use of the orthogonality conditions for the direction cosines: $a_{ik}a_{jk} = a_{ki}a_{kj} = \delta_{ij}$ (Section 2.3). For the nine equations written above, the determinant of the K'_{ij} and K_{ii} coefficients must equal zero for a nontrivial solution.

$$\begin{vmatrix} K'_{11} - K_{ii} & K'_{12} & K'_{13} \\ K'_{12} & K'_{22} - K_{ii} & K'_{23} \\ K'_{13} & K'_{23} & K'_{33} - K_{ii} \end{vmatrix} = 0.$$

Solving the determinant for the three roots K_{ii}, gives the magnitudes of the three principal dielectric constants of K_{11}, K_{22}, and K_{33}. To obtain the direction cosines a_{ij}, we substitute the K_{ii} values into the nine equations relating K'_{ij} to K_{ii} and make use of the orthogonality conditions as well. The process involves many arithmetic operations but can easily be solved by a computer.

The process just described used six measurements to obtain six unknowns. To obtain greater accuracy and to assess the experimental errors, it is useful to carry out the measurements in more than six different directions. A least squares refinement allows standard errors to be assigned to each of the principal dielectric constants and the three angles relating the principal axes to the measurement axes, and ultimately to the triclinic crystallographic axes.

It is a tedious task to specify the anisotropic properties of low symmetry crystals, which is one reason that relatively few such measurements have been carried out.

9.5 Geometric representation

When plotted as a function of direction, the dielectric constant K'_{11} is easily calculated from the principal dielectric constants.

$$K'_{11} = a_{11}^2 K_{11} + a_{12}^2 K_{22} + a_{13}^2 K_{33}.$$

Using spherical coordinates this becomes

$$K'_{11} = K_{11} \cos^2 \phi \sin^2 \theta + K_{22} \sin^2 \phi \sin^2 \theta + K_{33} \cos^2 \theta$$

To illustrate, we plot the room temperature dielectric constant of potassium dihydrogen phosphate (KDP = KH_2PO_4). Large crystals of KDP are easily grown from water solution and have been used as piezoelectric transducers and electro-optic light shutters. As shown in Fig. 9.10, the permittivity of KDP is highly anisotropic and strongly temperature dependent. At room temperature KDP is tetragonal, point group $\bar{4}2m$. The principal dielectric constants are

Dielectric constant

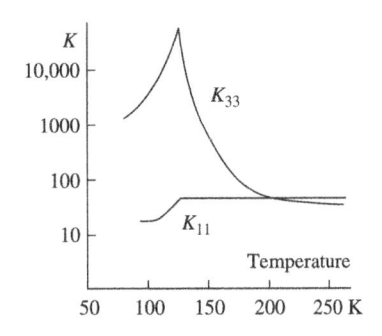

Fig. 9.10 Principal dielectric constants of KH$_2$PO$_4$ single crystals plotted as a function of temperature. Note change in sign of the anisotropy with $K_{11} > K_{33}$ at room temperature, and $K_{33} > K_{11}$ as the crystal cools toward the ferroelectric Curie point.

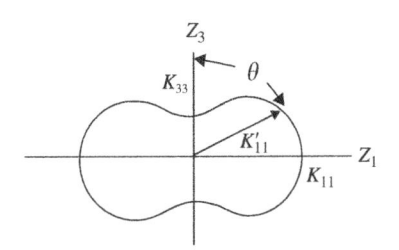

Fig. 9.11 The dielectric constant of KH$_2$PO$_4$ plotted as function of direction.

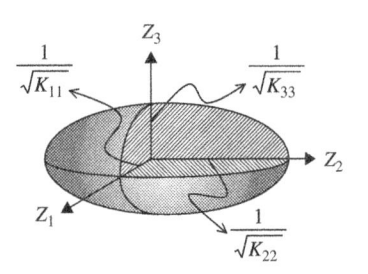

Fig. 9.12 By plotting $1/\sqrt{K}$ as a function of direction, the geometric surface representing the dielectric constant is an ellipsoid. The property coefficients can be negative for thermal expansion, magnetoelectricity, and certain other second rank tensors. In this case, hyperboloids may be generated.

$K_{11} = K_{22} = 60$ and $K_{33} = 24$. The geometric surface representing the longitudinal dielectric constant is given by

$$K'_{11} = 60 \sin^2 \theta + 24 \cos^2 \theta,$$

where θ is the angle between the measurement direction and Z_3, the $\bar{4}$ symmetry axis.

Fig. 9.11 shows the directional dependence of the dielectric constant at room temperature. The surface is cylindrically symmetric about Z_3 and resembles a doughnut without a hole in the center. As temperature is lowered, the surface becomes a sphere at the crossover point where $K_{11} = K_{33}$, and then develops into a long narrow peanut shape along Z_3.

There is another way of plotting symmetric second rank tensors like the dielectric constant. By plotting $1/\sqrt{K}$ rather than K, the surface is a simple quadric figure. To show that a quadric surface can be used to represent the dielectric constant, consider the general quadric equation referred to an arbitrary set of coordinates Z'_i.

$$S'_{11}Z'^2_1 + S'_{22}Z'^2_2 + S'_{33}Z'^2_3 + 2S'_{12}Z'_1Z'_2 + 2S'_{13}Z'_1Z'_3 + 2S'_{23}Z'_2Z'_3 = 1.$$

In shortened tensor form this is $S'_{ij}Z'_iZ'_j = 1$. The tensor character is demonstrated by inquiring how the coefficients S'_{ij} transform.

$$S'_{ij}Z'_iZ'_j = S'_{ij}a_{ik}Z_k a_{jl}Z_l = S_{kl}Z_k Z_l.$$

Thus $S_{kl} = a_{ik}a_{jl}S'_{ij}$ which is the way a second rank tensor transforms. When referred to principal axes, the quadric equation is

$$S_{11}Z^2_1 + S_{22}Z^2_2 + S_{33}Z^2_3 = 1.$$

Comparing this expression with the usual expression for an ellipsoid,

$$\frac{X^2}{a^2} + \frac{Y^2}{b^2} + \frac{Z^2}{c^2} = 1,$$

it is apparent that $S_{11} = 1/a^2$, etc. Therefore the intercept $a = 1/\sqrt{S_{11}}$. To obtain the quadric surface one plots $1/\sqrt{K}$ rather than K.

Since the principal dielectric constants are always positive numbers, the quadric surface is always an ellipsoid (Fig. 9.12), but for other second rank tensor properties like thermal expansion, the coefficient can be either positive or negative. Hyperboloid surfaces may be generated in these cases.

The quadric surfaces also make it easy to visualize the effect of symmetry on a physical property. The intrinsic symmetry of an ellipsoid is $2/m\,2/m\,2/m$, point group *mmm*. If the symmetry of the crystal is orthorhombic the twofold symmetry axes will be coincident with the principal axes. For monoclinic crystals only one principal axis is fixed, and for triclinic crystals, none. In tetragonal, trigonal, and hexagonal crystals, the quadric surface becomes an ellipsoid of revolution with a circular cross-section perpendicular to Z_3, the high symmetry direction. Cubic crystals have four threefold axes along body diagonal directions, and therefore there will be four circular cross-sections. Under this restriction the quadric surface representing the dielectric properties becomes a sphere.

9.7 Structure–property relationships

Problem 9.1
At 25°C a single-crystal (Single domain) of BaTiO$_3$ is tetragonal, point group 4mm.

a. Using Neumann's Law, deduce the dielectric constant matrix.
b. Measured values of the dielectric constants are

$$K_{11}^X = 4100 \quad K_{33}^X = 160 \quad (X = \text{constant stress})$$

and

$$K_{11}^x = 1970 \quad K_{33}^x = 109 \quad (x = \text{constant stress})$$

c. Why is K^X different from K^x? Discuss how they might be measured.
d. Plot $(K^X)'$ as a function of direction, showing the geometric representation. Replot as a quadric surface.

Problem 9.2
Muscovite mica, KAl$_2$[AlSi$_3$O$_{10}$](OH)$_2$, is a layer silicate that cleaves easily into thin transparent crystals. In the past, it has been used as a capacitor because of its heat resistance, high electric resistivity, and high breakdown strength. The hexagonal crystals are highly anisotropic, which a dielectric constant of 6.9 perpendicular to the layers, and 8.7 parallel to the layers. The dissipation factor (K''/K') is also very anisotropic. At 1 kHz, the loss factors are 0.0002 and 0.0980 in the perpendicular and parallel directions. Make drawings of K' and K'' as a function of direction and discuss the structure–property relationships.

9.6 Polycrystalline dielectrics

Many of the most useful dielectrics are polycrystalline ceramics with randomly oriented grains. These include barium titanate capacitors, alumina electronic packages, and porcelain high voltage insulators. To obtain the statistical average of the dielectric constant, we integrate the direction cosines over all angles:

$$\langle K'_{11} \rangle = K_{11}\langle a_{11}^2 \rangle + K_{22}\langle a_{12}^2 \rangle + K_{33}\langle a_{13}^2 \rangle$$

$$\langle a_{11}^2 \rangle = \frac{\int_{-1}^{+1} a_{11}^2 \, da_{11}}{\int_{-1}^{+1} da_{11}} = \frac{1}{3} = \langle a_{12}^2 \rangle = \langle a_{13}^2 \rangle.$$

The polycrystalline dielectric constant is simply the numerical average of the principal dielectric constants.

$$\langle K'_{11} \rangle = \frac{1}{3}(K_{11} + K_{22} + K_{33}).$$

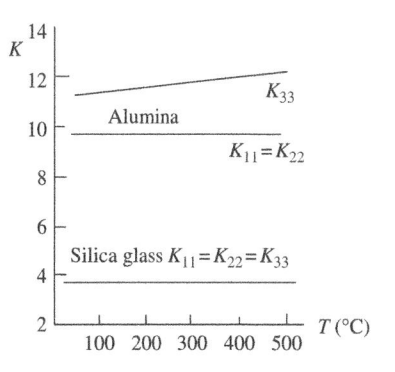

Fig. 9.13 Dielectric constants of Al$_2$O$_3$ crystals and SiO$_2$ glass.

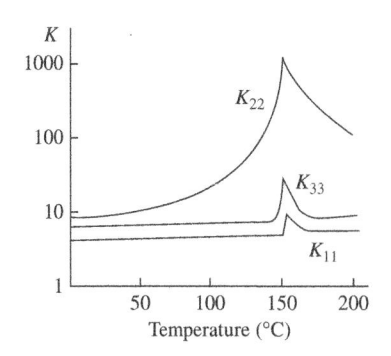

Fig. 9.14 Permittivity of ferroelectric NaNO$_2$ measured at 100 kHz.

9.7 Structure–property relationships

The dielectric constants of isotropic silica glass, a trigonal alumina crystal, and orthorhombic ferroelectric sodium nitrite are shown in Figs 9.13 and 9.14. Alumina and silica are typical of the low-K dielectrics used as insulators and electronic packages. The anisotropy is generally small and there is very little change in the permittivity over wide ranges in frequency and temperature. Sodium nitrite is a ferroelectric below 160°C with complex domain structure and hysteresis under high fields. The dielectric constant undergoes large changes with temperature and frequency as the molecular dipoles of the NO_2^- ions undergo rapid changes in orientation. Large anisotropies often occur in ferroelectric crystals.

The average dielectric constant of low-permittivity inorganic materials can be estimated using the Clausius–Mosotti equation. This involves assigning an atomic or molecular polarizability to each chemical species, and estimating the dielectric constant from the sum of the polarizabilities divided by the molar volume. The formula takes various forms depending on how the local electric fields are approximated, and how the polarizabilities are defined.

Working with dielectric data from a large number of simple oxides and fluorides, Shannon (1993) devised a predictive equation that generates K values accurate to about 0.5%. The structure–property relationship uses a table of dipolar polarizabilities derived from the Clausius–Mosotti equation:

$$\alpha_D = \frac{3}{4\pi}\left(V_m \frac{(K-1)}{(K+2)}\right).$$

V_m is the molar volume and K is the dielectric constant. Polarizability values for a number of cations and anions found in simple inorganic solids are listed in Table 9.2. It is interesting to note that the largest polarizabilities are from anions with large loosely bonded electron clouds, and from metals like Fe^{2+} and Ca^{2+} with more electrons than the lighter metals. The small, highly charged cations like B^{3+}, Be^{2+}, Si^{4+}, and Al^{3+} make only modest contributions to the dielectric constant. Fig. 9.15 shows how high (n^2) and low frequency dielectric constant depends on density. Oxides like silica with short, strong bonds, and low density have low dielectric constants while those with close packed structures and more ions per unit volume have larger permittivities.

The Clausius–Mosotti equation has also been used to predict polymer permittivities (Takahashi 1992). A listing of polymer components is given in Table 9.3 along with their molar polarizabilities a defined as the molar polarization divided by the molar volume ($a = P_m/V_m$). The molecular values

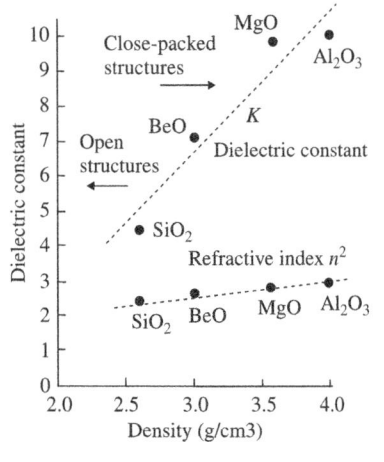

Fig. 9.15 Low-frequency dielectric constant K and high-frequency value (n^2) plotted as a function of density for simple oxide ceramics.

Table 9.2 Polarizabilities for cations and anions found in low-permittivity oxides and fluorides (Shannon 1993)

Anions	Cations
F^- 1.63 Å^3	B^{3+} 0.05 Å^3
O^{2-} 2.01	Be^{2+} 0.3
OH^- 2.18	Si^{4+} 0.85
	Al^{3+} 0.29
	Mg^{2+} 1.31
	Fe^{2+} 2.22
	Ca^{2+} 3.15

9.7 Structure–property relationships

Table 9.4 Average dielectric constants and tan δ values

	K	tan δ
Polyethylene	2.3	10^{-4}
Polypropylene	2.3	10^{-4}
Polystyrene	2.5	10^{-4}
Polymethylsiloxane	3.6	10^{-4}

Table 9.3 Molar polarizabilities for organic crystals and polymers (Takahashi 1992)

	a
$-CH_3$	0.24
$-CH_2-$	0.29
$-CH$	0.38
$-C_6H_4-$ (phenyl)	0.38
$-CH=CH-$	0.38
$-O-$	0.52
$-C=C-$	0.54
$-CO-O-$	0.82
$-N$ (amine)	0.82
$-CONH-$	1.04
$-COOH$	1.30
$-OH$	2.06

of a are obtained by averaging the a values for the components. To illustrate, for polyethylene $(-CH_2-CH_2\cdots)_n$, the molar polarizability is 0.29 and the expected dielectric constant is

$$K = \frac{1+2a}{1-a} = \frac{1.58}{0.71} = 2.23.$$

For polystyrene the molecular formula is

$$(-CH_2-CH-)$$
 with phenyl group on CH

and $\bar{a} = \frac{1}{3}(0.29 + 0.38 + 0.38) = 0.35$, giving $K = 2.6$. The experimental values for polyethylene and polystyrene are 2.3 and 2.5 (Table 9.4). Note that the groups containing the more ionic oxygen and hydroxyl ions give the largest polarizabilities.

Many physical properties measurements are accompanied by loss phenomena. The dielectric constant is a complex quantity $K^* = K' - iK''$ with real (K') and imaginary (K'') components. As shown in Fig. 9.16, the losses are sometimes highly anisotropic. The losses in rutile are anisotropic because of electrical conduction along the octahedral chains parallel to Z_3 (=[001]).

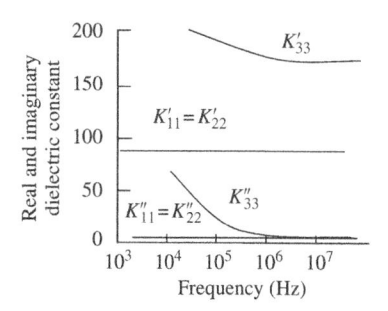

Fig. 9.16 Real and imaginary parts of the dielectric constant of rutile (TiO$_2$). Dielectric spectra are sometimes highly anisotropic.

10 Stress and strain

10.1 Mechanical stress — 72
10.2 Stress transformations — 74
10.3 Strain tensor — 75
10.4 Matrix transformation for strain — 77

Stress (force per unit area) and strain (change in length per unit length) are both symmetric second rank tensors like the dielectric constant, but they are *not* property tensors. Experimenters are at liberty to apply different types of forces to a specimen, therefore there is no reason that the stress tensors (and the resulting strain tensor) must conform to the crystal symmetry. Stress and strain tensors do not obey Neumann's Law. They are sometimes called field tensors to distinguish them from property tensors like the dielectric constant. Property tensors are relationships between field tensors.

For the same reason, electric and magnetic fields are first rank field tensors, as are magnetization and polarization. They do not obey the symmetry principles as first rank property tensors such as pyroelectricity or the magnetocaloric effect are required to do.

10.1 Mechanical stress

In arbitrary coordinate systems, the state of stress in a specimen is described by nine components of the stress tensor:

$$(X_{ij}) = \begin{pmatrix} X_{11} & X_{12} & X_{13} \\ X_{12} & X_{22} & X_{23} \\ X_{13} & X_{23} & X_{33} \end{pmatrix} \quad [\text{N/m}^2]$$

The first subscript refers to the direction of the force, the second to the normal to the face on which the force acts (Fig. 10.1). To prevent translational motion, each force is balanced by an equal and opposite force on the reverse side of the specimen. Stress component X_{22} is a tensile stress in which both the force and the normal are along Z_2, and X_{12} is a shear stress in which a force along Z_1 acts on a face normal to Z_2. For static equilibrium, the torques must be balanced, otherwise rotation occurs; this means that the stress tensor must be symmetric with $X_{12} = X_{21}$, $X_{13} = X_{31}$, and $X_{23} = X_{32}$. Thus the stress state is specified by six independent components: three tensile stresses X_{11}, X_{22}, and X_{33}, and three shear components X_{12}, X_{13}, and X_{23}.

For an arbitrary axial system (new axes) the general stress tensor can be rewritten as a 6×1 column matrix:

$$\begin{pmatrix} X'_{11} & X'_{12} & X'_{13} \\ X'_{12} & X'_{22} & X'_{23} \\ X'_{13} & X'_{23} & X'_{33} \end{pmatrix} = \begin{pmatrix} X'_1 = X'_{11} \\ X'_2 = X'_{22} \\ X'_3 = X'_{33} \\ X'_4 = X'_{23} \\ X'_5 = X'_{13} \\ X'_6 = X'_{12} \end{pmatrix}$$

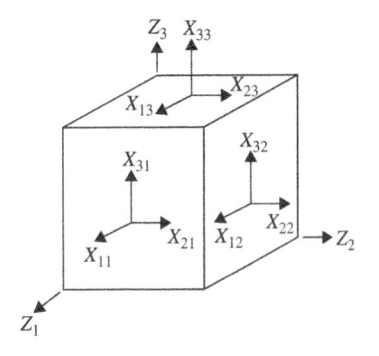

Fig. 10.1 Nine stress components (three tensile and six shear components) acting on a cube-shaped specimen. By convention, the directions shown in the drawing are taken as positive.

The first three components in the column matrix are tensile stresses along Z'_1, Z'_2, Z'_3, and the last three are shear stresses about Z'_1, Z'_2, Z'_3. Both the tensor and matrix forms are widely used in the literature.

When rotated to principal axes, the generalized stress becomes

$$\begin{pmatrix} X_{11} & 0 & 0 \\ 0 & X_{22} & 0 \\ 0 & 0 & X_{33} \end{pmatrix} = \begin{pmatrix} X_1 = X_{11} \\ X_2 = X_{22} \\ X_3 = X_{33} \\ 0 \\ 0 \\ 0 \end{pmatrix}$$

with no shear stress components. It should be remembered that the stress tensor is not a property tensor. Therefore the principal axes for stress will *not* coincide with symmetry axes unless the forces are applied in a symmetric manner.

There are several special forms of the stress tensor that are often used in experiments. A uniaxial *tensile* stress is

$$\begin{pmatrix} X_{11} & 0 & 0 \\ 0 & 0 & 0 \\ 0 & 0 & 0 \end{pmatrix} = \begin{pmatrix} X_1 = X_{11} \\ 0 \\ 0 \\ 0 \\ 0 \\ 0 \end{pmatrix}.$$

Biaxial stresses are represented by $\begin{pmatrix} X_{11} & 0 & 0 \\ 0 & X_{22} & 0 \\ 0 & 0 & 0 \end{pmatrix} = \begin{pmatrix} X_1 = X_{11} \\ X_2 = X_{22} \\ 0 \\ 0 \\ 0 \\ 0 \end{pmatrix}.$

Hydrostatic pressures are described by $\begin{pmatrix} -p & 0 & 0 \\ 0 & -p & 0 \\ 0 & 0 & -p \end{pmatrix} = \begin{pmatrix} -p \\ -p \\ -p \\ 0 \\ 0 \\ 0 \end{pmatrix}.$

A *pure shear* about Z_3 can be written in three equivalent ways

$$\begin{pmatrix} -X_{11} & 0 & 0 \\ 0 & X_{11} & 0 \\ 0 & 0 & 0 \end{pmatrix} = \begin{pmatrix} 0 & X_{12} & 0 \\ X_{12} & 0 & 0 \\ 0 & 0 & 0 \end{pmatrix} = \begin{pmatrix} 0 \\ 0 \\ 0 \\ 0 \\ 0 \\ X_6 \end{pmatrix},$$

where $X_{11} = X_{12} = X_6$. The first two forms are related by a 45° rotation about Z_3.

By convention, the tensile and shear stresses are positive if applied in the directions shown in Fig. 10.1. This means that a hydrostatic pressure, p, is a negative stress. One must be careful about signs because some authors use a different convention.

74 Stress and strain

Problem 10.1
Stress is force per unit area and is expressed in a number of different units including N/m^2, $dynes/cm^2$, Pascals, atmospheres, pounds per square inch, and mm of Hg. Convert 1 N/m^2 into these various systems.

10.2 Stress transformations

Like other second rank tensors, stress transforms as
$$X'_{ij} = a_{ik}a_{jl}X_{kl},$$
but how do the matrix coefficients transform? The 6×1 column matrices take the following form.

$$\begin{pmatrix} X'_1 \\ X'_2 \\ X'_3 \\ X'_4 \\ X'_5 \\ X'_6 \end{pmatrix} = \begin{pmatrix} \alpha_{11} & \alpha_{12} & \alpha_{13} & \alpha_{14} & \alpha_{15} & \alpha_{16} \\ \alpha_{21} & \alpha_{22} & \alpha_{23} & \alpha_{24} & \alpha_{25} & \alpha_{26} \\ \alpha_{31} & \alpha_{32} & \alpha_{33} & \alpha_{34} & \alpha_{35} & \alpha_{36} \\ \alpha_{41} & \alpha_{42} & \alpha_{43} & \alpha_{44} & \alpha_{45} & \alpha_{46} \\ \alpha_{51} & \alpha_{52} & \alpha_{53} & \alpha_{54} & \alpha_{55} & \alpha_{56} \\ \alpha_{61} & \alpha_{62} & \alpha_{63} & \alpha_{64} & \alpha_{65} & \alpha_{66} \end{pmatrix} \begin{pmatrix} X_1 \\ X_2 \\ X_3 \\ X_4 \\ X_5 \\ X_6 \end{pmatrix}.$$

The unknown α coefficients in the 6×6 matrix transformation are determined by writing out the tensor and matrix transformations and equating equivalent terms. For the tensile stress component $X'_{11} = X'_1$, the matrix and tensor expressions are
$$X'_1 = \alpha_{11}X_1 + \alpha_{12}X_2 + \cdots + \alpha_{16}X_6$$
$$X'_{11} = a^2_{11}X_{11} + a_{11}a_{12}aX_{12} + a_{12}a_{11}X_{21} + \cdots + a^2_{12}X_{22} + \cdots.$$

Therefore $\alpha_{11} = a^2_{11}$, $\alpha_{12} = a^2_{12}$, $\alpha_{16} = 2a_{11}a_{12}$. The other α coefficients are obtained in a similar way. The complete (α) matrix and reciprocal $(\alpha)^{-1}$ matrix are given in Table 10.1.

The (α) and $(\alpha)^{-1}$ matrices can be written in a more compact form. To find α_{mn} (or α^{-1}_{mn}) in terms of the direction cosines, rewrite the matrix coefficient subscripts in terms of the equivalent tensor subscripts:

$$1 = 11, \ 2 = 22, \ 3 = 33, \ 4 = 23, \ 5 = 31, \text{ and } 6 = 12.$$

Table 10.1 Transformation matrices for stresses and strains written in matrix form

(α)
$$\begin{pmatrix} (a^2_{11}) & (a^2_{12}) & (a^2_{13}) & (2a_{12}a_{13}) & (2a_{13}a_{11}) & (2a_{11}a_{12}) \\ (a^2_{21}) & (a^2_{22}) & (a^2_{23}) & (2a_{22}a_{23}) & (2a_{23}a_{21}) & (2a_{21}a_{22}) \\ (a^2_{31}) & (a^2_{32}) & (a^2_{33}) & (2a_{32}a_{33}) & (2a_{33}a_{31}) & (2a_{31}a_{32}) \\ (a_{21}a_{31}) & (a_{22}a_{32}) & (a_{23}a_{33}) & (a_{22}a_{33} + a_{23}a_{32}) & (a_{21}a_{33} + a_{23}a_{31}) & (a_{22}a_{31} + a_{21}a_{32}) \\ (a_{31}a_{11}) & (a_{32}a_{12}) & (a_{33}a_{13}) & (a_{12}a_{33} + a_{13}a_{32}) & (a_{13}a_{31} + a_{11}a_{33}) & (a_{11}a_{32} + a_{12}a_{31}) \\ (a_{11}a_{21}) & (a_{12}a_{22}) & (a_{13}a_{23}) & (a_{12}a_{23} + a_{13}a_{22}) & (a_{13}a_{21} + a_{11}a_{23}) & (a_{11}a_{22} + a_{12}a_{21}) \end{pmatrix}$$

(α^{-1})
$$\begin{pmatrix} (a^2_{11}) & (a^2_{21}) & (a^2_{31}) & (2a_{21}a_{31}) & (2a_{31}a_{11}) & (2a_{11}a_{21}) \\ (a^2_{12}) & (a^2_{22}) & (a^2_{32}) & (2a_{22}a_{32}) & (2a_{32}a_{12}) & (2a_{12}a_{22}) \\ (a^2_{13}) & (a^2_{23}) & (a^2_{33}) & (2a_{23}a_{33}) & (2a_{33}a_{13}) & (2a_{13}a_{23}) \\ (a_{12}a_{13}) & (a_{22}a_{23}) & (a_{32}a_{33}) & (a_{22}a_{33} + a_{32}a_{23}) & (a_{12}a_{33} + a_{32}a_{13}) & (a_{22}a_{13} + a_{12}a_{23}) \\ (a_{13}a_{11}) & (a_{23}a_{21}) & (a_{33}a_{31}) & (a_{21}a_{33} + a_{31}a_{23}) & (a_{31}a_{13} + a_{11}a_{33}) & (a_{11}a_{23} + a_{21}a_{13}) \\ (a_{11}a_{12}) & (a_{21}a_{22}) & (a_{31}a_{32}) & (a_{21}a_{32} + a_{31}a_{22}) & (a_{31}a_{12} + a_{11}a_{32}) & (a_{11}a_{22} + a_{21}a_{12}) \end{pmatrix}$$

The elements in the (α) and $(\alpha)^{-1}$ matrices are then given by the expressions

$$\alpha_{mn} = \alpha_{ijkl} = a_{ik}a_{jl} + (1 - \delta_{kl})a_{il}a_{jk}$$

and

$$\alpha_{mn}^{-1} = \alpha_{ijkl}^{-1} = a_{ki}a_{lj} + (1 - \delta_{kl})a_{kj}a_{li}.$$

As examples,

$$\alpha_{56} = \alpha_{3112} = a_{31}a_{12} + (1 - 0)a_{32}a_{11} = a_{31}a_{12} + a_{32}a_{11}$$

$$\alpha_{23}^{-1} = \alpha_{2233}^{-1} = a_{32}a_{32} + (1 - 1)a_{32}a_{32} = a_{32}^2.$$

Problem 10.2
A crystal is subjected to a tensile stress of 1 N/m² along Z_1 and a compressive stress of 2 N/m² along Z_2. A new coordinate system Z_i' is chosen. It is related to the original coordinate by a counterclockwise rotation of $\theta°$ about Z_3.

Write out the new stress components X_{ij}' in tensor form. Using the α matrix, repeat the exercise in matrix form to obtain the stress components of X_i'. Plot X_1', X_2', and X_6' as a function of the angle θ using the numerical values given above.

10.3 Strain tensor

Strain refers to the fractional change in shape of a specimen. It is a dimensionless quantity that refers to the change in length per unit length.

In order to describe strain, consider a solid in an orthogonal coordinate system Z_1, Z_2, Z_3 with a fixed origin (Fig. 10.2(a)). All other points can be displaced.

Strain is a symmetric second rank tensor like stress. The strain tensor relates two vectors: displacement u_i and coordinate Z_j. Written in differential form the strain is

$$(x_{ij}) = \frac{\delta u_i}{\delta Z_j} = \begin{pmatrix} x_{11} & x_{12} & x_{13} \\ x_{12} & x_{22} & x_{23} \\ x_{13} & x_{23} & x_{33} \end{pmatrix}.$$

The strain is symmetric ($x_{ij} = x_{ji}$) to eliminate body rotations that are not part of the shape change. The meaning of x_{11} and $x_{12} = x_{21}$ is illustrated in Fig. 10.2(b) and (c). x_{11} is a tensile strain and $x_{12} = x_{21}$ is a shear strain about Z_3.

As an illustration of an actual strain calculation, consider the object shown in Fig. 10.3(a) plotted in the Z_1–Z_2 plane. The initial coordinates of four points are listed. How will the shape change when a strain takes place? Let the strain be

$$(x_{ij}) = \begin{pmatrix} 0.1 & -0.1 & 0 \\ -0.1 & 0.2 & 0 \\ 0 & 0 & 0 \end{pmatrix}.$$

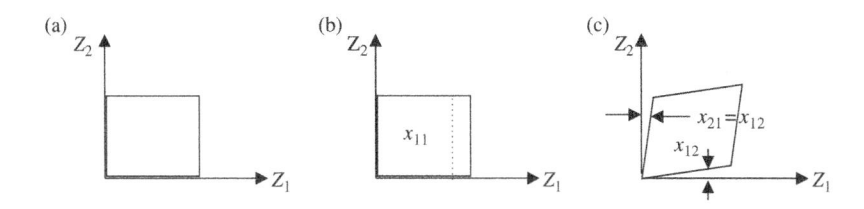

Fig. 10.2 Drawing illustrating (a) an unstrained solid, (b) tensile strain x_{11}, and (c) shear strain $x_{12} = x_{21}$.

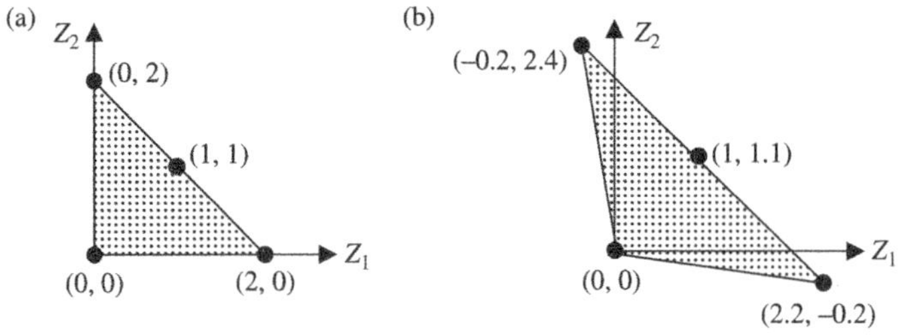

Fig. 10.3 A two-dimensional object illustrating how shape changes for a given strain.

Remembering that the origin is fixed, integration of the differential strain equation gives displacements

$$u_i = x_{ij} Z_j.$$

For the point at $Z_1 = 2$ and $Z_2 = 0$, the displacements are $u_1 = 0.1(2) - 0.1(0)$, $u_2 = -0.1(2) + 0.2(0)$, and the strained coordinates $Z_1 = 2.2$ and $Z_2 = -0.2$. Displacements for the other three points are illustrated in Fig. 10.3(b).

Problem 10.3
Verify the new coordinates for the other points in Fig. 10.3.

As with the stress tensor, the generalized strain tensor can be rewritten as a 6×1 matrix.

$$\begin{pmatrix} x'_{11} & x'_{12} & x'_{13} \\ x'_{12} & x'_{22} & x'_{23} \\ x'_{13} & x'_{23} & x'_{33} \end{pmatrix} = \begin{pmatrix} x'_1 = x'_{11} \\ x'_2 = x'_{22} \\ x'_3 = x'_{33} \\ x'_4 = 2x'_{23} \\ x'_5 = 2x'_{13} \\ x'_6 = 2x'_{12} \end{pmatrix},$$

where x'_1, x'_2 and x'_3 are the tensile strains along an arbitrary set of axes Z'_1, Z'_2, and Z'_3. The remaining three strain coefficients, x'_4, x'_5, and x'_6, are shear strains about Z'_1, Z'_2, and Z'_3, respectively. Note the factors of two that appear between shear tensor coefficients and matrix coefficients, as shown in Fig. 10.2(c),

$$x'_6 = x'_{12} + x'_{21} = 2x'_{12}.$$

Factors of two are encountered for strains but not for stresses.

When transformed from measurement axes (Z'_i) to principal axes (Z_i), the strain tensor has no shear components. The principal axes are not necessarily along symmetry directions since strain is not a property tensor. It is not covered by Neumann's Principle. For principal axis

$$\begin{pmatrix} x_{11} & 0 & 0 \\ 0 & x_{22} & 0 \\ 0 & 0 & x_{33} \end{pmatrix} = \begin{pmatrix} x_1 \\ x_2 \\ x_3 \\ 0 \\ 0 \\ 0 \end{pmatrix}.$$

Volume changes are easily visualized from principal axes (Fig. 10.4).

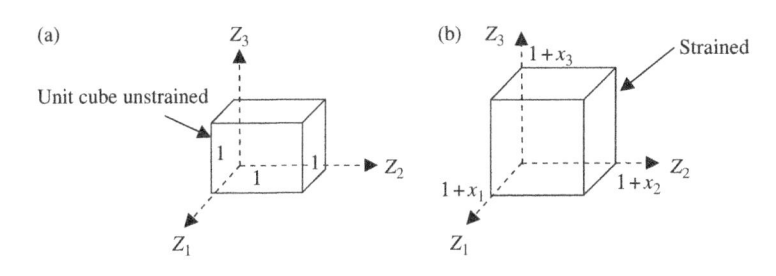

Fig. 10.4 A unit cube in (a) unstrained and (b) strained states referred to principal axes.

The volume change between the strained and unstrained states is
$$\Delta V = (1 + x_{11})(1 + x_{22})(1 + x_{33}) - 1 \cong x_{11} + x_{22} + x_{33}$$
for small strains. The fractional volume change for the unit cube is
$$\frac{\Delta V}{V} = x_{11} + x_{22} + x_{33}.$$

10.4 Matrix transformation for strain

In tensor form strain transforms as a symmetric second rank tensor
$$x'_{ij} = a_{ik} a_{jl} x_{kl}.$$
Stress also transforms as a symmetric second rank tensor, and in matrix form it transforms according to the α matrix (Section 10.2):

$$\begin{array}{ccc} 6 \times 1 & & 6 \times 6 \quad 6 \times 1 \\ (X') & = & (\alpha) \quad (X) \end{array}$$

To determine how the strain matrix transforms, it is helpful to recall that the product of stress and strain is mechanical energy density, W. Energy density is a scalar quantity. In matrix form,

$$W = x_i X_i = x_1 X_1 + x_2 X_2 + x_3 X_3 + x_4 X_4 + x_5 X_5 + x_6 X_6$$

$$= (x_1 \ x_2 \ x_3 \ x_4 \ x_5 \ x_6) \begin{pmatrix} X_1 \\ X_2 \\ X_3 \\ X_4 \\ X_5 \\ X_6 \end{pmatrix} = (x)_t (X).$$

Since energy density is a scalar, it has the same value in both the new and old coordinate systems ($W' = W$).

$$W = (x)_t (X) = (x)_t (\alpha)^{-1} (\alpha)(X) = (x)_t (\alpha)^{-1} (X')$$
$$= W' = (x')_t (X').$$

Therefore $(x')_t = (x)_t (\alpha)^{-1}$.

Taking the transpose of both sides, this becomes
$$((x')_t)_t = (x') = [(x)_t (\alpha)^{-1}]_t = (\alpha)_t^{-1} (x).$$

Therefore the strain matrix transforms as
$$(x') = (\alpha)_t^{-1} (x) \quad \text{from old to new, and as}$$
$$(x) = (\alpha)_t (x') \quad \text{from new to old.}$$

The transpose matrices $(\alpha)_t$ and $(\alpha)_t^{-1}$ are easily obtained from Table 10.1 by interchanging rows and columns in the (α) and $(\alpha)^{-1}$ matrices. The stress and strain matrix transformations are often used in working with higher rank tensor properties such as piezoelectricity (Chapter 12) and elasticity (Chapter 13).

Problem 10.4
Mathematically, stress and strain are symmetric second rank tensors like the dielectric constant. This means that a strain tensor such as

$$(x') = \begin{pmatrix} x'_{11} & x'_{12} & x'_{13} \\ x'_{12} & x'_{22} & x'_{32} \\ x'_{13} & x'_{23} & x'_{33} \end{pmatrix} = \begin{pmatrix} 5 & 1 & 2 \\ 1 & 4 & -1 \\ 2 & -1 & 3 \end{pmatrix} \times 10^{-3}$$

can be diagonalized in principal axes to

$$(x) = \begin{pmatrix} x_{11} & 0 & 0 \\ 0 & x_{22} & 0 \\ 0 & 0 & x_{33} \end{pmatrix},$$

where

$$(x')(a) = (a)(x).$$

a. Find the three principal strains x_{11}, x_{22}, and x_{33}.
b. Find the nine direction cosines $a_{11}, a_{12}, a_{13}, \ldots, a_{33}$ that relate the measurement axes to the principal axes.
c. What is the fractional change in volume for this set of strains?

The procedure for doing this was outlined in the previous chapter, Section 9.4.

Thermal expansion

11

When a material is heated uniformly it undergoes a strain described by the relationship

$$x_{ij} = \alpha_{ij}\Delta T,$$

where α_{ij} are the thermal expansion coefficients and ΔT is the change in temperature. Room temperature thermal expansion coefficients range from about 10^{-6}/K for an oxide like silica glass to 10^{-3}/K for an elastomeric polymer. Thermal expansion coefficients are often a strong function of temperature, as shown in Fig. 11.1. Simple linear relations are insufficient to describe thermal expansion over a wide temperature range. A power series consisting of terms in ΔT, $(\Delta T)^2$ and higher order terms can be used to describe this thermal expansion over extended temperature ranges. In hexagonal zinc oxide both coefficients are slightly negative at low temperatures and exhibit increasing anisotropy at high temperatures. The negative thermal expansion coefficients are attributed to the low-energy transverse vibrations that dominate at very low temperatures.

11.1	Effect of symmetry	79
11.2	Thermal expansion measurements	81
11.3	Structure–property relations	82
11.4	Temperature dependence	85

11.1 Effect of symmetry

Thermal expansion relates a second rank tensor (strain) to a scalar (temperature change). It is a symmetric second rank tensor because the strain tensor

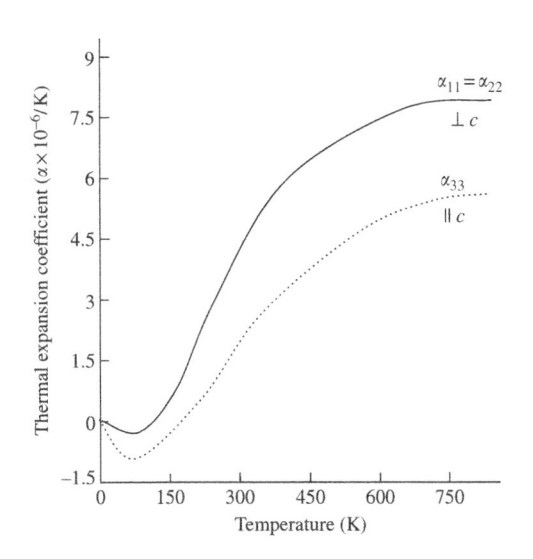

Fig. 11.1 Zinc oxide is a hexagonal crystal with tetrahedrally bonded zinc and oxygen atoms. The thermal expansion coefficients approach zero at 0 K. Anisotropy also changes sign at low temperatures with both coefficients becoming slightly negative.

80 Thermal expansion

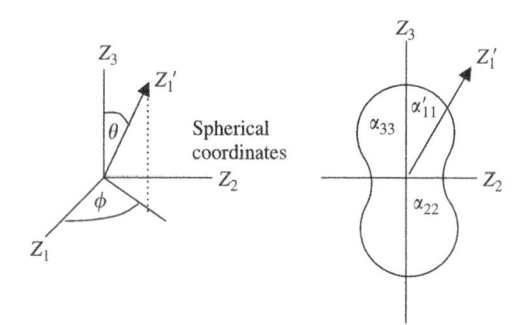

Fig. 11.2 Anisotropy surface for the thermal expansion coefficient of low symmetry crystals.

is symmetric. Therefore thermal expansion and the dielectric constant are the same type of tensor, and the effect of symmetry is the same. This means that for the general case (a triclinic crystal), six measurements are required to find the three principal thermal expansion coefficients and the three angles needed to orient the principal axes. The appropriate matrices for other symmetries are given in Table 9.1. Four measurements are needed for monoclinic crystals, three for orthorhombic, two for trigonal, tetragonal, and hexagonal crystals, and only one for cubic crystals.

When referred to principal axes, the thermal expansion coefficient in an arbitrary direction is

$$\alpha'_{11} = \alpha_{11} \cos^2\phi \sin^2\theta + \alpha_{22} \sin^2\phi \sin^2\theta + \alpha_{33} \cos^2\theta,$$

where θ and ϕ are the spherical coordinate angles. When all three principal coefficients are positive numbers, the resulting surface is shaped like a peanut (Fig. 11.2). For trigonal, tetragonal, and hexagonal crystals, the thermal expansion coefficient surface is

$$\alpha'_{11} = \alpha_{11} \sin^2\theta + \alpha_{33} \cos^2\theta$$

and for cubic crystals the surface is a sphere of radius $\alpha'_{11} = \alpha_{11}$.

Unlike the dielectric constant, thermal expansion coefficients can be positive, negative, or both positive and negative. Trigonal calcite ($CaCO_3$) is a good example. At room temperature, the thermal expansion coefficients are $\alpha_{11} = \alpha_{22} = -5.6 \times 10^{-6}$/K and $\alpha_{33} = +25 \times 10^{-6}$/K. The resulting surface (Fig. 11.3) has both positive and negative lobes. In between the two lobes is a cone of zero thermal expansion given by $\alpha'_{11} = 0 = \alpha_{11} \sin^2\theta + \alpha_{33} \cos^2\theta$ or $\tan^2\theta = -\alpha_{33}/\alpha_{11}$. The zero expansion angle is $\theta = 65°$.

Problem 11.1

Al_2TiO_5 is orthorhombic, point group *mmm*. The structure contains chains of TiO_6 octahedra along the c-axis, the direction of negative thermal expansion. Measured values along the principal axes are $\alpha_{11} = 9.51$, $\alpha_{22} = 19$, $\alpha_{33} = -1.4$, all in units of 10^{-6}/K. Plot the thermal expansion as a function for direction in the (100), (010), and (001) planes. Replot this data as a quadratic surface. The large anisotropy in α leads to intergranular microfracture and low mechanical strength in aluminum titanate ceramics.

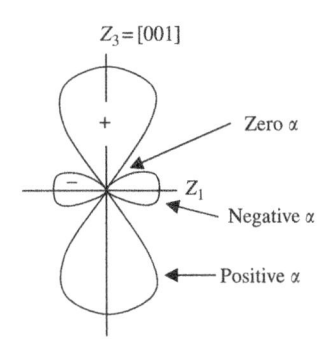

Fig. 11.3 Thermal expansion surface of calcite with circular symmetry about Z_3, the trigonal axis. The maximum expansion is perpendicular to the flat carbonate groups of the structure.

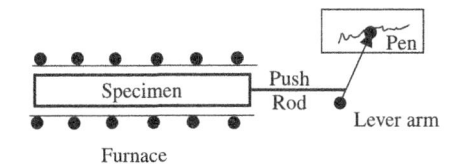

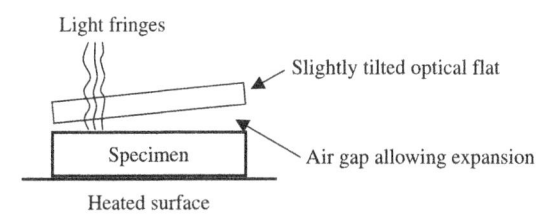

Fig. 11.4 Lever-arm dilatometers and optical interferometers are two of the experimental methods used to measure thermal expansion.

11.2 Thermal expansion measurements

Schematic illustrations of two of the classical techniques are shown in Fig. 11.4. Ceramists use push-rod dilatometers to measure the expansion of heated specimens inside a furnace. The lever arm amplifies the thermal motion. Physicists traditionally favored optical techniques. Small changes in thickness of a heated specimen were measured by counting light fringes formed by optical interference across a small air gap.

The classical techniques are sufficient for isotropic polycrystalline specimens or cubic crystals but low symmetry crystals require up to six different crystal orientations. Fortunately there is another method to measure the thermal expansion coefficients of anisotropic materials. The single crystal coefficients can all be determined from X-ray powder patterns.

The interplanar d-spacing are given by Bragg's Law

$$\lambda = 2d \sin \theta,$$

where θ is the Bragg angle and λ is the X-ray wavelength. Differentiating this equation with respect to temperature gives

$$\frac{dd}{dT} = -\frac{\lambda}{2} \frac{\cos \theta}{\sin^2 \theta} \frac{d\theta}{dT}$$

from which the thermal expansion coefficient

$$\alpha_d = \frac{1}{d} \frac{dd}{dT} = -\cot \theta \frac{d\theta}{dT}.$$

The shifts in Bragg angle, $d\theta/dT$, are determined by recording the powder pattern at two different temperatures (Fig. 11.5).

After determining the angular shifts and computing the thermal expansion coefficients for each reflection, the coefficients are plotted as a function of direction. The measured values α_d refer to the directions perpendicular to each Bragg plane (hkl). Thus for the $\bar{2}13$ reflection, the α_d value is plotted for the direction normal to $(\bar{2}13)$.

If, instead of plotting α_d, one plots $\pm 1/\sqrt{\alpha_d}$, then a representation quadric surface is obtained (see Section 9.5). A sample surface for a monoclinic crystal is shown in Fig. 11.6.

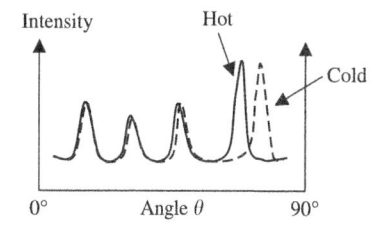

Fig. 11.5 An X-ray powder pattern recorded at two different temperatures showing the shifts in θ from which thermal expansion coefficients are calculated. The shifts are especially large in the back reflection region near 90°.

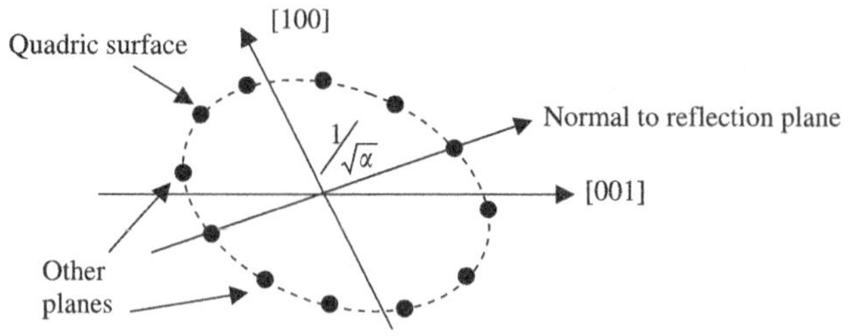

Fig. 11.6 Thermal expansion ellipsoid for a monoclinic crystal. By plotting $1/\sqrt{\alpha}$ for various planes, the principal axes can be identified.

The major and minor axes of the ellipse correspond to two of the principal axes. For a monoclinic crystal, the third principal axis is along the [010] direction.

The volume expansivity β is defined as the fractional change in volume for a 1° change in temperature: $\beta = (1/V)(dV/dT)$. When referred to principal axes,

$$\beta = \alpha_{11} + \alpha_{22} + \alpha_{33}.$$

The temperature dependence of the density, ρ, is

$$\frac{1}{\rho}\frac{d\rho}{dT} = -\beta = -(\alpha_{11} + \alpha_{22} + \alpha_{33}).$$

Problem 11.2
As pointed out in Section 6.2, there is a thermodynamic relationship between the linear thermal expansion coefficient and the piezocaloric effect. A piezocaloric experiment is performed on calcite (CaCO$_3$) using two plates cut parallel and perpendicular to the threefold symmetry axis. From the values of α_{11} and α_{33} listed in Table 11.1, calculate the changes in entropy and temperature when the plates are subjected to a compressive stress of 1000 N/m^2. Use the Law of Dulong and Petit in estimating the specific heat. The density of calcite is 2.71 g/cm^3.

11.3 Structure–property relations

Strong interatomic forces are associated with low thermal expansion, weak forces with high expansion (Fig. 11.7). The room temperature thermal expansion coefficients of cubic inorganic crystals illustrates the trend. Fig. 11.8 shows the thermal expansion coefficients of six compounds with univalent and divalent ions. In every case the divalent compounds have stronger bonds and smaller expansivities. The product of the charges differs by a factor of four and so do the thermal expansion coefficients.

Megaw has shown that α is inversely proportional to the Pauling bond strength, defined as the cation valence divided by its coordination number. For cubic ZrO$_2$ the Zr–O bond strength is $4/8 = 0.5$. Fig. 11.8 shows the thermal expansion coefficients plotted against bond strength. The bond strength is approximately proportional to $\alpha^{-1/2}$.

Although α depends mainly on bond strength, there are variations among isomorphous crystals. Thermal expansion coefficients for alkali halides with the

11.3 Structure–property relations

Table 11.1 Thermal expansion coefficients near room temperature in units of $10^{-6}/K$

Cubic crystals	α		
Diamond (C)	1.4		
Silicon (Si)	4.2		
Germanium (Ge)	5.9		
Copper (Cu)	17		
Silver (Ag)	20		
Gold (Au)	15		
Iron (Fe)	12		
Platinum (Pt)	8.3		
Tungsten (W)	4.3		
Hexagonal crystals	α_{11}	α_{33}	
Magnesium (Mg)	27	28	
Zinc (Zn)	14	61	
Cadmium (Cd)	19	48	
Magnesium Hydroxide (Mg(OH)$_2$)	11	45	
Tetragonal crystals	α_{11}	α_{33}	
Tin (Sn)	46	22	
Titanium Oxide (TiO$_2$)	7.1	9.2	
Trigonal crystals	α_{11}	α_{33}	
Calcium Carbonate (CaCO$_3$)	−3.8	21	
Sodium Nitrate (NaNO$_3$)	11	120	
Tellurium (Te)	28	−1.7	
Antimony (Sb)	8.2	16	
Aluminum Oxide (Al$_2$O$_3$)	5.4	6.6	
Orthorhombic crystals	α_{11}	α_{22}	α_{33}
Iodine (I$_2$)	133	95	35
Lead Chloride (PbCl$_2$)	34	39	17

rocksalt structure range from NaF $34 \times 10^{-6}/K$ to LiI $56 \times 10^{-6}/K$. Radius-ratio appears to be important since α is largest for LiI, LiBr, LiCl, NaI, and NaBr where r_+/r_- is small. Anion–anion repulsion loosens the structure making expansion easier.

Thermal expansion of layer-type crystals is largest normal to the layer. In hydrocarbons and other planar molecular crystals, expansion coefficients for crystalline benzene (−193°C to 3°C) are 11.9, 10.6, and $22.1 \times 10^{-5}/K$ parallel to the a, b, and c orthorhombic axes. The C$_6$H$_6$ molecules in benzene lie close to the (001) plane, perpendicular to the direction of largest expansion. The same is true of naphthalene (C$_{10}$H$_8$) and anthracene (C$_{14}$H$_{10}$). Expansion coefficients are greater for crystals with small molecules than their larger homologs. The average α values for benzene, naphthalene, and anthracene are 14.7, 12.7, and $8.0 \times 10^{-5}/K$. Because of the weak bonding between molecules, the thermal expansion coefficient of organic crystals are an order of magnitude larger than those of metals and ceramics.

Anisotropy in inorganic crystals follow similar trends. Calcite is an ionic crystal consisting Ca^{2+} cations and (CO$_3$)$^{2-}$ anions. The flat triangular carbonate groups are perpendicular to the trigonal c-axis, the direction of maximum

Rocksalt structure
NaCl $\alpha = 40 \times 10^{-6}/K$
MgO $\alpha = 10 \times 10^{-6}$

Fluorite structure
CaF$_2$ $\alpha = 19 \times 10^{-6}$
ZrO$_2$ $\alpha = 4.5 \times 10^{-6}$

Zincblende structure
CuBr $\alpha = 19 \times 10^{-6}$
ZnS $\alpha = 4.5 \times 10^{-6}$

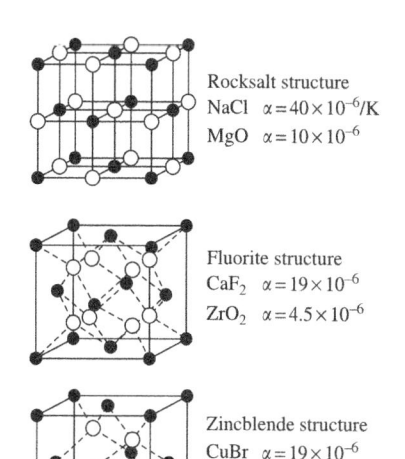

Fig. 11.7 For simple ionic structures like rocksalt and fluorite, higher valence compounds have lower thermal expansion coefficients.

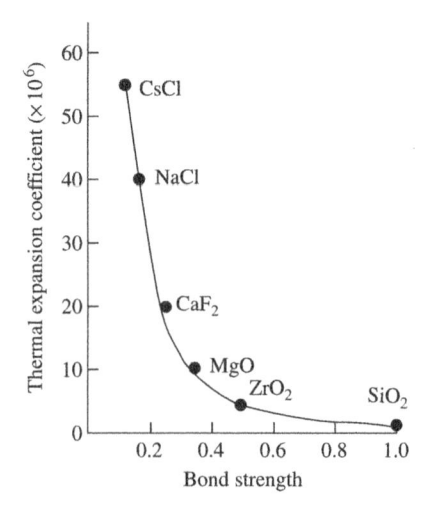

Fig. 11.8 Thermal expansion coefficients are inversely proportional to Pauling bond strength.

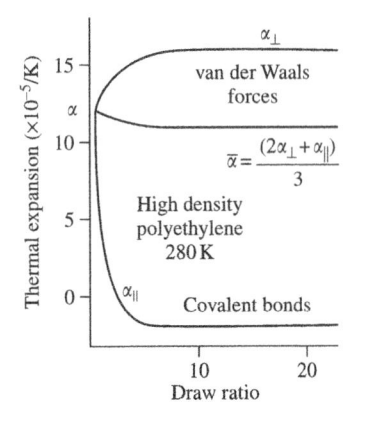

Fig. 11.9 Thermal expansion coefficients of polyethylene fibers measured parallel and perpendicular to the draw axis. As the $(CH_2)_n$ chains align under stress, the bonding and thermal expansion becomes very anisotropic.

thermal expansion (Fig. 11.3). Table 11.1 lists thermal expansion coefficients for a number of isotropic and anisotropic crystals.

Layer structures like antimony, brucite ($Mg(OH)_2$), and graphite also have the largest thermal expansion coefficients α_{33} perpendicular to the layers. Tellurium and tin have strong bonding along the c-axis making $\alpha_{11} > \alpha_{33}$. Metals also show similar anisotropy effects. Metallic zinc has a distorted hexagonal close-packed structure with the shortest bonds in the (001) plane. The thermal expansion coefficient in the perpendicular direction α_{33} is much greater than α_{11}.

Thermal expansion depends on the strength of the chemical bonds in different directions. Polymers form very long molecules with covalent bonds in the chain direction and van der Waals forces in the perpendicular directions between adjacent chains. A large anisotropy in thermal expansion is therefore expected. X-ray measurements on crystalline polyethylene along the orthorhombic axes gave $\alpha_a = 20 \times 10^{-5}/K$, $\alpha_b = 6.4 \times 10^{-5}$, and $\alpha_c = -1.3 \times 10^{-5}$. The negative value α_c along the chain axis is somewhat surprising but has been verified in other polymers. Thermal agitation gives rise to lateral vibrations in the chain which produce an effective contraction by bending motions.

In an isotropic polymer the chains are entangled in random directions, so that thermal expansion is controlled by the weak bonds between chains. When drawn into fibers, however, the polymer becomes anisotropic with a large decrease in α parallel to the fiber axis and modest increase in the perpendicular directions (Fig. 11.9).

Thermal expansion anisotropy leads to problems in polycrystalline ceramics and metals. Under thermal cycling, neighboring grains expand differently, leading to stresses at the grain boundaries. Ceramics made of cubic materials do not experience this problem because grains expand and contract uniformly as temperature changes. To relieve the intergranular stresses in anisotropic materials, it is sometimes possible to choose a composition that is accidentally isotropic. Consider the trigonal $Al_{2-x}Cr_xO_3$ grains in a ruby ceramic. The physical properties of a solid solution change smoothly from one end member to the other. In the Al_2O_3–Cr_2O_3 solid solution the thermal expansion coefficients

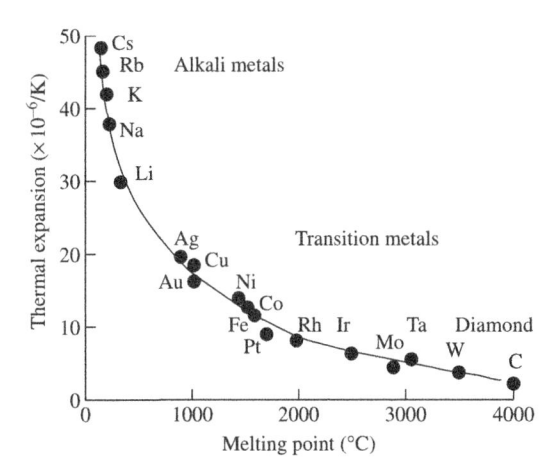

Fig. 11.10 Melting points and room temperature thermal expansion coefficients for a number of different chemical elements.

change from a positive ($\alpha_{33} > \alpha_{11}$) to a negative ($\alpha_{33} < \alpha_{11}$) anisotropy. In between there is an isotropic composition where $\alpha_{33} = \alpha_{11}$ and the intergranular stresses are minimized. This is an interesting bit of tensor engineering that improves thermal shock resistance.

11.4 Temperature dependence

There is an inverse relationship between thermal expansion and melting point, as shown in Fig. 11.10. The product αT_m is approximately constant. Low melting elements like the alkali metals have much larger coefficients than the refractory transition metal elements.

The crystal structures of MgO, BeO, Al_2O_3, $MgAl_2O_4$, and $BeAl_2O_4$ are all based on close-packed oxygen lattices and all exhibit fairly large thermal expansion coefficients, $5-10 \times 10^{-6}$/K. The effect of temperature is to increase thermal vibration, and in close-packed structures this results in atoms vibrating against one another since they are in close contact. The situation is more complicated in open structures where two additional effects can occur. The atoms can vibrate anisotropically toward open spaces in the structure, resulting in low thermal expansion coefficients. Thus many open structure oxides have small thermal expansion coefficients; compounds like spodumene (2×10^{-6}/K) are therefore useful because of their thermal shock resistance. Second, there can be cooperative rotational effects that lead to a rapid change in thermal expansion coefficients with temperature, as in quartz. The small expansivity of silica glass ($<10^{-6}$/K) has been attributed to the fact that the densities are low and also that cooperative rotations are not possible in amorphous materials.

Thermal expansion coefficients often change dramatically with temperature, especially when phase transformations are involved. The correlation between thermal expansion and melting points has already been mentioned. Refractory compounds like silica glass have small thermal expansion coefficients but the crystalline forms of silica (cristobalite, tridymite, and quartz) all have displacive phase transformations that can lead to thermal shock in silica refractories. Silica ceramics must be heated or cooled slowly below 600°C where the transitions

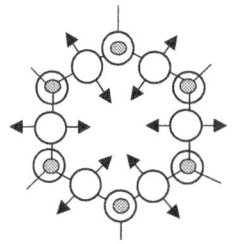

Fig. 11.11 Silica rings consist of tetrahedrally coordinated silicon atoms linked together by oxygens. Open spaces in the silicate structures leads to transverse motions of the oxygen atom causing large displacive phase transformations and unusual thermal expansion effects.

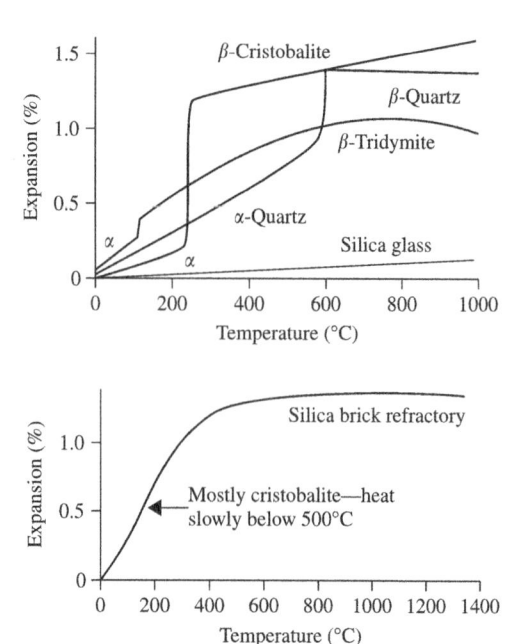

Fig. 11.12 Thermal expansion of various forms of silica and silica brick.

occur. At higher temperatures the crystalline phases have small α values similar to silica glass and the silica bricks are more resistant to thermal shock. The rapid increase in volume at the phase α–β phase transformations can be understood as an opening up of the silicate rings. At room temperature the rings (Fig. 11.11) are partially crumpled but the increase in thermal motion straightens out the silicate rings. The transverse vibrations of oxygen also explain the peculiar behavior at very low temperatures. Silica has a negative thermal expansion coefficient near 0 K because the lowest energy vibration modes are the transverse motions of the oxygen atoms. This leads to a crumpling of the silicate rings and thermal shrinkage. Zinc oxide (Fig. 11.1) shows a similar effect.

Problem 11.3
Using the data for silica brick refractories in Figure 11.12, estimate the thermal expansion coefficient α as a function of temperature from 100°C to 1400°C. Express the dependence of α on T as a power series in ΔT, $(\Delta T)^2$, $(\Delta T)^3$, etc., and derive a set of coefficients which fit the measured values of α over this temperature range.

Piezoelectricity

12

The prefix "piezo" (pronounced pie-ease-o) comes from the Greek word for pressure or mechanical force. Piezoelectricity refers to the linear coupling between mechanical stress and electric polarization (the direct piezoelectric effect) or between mechanical strain and applied electric field (the converse piezoelectric effect). The equivalence between the direct and converse effects was established earlier using thermodynamic arguments (Section 6.2).

The principal piezoelectric coefficient, d, relates polarization, P, to stress, X, in the direct effect ($P = dX$) and strain, x, to electric field E ($x = dE$). Thus the units of d are [C/N] or [m/V] which are equivalent to one another. Typical sizes for useful piezoelectric materials range from about 1 pC/N for quartz crystals to about 1000 pC/N for PZT (lead zirconate titanate) ceramics.

12.1	Tensor and matrix formulations	87
12.2	Matrix transformations and Neumann's Law	89
12.3	Piezoelectric symmetry groups	91
12.4	Experimental techniques	93
12.5	Structure–property relations	94
12.6	Hydrostatic piezoelectric effect	97
12.7	Piezoelectric ceramics	99
12.8	Practical piezoelectrics: Quartz crystals	100

12.1 Tensor and matrix formulations

To understand how the piezoelectric effect varies with direction and how it is affected by symmetry, it is necessary to determine how piezoelectric coefficients transform between coordinate systems. Since polarization is a vector and stress a second rank tensor, the physical property relating these two variables must involve three directions:

$$P_j = d_{jkl} X_{kl}.$$

In the new coordinate system

$$P'_i = a_{ij} P_j = a_{ij} d_{jkl} X_{kl}.$$

Transforming the stress to the new coordinate system gives

$$P'_i = a_{ij} d_{jkl} a_{mk} a_{nl} X'_{mn} = d'_{imn} X'_{mn}.$$

Thus piezoelectricity transforms as a polar third rank tensor.

$$d'_{imn} = a_{ij} a_{mk} a_{nl} d_{jkl}.$$

In general there are $3^3 = 27$ tensor components, but because the stress tensor is symmetric ($X_{ij} = X_{ji}$), only 18 of the components are independent. Therefore the piezoelectric effect can be described by a 6×3 matrix. The matrix form of the piezoelectric effect uses only two subscripts: $P_i = d_{ij} X_j$ where $i = 1$–3, and $j = 1$–6. Written out it is

$$\begin{pmatrix} P_1 \\ P_2 \\ P_3 \end{pmatrix} = \begin{pmatrix} d_{11} & d_{12} & d_{13} & d_{14} & d_{15} & d_{16} \\ d_{21} & d_{22} & d_{23} & d_{24} & d_{25} & d_{26} \\ d_{31} & d_{32} & d_{33} & d_{34} & d_{35} & d_{36} \end{pmatrix} \begin{pmatrix} X_1 \\ X_2 \\ X_3 \\ X_4 \\ X_5 \\ X_6 \end{pmatrix}.$$

To determine the relationships between the tensor coefficients d_{ijk} and the matrix coefficient d_{ij}, it is helpful to write out the equations and identify corresponding coefficients. To illustrate, write out P_1 first in tensor form and then in matrix form.

$$P_1 = d_{1jk}X_{jk} = d_{111}X_{11} + d_{112}X_{12} + d_{113}X_{13} + d_{121}X_{21} + d_{122}X_{22}$$
$$+ d_{123}X_{23} + d_{131}X_{31} + d_{132}X_{32} + d_{133}X_{33}.$$

And in matrix form,

$$P_1 = d_{1j} = d_{11}X_1 + d_{12}X_2 + d_{13}X_3 + d_{14}X_4 + d_{15}X_5 + d_{16}X_6.$$

Identifying corresponding stresses and piezoelectric coefficients in the two expressions for P_1:

$$X_{11} = X_1, \quad X_{22} = X_2, \quad X_{33} = X_3, \quad X_{12} = X_{21} = X_6,$$
$$X_{13} = X_{31} = X_5, \quad \text{and} \quad X_{23} = X_{32} = X_4.$$
$$d_{111} = d_{11}, \quad d_{122} = d_{12}, \quad d_{133} = d_{13}, \quad d_{123} + d_{132} = 2d_{123} = d_{14},$$
$$d_{131} + d_{113} = 2d_{113} = d_{15}, \quad d_{112} + d_{121} = 2d_{112} = d_{16}.$$

Factors of two enter the matrix–tensor relations for piezoelectric coefficients involving shear stresses or shear strains. If $j = 1, 2, 3$ matrix coefficient $d_{ij} = d_{ijj}$. If $j = 4, 5, 6$, then $d_{ij} = 2d_{ikl}$.

For the converse piezoelectric effect, strain (x) is related to electric field (E). In tensor form, strain is a second rank tensor, and electric field a first rank tensor. Therefore the converse piezoelectric coefficient is a third rank tensor.

$$x_{ij} = d_{ijk}E_k.$$

It was shown earlier (Section 6.2) that the direct and converse piezoelectric coefficients are equal based on thermodynamic arguments.

In matrix form the converse coefficient matrix is the transpose of the direct effect matrix

$$(x) = (d)_t(E).$$

Written out, the converse effect is

$$\begin{pmatrix} x_1 \\ x_2 \\ x_3 \\ x_4 \\ x_5 \\ x_6 \end{pmatrix} = \begin{pmatrix} d_{11} & d_{21} & d_{31} \\ d_{12} & d_{22} & d_{32} \\ d_{13} & d_{23} & d_{33} \\ d_{14} & d_{24} & d_{34} \\ d_{15} & d_{25} & d_{35} \\ d_{16} & d_{26} & d_{36} \end{pmatrix} \begin{pmatrix} E_1 \\ E_2 \\ E_3 \end{pmatrix}.$$

12.2 Matrix transformations and Neumann's Law

In matrix form the direct piezoelectric effect is $(P) = (d)(X)$. When written in tensor form the piezoelectric coefficients transform as a third rank tensor, but how do they transform in matrix notation? Here we make use of the matrix transformations for stress (Section 10.2):

$$(X') = (\alpha)(X) \quad \text{and} \quad (X) = (\alpha)^{-1}(X'),$$

where the (α) and $(\alpha)^{-1}$ matrices are functions of the direction cosines a_{ij} (Table 10.1). For the direct piezoelectric effect, the matrix transformation is derived as follows.

$$(P') = (a)(P) = (a)(d)(X) = (a)(d)(\alpha)^{-1}(X')$$
$$= (d')(X').$$

Therefore the matrix transformation in going from the old to the new coordinate system is

$$(d') = (a)(d)(\alpha)^{-1}.$$

Premultiplying this equation by $(a)^{-1}$ and postmutiplying by (α) gives the transformation from new to old.

$$(d) = (a)^{-1}(d')(\alpha).$$

To illustrate how the matrix transformation is used, we derive the piezoelectric coefficients for monoclinic crystals belonging to point group 2. By convention, the twofold symmetry axis is along the crystallographic [010] direction and is labeled Z_2 for the physical properties. The direction cosine matrix for 180° rotation about Z_2 is

$$(a) = \begin{pmatrix} -1 & 0 & 0 \\ 0 & 1 & 0 \\ 0 & 0 & -1 \end{pmatrix}.$$

Since this is a symmetry element for sucrose and other crystals belonging to point group 2, the piezoelectric matrix must remain the same after the transformation (Neumann's Law). We carry out this transformation in matrix form.

$$(d') = (a)(d)(\alpha)^{-1}$$

$$(d') = (a) \begin{pmatrix} d_{11} & d_{12} & d_{13} & d_{14} & d_{15} & d_{16} \\ d_{21} & d_{22} & d_{23} & d_{24} & d_{25} & d_{26} \\ d_{31} & d_{32} & d_{33} & d_{34} & d_{35} & d_{36} \end{pmatrix} \begin{pmatrix} 1 & 0 & 0 & 0 & 0 & 0 \\ 0 & 1 & 0 & 0 & 0 & 0 \\ 0 & 0 & 1 & 0 & 0 & 0 \\ 0 & 0 & 0 & -1 & 0 & 0 \\ 0 & 0 & 0 & 0 & 1 & 0 \\ 0 & 0 & 0 & 0 & 0 & -1 \end{pmatrix}.$$

The reciprocal (α) matrix was obtained by substituting the a_{ij} direction cosines for a twofold rotation in the $(\alpha)^{-1}$ matrix in Table 10.1.

$$(d') = \begin{pmatrix} -1 & 0 & 0 \\ 0 & 1 & 0 \\ 0 & 0 & -1 \end{pmatrix} \begin{pmatrix} d_{11} & d_{12} & d_{13} & -d_{14} & d_{15} & -d_{16} \\ d_{21} & d_{22} & d_{23} & -d_{24} & d_{25} & -d_{26} \\ d_{31} & d_{32} & d_{33} & -d_{34} & d_{35} & -d_{36} \end{pmatrix}$$

$$= \begin{pmatrix} -d_{11} & -d_{12} & -d_{13} & +d_{14} & -d_{15} & +d_{16} \\ +d_{21} & +d_{22} & +d_{23} & -d_{24} & +d_{25} & -d_{26} \\ -d_{31} & -d_{32} & -d_{33} & +d_{34} & -d_{35} & +d_{36} \end{pmatrix}.$$

Equating this transformed matrix (d') to the original piezoelectric matrix (d) shows that 10 of the coefficients are zero. The piezoelectric matrix for point group 2 has eight nonzero coefficients and all eight are different. Therefore a minimum of eight piezoelectric measurements will be required for these crystals.

Sucrose ($C_{12}H_{22}O_{11}$) and triglycine sulfate (TGS = $(NH_2CH_2COOH)_3 \cdot H_2SO_4$) are monoclinic crystals belonging to point group 2. Their piezoelectric coefficients are given in Table 12.1. TGS is ferroelectric with active domain walls but sucrose (common table sugar) is not. Note that the some of piezoelectric coefficients of ferroelectric TGS are much larger than those of sucrose.

Piezoelectric matrices for other symmetry groups are derived in a similar manner. An important case is that of a poled ferroelectric ceramic. The poling process is carried out with a strong DC field at elevated temperatures where domain walls move more easily. The symmetry of a uniform DC field is ∞m, one of the Curie group textures.

Point group ∞m has an ∞-fold axis parallel to Z_3 and an infinite number of mirror planes parallel to Z_3. Three of these mirror planes are perpendicular to Z_1, perpendicular to Z_2, and at 45° to Z_1 and Z_2. Using these three mirrors and the direct inspection method for tensor coefficients, we derive the piezoelectric matrix for poled ferroelectric ceramics by the direct inspection method.

For the mirror perpendicular to Z_1, $1 \to -1$, $2 \to 2$, $3 \to 3$. Therefore for coefficient d_{111}, $111 \to -111$, which means that according to Neumann's Principle, $d_{111} = 0$. In matrix form $d_{11} = d_{111} = 0$. Carrying out the same procedure for other tensor coefficients, it is obvious that $d_{ijk} = 0$ whenever there is an odd number of 1s in the subscript. Thus, for example, $d_{123} = 0 = \frac{1}{2}d_{14}$.

Table 12.1 Piezoelectric coefficients of sucrose and triglycine sulfate measured in units of pC/N at room temperature

Coefficient	Sucrose	Triglycine sulfate
d_{21}	1.48	23.6
d_{22}	−3.42	7.9
d_{23}	0.74	25.3
d_{14}	1.25	2.8
d_{16}	−2.42	−4.6
d_{25}	−0.87	24.3
d_{34}	−4.22	−3.2
d_{36}	0.42	2.8

The resulting matrix for $m \perp Z_1$ is

$$\begin{pmatrix} 0 & 0 & 0 & 0 & d_{15} & d_{16} \\ d_{21} & d_{22} & d_{23} & d_{24} & 0 & 0 \\ d_{31} & d_{32} & d_{33} & d_{34} & 0 & 0 \end{pmatrix}.$$

The second symmetry element is $m \perp Z_2$ for which $1 \to 1$, $2 \to -2$, $3 \to 3$. In this case all the remaining tensor coefficients with an odd number of 2s go to zero. Thus $d_{222} = 0 = d_{22}$ and $d_{122} = \frac{1}{2}d_{16} = 0$. The remaining matrix coefficients are

$$\begin{pmatrix} 0 & 0 & 0 & 0 & d_{15} & 0 \\ 0 & 0 & 0 & d_{24} & 0 & 0 \\ d_{31} & d_{32} & d_{33} & 0 & 0 & 0 \end{pmatrix}.$$

The third symmetry element is a mirror plane at 45° to X_1 and X_2. Such a transformation takes $1 \to 2$, $2 \to 1$, and $3 \to 3$. For the remaining tensor coefficients $113 \to 223$, or in matrix notation, $d_{15} = d_{24}$. Similarly $311 \to 322$, and $d_{31} = d_{32}$. The final coefficient d_{33} is left unchanged.

For poled ferroelectric ceramics and other materials belonging to point group ∞m, the piezoelectric matrix has five coefficients, three of which are independent:

$$\begin{pmatrix} 0 & 0 & 0 & 0 & d_{15} & 0 \\ 0 & 0 & 0 & d_{15} & 0 & 0 \\ d_{31} & d_{31} & d_{33} & 0 & 0 & 0 \end{pmatrix}.$$

12.3 Piezoelectric symmetry groups

Applying Neumann's Principle to the 32 crystal classes and seven Curie groups leads to the matrices listed in Table 12.2. Piezoelectricity is a null property, which means that the piezoelectric effect disappears for certain symmetry groups. Eleven of the crystal classes and four of the Curie groups are non-piezoelectric. All but two of the nonpiezoelectric groups are centrosymmetric. The presence of a center of symmetry eliminates all piezoelectric coefficients.

Table 12.2 Piezoelectric matrices for the crystallographic and limiting point groups

Point groups $\bar{1}, 2/m, mmm, \bar{3}, \bar{3}m, 4/m,$ $4/mmm, 6/m, 6/mmm, m3, 432,$ $m3m, \infty/m, \infty/mm, \infty\infty, \infty\infty m$	$\begin{pmatrix} 0 & 0 & 0 & 0 & 0 & 0 \\ 0 & 0 & 0 & 0 & 0 & 0 \\ 0 & 0 & 0 & 0 & 0 & 0 \end{pmatrix}$
Point group 1	$\begin{pmatrix} d_{11} & d_{12} & d_{13} & d_{14} & d_{15} & d_{16} \\ d_{21} & d_{22} & d_{23} & d_{24} & d_{25} & d_{26} \\ d_{31} & d_{32} & d_{33} & d_{34} & d_{35} & d_{36} \end{pmatrix}$ (18)
Point group 2	$\begin{pmatrix} 0 & 0 & 0 & d_{14} & 0 & d_{16} \\ d_{21} & d_{22} & d_{23} & 0 & d_{25} & 0 \\ 0 & 0 & 0 & d_{34} & 0 & d_{36} \end{pmatrix}$ (8)

Continued overleaf

Table 12.2 (*Continued*)

Point group m
$$\begin{pmatrix} d_{11} & d_{12} & d_{13} & 0 & d_{15} & 0 \\ 0 & 0 & 0 & d_{24} & 0 & d_{26} \\ d_{31} & d_{32} & d_{33} & 0 & d_{35} & 0 \end{pmatrix}$$
(10)

Point group $mm2$
$$\begin{pmatrix} 0 & 0 & 0 & 0 & d_{15} & 0 \\ 0 & 0 & 0 & d_{24} & 0 & 0 \\ d_{31} & d_{32} & d_{33} & 0 & 0 & 0 \end{pmatrix}$$
(5)

Point group 222
$$\begin{pmatrix} 0 & 0 & 0 & d_{14} & 0 & 0 \\ 0 & 0 & 0 & 0 & d_{25} & 0 \\ 0 & 0 & 0 & 0 & 0 & d_{36} \end{pmatrix}$$
(3)

Point group 3
$$\begin{pmatrix} d_{11} & -d_{11} & 0 & d_{14} & d_{15} & -2d_{22} \\ -d_{22} & d_{22} & 0 & d_{15} & -d_{14} & -2d_{11} \\ d_{31} & d_{31} & d_{33} & 0 & 0 & 0 \end{pmatrix}$$
(6)

Point group 32
$$\begin{pmatrix} d_{11} & -d_{11} & 0 & d_{14} & 0 & 0 \\ 0 & 0 & 0 & 0 & -d_{14} & -2d_{11} \\ 0 & 0 & 0 & 0 & 0 & 0 \end{pmatrix}$$
(2)

Point group $3m$
$$\begin{pmatrix} 0 & 0 & 0 & 0 & d_{15} & -2d_{22} \\ -d_{22} & d_{22} & 0 & d_{15} & 0 & 0 \\ d_{31} & d_{31} & d_{33} & 0 & 0 & 0 \end{pmatrix}$$
(4)

Point group $4, 6, \infty$
$$\begin{pmatrix} 0 & 0 & 0 & d_{14} & d_{15} & 0 \\ 0 & 0 & 0 & d_{15} & -d_{14} & 0 \\ d_{31} & d_{31} & d_{33} & 0 & 0 & 0 \end{pmatrix}$$
(4)

Point group $\bar{4}$
$$\begin{pmatrix} 0 & 0 & 0 & d_{14} & d_{15} & 0 \\ 0 & 0 & 0 & -d_{15} & d_{14} & 0 \\ d_{31} & -d_{31} & 0 & 0 & 0 & d_{36} \end{pmatrix}$$
(3)

Point group $4mm, 6mm, \infty m$
$$\begin{pmatrix} 0 & 0 & 0 & 0 & d_{15} & 0 \\ 0 & 0 & 0 & d_{15} & 0 & 0 \\ d_{31} & d_{31} & d_{33} & 0 & 0 & 0 \end{pmatrix}$$
(3)

Point group $422, 622, \infty 2$
$$\begin{pmatrix} 0 & 0 & 0 & d_{14} & 0 & 0 \\ 0 & 0 & 0 & 0 & -d_{14} & 0 \\ 0 & 0 & 0 & 0 & 0 & 0 \end{pmatrix}$$
(1)

Point group $\bar{4}2m$
$$\begin{pmatrix} 0 & 0 & 0 & d_{14} & 0 & 0 \\ 0 & 0 & 0 & 0 & d_{14} & 0 \\ 0 & 0 & 0 & 0 & 0 & d_{36} \end{pmatrix}$$
(2)

Point group $\bar{6}$
$$\begin{pmatrix} d_{11} & -d_{11} & 0 & 0 & 0 & -2d_{22} \\ -d_{22} & d_{22} & 0 & 0 & 0 & -2d_{11} \\ 0 & 0 & 0 & 0 & 0 & 0 \end{pmatrix}$$
(2)

Point group $\bar{6}m2$
$$\begin{pmatrix} 0 & 0 & 0 & 0 & 0 & -2d_{22} \\ -d_{22} & d_{22} & 0 & 0 & 0 & 0 \\ 0 & 0 & 0 & 0 & 0 & 0 \end{pmatrix}$$
(1)

Point group $\bar{4}3m, 23$
$$\begin{pmatrix} 0 & 0 & 0 & d_{14} & 0 & 0 \\ 0 & 0 & 0 & 0 & d_{14} & 0 \\ 0 & 0 & 0 & 0 & 0 & d_{14} \end{pmatrix}$$
(1)

12.4 Experimental techniques

The number of independent piezoelectric coefficients varies from 18 for triclinic group 1 to one for crystals in cubic classes $\bar{4}3m$ and 23. The piezoelectric constants can be measured under static or quasistatic conditions with reasonable accuracy, although the precision is inferior to the resonance method. The Berlincourt d_{33} Meter is a widely used instrument that applies a vibrating mechanical force to the sample and simultaneously to a standard piezoelectric with a known d_{33} coefficient. By comparing the electric signals from the two samples, the d_{33} coefficient of the sample is obtained. Other coefficients are measured by using crystal plates of various orientations.

The resonance method is the most widely used measurement technique. Using the schematic circuit in Fig. 12.1, the impedance of the sample is monitored as function of frequency. The piezoelectric coefficients couple the electric field to mechanical strain leading to various resonant motions. Five of these resonances for poled piezoelectric ceramics are illustrated in Fig. 12.2. Two are length extensional modes generated with slender bars, two use poled disks, and the fifth is a thickness shear mode. The three piezoelectric coefficients are obtained by measuring the resonant frequencies, together with the sample dimensions, dielectric properties, and elastic constants. Details of the measurements are given in the IEEE Standards on Piezoelectric Crystals.

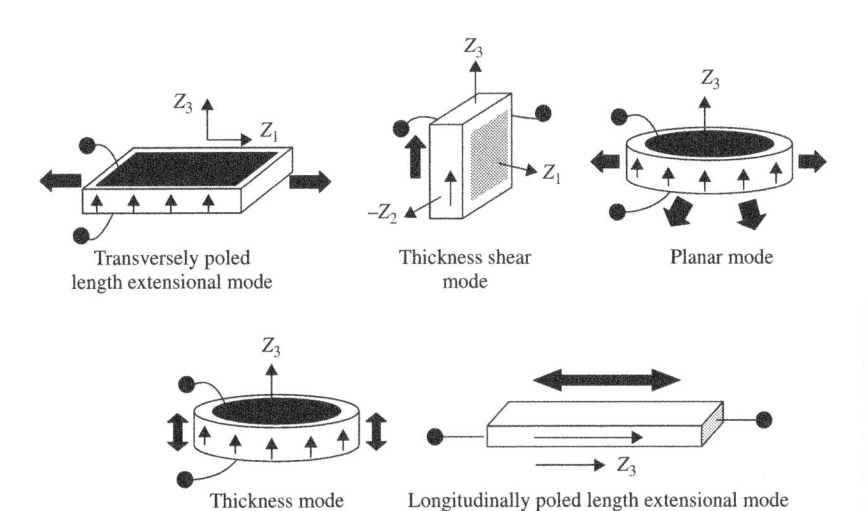

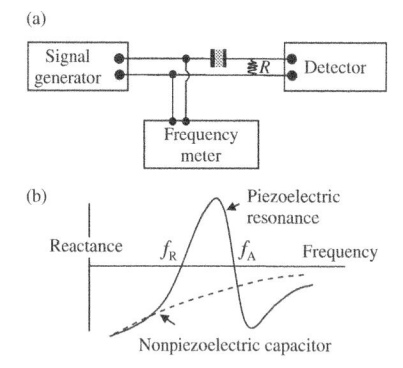

Fig. 12.1 (a) Electronic network for determining the resonant (f_R) and antiresonant (f_A) frequencies of a piezoelectric resonator. (b) The reactance curve associated with a piezoelectric resonance.

Problem 12.1

At room temperature KNbO$_3$ belongs to orthorhombic point group $mm2$ which has five independent piezoelectric coefficients. For a crystal plate oriented perpendicular to an arbitrary direction Z'_3, the piezoelectric coefficient in the thickness direction is d'_{33}. Determine d'_{33} for KNbO$_3$ in terms of the five piezoelectric constants and the orientation angles θ and ϕ of the plate.

The Berlincourt d_{33} Meter described in Section 12.4 works well for strong piezoelectrics like KNbO$_3$. Which of its piezoelectric coefficients can be determined by this method? What orientations should be used?

Fig. 12.2 Resonant modes used in determining the piezoelectric coefficients of poled ferroelectric ceramics. The transversely poled extensional mode and planar mode are governed by d_{31}, the shear mode gives d_{15}, and d_{33} can be measured by either the longitudinally poled extensional mode or the thickness mode. All five modes are used in piezoelectric devices as well.

12.5 Structure–property relations

Data for a number of piezoelectric crystals and poled ceramics are collected in Table 12.3.

All piezoelectric coefficients are zero in centric crystals and untextured polycrystalline materials. In some acentric crystals the causes of piezoelectricity can be readily identified from molecular shapes. $NiSO_3 \cdot 6H_2O$ contains Ni^{2+} ions in octahedral coordination with six water molecules and pyramidal sulfite groups (Fig. 12.3). In nickel sulfite hexahydrate, all the sulfite groups point in the same direction with the trigonal axis of the crystal coinciding with the symmetry of the molecule. It is obvious why the crystal is piezoelectric when squeezed in this direction.

Crystals containing tetrahedral groups are often piezoelectric as well. Zincite (ZnO), zincblende (ZnS), and quartz (SiO_2) are examples. The symmetry of a

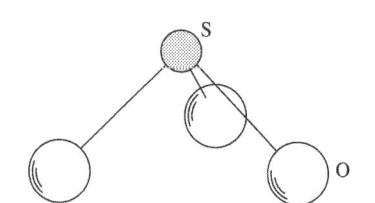

Fig. 12.3 Pyramidal groups such as the sulfite ion with polar symmetry sometimes play an important role in piezoelectric and pyroelectric crystals.

Table 12.3 Piezoelectric strain coefficients in pC/N

Cubic ($\bar{4}3m$)	d_{14}				
$Bi_{12}SiO_{20}$	40				
$NaClO_3$	1.7				
GaAs	2.6				
ZnS	3.2				
Hexagonal (6mm)	d_{31}	d_{33}	d_{15}		
ZnO	−5.0	12.4	−8.3		
CdS	−5.2	10.3	−14.0		
AlN	−2.0	5.0	4.0		
Tetragonal (4mm)	d_{31}	d_{33}	d_{15}		
$BaTiO_3$	−34.5	85.6	392		
$PbTiO_3$	−25	117	62		
Tetragonal ($\bar{4}2m$)	d_{14}	d_{36}			
KH_2PO_4	1.3	21			
$NH_4H_2PO_4$	1.8	48			
Tetragonal (422)	d_{14}				
TeO_2	8.1				
Trigonal (3m)	d_{31}	d_{22}	d_{33}	d_{15}	
$LiNbO_3$	−1.0	21	16	74	
$LiTaO_3$	−3.0	9.0	9.0	26	
Tourmaline	−0.3	−0.3	−1.8	−3.6	
Trigonal (32)	d_{11}	d_{14}			
α-Quartz	2.3	−0.67			
Orthorhombic (222)	d_{14}	d_{25}	d_{36}		
Rochelle salt	2300	−56	12		
Orthorhombic (mm2)	d_{31}	d_{32}	d_{33}	d_{15}	d_{24}
$PbNb_2O_6$	−43	24	60	180	170
Poled ceramics (∞m)	d_{31}	d_{33}	d_{15}		
$BaTiO_3$	−78	190	260		
$Pb(Zr, Ti)O_3$					
PZT-5H	−274	593	741		
PZT-8	−27	225	330		
$K_{0.5}Na_{0.5}NbO_3$	−51	127	306		

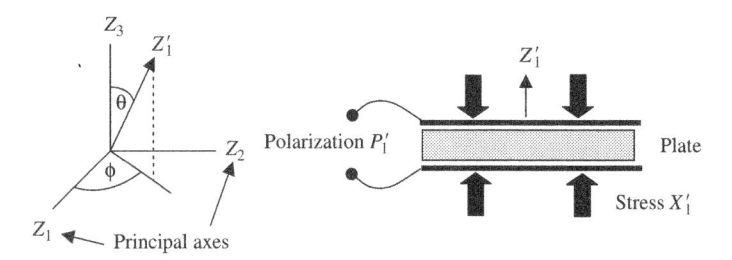

Fig. 12.4 Orientation of the longitudinal piezoelectric experiment in an arbitrary direction.

regular tetrahedron is $\bar{4}3m$, a noncentrosymmetric point group. It is interesting to examine the piezoelectric surface for point group $\bar{4}3m$ and to speculate on the size of the piezoelectric coefficient.

For the longitudinal piezoelectric surface each radius vector has a length proportional to the charge per unit area generated by a stress in the same direction. Imagine a plate cut from a crystal and electroded on its major faces (Fig. 12.4).

Direction Z'_1 is normal to the electroded surfaces and parallel to the stress direction. In tensor notation the piezoelectric coefficient in this direction is $d'_{111} = a_{1i}a_{1j}a_{1k}d_{ijk}$ where d_{ijk} are the measured coefficients referred to the principal axes Z_1, Z_2, Z_3. For point group $\bar{4}3m$ (see Table 12.1), all the piezoelectric coefficients are zero except $d_{14} = d_{25} = d_{36}$. In tensor notation the nonzero coefficients are

$$d_{123} = d_{132} = d_{213} = d_{231} = d_{312} = d_{321}.$$

Therefore

$$d'_{111} = 6a_{11}a_{12}a_{13}d_{123} = 3d_{14}\cos\theta \sin^2\theta \sin\phi \cos\phi.$$

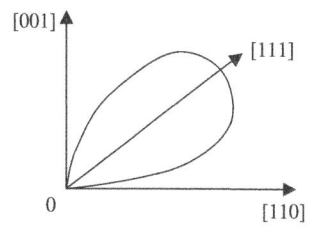

Fig. 12.5 Cross-section of piezoelectric surface for cubic crystals belonging to point group $\bar{4}3m$.

Plotting out the surface gives four lobes pointing along the [111], [$\bar{1}\bar{1}$1] [1$\bar{1}\bar{1}$], and [$\bar{1}$1$\bar{1}$] cube axes. The four lobes resemble the tetrahedral sp^3 hybrid bonds found in chemistry. A cross-section of one of the lobes is shown in Fig. 12.5.

Zincblende and other crystals belonging to this point group are not piezoelectric when stressed along the $\langle 100 \rangle$ cube edges or the $\langle 110 \rangle$ face diagonals.

Many piezoelectric crystals have rather complicated structures, but zincblende (cubic ZnS) is relatively simple. It has a diamond-like structure with both zinc and sulfur in tetrahedral coordination (Fig. 12.6(a)). As pointed out previously, the point group is $\bar{4}3m$ with just one independent piezoelectric coefficient d_{14}. When subject to a shear stress X_4 acting about $Z_1 = [100]$, the crystal structure shears, deforming the tetrahedra (Fig. 12.6(b)). As a result two of the sulfur atoms near each zinc move closer, and two move further away. To maintain four equal bonds, the zinc ion moves in the Z_1 direction creating a polarization P_1. This is the underlying mechanism of piezoelectric coefficient d_{14}.

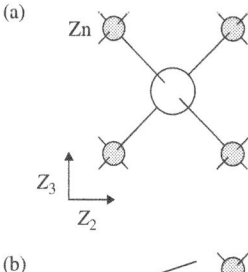

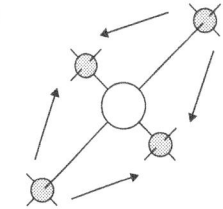

Fig. 12.6 A ZnS$_4$ tetrahedron in zincblende in the unstressed state (a) and then under a shear stress X_4 (b). To equalize the four bondlengths in the sheared state, the divalent zinc atom moves in the Z_1 direction, creating piezoelectric coefficients d_{14}.

Polymers—both natural and synthetic—are sometimes piezoelectric because of the chain configurations and polar side groups. In its β form,

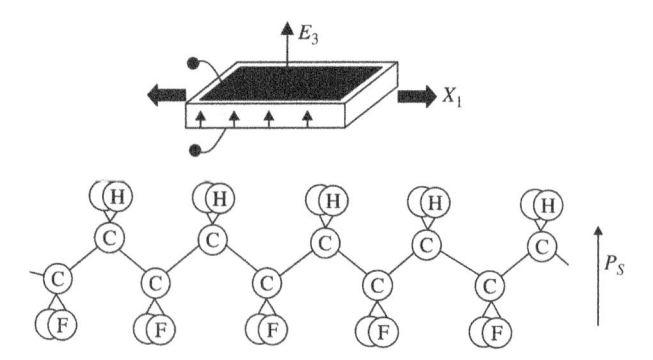

Fig. 12.7 Stretched molecules of PVDF aligned in a strong poling field have an orthorhombic texture and a strong piezoelectric effect.

polyvinylidene fluoride (PVDF = $(CH_2CF_2)_n$) is strongly piezoelectric when mechanically stretched and electrically poled (Fig. 12.7). With the electric field along Z_3 and the stretching force along Z_1, the polymer develops an orthorhombic texture in polar point group $mm2$. This is an interesting example of *Curie's Principle of Symmetry Superposition*. The symmetry of the electric field is Curie group ∞m, and that of a tensile stress is ∞/mm. When the two ∞-fold axes are at right angles to one another along Z_3 and Z_1, respectively, there are three symmetry elements that are common to the two groups: mirror planes perpendicular to Z_1 and Z_2 and a twofold rotation axis parallel to Z_3. The resulting symmetry group, orthorhombic $mm2$, has five independent piezoelectric coefficients: d_{31}, d_{32}, d_{33}, d_{15}, and d_{24}. Measured values for the coefficients are $d_{31} = 20$ pC/N, $d_{32} = 2$, $d_{33} = -30$, and d_{15}, d_{24} are small.

The atomistic origin of these effects can be visualized from the molecular structure (Fig. 12.7). Under tensile force, the carbon backbone of the PVDF polymer will align along Z_1, but electric dipoles in this direction are small. The dipoles are associated with the hydrogen and fluorine side groups. A DC field in the Z_3 direction aligns the dipoles with the protons parallel to Z_3 and the fluorine ions parallel to $-Z_3$.

Natural polymers such as cellulose also exhibit weak piezoelectric effects. There are no strong poling fields in this case but the spiral nature of the molecules together with the natural texture of a tree again imparts orthorhombic symmetry to pieces of wood. As shown in Fig. 12.8, the three principal axes are radial (Z_1), tangential to the tree rings (Z_2), and parallel to the height (Z_3) of the tree. Because of the helical nature of the cellulose molecules and the orthorhombic texture, wood has been assigned to point group 222, which has three piezoelectric coefficients. Measurements carried out on rectangular pieces of birch gave shear coefficients $d_{14} = -0.18$, $d_{25} = 0.30$, $d_{36} = 0.07$ pC/N, more than an order of magnitude smaller than engineering materials. The origin of the weak piezoelectric effects in living systems (including our bodies) and the role it may play in growth and healing mechanisms has been a subject of much speculation.

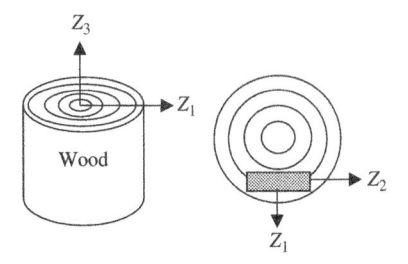

Fig. 12.8 The texture of wood. The ring structure of a large tree has orthorhombic symmetry.

Problem 12.2
Zinc oxide crystals belong to point group $6mm$ with three independent piezoelectric coefficients d_{31}, d_{33}, and d_{15}. Using the information in Tables 12.2 and 12.3, plot the longitudinal piezoelectric coefficient $d'_{11} = d'_{111}$ as a function of direction in the Z_1–Z_2 and Z_1–Z_3 planes.

12.6 Hydrostatic piezoelectric effect

Hydrostatic stress is a scalar (zero-rank tensor) quantity like temperature. Therefore, the hydrostatic piezoelectric effect is analogous to the pyroelectric effect (Chapter 8) with the same symmetry restrictions. Only ten of the 32 crystal classes and two of the seven Curie groups exhibit the hydrostatic piezoelectric effect.

As discussed in Section 10.1, the nonzero stress components associated with a change in pressure are $X_{11} = X_{22} = X_{33} = -p$. The resulting components of electrical polarization are,

$$P_1 = -(d_{11} + d_{12} + d_{13})p$$
$$P_2 = -(d_{21} + d_{22} + d_{23})p$$
$$P_3 = -(d_{31} + d_{32} + d_{33})p.$$

These expressions simplify for the various symmetry groups. For a poled ceramic (group ∞m),

$$P_1 = P_2 = 0 \quad \text{and} \quad P_3 = -(2d_{31} + d_{33})p.$$

The vector sum of the polarization is the hydrostatic piezoelectric effect, $P = d_h p$, where d_h is the hydrostatic charge coefficient. Also of interest is the hydrostatic voltage coefficient $g_h = d_h/\varepsilon$, where ε is the dielectric permittivity. The $d_h g_h$ product is often used as a figure of merit for underwater hydrophones.

Hydrostatic piezoelectric coefficients for a number of crystals and poled ceramics are given in Table 12.4. Since the symmetry requirements for pyroelectricity and the hydrostatic piezoelectric effect are identical, all the materials are also pyroelectric.

For the purposes of discussion, they can be divided into ferroelectric pyroelectrics and ordinary (nonferroelectric) pyroelectrics. As shown in Table 12.4, the ferroelectrics have substantial d_h coefficients but the g_h values are not very big because of their large permittivities.

Ordinary pyroelectrics can be further subdivided into water-soluble pyroelectrics and oxide pyroelectrics. Oxides and sulfides with the wurtzite structure (Fig. 8.4) have very small hydrostatic piezoelectric effects. The wurtzite crystal structure is based on a hexagonal close-packed anion lattice with cations in tetrahedral interstices. Compared to the other pyroelectrics, the atomic bonding in wurtzite is very isotropic. It is not surprising, therefore, that under hydrostatic pressure they deform isotropically, leading to very small piezoelectric effects.

Silicate pyroelectrics have somewhat larger hydrostatic coefficients than the wurtzite group. Tourmaline is a complex borosilicate mineral containing tetrahedral SiO_4 groups. The silica tetrahedra are arranged in Si_6O_{18} rings oriented perpendicular to the pyroelectric axis. This imparts an anisotropy to the structure not found in the wurtzite group, but the silicate and borate groups are linked together by Al^{3+} and Mg^{2+} ions that also form fairly strong chemical bonds and hence tourmaline is not as anisotropic as some other crystals.

More anisotropic structures are found among the water-soluble pyroelectrics. Lithium sulfate monohydrate ($Li_2SO_4 \cdot H_2O$) is an important example with an extremely large g_h coefficient, so large that the crystals have been used as hydrostatic pressure sensors. The crystal structure of lithium sulfate contains Li$^+$ cations, tetrahedral SO$_4^{2-}$ anions, and water molecules. Ionic bonds between cations and anions extend in all directions in the crystals but the hydrogen bonding between water molecules extends only along b, the unique polar axis (Fig. 12.9(a)). Tension in this direction produces a large electric polarization. Short hydrogen bonds like those in lithium sulfate make an important contribution to the piezoelectric effect because the proton position changes as the oxygen–oxygen is stretched. In short hydrogen bonds the proton is midway between the oxygens, whereas the proton is asymmetrically positioned in long H bonds. Mechanical stress, therefore, directly affects the dipole moments of the water molecules, producing electric polarization along the b-axis and an unusually large d_{22} coefficient in lithium sulfate. The large piezoelectric effect together with a small dielectric constant gives it the largest hydrostatic voltage coefficient (g_h) of any material, including ferroelectrics.

Because of their large polarizabilities, ferroelectrics also have large piezoelectric constants but the hydrostatic coefficients are not large for those with nearly symmetric crystal structures. BaTiO$_3$, (Na, K)NbO$_3$ and Pb(Zr, Ti)O$_3$ have the perovskite structure that has a close-packed array of oxygens and large

Table 12.4 Hydrostatic piezoelectric coefficients for a number of materials. For a given pressure, d_h measures the electric polarization, and g_h the open-circuit electric field. d_h is expressed in units of 10^{-12} C/N and g_h in 10^{-3} m^2/C

	d_h	g_h
Water-soluble pyroelectrics		
Ethylene diamine tartrate (EDT)		
C$_2$H$_4$(NH$_3$)$_2$ C$_4$H$_4$O$_6$	1.0	15
Lithium sulphate monohydrate (LH)		
Li$_2$SO$_4 \cdot$ H$_2$O	16.4	180
Others	<4.0	<100
Pyroelectric silicate minerals		
Tourmaline		
(Na, Ca)(Mg, Fe)$_3$B$_3$Al$_6$Si$_6$(O, OH, F)$_{31}$	2.5	38
Others	<3.0	<30
Wurtzite-family pyroelectrics		
BeO, ZnO, CdS, CdSe	<0.2	<3
Ferroelectric single crystals		
Barium titanate		
BaTiO$_3$	16.6	11
Triglycine sulfate (TGS)		
(NH$_2$CH$_2$COOH)$_3$H$_2$SO$_4$	8.0	30
Antimony sulfur iodide (10°C)		
SbSI	1100	14
Lithium niobate		
LiNbO$_3$	14.5	57
Poled ferroelectric ceramics		
Barium titanate		
BaTiO$_3$	34.0	2
Lead niobate		
PbNb$_2$O$_6$	67.0	34
Lead zirconate titanate (PZT)		
Pb(Ti, Zr)O$_3$	20–50	2–9
Sodium potassium niobate		
(Na, K) NbO$_3$	40	10

cations. The LiNbO$_3$ structure also has close-packed oxygens. The hydrostatic piezoelectric coefficients for these materials are small compared to antimony sulfur iodide (SbSI) that has the largest d_h coefficient in Table 12.4.

The structure of SbSI (Fig. 12.9(b)) is very anisotropic with covalently-bonded chains parallel to the polar axis. Neighboring chains are only weakly bonded by ionic or van der Waals forces. Crystals of SbSI cleave readily parallel to the polar c-axis. Under tensile force parallel to c, the crystals develop a large piezoelectric polarization similar in origin to that in lithium sulfate. The antimony cations displace relative to the anions causing the polarization. Piezoelectric effects in the perpendicular directions are much smaller because of the loose packing of chains. The g_h coefficient of SbSI is small because of its large permittivity.

In summary, the best piezoelectrics for hydrostatic sensors are those with anisotropic structures and a molecular mechanism for piezoelectricity. This structure–property relation is the basic idea behind an important family of transducers made from poled ferroelectric ceramics and elastically-compliant polymers. Parallel fibers of PZT embedded in a polymer matrix give larger d_h and g_h coefficients than solid PZT.

12.7 Piezoelectric ceramics

For the past 50 years PZT (PbZr$_{1-x}$Ti$_x$O$_3$) has been the workhorse of transducer technology. Underwater sonar, biomedical ultrasound, multilayer actuators for fuel injection, piezoelectric printers, and bimorph pneumatic valves all make use of poled PZT ceramics. PZT is one of a number of ferroelectric substances crystallizing with the perovskite structure (Fig. 12.10). Lead atoms appear at the corners of the unit cell and oxygens at the face centers. Together they make up a cubic close-packed array, having a lattice parameter near 4 Å. Octahedrally coordinated titanium or zirconium ions are located at the center of the unit cell.

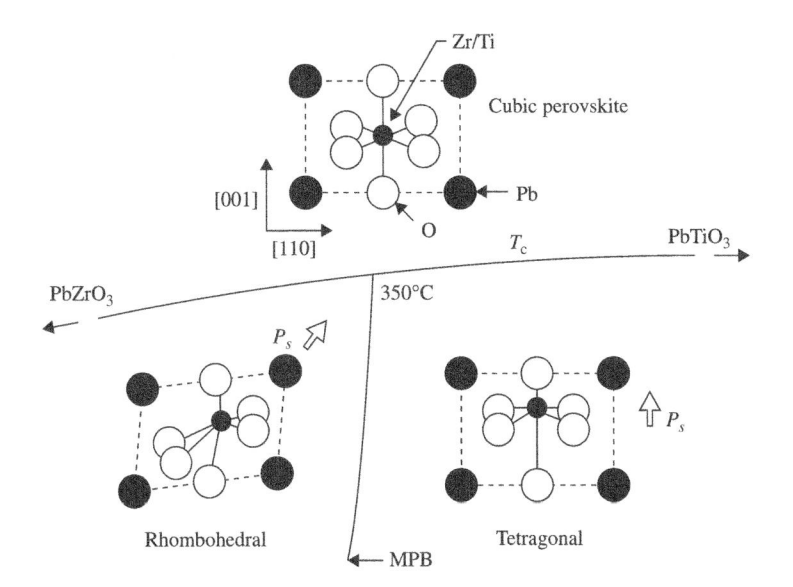

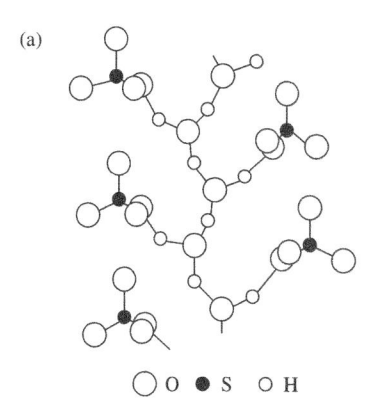

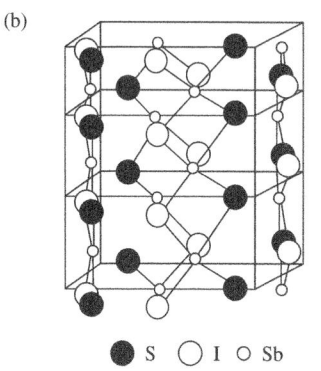

Fig. 12.9 (a) In lithium sulfate monohydrate the water molecules form a chain of hydrogen bonds along the polar axis, but are not linked laterally. (b) Ferroelectric SbSI also has a chain structure and great hydrostatic sensitivity. Both materials provide the parallel connectivity required for large longitudinal d coefficients and small transverse piezoelectric effects.

Fig. 12.10 A portion of the PbZrO$_3$–PbTiO$_3$ phase diagram showing the structure change at the Curie temperature (T_c) and the morphotropic phase boundary. Compositions near the morphotropic phase boundary (MPB) have 14 possible orientations of the polar axis in the coexisting tetragonal and rhombohedral domain structures.

On cooling from high temperature, the cubic unit cell undergoes a displacive phase transformation with atomic displacements of about 0.1 Å. For Ti-rich compositions, the point symmetry changes from cubic $m3m$ to tetragonal $4mm$ at the Curie temperature. The tetragonal state with its polar axis along [001] = Z_3 persists to 0 K. The structure changes are shown in Fig. 12.10. Zr-rich compositions prefer rhombohedral point group $3m$ with the polar axis along a [111] direction.

To enhance the piezoelectric properties of PZT ceramics, compositions near a phase transition are chosen. The compositions that pole best lie near the morphotropic boundary between the rhombohedral and tetragonal distortions of the perovskite structure. At the boundary there are 14 possible poling directions, six ⟨100⟩ directions in the tetragonal state and eight ⟨111⟩ directions in the rhombohedral phase. Domain wall movements and phase changes between the rhombohedral and tetragonal phases occur during the poling process.

As pointed out in Section 12.1, electrically poled ceramics belong to symmetry group ∞m with five piezoelectric coefficients $d_{31} = d_{32}$, d_{33}, and $d_{15} = d_{24}$. Both intrinsic and extrinsic contributions to these piezoelectric coefficients are important. The intrinsic effects coming from the distortions of the crystal structure under mechanical stress are pictured in Fig. 12.11. Under a tensile stress parallel to the dipole moment, there is an enhancement of the polarization along [001] = Z_3. When a tensile stress is applied perpendicular to the dipole, the polarization along Z_3 is reduced. These are the d_{33} and d_{31} effects, respectively. When the dipole is tilted by shear stress, charges appear on the side faces, the d_{15} coefficient. There are extrinsic contributions to the piezoelectric coefficients as well. For PZT compositions near the morphotropic phase boundary, these can be very large, often involving reversible domain wall motions and phase changes. Domain wall contribution to the piezoelectric and dielectric properties of PZT are controlled by doping the ceramic with higher- or lower-valent ions. Substituting a small amount of Nb^{5+} for Ti^{4+} gives a soft-PZT with easily movable domain walls and a large piezoelectric coefficient. Hard PZTs are made by adding Fe^{3+} for Ti^{4+}. High-power transducers use hard PZT because they will not depole under high fields in the reverse directions. Soft PZT is preferred for hydrophones and other sensors. In Table 12.2, the poled ceramics labeled PZT-8 and PZT-5H are representative of hard- and soft-PZT, respectively.

The multimillion-dollar market for PZT actuators and transducers includes multilayer d_{33} thickness mode actuators, d_{31} transverse mode transducers, and various bender types. Typical applications, forces, and displacements for these three families of actuators are shown in Fig. 12.12. High power PZT transmitters for underwater sonar systems operate in the kHz range and are up to a meter in size. Biomedical ultrasonic transducers are much smaller and operate in the MHz range.

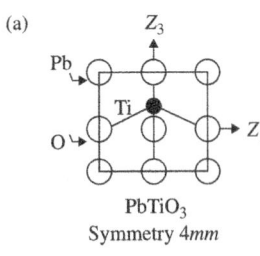

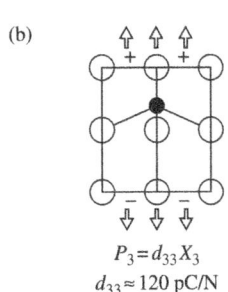

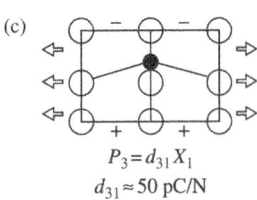

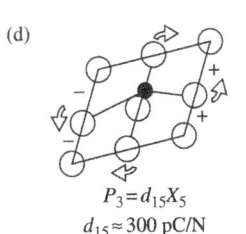

Fig. 12.11 Structure–property relations for the intrinsic piezoelectric effect in PbTiO$_3$. In the unstressed state there is an electric dipole associated with the off-center shift of the titanium atom. Under stress, this dipole can be increased (d_{33}), decreased (d_{31}), or tilted (d_{15}).

(a) PbTiO$_3$ Symmetry $4mm$

(b) $P_3 = d_{33}X_3$
$d_{33} \approx 120$ pC/N

(c) $P_3 = d_{31}X_1$
$d_{31} \approx 50$ pC/N

(d) $P_3 = d_{15}X_5$
$d_{15} \approx 300$ pC/N

12.8 Practical piezoelectrics: Quartz crystals

Piezoelectric quartz crystals are used to control frequency and time standards in electronic circuits. Despite its modest piezoelectric coefficients, quartz has dominated this market for many years because of its low losses (high Q) and zero-temperature coefficient cuts (Fig. 12.13). In addition to wristwatches and

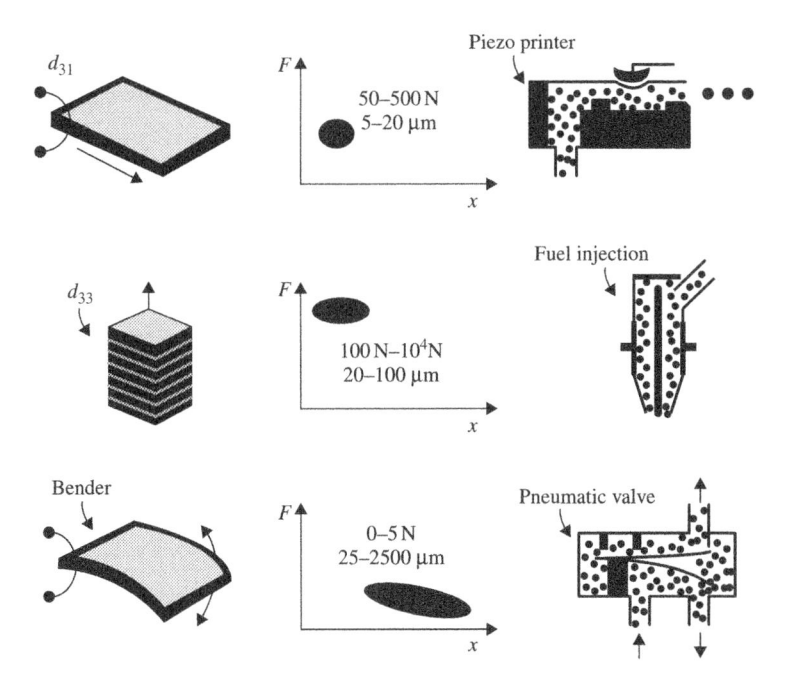

Fig. 12.12 Ceramic multilayer actuators consist of thin layers of piezoelectric ceramic and metal electrodes. In contrast to traditional piezoelectrics, even low voltages produce large forces and substantial displacements. A tradeoff exists between force and displacement. The multilayer stack utilizing the d_{33} coefficient give kilonewton forces capable of pushing heavy weights through small distances. Bimorph benders make use of the smaller transverse of d_{31} coefficients to give larger displacements in the millimeter range, but only small forces.

other time-keeping devices, quartz is used in selective band-pass filters for long-distance telephone lines and other broadband carrier frequency systems.

Quartz belongs to crystal class 32 with five nonzero piezoelectric coefficients. Only two of the coefficients are independent. The converse piezoelectric effect is given by the matrix relation (see Table 12.1):

$$\begin{pmatrix} x_1 \\ x_2 \\ x_3 \\ x_4 \\ x_5 \\ x_6 \end{pmatrix} = \begin{pmatrix} d_{11} & 0 & 0 \\ -d_{11} & 0 & 0 \\ 0 & 0 & 0 \\ d_{14} & 0 & 0 \\ 0 & -d_{14} & 0 \\ 0 & -2d_{11} & 0 \end{pmatrix} \begin{pmatrix} E_1 \\ E_2 \\ E_3 \end{pmatrix}.$$

For right-handed quartz, $d_{11} = 2.27$ and $d_{14} = -0.67$ pC/N.

The three principal axes (Fig. 12.12) are chosen such that the threefold symmetry axis is along Z_3, the twofold axis along Z_1, and Z_2 is perpendicular to both Z_1 and Z_3. In the quartz literature, plates cut perpendicular to Z_1, Z_2, and Z_3 are known as X-, Y-, and Z-cuts, respectively. The vibration modes set up by the five piezoelectric coefficients are pictured in Fig. 12.13.

The AT- and BT-cuts shown in Fig. 12.13 are two of the commonly used orientations for quartz oscillator plates. For frequency- and time-standards it is important that the resonant frequency does not change with temperature. Both the AT- and BT-cuts are rotated Y-cuts that have been rotated about $X(=Z_1)$, the twofold symmetry axis. By rotating in the Y–Z plane, the field component E_1 remains zero, thus avoiding excitation of the X-cut modes shown in Fig. 12.14. The AT- and BT-cut modes are thickness shear modes driven by piezoelectric coefficient d'_{26}, where

$$x'_6 = d'_{26} E'_2.$$

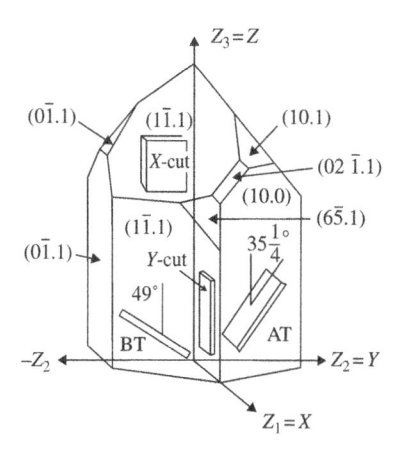

Fig. 12.13 Right-handed quartz crystal showing the orientation of the X-, Y-, AT-, and BT-cuts.

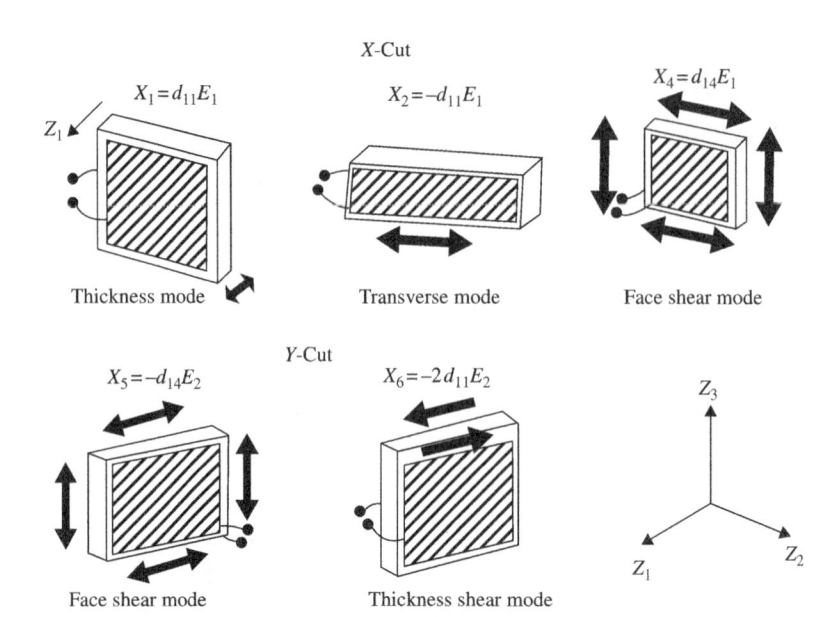

Fig. 12.14 Electrically driven vibration modes of X- and Y-cut quartz. There are no active modes for the Z-cut.

For the rotated cuts (Fig. 12.13) the axis Z_1 remains unchanged while Z_2 and Z_3 are rotated through an angle θ to two new axes Z_2' and Z_3'. The shear motion x_6' is about Z_3' while the field E_2' is in the Z_2' direction.

Further discussion is required to explain how the orientation angles of the AT- and BT-cuts were determined. The resonant frequency depends on the sample dimensions, density, and elastic constants. We will return to this discussion in the next chapter on elasticity.

Problem 12.3
Compute the sign and magnitude of coefficient d_{26}' in terms of θ, d_{11}, and d_{14}. Since only one coefficient is needed, it is easiest to carry out the calculation in tensor notation. Substitute the angles for the AT- and BT-cuts from Fig. 12.13 to evaluate the magnitudes of the d_{26}' coefficients.

Elasticity

13

All solids change shape under mechanical force. Under small stresses, the strain x is related to stress X by Hooke's Law $(x) = (s)(X)$, or the converse relationship $(X) = (c)(x)$. The elastic compliance coefficients (s) are generally reported in units of m^2/N, and the stiffness coefficients (c) in N/m^2. For a fairly stiff material like a metal or a ceramic, c is about 10^{11} N/m^2 = 10^{12} dynes/cm^2 = 100 GPa = 0.145×10^8 PSI. Hooke's Law is a linear relation between stress and strain, and does not describe the elastic behavior at high stress levels that requires higher order elastic constants (Chapter 15). Irreversible phenomena such as plasticity and fracture occur at still higher stress levels.

13.1	Tensor and matrix coefficients	103
13.2	Tensor and matrix transformations	105
13.3	Stiffness–compliance relations	106
13.4	Effect of symmetry	107
13.5	Engineering coefficients and measurement methods	109
13.6	Anisotropy and structure–property relations	110
13.7	Compressibility	113
13.8	Polycrystalline averages	114
13.9	Temperature coefficients	116
13.10	Quartz crystal resonators	118

13.1 Tensor and matrix coefficients

Two directions are needed to specify stress (the direction of the force and the normal to the face on which the force acts), and two directions are needed to specify strain (the direction of the displacement and the orientation of the measurement axis). Thus there are four directions involved in measuring elastic stiffness, which is therefore a fourth rank tensor:

$$X_{ij} = c_{ijkl} x_{kl}.$$

The tensor transformation from old to new goes as follows.

$$X'_{ij} = a_{ik} a_{jl} X_{kl} = a_{ik} a_{jl} c_{klmn} x_{mn}$$
$$= a_{ik} a_{jl} c_{klmn} a_{om} a_{pn} x'_{op}$$
$$= c'_{ijop} x'_{op}.$$

Therefore the stiffness coefficient transforms as a polar fourth rank tensor involving the product of four direction cosines relating the new (primed) coordinate system to the old (unprimed) coordinate system:

$$c'_{ijop} = a_{ik} a_{jl} a_{om} a_{pn} c_{klmn}.$$

The directional subscripts all range from 1 to 3 so there are $(3)^4 = 81$ tensor components for the elastic stiffness or the elastic compliance. The number coefficients is considerably reduced by the fact that both stress and strain are symmetric second rank tensors for which $X_{ij} = X_{ji}$ and $x_{kl} = x_{lk}$. Therefore $c_{ijkl} = c_{jikl} = c_{ijlk} = c_{jilk}$. These equalities reduce the number of independent stiffness coefficients from 81 to 36, and make it possible to describe the elastic constants by a 6 × 6 matrix:

$$\begin{pmatrix} X_1 \\ X_2 \\ X_3 \\ X_4 \\ X_5 \\ X_6 \end{pmatrix} = \begin{pmatrix} c_{11} & c_{12} & c_{13} & c_{14} & c_{15} & c_{16} \\ c_{21} & c_{22} & c_{23} & c_{24} & c_{25} & c_{26} \\ c_{31} & c_{32} & c_{33} & c_{34} & c_{35} & c_{36} \\ c_{41} & c_{42} & c_{43} & c_{44} & c_{45} & c_{46} \\ c_{51} & c_{52} & c_{53} & c_{54} & c_{55} & c_{56} \\ c_{61} & c_{62} & c_{63} & c_{64} & c_{65} & c_{66} \end{pmatrix} \begin{pmatrix} x_1 \\ x_2 \\ x_3 \\ x_4 \\ x_5 \\ x_6 \end{pmatrix}.$$

The matrix can be further simplified by an energy argument. The product of stress and strain is the stored mechanical energy density. In differential form, the energy density is

$$dW = X_i dx_i = c_{ij} x_j dx_i$$
$$= c_{11} x_1 dx_1 + c_{12} x_2 dx_1 + c_{21} x_1 dx_2 + \cdots.$$

Taking the partial derivatives with respect to strain components x_1 and x_2,

$$\frac{\partial W}{\partial x_1} = c_{11} x_1 + c_{12} x_2 + \cdots + c_{16} x_6$$

$$\frac{\partial W}{\partial x_2} = c_{21} x_1 + c_{22} x_2 + \cdots + c_{26} x_6.$$

The second derivatives are

$$\frac{\partial^2 W}{\partial x_1 \partial x_2} = c_{12}$$

and

$$\frac{\partial^2 W}{\partial x_2 \partial x_1} = c_{21}$$

and since the order of differentiation does not affect the energy density,

$$c_{12} = c_{21}.$$

Applying the same argument to the other stiffness coefficients shows that they are all symmetric, $c_{ij} = c_{ji}$. Therefore only 21 independent elastic coefficients are needed to describe Hooke's Law. For symmetries higher than triclinic, the number is reduced even further.

Before proceeding further it is necessary to establish the relationships between the tensor coefficients c_{ijkl} and the matrix coefficients c_{ij}. Equivalences are established by writing out Hooke's Law first in tensor form, and then in matrix form, and identifying equivalent terms. In tensor form a tensile stress along Z_1 is

$$X_{11} = c_{1111} x_{11} + c_{1112} x_{12} + c_{1121} x_{21} + c_{1122} x_{22} + \cdots.$$

The corresponding terms in matrix form are

$$X_1 = c_{11} x_1 + c_{12} x_2 + \cdots + c_{16} x_6,$$

where the strains are $x_6 = x_{12} + x_{21} = 2x_{12}$, $x_1 = x_{11}$ and $x_2 = x_{22}$ (see Section 10.3). Therefore $c_{11} = c_{1111}$, $c_{12} = c_{1122}$, and $c_{16} = \frac{1}{2}(c_{1112} + c_{1121}) = c_{1112}$.

Carrying out the same procedure for the shear stress about Z_3, in tensor form, and remembering that $X_{12} = X_{21}$:

$$X_{12} = c_{1211} x_{11} + c_{1212} x_{12} + c_{1221} x_{21} + c_{1222} x_{22} + \cdots$$

$$X_{21} = c_{2111} x_{11} + c_{2112} x_{12} + c_{2121} x_{21} + c_{2122} x_{22} + \cdots.$$

The equivalent stress in matrix form (see Section 10.2) is
$$X_6 = c_{61}x_1 + c_{62}x_2 + \cdots + c_{66}x_6.$$
Identifying corresponding terms and remembering that $X_6 = X_{12} = X_{21}$ and $x_6 = x_{12} + x_{21}$, leads to the result that $c_{61} = c_{1211} = c_{2111}$, $c_{62} = c_{1222} = c_{2122}$, and $c_{66} = c_{1212} = c_{1221} = c_{2112} = c_{2121}$.

In summary, no factors of two or four are needed when converting stiffness coefficients from tensor to matrix form. This is *not* true for the compliance coefficients. In tensor form the compliances are given by
$$x_{ij} = s_{ijkl}X_{kl}.$$
And in matrix form
$$x_i = s_{ij}X_j.$$
Written out, a tensile strain along Z_1 is
$$x_{11} = s_{1111}X_{11} + s_{1112}X_{12} + s_{1121}X_{21} + s_{1122}X_{22} + \cdots$$
in tensor form, and
$$x_1 = s_{11}X_1 + s_{12}X_2 + \cdots + s_{16}X_6$$
in matrix form.

Since $x_{11} = x_1$, $X_{11} = X_1$, $X_{12} = X_{21} = X_6$, and $X_{22} = X_2$, it is apparent that $s_{11} = s_{1111}$, $s_{12} = s_{1122}$, and $s_{16} = s_{1112} + s_{1121} = 2s_{1112}$.

For shear strains about Z_3,
$$x_{12} = s_{1211}X_{11} + s_{1212}X_{12} + s_{1221}X_{21} + s_{1222}X_{22} + \cdots$$
$$x_{21} = s_{2111}X_{11} + s_{2112}X_{12} + s_{2121}X_{21} + s_{2122}X_{22} + \cdots$$
in tensor form. The corresponding matrix terms are
$$x_6 = s_{61}X_1 + s_{62}X_2 + \cdots + s_{66}X_6.$$
Remembering that $x_6 = x_{12} + x_{21} = 2x_{12}$, $X_1 = X_{11}$, $X_2 = X_{22}$, and $X_6 = X_{12} = X_{21}$, it is clear that $s_{61} = s_{1211} + s_{2111} = 2s_{1211}$, and that $s_{66} = s_{1212} + s_{1221} + s_{2112} + s_{2121} = 4s_{1212}$.

In general, $s_{ijkl} = s_{mn}$ when $m, n = 1, 2, 3$; $2s_{ijkl} = s_{mn}$ when m or $n = 4, 5, 6$; and $4s_{ijkl} = s_{mn}$ when $m, n = 4, 5, 6$. It is important to remember these factors of two or four when working with compliance coefficients.

13.2 Tensor and matrix transformations

In tensor form the stiffness and compliance coefficients transform as fourth rank polar tensors:
$$c'_{ijkl} = a_{im}a_{jn}a_{ko}a_{lp}c_{mnop}$$
and
$$s'_{ijkl} = a_{im}a_{jn}a_{ko}a_{lp}s_{mnop}.$$

Matrix transformations for stress and strain were discussed in Sections 10.2 and 10.3. In transforming between new (primed) and old (unprimed) coordinate systems, stress (X) and strain (x) transform as follows
$$(X') = (\alpha)(X),$$
$$(X) = (\alpha)^{-1}(X'),$$
$$(x') = (\alpha)_t^{-1}(x),$$

and
$$(x) = (\alpha)_t(x').$$

The 6 × 6 (α) matrices involve the direction cosine coefficients and are written out in Table 10.1.

Using these relationships, together with Hooke's Law, the compliance and stiffness transformations are easily obtained. For compliances,
$$(x') = (\alpha)_t^{-1}(x) = (\alpha)_t^{-1}(s)(X) = (\alpha)_t^{-1}(s)(\alpha)^{-1}(X')$$
$$= (s')(X').$$

Therefore $(s') = (\alpha)_t^{-1}(s)(\alpha)^{-1}$. In a similar way it can be shown that
$$(s) = (\alpha)_t(s')(\alpha),$$
$$(c') = (\alpha)(c)(\alpha)_t,$$

and
$$(c) = (\alpha)^{-1}(c')(\alpha)_t^{-1}.$$

13.3 Stiffness-compliance relations

In matrix form the compliance coefficients are given by
$$(x) = (s)(X).$$
Premultiplying both sides by the reciprocal compliance matrix
$$(s)^{-1}(x) = (s)^{-1}(s)(X) = (X) = (c)(x).$$
Therefore $(s)^{-1} = (c)$ or $(c)(s) = 1$.

Multiplying out these symmetric 6 × 6 matrices leads to a set of linear equations of which these are the first two:
$$c_{11}s_{11} + c_{12}s_{12} + c_{13}s_{13} + c_{14}s_{14} + c_{15}s_{15} + c_{16}s_{16} = 1$$
$$c_{11}s_{12} + c_{12}s_{22} + c_{13}s_{23} + c_{14}s_{24} + c_{15}s_{25} + c_{16}s_{26} = 0.$$

In shorthand notation the relationships between the matrix stiffness and compliance coefficients can be written as
$$c_{ij}s_{jk} = \delta_{ik} = 1 \quad \text{when } i = k$$
$$= 0 \quad \text{when } i \neq k.$$

For tensor notation the corresponding equations are
$$c_{ijkl}s_{klmn} = \delta_{im}\delta_{jn}.$$

For triclinic crystals where all 36-matrix coefficients are nonzero, it is necessary to solve the equations in full, but the relationships between stiffness and compliance coefficients are greatly simplified for higher symmetry materials.

For cubic crystals there are three independent elastic constants.
$$c_{11} = (s_{11} + s_{12})/(s_{11} - s_{12})(s_{11} + 2s_{12})$$
$$c_{12} = -s_{12}/(s_{11} - s_{12})(s_{11} + 2s_{12})$$
$$c_{44} = 1/s_{44}.$$

Hexagonal crystals have five independent coefficients related by the equations:

$$c_{11} + c_{12} = s_{33}/S$$
$$c_{11} - c_{12} = 1/(s_{11} - s_{12})$$
$$c_{13} = -s_{13}/S$$
$$c_{33} = (s_{11} + s_{12})/S$$
$$c_{44} = 1/s_{44},$$

where $S = s_{33}(s_{11} + s_{12}) - 2s_{13}^2$.

Problem 13.1
Using the appropriate matrix from Table 13.1, derive the corresponding stiffness-compliance relations for crystals belonging to tetragonal point group 4/*mmm*.

13.4 Effect of symmetry

To demonstrate the way in which Neumann's Principle simplifies the elastic constant matrix, consider the tetragonal point group $4/m$. The mineral scheelite ($CaWO_4$) belongs to this symmetry class. There are two independent symmetry elements, a fourfold rotation axis parallel to Z_3 and a mirror plane perpendicular to Z_3.

Working with tensor notation and the direct inspection method, $4 \parallel Z_3$ takes $1 \to 2, 2 \to -1, 3 \to 3$. Beginning with compliance coefficient s_{1111} and then proceeding to others, it can be seen that under the symmetry transformation

$1111 \to 2222$	$(s_{11} = s_{22})$
$2222 \to 1111$	$(s_{22} = s_{11})$
$3333 \to 3333$	(s_{33})
$1122 \to 2211$	$(s_{12} = s_{21})$
$1133 \to 2233$	$(s_{13} = s_{23})$
$1212 \to 2121$	(s_{66})
$1313 \to 2323$	$(s_{55} = s_{44})$
$1233 \to -2133$	$(s_{63} = -s_{63} = 0)$
$2321 \to 1312$	$(s_{46} = s_{56})$
$1312 \to -2321$	$(s_{56} = -s_{46} = -s_{56} = 0)$
$1112 \to -2221$	$(s_{16} = -s_{26})$
$\vdots$	

The corresponding matrix coefficients are given in parentheses. The resulting compliance matrix is

$$\begin{pmatrix} s_{11} & s_{12} & s_{13} & 0 & 0 & s_{16} \\ s_{12} & s_{11} & s_{13} & 0 & 0 & -s_{16} \\ s_{13} & s_{13} & s_{33} & 0 & 0 & 0 \\ 0 & 0 & 0 & s_{44} & 0 & 0 \\ 0 & 0 & 0 & 0 & s_{44} & 0 \\ s_{16} & -s_{16} & 0 & 0 & 0 & s_{66} \end{pmatrix}.$$

Table 13.1 Elastic constant matrices for the 32 crystal classes and seven limiting groups

Triclinic
1, $\bar{1}$

$$\begin{pmatrix} c_{11} & c_{12} & c_{13} & c_{14} & c_{15} & c_{16} \\ c_{12} & c_{22} & c_{23} & c_{24} & c_{25} & c_{26} \\ c_{13} & c_{23} & c_{33} & c_{34} & c_{35} & c_{36} \\ c_{14} & c_{24} & c_{34} & c_{44} & c_{45} & c_{46} \\ c_{15} & c_{25} & c_{35} & c_{45} & c_{55} & c_{56} \\ c_{16} & c_{26} & c_{36} & c_{46} & c_{56} & c_{66} \end{pmatrix}$$

Monoclinic
2, m, $2/m$

$$\begin{pmatrix} c_{11} & c_{12} & c_{13} & 0 & c_{15} & 0 \\ c_{12} & c_{22} & c_{23} & 0 & c_{25} & 0 \\ c_{13} & c_{23} & c_{33} & 0 & c_{35} & 0 \\ 0 & 0 & 0 & c_{44} & 0 & c_{46} \\ c_{15} & c_{25} & c_{35} & 0 & c_{55} & 0 \\ 0 & 0 & 0 & c_{46} & 0 & c_{66} \end{pmatrix}$$

Orthorhombic
222, $mm2$, mmm

$$\begin{pmatrix} c_{11} & c_{12} & c_{13} & 0 & 0 & 0 \\ c_{12} & c_{22} & c_{23} & 0 & 0 & 0 \\ c_{13} & c_{23} & c_{33} & 0 & 0 & 0 \\ 0 & 0 & 0 & c_{44} & 0 & 0 \\ 0 & 0 & 0 & 0 & c_{55} & 0 \\ 0 & 0 & 0 & 0 & 0 & c_{66} \end{pmatrix}$$

Tetragonal
4, $\bar{4}$, $4/m$

$$\begin{pmatrix} c_{11} & c_{12} & c_{13} & 0 & 0 & c_{16} \\ c_{12} & c_{11} & c_{13} & 0 & 0 & -c_{16} \\ c_{13} & c_{13} & c_{33} & 0 & 0 & 0 \\ 0 & 0 & 0 & c_{44} & 0 & 0 \\ 0 & 0 & 0 & 0 & c_{44} & 0 \\ c_{16} & -c_{16} & 0 & 0 & 0 & c_{66} \end{pmatrix}$$

Tetragonal
$4mm$, $\bar{4}2m$, 422, $4/mmm$

$$\begin{pmatrix} c_{11} & c_{12} & c_{13} & 0 & 0 & 0 \\ c_{12} & c_{11} & c_{13} & 0 & 0 & 0 \\ c_{13} & c_{13} & c_{33} & 0 & 0 & 0 \\ 0 & 0 & 0 & c_{44} & 0 & 0 \\ 0 & 0 & 0 & 0 & c_{44} & 0 \\ 0 & 0 & 0 & 0 & 0 & c_{66} \end{pmatrix}$$

Trigonal
3, $\bar{3}$

$$\begin{pmatrix} c_{11} & c_{12} & c_{13} & c_{14} & -c_{25} & 0 \\ c_{12} & c_{11} & c_{13} & -c_{14} & c_{25} & 0 \\ c_{13} & c_{13} & c_{33} & 0 & 0 & 0 \\ c_{14} & -c_{14} & 0 & c_{44} & 0 & c_{25} \\ -c_{25} & c_{25} & 0 & 0 & c_{44} & c_{14} \\ 0 & 0 & 0 & c_{25} & c_{14} & \frac{1}{2}(c_{11}-c_{12}) \end{pmatrix}$$

Trigonal
32, $3m$, $\bar{3}m$

$$\begin{pmatrix} c_{11} & c_{12} & c_{13} & c_{14} & 0 & 0 \\ c_{12} & c_{11} & c_{13} & -c_{14} & 0 & 0 \\ c_{13} & c_{13} & c_{33} & 0 & 0 & 0 \\ c_{14} & -c_{14} & 0 & c_{44} & 0 & 0 \\ 0 & 0 & 0 & 0 & c_{44} & c_{14} \\ 0 & 0 & 0 & 0 & c_{14} & \frac{1}{2}(c_{11}-c_{12}) \end{pmatrix}$$

Hexagonal
6, $\bar{6}$, $6/m$, 622, $6mm$, $\bar{6}m2$, $6/mmm$
Curie groups
∞, ∞m, ∞/m, $\infty 2$, ∞/mm

$$\begin{pmatrix} c_{11} & c_{12} & c_{13} & 0 & 0 & 0 \\ c_{12} & c_{11} & c_{13} & 0 & 0 & 0 \\ c_{13} & c_{13} & c_{33} & 0 & 0 & 0 \\ 0 & 0 & 0 & c_{44} & 0 & 0 \\ 0 & 0 & 0 & 0 & c_{44} & 0 \\ 0 & 0 & 0 & 0 & 0 & \frac{1}{2}(c_{11}-c_{12}) \end{pmatrix}$$

Cubic
23, $m3$, 432, $\bar{4}3m$, $m3m$

$$\begin{pmatrix} c_{11} & c_{12} & c_{12} & 0 & 0 & 0 \\ c_{12} & c_{11} & c_{12} & 0 & 0 & 0 \\ c_{12} & c_{12} & c_{11} & 0 & 0 & 0 \\ 0 & 0 & 0 & c_{44} & 0 & 0 \\ 0 & 0 & 0 & 0 & c_{44} & 0 \\ 0 & 0 & 0 & 0 & 0 & c_{44} \end{pmatrix}$$

Continued overleaf

Table 13.1 (*Continued*)

Curie groups $\infty\infty$, $\infty\infty m$	$\begin{pmatrix} c_{11} \\ c_{12} \\ c_{12} \\ 0 \\ 0 \\ 0 \end{pmatrix}$	$\begin{matrix} c_{12} \\ c_{11} \\ c_{12} \\ 0 \\ 0 \\ 0 \end{matrix}$	$\begin{matrix} c_{12} \\ c_{12} \\ c_{11} \\ 0 \\ 0 \\ 0 \end{matrix}$	$\begin{matrix} 0 \\ 0 \\ 0 \\ \frac{1}{2}(c_{11}-c_{12}) \\ 0 \\ 0 \end{matrix}$	$\begin{matrix} 0 \\ 0 \\ 0 \\ 0 \\ \frac{1}{2}(c_{11}-c_{12}) \\ 0 \end{matrix}$	$\begin{matrix} 0 \\ 0 \\ 0 \\ 0 \\ 0 \\ \frac{1}{2}(c_{11}-c_{12}) \end{matrix}$	

The second symmetry element in point group $4/m$ is the mirror plane perpendicular to Z_3. For $m \perp Z_3$, $1 \rightarrow 1$, $2 \rightarrow 2$, $3 \rightarrow -3$. This means that any tensor coefficient such as s_{1312} or s_{1333} with an odd number of 3s will disappear. Beginning with the already simplified matrix for $4 \parallel Z_3$, none of the remaining nonzero coefficients is affected by the mirror operation. Therefore there is no further simplification of the compliance matrix. Point group $4/m$ has seven independent elastic constants: s_{11}, s_{12}, s_{13}, s_{33}, s_{44}, s_{16}, and s_{66}. For scheelite and other crystals belonging to $4/m$, at least seven measurements are required.

Elastic constant matrices for other crystallographic point groups and for the limiting groups are given in Table 13.1. The proofs are carried out by tensor or matrix transformations, or by the direct inspection method, using Neumann's Principle. The number of independent elastic constant range from 21 for triclinic crystals down to 2 for isotropic bodies such as glass or polycrystalline metals and ceramics.

13.5 Engineering coefficients and measurement methods

For isotropic groups $\infty\infty m$ and $\infty\infty$, there are just two independent elastic compliance coefficients s_{11} and s_{12}. The shear coefficient $s_{44} = 2(s_{11} - s_{12})$. In engineering texts the elastic properties of isotropic materials are usually described by Young's Modulus E, Poisson's Ratio ν, and the Rigidity Modulus G. The basic experiments are illustrated in Fig. 13.1. Young's Modulus is change in length per unit length for a tensile stress: $x_1 = s_{11}X_1 = (1/E)X_1$. Poisson's Ratio is also measured under a tensile stress X_1. It is the ratio of the transverse contraction to the longitudinal elongation: $\nu = |x_1|/|x_2| = -s_{12}/s_{11}$. The Rigidity Modulus G is a measure of shear strain under shear stress. For shear about Z_1, $x_4 = s_{44}X_4 = (1/G)X_4$. Therefore $G = 1/s_{44} = 1/2(s_{11}-s_{12})$. Since there are only two independent coefficients for isotropic bodies, E, ν, and G are interrelated. The relationship is $G = E/2(1+\nu)$.

The most commonly used experimental techniques for measuring elastic constants can be classified as static and dynamic methods. The optical experiment illustrated in Fig. 13.2 is a typical static method carried out under slowly applied bending stress. A long thin bar cut in the Z'_1 direction is deformed in a four-point bending experiment through a small angle θ. The resulting deflection can be measured using a collimated light beam reflected from mirrors attached to the sample. The radius of curvature $R = l/\theta$ where l is the separation of the two mirrors and θ is the bending angle expressed in radians. The bending moment $B = F/a$ where F is the applied force and a is the separation between loading points. Young's Modulus $E = 1/s'_{11}$ is obtained from R, B, and I, the moment of

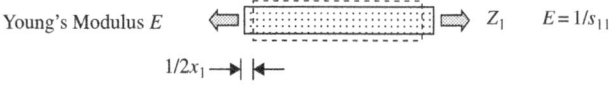

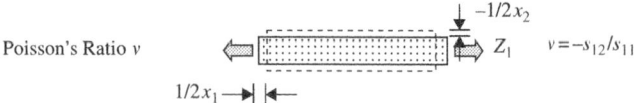

Fig. 13.1 Engineering coefficients for isotropic solids, Young's Modulus is a measure of longitudinal elongation and Poisson's Ratio the transverse contraction. The Rigidity Modulus measures the shear stiffness.

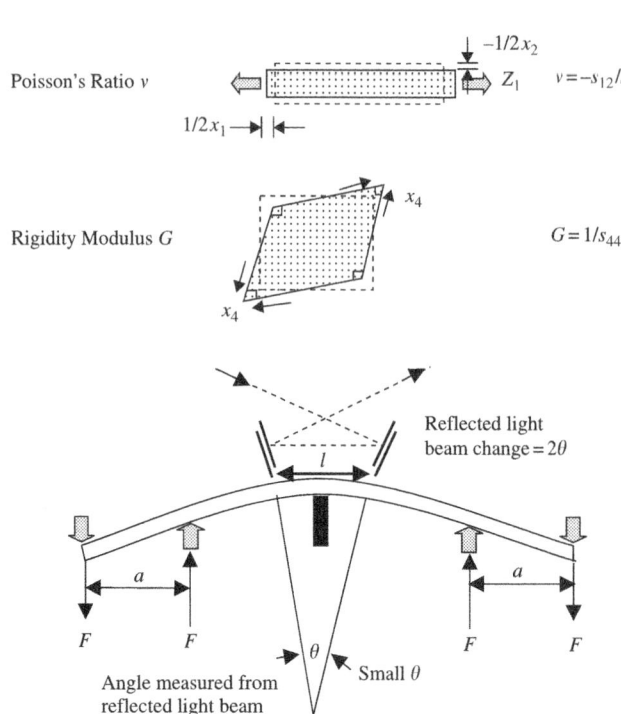

Fig. 13.2 A static optical method used for measuring bending strains. Long optical paths provide the accuracy necessary to measure small strains. Anisotropic elastic constants are obtained by using samples cut along different crystal directions.

inertia of the bar: $E = BR/I = 1/s'_{11}$. For crystals and other anisotropic materials the bars are cut in various orientations to evaluate the elastic compliance coefficients.

Typical of the dynamic methods used to measure elastic constants is the pulse propagation experiment in Fig. 13.3. The velocity v of longitudinal and transverse acoustic waves are measured for various crystal orientations to obtain the elastic stiffness coefficients. As explained later in the chapter on acoustics (Chapter 23), the wave velocity $v = \sqrt{c/\rho}$ where c is the stiffness coefficient governing the acoustic vibration and ρ is the crystal density. In the experiment, a piezoelectric quartz crystal is used to launch the wave and to detect the return signal reflected from the opposite side of the sample. The converse and direct piezoelectric effect in quartz are used to generate the wave and to sense the return reflection. An oscilloscope measures the time t required to traverse the specimen. The velocity $v = 2d/t$ where d is the specimen thickness. X-cut quartz is used to launch longitudinal waves and AC-cuts for transverse waves. The AC-cut is described in Section 13.10.

13.6 Anisotropy and structure–property relations

Beginning from either the matrix or tensor transformation equations, the longitudinal stiffness coefficient can be derived as a function of crystallographic

Fig. 13.3 Dynamic pulse propagation method used to measure elastic stiffness coefficients. An acoustic wave is launched into specimen by applying an electric pulse to a piezoelectric quartz crystal. The quartz plate later detects the wave reflected from the opposite side of the specimen. Roundtrip time t is measured with an oscilloscope to give the acoustic velocity and the elastic constant.

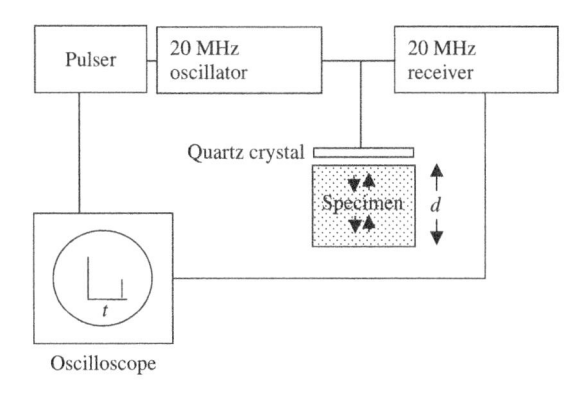

Table 13.2 Elastic stiffness coefficients (in units of 10^{11} N/m^2) for several alkali metal crystals, semiconductors, and alkali halide crystals all with cubic structure (A is the anisotropy factor $2c_{44}/(c_{11} - c_{12})$)

	c_{11}	c_{12}	c_{14}	A
Alkali metals (body-centered cubic)				
Li	0.135	0.114	0.088	8.4
Na	0.074	0.062	0.042	7.2
K	0.037	0.031	0.019	6.7
Semiconductors (diamond structure)				
C	10.20	2.50	4.92	1.3
Si	1.66	0.64	0.80	1.6
Ge	1.30	0.49	0.67	1.7
Alkali halides (NaCl structure)				
NaCl	0.485	0.125	0.127	0.7
KCl	0.405	0.066	0.063	0.37
RbCl	0.363	0.062	0.047	0.31

direction. With stress and strain both measured along an arbitrary direction Z'_1, the stiffness is

$$c'_{11} = c'_{1111} = a_{1i}a_{1j}a_{1k}a_{1l}c_{ijkl},$$

where the direction cosines a_{1i} relate Z'_1 to the principal axes. For cubic crystals,

$$c'_{11} = c_{11} - 2(c_{11} - c_{12} - 2c_{44})(a_{11}^2 a_{12}^2 + a_{11}^2 a_{13}^2 + a_{12}^2 a_{13}^2).$$

Note that fourth rank tensor properties like the elastic constants are anisotropic even in cubic crystals. The only exception is when $c_{11} = c_{12} + 2c_{44}$. The quantity $A = 2c_{44}/(c_{11} - c_{12})$ is often referred to as the anisotropy factor. When $A = 1$, $c'_{11} = c_{11}$ for all directions. If $A < 1$ the crystal is stiffest along $\langle 100 \rangle$ cube axes, and when $A > 1$ it is stiffest along the $\langle 111 \rangle$ body diagonals.

The elastic constants for three types of cubic crystals are listed in Table 13.2. Note the major changes in the anisotropy factor A. The body-centered cubic alkali metals are much stiffer along $\langle 111 \rangle$ directions, as are the column IV elements with the diamond structure. In both structures the nearest-neighbor bonds are also in $\langle 111 \rangle$ directions. For alkali halide crystals with the rocksalt structure, the cation–anion bonds are oriented along $\langle 100 \rangle$ directions. These are usually the directions of greatest elastic stiffness as well (Fig. 13.4).

Similar trends can be seen in crystals of lower symmetry. Chain silicates (pyroxenes) and double chain silicates (amphiboles) crystallize in monoclinic

112   *Elasticity*

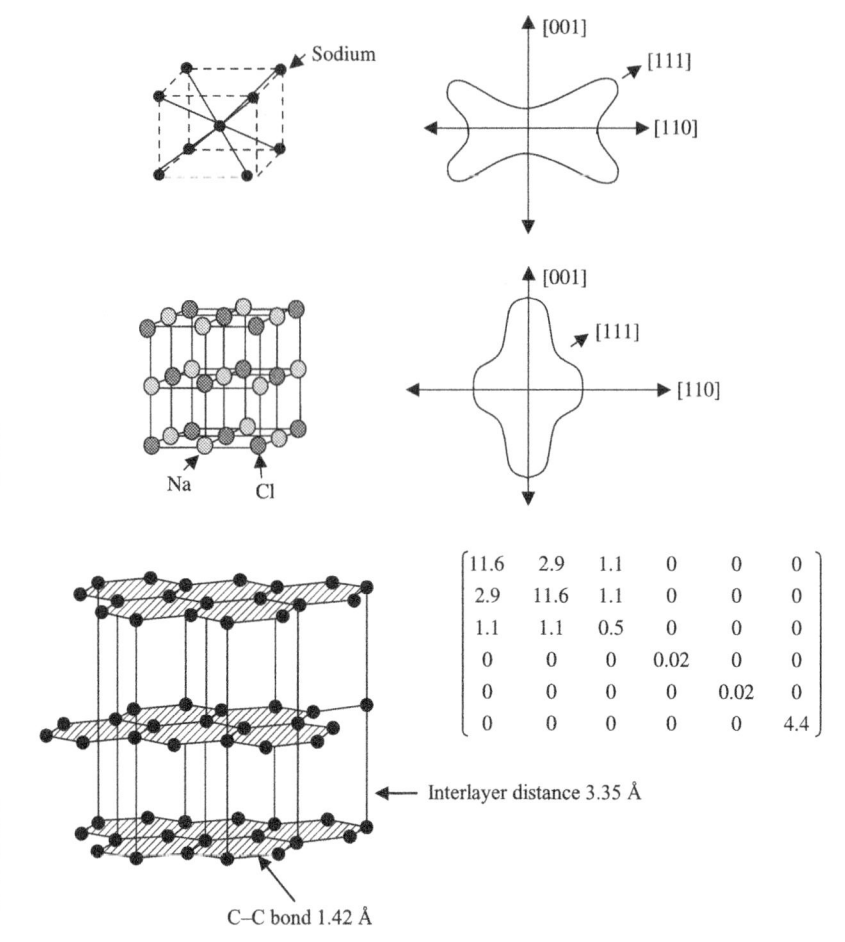

Fig. 13.4 Crystal structures and stiffness surfaces for sodium and sodium chloride. The stiffest directions are generally aligned with near-neighbor bonds.

$$\begin{bmatrix} 11.6 & 2.9 & 1.1 & 0 & 0 & 0 \\ 2.9 & 11.6 & 1.1 & 0 & 0 & 0 \\ 1.1 & 1.1 & 0.5 & 0 & 0 & 0 \\ 0 & 0 & 0 & 0.02 & 0 & 0 \\ 0 & 0 & 0 & 0 & 0.02 & 0 \\ 0 & 0 & 0 & 0 & 0 & 4.4 \end{bmatrix}$$

Fig. 13.5 Graphite (C) has a very anisotropic crystal structure with strong bonds in the (001) plane and weak interatomic forces between layers. As a result, c_{11} is much larger than c_{33}, and c_{66} is much bigger than c_{44}. Elastic stiffnesses are in units of 10^{11} N/m².

or orthorhombic structures with silicate chains parallel to the $Z_3 = [001]$ directions. These crystals tend to be stiffest along the chain directions. In micas and other layer silicates, the silica tetrahedra are bonded together in layers perpendicular to $Z_3 = [001]$. The bonding between layers is relatively weak. For this reason c_{11} and c_{22} are about three times larger than c_{33}.

Graphite is even more anisotropic than mica. The crystal structure is hexagonal with strong covalent bonds (bondlength 1.42 Å) in the (001) plane. The bonding between the layers is very weak (interatomic distance 3.35 Å) leading to easy cleavage and extreme elastic anisotropy. The c_{11} and c_{22} coefficients are similar to those of diamond but c_{33} is much smaller. Elastic constants and crystal structure are shown in Fig. 13.5. Note that the anisotropy in shear stiffness is also very large with $c_{66} \gg c_{44}$.

A similar, but less dramatic, trend is found in hexagonal close-packed (HCP) metals (Table 13.3). In the ideal HCP structure each metal atom is bonded to 12 neighbors with equal bond lengths and a c/a ratio of 1.63. For most HCP metals the ratio is slightly smaller with slightly longer bond lengths in the (001) plane. As a result c_{33} is slightly larger than c_{11}. Cadmium and zinc are the exceptions. Their c/a ratios are much larger than other HCP metals, leading to six short strong bonds in the (001) plane. The elastic stiffness coefficients $c_{11} = c_{22}$ are much greater than c_{33}.

Table 13.3 Elastic constants (in units of 10^{11} N/m^2) and lattice parameters (in Å) of hexagonal close-packed metals. Zn and Cd have strong bonds in the (001) plane giving large c/a ratios and small c_{33}/c_{11} ratios

Metal	c_{11}	c_{33}	c_{33}/c_{11}	a	c	c/a
Be	2.92	3.36	1.16	2.29	3.58	1.56
Cd	1.16	0.51	0.44	2.98	5.62	1.89
Co	3.07	3.58	1.17	2.51	4.07	1.62
Hf	1.81	1.97	1.09	3.20	5.06	1.58
Mg	0.60	0.62	1.03	3.21	5.21	1.62
Re	6.13	6.83	1.12	2.76	4.46	1.62
Ti	1.62	1.81	1.12	2.95	4.69	1.59
Zn	1.61	0.61	0.38	2.66	4.95	1.86
Zr	1.43	1.65	1.15	3.23	5.15	1.59

Polyethylene single crystals have enormous anisotropy in elastic stiffness. Parallel to the chain direction Young's Modulus ($1/s_{33}$) is 283 GPa at room temperature. The C–C covalent bonds in this direction give a Young's Modulus greater than that of steel (200 GPa). In the transverse direction the modulus of polyethylene crystal ($1/s_{11}$) is only about 7 GPa. Heavily drawn polyethylene fibers have similar anisotropy. The anisotropy remains large over a wide temperature range.

Problem 13.2
Ice crystals are hexagonal with five independent elastic compliance coefficients (units of 10^{-11} m^2/N): $s_{11} = 10.1$, $s_{12} = -4.1$, $s_{13} = -1.9$, $s_{33} = 8.3$, $s_{44} = 32.5$. The full matrix is given in Table 13.3.

Show that the elastic compliance surface for ice is cylindrically symmetric about the hexagonal axis ($Z_3 = [001]$) by deriving $s'_{1111} = s'_{11}$ in spherical coordinates θ and ϕ. Prove that
$$s'_{11} = s_{11} \sin^4 \theta + s_{33} \cos^4 \theta + (s_{44} + 2s_{13}) \sin^2 \theta \cos^2 \theta.$$
Plot Young's Modulus ($E = 1/s'_{11}$) as a function of θ in the Z_1–Z_3 plane.

13.7 Compressibility

The compressibility of a solid is defined as
$$K = -\frac{1}{V}\frac{dV}{dp},$$
where V is volume and p is hydrostatic pressure. As pointed out earlier (Section 10.4), the change in volume per unit volume is $x_{11} + x_{22} + x_{33} = x_{ii}$, the sum of the three longitudinal strains. To evaluate K for an anisotropic solid, it is only necessary to obtain x_{ii} from Hooke's Law.

$$x_{ii} = s_{iikl} X_{kl} = -s_{iikl}\, p\, \delta_{kl}$$

$$\begin{aligned} K = s_{iikk} &= s_{1111} + s_{1122} + s_{1133} + s_{2211} + s_{2222} + s_{2233} \\ &\quad + s_{3311} + s_{3322} + s_{3333} \\ &= s_{11} + s_{22} + s_{33} + 2s_{12} + 2s_{13} + 2s_{23}. \end{aligned}$$

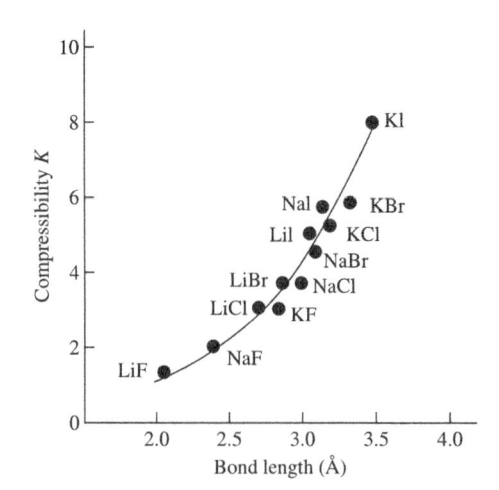

Fig. 13.6 Compressibilities of alkali halide crystals with the rocksalt structure. Crystals with long weak bonds are more easily compressed. K values in units of 10^{-11} m^2/N.

The compressibility can be quickly computed from the elastic compliance matrix by summing the nine coefficients in the upper left quadrant.

As might be expected, solids with long weak bonds generally have higher compressibilities than those with short strong bonds. This is nicely illustrated with the compressibilities of alkali halide crystals having the rocksalt structure (Fig. 13.6).

Problem 13.3
Calculate the elastic compliance coefficients and compressibilities of diamond, silicon, and germanium, and discuss the trends among these column IV elements.

13.8 Polycrystalline averages

If the single crystal coefficients are known, the elastic constants of polycrystalline solids can be estimated to within a few percent using the Voigt–Reuss–Hill method. Finite element calculations are required to obtain more accurate predictions.

Polycrystalline solids with large numbers of randomly oriented grains possess spherical symmetry, point group $\infty\infty m$. (If the grains all have the same handedness, as in a sugar cube, the appropriate group is $\infty\infty$.) In either case there are two independent elastic constants, c_{11} and c_{12}. The shear stiffness $c_{44} = \frac{1}{2}(c_{11} - c_{12})$. As pointed out in Section 13.5, the elastic properties of isotropic solids can also be described in terms of Young's Modulus E, Poisson's Ratio ν, and the Rigidity Modulus G.

The Voigt–Reuss–Hill method begins by calculating the Voigt averages for the average elastic stiffness $\langle c'_{11} \rangle$, $\langle c'_{12} \rangle$, and $\langle c'_{44} \rangle$. For the most general case of a triclinic crystal, there are 36 elastic constants, 21 of which are independent. In tensor form,

$$c'_{11} = c'_{1111} = a_{1i}a_{1j}a_{1k}a_{1l}c_{ijkl}$$
$$= c_{1111}a_{11}^4 + 2c_{1122}a_{11}^2 a_{12}^2 + 2c_{1133}a_{11}^2 a_{13}^2$$
$$+ 4c_{1123}a_{11}^2 a_{12} a_{13} + \cdots + 8\, c_{1312}a_{11}^2 a_{12} a_{13} + 4c_{1212}a_{11}^2 a_{12}^2.$$

In matrix form the average value is

$$\langle c'_{11}\rangle = c_{11}\langle a^4_{11}\rangle + 2c_{12}\langle a^2_{11}a^2_{12}\rangle + 2c_{13}\langle a^2_{11}a^2_{13}\rangle$$
$$+ 4c_{14}\langle a^2_{11}a_{12}a_{13}\rangle + \cdots + 8\,c_{56}\langle a^2_{11}a_{12}a_{13}\rangle + 4c_{66}\langle a^2_{11}a^2_{12}\rangle.$$

When averaged over all angles, the direction cosine a_{11} is

$$\langle a_{11}\rangle = \frac{\int_{-1}^{1} a_{11}\,da_{11}}{\int_{-1}^{1} da_{11}} = 0.$$

All odd powers of direction cosines average to zero. For the even powers, $\langle a^2_{11}\rangle = \frac{1}{3}$, $\langle a^4_{11}\rangle = \frac{1}{5}$, and $\langle a^2_{11}a^2_{12}\rangle = \frac{1}{15}$.

Therefore

$$\langle c'_{11}\rangle = \tfrac{3}{15}(c_{11} + c_{22} + c_{33}) + \tfrac{2}{15}(c_{12} + c_{13} + c_{23}) + \tfrac{4}{15}(c_{44} + c_{55} + c_{56}).$$

Similarly,

$$\langle c'_{12}\rangle = \tfrac{1}{15}(c_{11} + c_{22} + c_{33}) + \tfrac{4}{15}(c_{12} + c_{13} + c_{23}) - \tfrac{2}{15}(c_{44} + c_{55} + c_{66})$$

and

$$\langle c'_{44}\rangle = \tfrac{1}{15}(c_{11} + c_{22} + c_{33}) - \tfrac{1}{15}(c_{12} + c_{13} + c_{23}) + \tfrac{3}{15}(c_{44} + c_{55} + c_{66}).$$

Note that the Voigt averages satisfy the isotropy condition

$$\langle c'_{44}\rangle = \tfrac{1}{2}(\langle c'_{11}\rangle - \langle c'_{12}\rangle).$$

To obtain the Voigt average for Young's Modulus $E_V = 1/s_{11}$, convert s_{11} to stiffness coefficients for an isotropic solid (Section 13.3 and Table 13.1).

$$s_{11} = (c_{11} + c_{12})/(c_{11} - c_{12})(c_{11} + 2c_{12}).$$

Therefore $E_V = (\langle c'_{11}\rangle - \langle c'_{12}\rangle)(\langle c'_{11}\rangle + 2\langle c'_{12}\rangle)/\langle c'_{11}\rangle + \langle c'_{12}\rangle$. The Voigt average for the Rigidity Modulus $G_V = 1/\langle s'_{44}\rangle$ and Poisson's Ratio is $\nu_V = (E_V/2G_V) - 1$. Numerical values for E_V, G_V, and ν_V are evaluated from the single crystal stiffness coefficients.

The second part of the Voigt–Reuss–Hill averaging procedure is the Reuss average of the compliance coefficients, beginning with the tensor transformation

$$s'_{ijkl} = a_{im}a_{jn}a_{ko}a_{lp}s_{mnop}.$$

Following the same procedure used for the stiffness coefficients gives the Reuss averages:

$$E_R = 15/[5(s_{11} + s_{22} + s_{33}) + 2(s_{12} + s_{13} + s_{23}) + (s_{44} + s_{55} + s_{66})]$$
$$G_R = 15/[4(s_{11} + s_{22} + s_{33}) - 4(s_{12} + s_{13} + s_{23}) + 3(s_{44} + s_{55} + s_{66})]$$
$$\nu_R = (E_R/2G_R) - 1.$$

Neither the Reuss average or the Voigt average is quite right. The Reuss averaging procedure assumes the polycrystalline specimen is in a condition of uniform stress, while the Voigt method assumes uniform strain. In fact there are bound to be variations in stress and strain throughout the sample, especially near grain boundaries where mismatched orientations occur. The truth lies between these two extremes: this is an average of the average.

To illustrate, the Voigt and Reuss averages for three metals are listed in Table 13.4, along with the observed experimental value. In every case the Voigt values are too high, and Reuss moduli too low, but the average of the averages

Table 13.4 Experimental values of (in 10^{11} N/m^2) Young's Modulus (E) and Rigidity Modulus (G) for polycrystalline copper, gold, and iron. Averages calculated from single crystal elastic constants by the Voigt (E_V and G_V) and Reuss (E_R and G_R) methods are shown for comparison. The experimental values lie about halfway between the calculated results

Moduli	Copper	Gold	α-Iron
E_R	1.09	0.69	1.93
E	1.23	0.79	2.13
E_V	1.44	0.87	2.29
G_R	0.40	0.24	0.74
G	0.46	0.28	0.83
G_V	0.54	0.31	0.86

Table 13.5 A comparison of stiff, low-density ceramics with common metals, glass and wood (E values expressed in 10^{11} N/m^2)

	Specific gravity	Young's Modulus	Ratio
Iron	7.8	2.10	0.25
Titanium	4.5	1.20	0.25
Aluminum	2.7	0.73	0.25
Common glass	2.5	0.70	0.26
Spruce wood	0.5	0.13	0.25
AlN	3.3	3.4	1.03
Al$_2$O$_3$	4.0	3.8	0.95
Boron	2.3	4.1	1.80
C whiskers	2.3	7.5	3.30
SiC	3.2	5.1	1.6
Si$_3$N$_4$	3.2	3.8	1.2

is very close. This is the Voigt–Reuss–Hill procedure for estimating the elastic properties of polycrystalline materials from the single crystal values.

Problem 13.4
Using the Voigt–Reuss–Hill procedure, estimate the Young's Modulus, E, the Rigidity Modulus, G, and the Poisson Ratio, ν, for polycrystalline beryllium. Single crystal stiffness coefficients are listed in Table 13.3. Beryllium is one of the stiffest of metals and has a superb stiffness/weight ratio.

For many engineering applications, it is not the stiffness that is important, but the stiffness per unit weight. It is here that ceramics excel with much greater stiffnesses than common construction materials such as steel and wood (Table 13.5).

13.9 Temperature coefficients

Most materials become more compliant with increasing temperature as thermal vibrations increase in amplitude, leading to longer interatomic distances, and weaker chemical bonds. The temperature coefficients of elastic

Table 13.6 Temperature coefficients of the elastic stiffnesses for several cubic crystals (the fractional change in stiffness with increasing temperature is expressed in units of 10^{-4}/K, and is measured near room temperature)

	Tc_{11}	Tc_{12}	Tc_{44}
Diamond structure			
Carbon	−0.137	−0.57	−0.125
Silicon	−0.81	−1.1	−0.63
Germanium	−1.2	−1.1	−1.15
Body-centered cubic			
Iron	−1.7	−0.8	−1.5
Sodium	−6.3	−5.6	−17
Face-centered cubic			
Aluminum	−3.1	−1.3	−4.45
Copper	−2.01	−1.24	−3.33
Gold	−1.8	−1.5	−3.0

stiffness, defined as $T_c = (1/c)(dc/dT)$ is generally a negative number. It ranges from about 10^{-5}/K in strongly bonded solids like diamond to 10^{-2}/K in molecular solids with low melting points.

Typical values for cubic metals and semiconductor crystals are listed in Table 13.6.

Changes in the elastic stiffnesses of diamond, aluminum, and anthracene (Fig. 13.7) illustrate typical behavior over a wide temperature range. The coefficients of diamond vary little with temperature compared to those of weakly bonded metals and molecular crystals.

Phase transformations cause rapid changes in elastic stiffness, and often these changes are highly anisotropic depending on the modes of vibration associated with the transformation. The melting of tin (Fig. 13.8) and the ferroelectric phase transformation in barium titanate are good examples. The common white tin structure is tetragonal, point group $4/mmm$. Each Sn atom is bonded to six Sn neighbors, four in a distorted tetrahedron at 3.02 Å and two others at 3.17 Å along the c-axis. The two additional neighbors stiffen the structure along Z_3 and lower the compliance coefficient s_{33}. As temperature increases toward the melting point of 505 K, the thermal oscillations lead to further softening of s_{11}.

Barium titanate has three displacive phase transformations at 130°C, 0°C, and −90°C. Around room temperature in the ferroelectric tetragonal phase (point group $4mm$), the unit cell is elongated along the polar c-axis. As a result, s_{33} is much larger than s_{11}. It grows especially large near the Curie temperature where the c lattice parameter changes rapidly with temperature. On cooling toward the tetragonal–orthorhombic transition at 0°C, s_{44} softens dramatically as the polar axis swings from [001] to [011].

The elastic properties of polymers are also strongly temperature dependent. Amorphous polymers like polyisobutylene (Fig. 13.9) show a pronounced glass transition in which the shear stiffness increases by three orders of magnitude on cooling below −50°C. The polymer loses its rubber-like properties as it freezes into a brittle glass-like state. Cross-linking between chains greatly stiffens polymeric materials, increasing Young's Modulus from about 10^8 N/m^2 in polyethylene to 10^{12} N/m^2 in diamond.

118 *Elasticity*

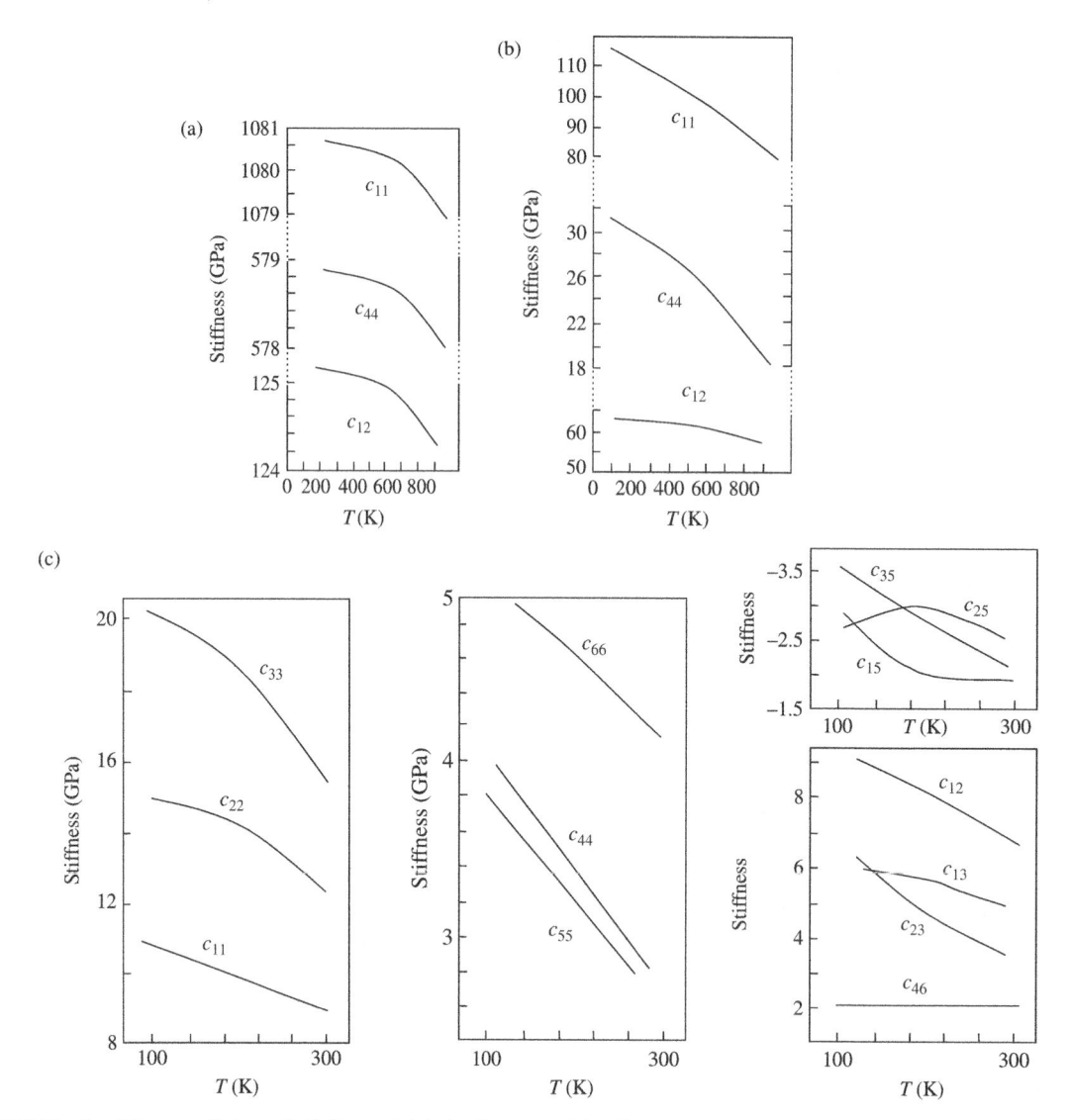

Fig. 13.7 Elastic stiffness coefficients of (a) diamond, (b) aluminum, and (c) anthracene plotted as a function of temperature. Stiffnesses are expressed in GPa= 10^9 N/m². Anthracene ($C_{14}H_{10}$) belongs to monoclinic point group $2/m$ and is highly anisotropic because of its aromatic structure. The weak van der Waals bonding between molecules leads to low stiffness that softens rapidly with increasing temperature.

13.10 Quartz crystal resonators

Vibrating quartz crystals are used as time and frequency standards in wristwatches, radio oscillators, and computers. The optimization of these resonant devices present many interesting applications in crystal physics. As pointed out in Section 12.8, quartz is piezoelectric with a very high mechanical Q leading to very sharp resonances. The five basic vibration modes excited by electric fields were pictured in Fig. 12.13. Here we discuss two families of interesting rotated Y-cuts: (1) the AT- and BT-cuts with zero temperature coefficients, and (2) the AC- and BC-cuts used to generate pure shear waves. Both types of resonators use thickness shear vibrations driven by the largest piezoelectric coefficient $d_{26} = -2d_{11}$.

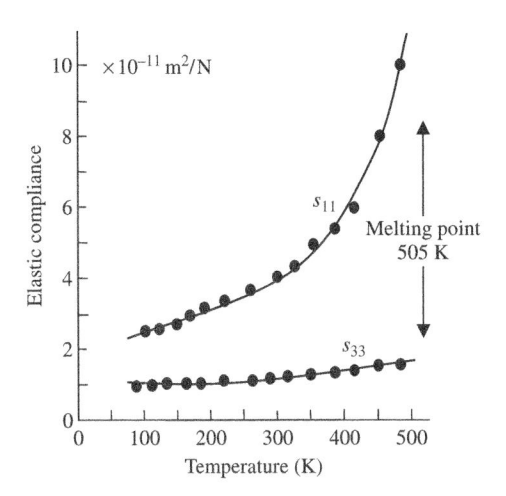

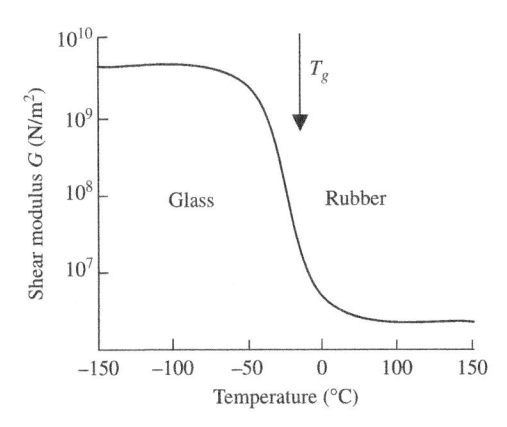

Fig. 13.8 Tin has a very low melting point and is an important constituent of solder. Compliance coefficient s_{11} softens markedly as temperature approaches the melting point. Coefficient s_{33} remains rather stiff because of the two near neighbors along the $[001] = Z_3$ axis.

Fig. 13.9 Rigidity Modulus of polyisobutylene stiffens rapidly as it cools from the rubbery state at room temperature to the more rigid glassy state. T_g is the glass transition temperature.

13.10.1 AT- and BT-cuts

The resonant frequency for the thickness shear mode is given by

$$f = \frac{1}{2t}\sqrt{\frac{c'_{66}}{\rho}},$$

where t is thickness of the quartz plate (Fig. 12.13) and ρ is the density. Elastic stiffness coefficient c'_{66} governs shear motions about Z'_3 for the rotated Y-cut. The engineering objective is to find an orientation angle θ for which the resonant frequency does not drift with changes in temperature. The temperature derivative of f is

$$\frac{1}{f}\frac{df}{dT} = -\frac{1}{t}\frac{dt}{dT} + \frac{1}{2c'_{66}}\frac{dc'_{66}}{dT} - \frac{1}{2\rho}\frac{d\rho}{dT}.$$

To make $df/dT = 0$, requires that the three terms governed by t, c'_{66}, and ρ have off-setting temperature changes.

The changes in the thickness t and the density ρ are controlled by the thermal expansion coefficients. Quartz is trigonal so there are two independent thermal expansion coefficients $\alpha_{11} = \alpha_{22}$ and α_{33}. The room temperature thermal expansion coefficients are $\alpha_{11} = 14.3 \times 10^{-6}$/K and $\alpha_{33} = 7.8 \times 10^{-6}$/K.

Table 13.7 Elastic stiffnesses (in 10^{11} N/m^2) and temperature coefficients (in 10^{-4}/K) for quartz

c_{11}	0.860	Tc_{11}	−0.465
c_{12}	0.051	Tc_{12}	−33
c_{13}	0.105	Tc_{13}	−7
c_{14}	0.183	Tc_{14}	−9
c_{33}	1.070	Tc_{33}	−2.05
c_{44}	0.590	Tc_{44}	−1.66
c_{66}	0.410	Tc_{66}	+1.64

As shown in Section 10.2, the temperature derivative of the density is

$$\frac{1}{\rho}\frac{d\rho}{dT} = -(2\alpha_{11} + \alpha_{33}) = -36.4 \times 10^{-6}/\text{K}.$$

For the thickness shear mode the controlling dimension is the thickness t measured in the Z_2' direction. The temperature change comes from thermal expansion coefficient α_{22}':

$$\frac{1}{t}\frac{dt}{dT} = \alpha_{22}' = \alpha_{11}\cos^2\theta + \alpha_{33}\sin^2\theta.$$

Together these two terms are rather small compared to the temperature derivatives of the elastic constants ($T_c = (1/c)(dc/dT)$). Quartz belongs to trigonal point group 32 with seven independent stiffness coefficients, each with a different temperature coefficient (Table 13.7).

c_{66}' is obtained by rotating the coordinate system about the $X = Z_1 = [100]$ axis. In tensor notation

$$c_{66}' = c_{1212}' = a_{1i}a_{2j}a_{1k}a_{2l}c_{ijkl}$$

$$(a) = \begin{pmatrix} 1 & 0 & 0 \\ 0 & \cos\theta & \sin\theta \\ 0 & -\sin\theta & \cos\theta \end{pmatrix}$$

$$c_{66}' = c_{1313}\sin^2\theta + c_{1312}\sin\theta\cos\theta + c_{1213}\cos\theta\sin\theta + c_{1212}\cos^2\theta.$$

Converting back to matrix form and noting that $c_{55} = c_{44}$ and $c_{56} = c_{65} = c_{14}$ for point group 32, $c_{66}' = c_{44}\sin^2\theta + 2c_{14}\sin\theta\cos\theta + c_{66}\cos^2\theta$.

Taking the temperature derivative of c_{66}' and substituting the values into the equation for the temperature derivative of the resonant frequency gives

$$\frac{1}{f}\frac{df}{dT} = -\alpha_{11}\cos^2\theta - \alpha_{33}\sin^2\theta + \frac{1}{2}(2\alpha_{11} + \alpha_{33})$$

$$+ \frac{(1/2)(c_{44}Tc_{44}\sin^2\theta + 2c_{14}Tc_{14}\sin\theta\cos\theta + c_{66}Tc_{66}\cos^2\theta)}{c_{44}\sin^2\theta + 2c_{14}\sin\theta\cos\theta + c_{66}\cos^2\theta}.$$

To find the orientation of the zero temperature coefficient cuts, set $(1/f)(df/dT) = 0$ and solve for θ. The equation has two roots, $\theta = -35°$ and $+49°$ corresponding to the widely used AT- and BT-cuts.

It should be pointed out that zero temperature coefficient cuts are not possible for many piezoelectric crystals. The positive temperature coefficient Tc_{66} makes it possible for quartz. Tc coefficients are generally negative since most crystals soften elastically with increasing temperature, but quartz has a displacive phase transformation at 573°C, the so-called α–β transition. As explained later in the chapter on Ferroic Crystals, the α–β transformation involves rotational changes in the crystal structure. The stiffening of c_{66} is one of the results.

13.10.2 AC- and BC-cuts

The second example concerns mode coupling in quartz and the identification of crystal orientations with pure shear motions. As pointed out in Section 13.5, AC-cut quartz plates are used to generate pure shear waves.

Interfering vibration modes arise either from electromechanical coupling or from mechanical coupling. For the AT- and BT-cuts an electric field E'_2 excites a thickness shear vibration x'_6 through piezoelectric coefficient d'_{26}. But E'_2 also excites a face shear vibration through d_{25} (see Fig. 12.14). This mode has a much lower frequency but care must be exercised that the harmonics of the face shear mode do not come close to the thickness shear. Other piezoelectric vibrations can be excited by E_1. Rotated Y-cuts with $E_1 = 0$ are used to avoid these modes.

Mechanical coupling between modes takes place through elastic constants which couple one motion to another. For example, stiffness coefficient c_{12} couples motions in the Z_1 and Z_2 directions. For thickness shear motions in rotated Y-cuts, a shear strain x'_6 is excited by E'_2 through piezoelectric coefficient d'_{26}. The strain component x'_6 is coupled to stress component X'_6 through c'_{66}. It is also coupled to stress components X'_1, X'_2, X'_3, X'_4, and X'_5 through elastic constants c'_{16}, c'_{26}, c'_{36}, c'_{46}, and c'_{56}. The stresses generated through these five stiffness coefficients will excite additional vibrations coupled to the shear motion about Z'_3.

Problem 13.5
To avoid coupled modes of motion, it is necessary to make $c'_{16} = c'_{26} = c'_{36} = c'_{46} = c'_{56} = 0$. This can be done for quartz if the angle θ is chosen correctly. As an exercise, evaluate these five coefficients for a Y-cut rotated through an angle θ and determine the critical angles for AC- and BC-cuts. These are the orientation angles for the piezoelectric transducers that are widely used in acoustics to generate transversely-polarized waves.

14 Magnetic phenomena

14.1 Basic ideas and units 122
14.2 Magnetic structures and time reversal 124
14.3 Magnetic point groups 125
14.4 Magnetic axial vectors 130
14.5 Saturation magnetization and pyromagnetism 131
14.6 Magnetic susceptibility and permeability 134
14.7 Diamagnetic and paramagnetic crystals 135
14.8 Susceptibility measurements 137
14.9 Magnetoelectricity 138
14.10 Piezomagnetism 142
14.11 Summary 146

In this chapter we deal with a number of magnetic properties and their directional dependence: pyromagnetism, magnetic susceptibility, magnetoelectricity, and piezomagnetism. In the course of dealing with these properties, two new ideas are introduced: magnetic symmetry and axial tensors.

14.1 Basic ideas and units

Moving electric charge generates magnetic fields and magnetization. Macroscopically, an electric current i flowing in a coil of n turns per meter produces a magnetic field $H = ni$ amperes/meter [A/m]. On the atomic scale, magnetization arises from unpaired electron spins and unbalanced electronic orbital motion.

The weber [Wb] is the basic unit of magnetic charge m. The force between two magnetic charges m_1 and m_2 is

$$F = \frac{m_1 m_2}{4\pi \mu_0 r^2} \quad [\text{N}],$$

where r is the separation distance and μ_0 ($= 4\pi \times 10^{-7}$ H/m) is the permeability of vacuum. In a magnetic field H, magnetic charge experiences a force $F = mH$ [N]. North and south poles (magnetic charges) separated by a distance r create magnetic dipole moments mr [Wb m]. Magnetic dipole moments provide a convenient way of picturing the atomistic origins arising from moving electric charge.

Magnetization (I) is the magnetic dipole moment per unit volume and is expressed in units of Wb m/m^3 = Wb/m^2. The magnetic flux density ($B = I + \mu_0 H$) is also in Wb/m^2 and is analogous to the electric displacement D.

All materials respond to magnetic fields, producing a magnetization $I = \chi H$, and a magnetic flux density $B = \mu H$ where χ is the magnetic susceptibility and μ is the magnetic permeability. Both χ and μ are in henries/m (H/m). The permeability $\mu = \chi + \mu_0$ and is analogous to electric permittivity. χ and μ are sometimes expressed as dimensionless quantities ($\bar{\chi}$ and $\bar{\mu}$) like the dielectric constant, where $\bar{\chi} = \chi/\mu_0$ and $\bar{\mu} = \mu/\mu_0$. Other magnetic properties will be defined later in the chapter.

A schematic view of the submicroscopic origins of magnetic phenomena is presented in Fig. 14.1. Most materials are diamagnetic with only a weak magnetic response induced by an applied magnetic field. In most compounds the electrons are paired in bonding, but magnetic fields cause small changes in orbital motion that results in a small negative susceptibility ($\bar{\chi} \approx 10^{-5}$).

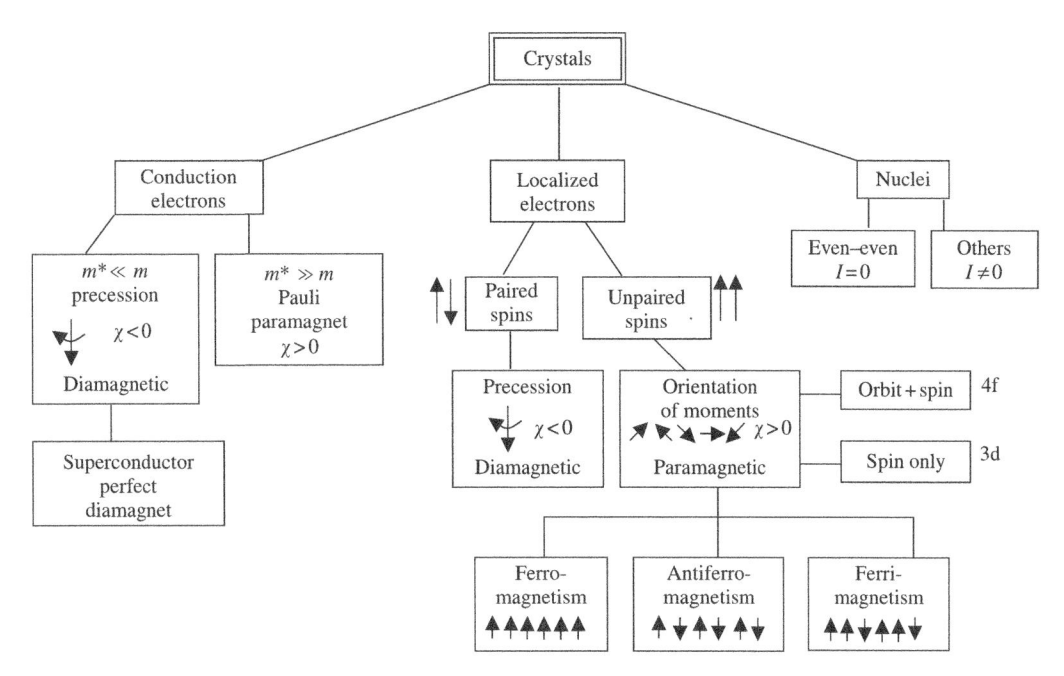

Fig. 14.1 Origins of magnetic phenomena in crystals. Electric currents come from both free and bound electrons and from atomic nuclei. The largest effects are in ferromagnetic, ferrimagnetic, and superconductor materials.

Induced currents enclose larger areas in aromatic molecules, giving larger effects. In a field gradient diamagnetic materials experience a force driving them out of the field.

Magnetization is also linearly proportional to field in *paramagnetic* materials but $\bar{\chi}$ is positive and usually larger, about 10^{-4}. Paramagnetism is common in dilute transition-metal salts, in which the metal ions with unpaired electrons interact with one another only weakly. The spins are randomly oriented at elevated temperatures but align slightly when a field is applied. Alignment becomes more difficult at high temperatures causing a decrease in susceptibility with temperature following the Curie Law. *Superparamagnets* contain clusters of paramagnetic ions. Exchange interactions are strong within a cluster, but weak between clusters. Yet another type of paramagnetism is found in metals where the conduction electrons create temperature independent Pauli paramagnetism. The effect is caused by small changes in the band structure for electrons of opposite spin.

When spins of neighboring atoms interact appreciably, three types of ordered configurations occur: antiferromagnetism, ferromagnetism, and ferrimagnetism. All three show Curie–Weiss Law behavior at high temperatures in the paramagnetic region. On cooling, the materials undergo a phase transition to a state in which the atomic dipoles are aligned, even in the absence of an applied field.

Antiferromagnetism is the most common of the three phenomena. In a simple collinear antiferromagnet, adjacent spins are aligned in antiparallel directions, producing zero net moment at zero field. Canted, spiral, and other more complicated antiferromagnetic arrays have also been observed. The magnetic susceptibility is small and field-independent, with a pronounced maximum near the transition temperature, called the Néel point. A few antiferromagnets such as $FeCl_2$ undergo spin reversal under applied fields, converting to ferromagnets with all spins aligned. Dysprosium and other materials with antiferromagnetic–ferromagnetic transitions exhibit peculiar field dependence, causing them to be classified as *metamagnets*.

The most useful magnetic materials are *ferromagnets* and ferrimagnets. Both possess a spontaneous magnetization that shows hysteresis under applied fields and disappears at the Curie temperature. The ordered spin array of a ferromagnet such as Co consists of parallel spins. Ferromagnets are rare among oxides, with only CrO_2, EuO, and a few other examples. In a simple ferrimagnet, neighboring spins are antiparallel but are either unequal in size or unequal in number. A number of important ferrimagnetic materials are found in the spinel and garnet families.

Parasitic ferromagnetism is an effect found in $\alpha Fe_2 O_3$ and certain other antiferromagnets. A weak ferromagnetism caused by canting of the spins is permitted for certain magnetic point groups. Parasitic ferromagnetism disappears at the Néel temperature.

14.2 Magnetic structures and time reversal

Thousands of ordered magnetic structures have been determined by neutron diffraction and various resonance techniques. As shown in Fig. 14.2, some are simple collinear structures with all the magnetic moments aligned in parallel or antiparallel configurations, but others have more complex configurations with spins arranged in canted or helical structures.

Geometric representations are helpful in determining the effect of symmetry on physical properties. In picturing magnetic structures, atomic magnetic moments are often visualized as arrows, but this is somewhat misleading. Magnetization arises from moving electric charge so that a current loop is a more meaningful symbol. It is helpful, however, to retain the arrow indicating the magnetization direction as an aid in visualizing the orientation of the current loop.

The transformations of a current loop under various symmetry operations are shown in Fig. 14.3. A twofold axis reverses the direction of atomic moments oriented perpendicular to the axis but does not affect the parallel components. The reverse is true for mirror planes: magnetic moments parallel to the mirror are reversed by the reflection operation, whereas perpendicular components are unaffected. An inversion center leaves the moment unaltered, regardless of orientation.

Time reversal, a nonspatial symmetry operation, is also used in describing magnetic structures. Reversing time reverses the direction of current flow, reversing the direction of magnetization. In describing magnetic structures the time reversal operator often occurs in combination with the geometrical symmetry elements. As an example, consider two identical atoms whose positions are related by a twofold rotation axis (Fig. 14.4). The magnetic moments associated with these atoms may also obey the rotational operation, or they may not, depending the orientation of the moments. If both spins are parallel to the twofold axis (Fig. 14.4(a)), the twofold axis is retained. If they are collinear with the twofold axis and antiparallel to one another (Fig. 14.4(b)), the twofold axis is

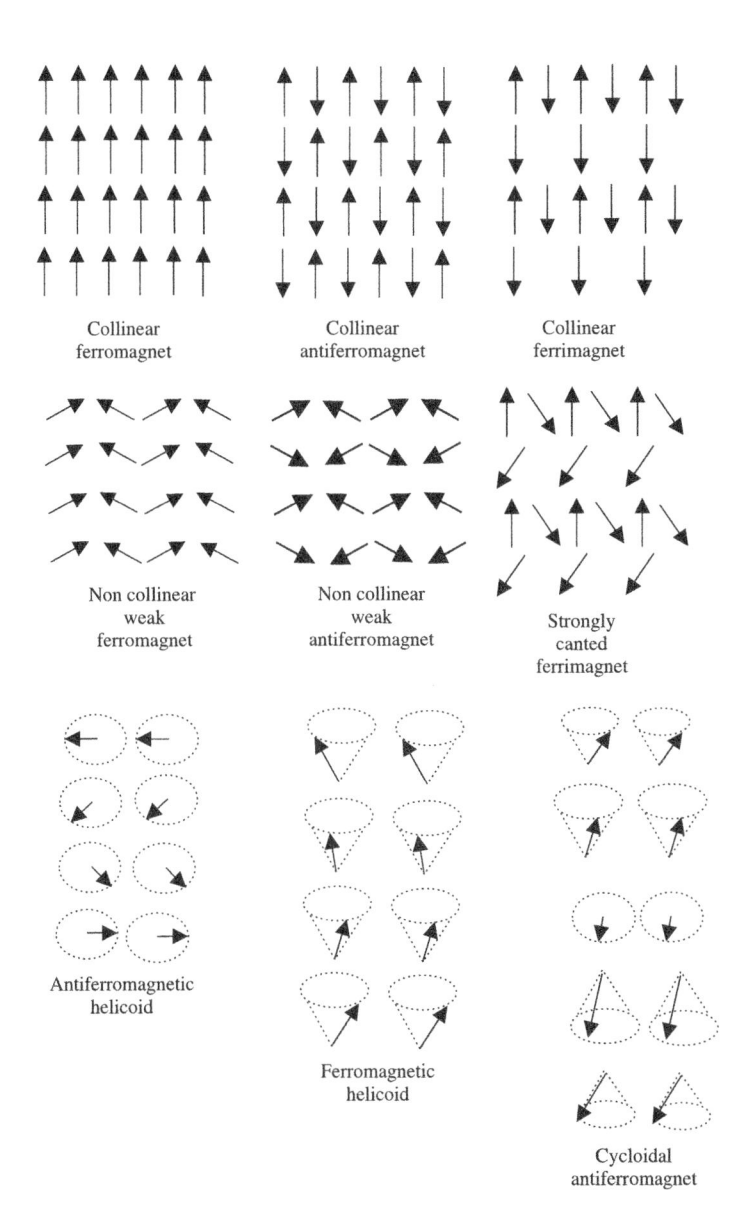

Fig. 14.2 Representative magnetic structures for ferromagnetic, ferrimagnetic, and antiferromagnetic materials. Each can be assigned to a magnetic symmetry group.

almost retained. In this case, reversing the spin retains the symmetry element. Spin reversal is brought about by time reversal, so that the two moments in Fig.14.4(b) are related by a twofold rotation accompanied by time reversal, designated 2′. The spins need not be parallel or antiparallel to the symmetry axis to maintain 2 or 2′ symmetry operators. The spin configurations shown in Fig. 14.4(c) and (d) also possess twofold axes.

14.3 Magnetic point groups

Magnetic point groups govern magnetic properties, just as nonmagnetic properties are subject to Neumann's Principle and crystallographic point

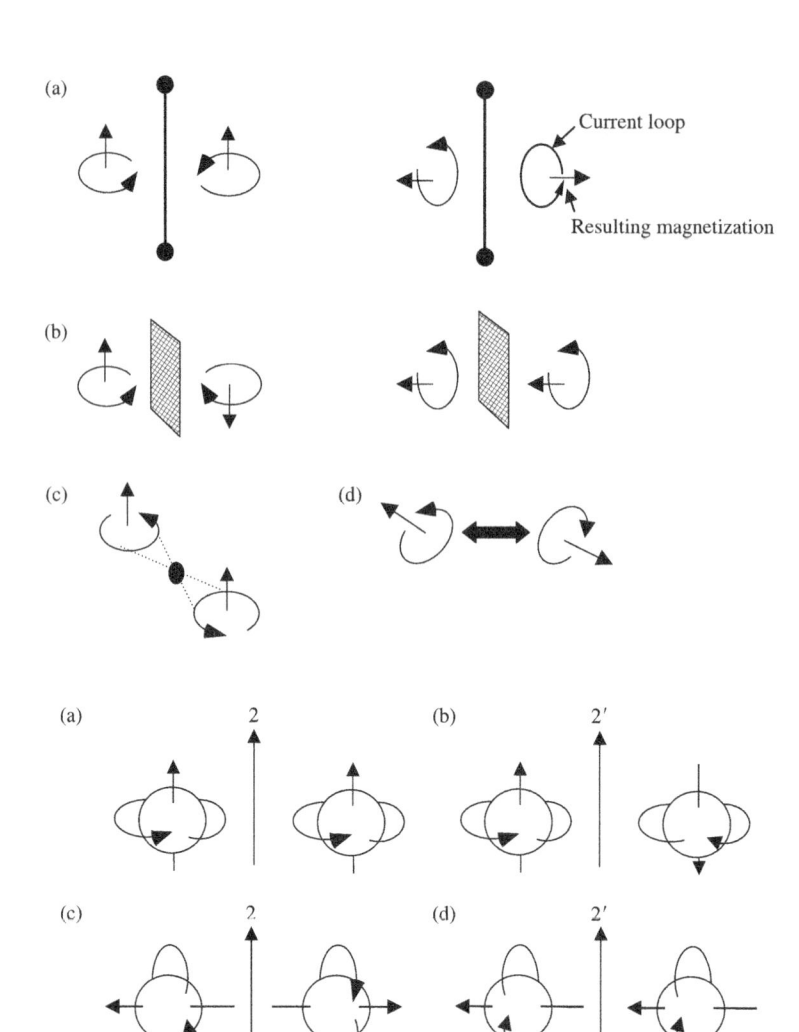

Fig. 14.3 Transformation of a current loop under (a) twofold rotation, (b) reflection, (c) inversion, and (d) time reversal. Note that for rotation and reflection, the transformation depends on the orientation of the loop relative to the symmetry element.

Fig. 14.4 Spin configurations in (a) and (c) possess twofold symmetry, while in (b) and (d) the rotational symmetry is accompanied by time reversal.

symmetry. Introduction of the time reversal operator increases the number of classes by adding 90 additional magnetic point groups.

To illustrate, we shall derive the magnetic point groups associated with $2/m$, one of the 32 crystallographic point groups. This a monoclinic point group with a twofold symmetry axis parallel to $Z_2 = [010]$, and a mirror plane perpendicular to Z_2. Point group $2/m$ also contains a center of symmetry ($\bar{1}$) that does not appear in the symbol. Orthoclase feldspar ($KAlSi_3O_8$) and other nonmagnetic crystals belonging to $2/m$ contain six types of symmetry: $2, 2', m, m', \bar{1}$, and $\bar{1}'$. They possess regular crystallographic symmetry elements and time-reversed symmetry elements. Two atoms in the feldspar structure related by the mirror plane m are also related by m' (reflection plus time reversal). Reversing time reverses spin directions, but since spin-up and spin-down electron distributions are identical in nonmagnetic crystals, both m and m' are present. This is not true in materials with long-range magnetic order. One symmetry element may be obeyed, but not both.

When discussing magnetic symmetry, it is necessary to distinguish the crystallographic point groups from the magnetic point groups. The crystallographic

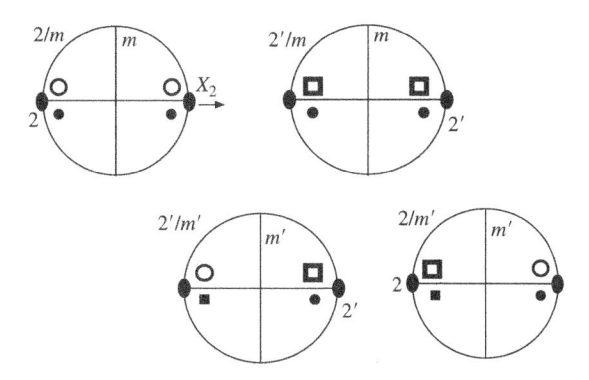

Fig. 14.5 Stereograms of the four magnetic point groups associated with crystallographic point group $2/m1'$. Solid and open symbols represent equivalent points in the northern and southern hemispheres. Square and circular symbols represent points related by time reversal. Both time-reversed and nontime-reversed symmetry elements are present in $2/m1'$ so there are twice the number of symmetry elements in the crystallographic group.

group is written as $2/m1'$ which indicates that it contains both the regular and the time-reversed symmetry elements. When written as $2/m$ it contains 2, m, and $\bar{1}$, but not $2'$, m', and $\bar{1}'$. $2/m$ is one of the magnetic point groups associated with crystallographic group $2/m1'$. As shown in Fig. 14.5 there are three more: $2/m'$, $2'/m$, and $2'/m'$. They are easily deduced from $2/m$ by systematic substitution of time-reversed symmetry elements.

Examples of many of the magnetic point groups have been determined by neutron diffraction (Oles). Referring to the four monoclinic groups in Fig. 14.5, ferromagnetic Fe_3Se_4 belongs to $2'/m'$, while the low temperature magnetic structures of antiferromagnetic DyOOH and ErOOH are in $2/m'$ and $2'/m$, respectively. Between 250 and 950 K, hematite (α-Fe_2O_3) is a weak ferrimagnet in magnetic point group $2/m$.

A full listing of the magnetic point groups and their associated physical properties is given in Table 14.1.

Problem 14.1
The ninety magnetic point groups can be deduced using stereographic projections. Derive the nine magnetic point groups belonging to the orthorhombic system. Begin with the three that do not contain time reversal (222, $mm2$, mmm) and generate six more. Label each group and indicate which are pyromagnetic.

Problem 14.2
Derive the magnetic Curie groups and list their generating symmetry elements.

Examples of four magnetic structures and their magnetic point groups are shown in Figs. 14.6 and 14.7. Common α-Fe is ferromagnetic with magnetic moments aligned parallel to the [001] direction. The body-centered cubic crystal structure (point group $m3m1'$ above T_c) distorts very slightly on cooling through the magnetic transition where the spins become ordered and the symmetry changes to magnetic point group $4/mm'm'$.

Cobalt is also ferromagnetic with magnetic spin along [001]. The crystal structure is hexagonal close-packed (Fig. 14.6(b)), point group $6/mmm1'$ in the high temperature paramagnetic state, which changes to magnetic group $6/mm'm'$ when the spins align. In this case the metal is hexagonal in both paramagnetic and ferromagnetic phases.

128 *Magnetic phenomena*

Table 14.1 Ninety magnetic point groups. Number of nonzero matrix coefficients for pyromagnetism, magnetoelectricity, and piezomagnetism are listed along with the number of independent coefficients in parentheses (the symmetry elements required to generate the group are also included)

	Pyromagnetic	Magnetoelectric	Piezomagnetic	Generating elements
Triclinic				
1	3(3)	9(9)	18(18)	1
$\bar{1}$	3(3)	0	18(18)	$\bar{1}$
$\bar{1}'$	0	9(9)	0	$\bar{1}'$
Monoclinic				
2	1(1)	5(5)	8(8)	$2 \parallel Z_2$
$2'$	2(2)	4(4)	10(10)	$2' \parallel Z_2$
m	1(1)	4(4)	8(8)	$m \perp Z_2$
m'	2(2)	5(5)	10(10)	$m' \perp Z_2$
$2/m$	1(1)	0	8(8)	$2 \parallel Z_2, m \perp Z_2$
$2'/m'$	2(2)	0	10(10)	$2' \parallel Z_2, m' \perp Z_2$
$2/m'$	0	5(5)	0	$2 \parallel Z_2, m' \perp Z_2$
$2'/m$	0	4(4)	0	$2' \parallel Z_2, m \perp Z_2$
Orthorhombic				
222	0	3(3)	3(3)	$2 \parallel Z_2, 2 \parallel Z_3$
$2'2'2$	1(1)	2(2)	5(5)	$2' \parallel Z_2, 2 \parallel Z_3$
$mm2$	0	2(2)	3(3)	$m \perp Z_2, 2 \parallel Z_3$
$m'm'2$	1(1)	3(3)	5(5)	$m' \perp Z_2, 2 \parallel Z_3$
$m'm2'$	1(1)	2(2)	5(5)	$m \perp Z_2, 2' \parallel Z_3$
mmm	0	0	3(3)	$m \perp Z_1, m \perp Z_2, m \perp Z_3$
$m'm'm$	1(1)	0	5(5)	$m' \perp Z_1, m' \perp Z_2, m \perp Z_3$
$m'm'm'$	0	3(3)	0	$m' \perp Z_1, m' \perp Z_2, m' \perp Z_3$
$m'mm$	0	2(2)	0	$m' \perp Z_1, m \perp Z_2, m \perp Z_3$
Trigonal				
3	1(1)	5(3)	13(6)	$3 \parallel Z_3$
$\bar{3}$	1(1)	0	13(6)	$\bar{3} \parallel Z_3$
$\bar{3}'$	0	5(3)	0	$\bar{3}' \parallel Z_3$
32	0	3(2)	5(2)	$2 \parallel Z_1, 3 \parallel Z_3$
$32'$	1(1)	2(1)	8(4)	$2' \parallel Z_1, 3 \parallel Z_3$
$3m$	0	2(1)	5(2)	$m \perp Z_1, 3 \parallel Z_3$
$3m'$	1(1)	3(2)	8(4)	$m' \perp Z_1, 3 \parallel Z_3$
$\bar{3}m$	0	2(1)	5(2)	$m \perp Z_1, \bar{3} \parallel Z_3$
$\bar{3}m'$	1(1)	0	8(4)	$m' \perp Z_1, \bar{3} \parallel Z_3$
$\bar{3}'m'$	0	3(2)	0	$m' \perp Z_1, \bar{3}' \parallel Z_3$
$\bar{3}'m$	0	0	0	$m \perp Z_1, \bar{3}' \parallel Z_3$
Tetragonal				
4	1(1)	5(3)	7(4)	$4 \parallel Z_3$
$4'$	0	4(2)	7(4)	$4' \parallel Z_3$
$\bar{4}$	1(1)	4(2)	7(4)	$\bar{4} \parallel Z_3$
$\bar{4}'$	0	5(3)	7(4)	$\bar{4}' \parallel Z_3$
$4/m$	1(1)	0	7(4)	$m \perp Z_3, 4 \parallel Z_3$
$4'/m$	0	0	7(4)	$m \perp Z_3, 4' \parallel Z_3$
$4/m'$	0	5(3)	0	$m' \perp Z_3, 4 \parallel Z_3$
$4'/m'$	0	4(2)	0	$m' \perp Z_3, 4' \parallel Z_3$
422	0	3(2)	2(1)	$2 \parallel Z_1, 4 \parallel Z_3$
$4'22$	0	2(1)	3(2)	$2 \parallel Z_1, 4' \parallel Z_3$
$42'2'$	1(1)	2(1)	5(3)	$2' \parallel Z_1, 4 \parallel Z_3$
$4mm$	0	2(1)	2(1)	$m \perp Z_1, 4 \parallel Z_3$
$4'mm'$	0	2(1)	3(2)	$m \perp Z_1, 4' \parallel Z_3$
$4m'm'$	1(1)	3(2)	5(3)	$m' \perp Z_1, 4 \parallel Z_3$
$\bar{4}2m$	0	2(1)	2(1)	$2 \parallel Z_1, \bar{4} \parallel Z_3$
$\bar{4}'2m'$	0	3(2)	3(2)	$2 \parallel Z_1, \bar{4}' \parallel Z_3$
$\bar{4}'2'm$	0	2(1)	3(2)	$2' \parallel Z_1, \bar{4}' \parallel Z_3$
$\bar{4}2'm'$	1(1)	2(1)	5(3)	$2' \parallel Z_1, \bar{4} \parallel Z_3$
$4/mmm$	0	0	2(1)	$m \perp Z_1, m \perp Z_3, 4 \parallel Z_3$

Continued overleaf

Table 14.1 (*Continued*)

	Pyromagnetic	Magnetoelectric	Piezomagnetic	Generating elements
Tetragonal				
$4'/mmm'$	0	0	3(2)	$m \perp Z_1, m \perp Z_3, 4' \parallel Z_3$
$4/mm'm'$	1(1)	0	5(3)	$m' \perp Z_1, m \perp Z_3, 4 \parallel Z_3$
$4/m'm'm'$	0	3(2)	0	$m' \perp Z_1, m' \perp Z_3, 4 \parallel Z_3$
$4/m'mm$	0	2(1)	0	$m \perp Z_1, m' \perp Z_3, 4 \parallel Z_3$
$4'/m'mm'$	0	2(1)	0	$m \perp Z_1, m' \perp Z_3, 4' \parallel Z_3$
Hexagonal				
6	1(1)	5(3)	7(4)	$6 \parallel Z_3$
$6'$	0	0	6(2)	$6' \parallel Z_3$
$\bar{6}$	1(1)	0	7(4)	$\bar{6} \parallel Z_3$
$\bar{6}'$	0	5(3)	6(2)	$\bar{6}' \parallel Z_3$
$6/m$	1(1)	0	7(4)	$m \perp Z_3, 6 \parallel Z_3$
$6'/m'$	0	0	6(2)	$m' \perp Z_3, 6' \parallel Z_3$
$6/m'$	0	5(3)	0	$m' \perp Z_3, 6 \parallel Z_3$
$6'/m$	0	0	0	$m \perp Z_3, 6' \parallel Z_3$
622	0	3(2)	2(1)	$2 \parallel Z_1, 6 \parallel Z_3$
$6'22'$	0	0	3(1)	$2 \parallel Z_1, 6' \parallel Z_3$
$62'2'$	1(1)	2(1)	5(3)	$2' \parallel Z_1, 6 \parallel Z_3$
6mm	0	2(1)	2(1)	$m \perp Z_1, 6 \parallel Z_3$
$6'mm'$	0	0	3(1)	$m \perp Z_1, 6' \parallel Z_3$
$6m'm'$	1(1)	3(2)	5(3)	$m' \perp Z_1, 6 \parallel Z_3$
$\bar{6}m2$	0	0	2(1)	$m \perp Z_1, \bar{6} \parallel Z_3$
$\bar{6}'m'2$	0	3(2)	3(1)	$m' \perp Z_1, \bar{6}' \parallel Z_3$
$\bar{6}'m2'$	0	2(1)	3(1)	$m \perp Z_1, \bar{6}' \parallel Z_3$
$\bar{6}m'2'$	1(1)	0	5(3)	$m' \perp Z_1, \bar{6} \parallel Z_3$
$6/mmm$	0	0	2(1)	$m \perp Z_1, m \perp Z_3, 6 \parallel Z_3$
$6'/m'mm'$	0	0	3(1)	$m \perp Z_1, m' \perp Z_3, 6' \parallel Z_3$
$6/mm'm'$	1(1)	0	5(3)	$m' \perp Z_1, m \perp Z_3, 6 \parallel Z_3$
$6/m'm'm'$	0	3(2)	0	$m' \perp Z_1, m' \perp Z_3, 6 \parallel Z_3$
$6/m'mm$	0	2(1)	0	$m \perp Z_1, m' \perp Z_3, 6 \parallel Z_3$
$6'/mmm'$	0	0	0	$m \perp Z_1, m \perp Z_3, 6' \parallel Z_3$
Cubic				
23	0	3(1)	3(1)	$2 \parallel Z_1, 3 \parallel [111]$
m3	0	0	3(1)	$m \perp Z_1, 3 \parallel [111]$
$m'3$	0	3(1)	0	$m' \perp Z_1, 3 \parallel [111]$
432	0	3(1)	0	$4 \parallel Z_1, 3 \parallel [111]$
$4'32$	0	0	3(1)	$4' \parallel Z_1, 3 \parallel [111]$
$\bar{4}3m$	0	0	0	$\bar{4} \parallel Z_1, 3 \parallel [111]$
$\bar{4}'3m'$	0	3(1)	3(1)	$\bar{4}' \parallel Z_1, 3 \parallel [111]$
m3m	0	0	0	$m \perp Z_1, 3 \parallel [111], m \perp [110]$
$m3m'$	0	0	3(1)	$m \perp Z_1, 3 \parallel [111], m' \perp [110]$
$m'3m'$	0	3(1)	0	$m' \perp Z_1, 3 \parallel [111], m' \perp [110]$
$m'3m$	0	0	0	$m' \perp Z_1, 3 \parallel [111], m \perp [110]$

Transition metal oxides are often antiferromagnetic because of strong superexchange interactions through neighboring oxygen atoms. Hematite (α-Fe$_2$O$_3$) and chrome oxide (Cr$_2$O$_3$) are corundum-family oxides with rhombohedral (trigonal) unit cells (Fig. 14.6(c) and (d)). The corundum (α-Al$_2$O$_3$) structure belongs to point group $\bar{3}m1'$. Below T_N, the Neel temperature, α-Fe$_2$O$_3$ exhibits a weak parasitic ferromagnetism between $-20°$C and $675°$C. In this temperature range the atomic moments lie in the basal (001) plane (point group $2/m$). Parasitic ferromagnetism occurs when a small ferromagnetic component can be developed without violating the point group symmetry. Point group $2/m$ is a subgroup of ∞/mm', the Curie group symmetry associated with the magnetization vector. Weak ferromagnetism occurs along the twofold rotation axis of $2/m$ in the crystallographic (001) plane. The canted spin arrangement corresponds to the noncollinear weak ferromagnet pattern in Fig. 14.2 Below $-20°C$ the spins in hematite are aligned along [001] in an antiparallel configuration (magnetic point group $\bar{3}m$, Fig. 14.7(c)). Weak ferromagnetism does not occur in this point group. In antiferromagnetic Cr_2O_3 the spins are also along [001] but arranged in different pattern conforming to point group $\bar{3}'m'$, Fig. 14.7(d). Stereograms for the four magnetic point groups are shown in Fig. 14.7.

In all the examples just discussed the symmetry of the crystal structure is modified by introducing a magnetic dipole array. When a spontaneous magnetization is superposed on crystallographic symmetry, the point group is changed to one of the magnetic subgroups, depending on the direction of the magnetization. The symmetry group of magnetization is ∞/mm', and the magnetic groups of α-Fe, Co, Ni are all subgroups of ∞/mm'.

14.4 Magnetic axial vectors

Electric and magnetic phenomena are related to one another through Maxwell's Equations. The equations are written out in vector and tensor notation in Table 14.2. From the first two equations it is apparent that electric and magnetic variables are related through vector products. In tensor form the vector product is written as the rotation tensor ε_{ijk} which is equal to zero unless all three subscripts are different ($i \neq j \neq k \neq i$); $\varepsilon_{ijk} = +1$ if i, j, k are cyclic, and $\varepsilon_{ijk} = -1$ if they are anticyclic. In other words the vector product changes sign when the handedness changes. Transformations such as inversion or reflection change the sign of a vector product. The vector product of two polar vectors creates an axial vector.

From the first Maxwell equation, the magnetic flux density B_i is related to the cross product of the spatial derivative of the electric field E_k with respect to coordinate Z_j. Since the electric field E and coordinate Z both transform as polar first rank tensors (vectors), and the time t is a scalar, then B must transform as an axial first rank tensor (axial vector). In tensor form,

$$B'_i = |a|a_{ij}B_j$$

and since B, I, and H are all related by the constitutive equation, they are all axial vectors.

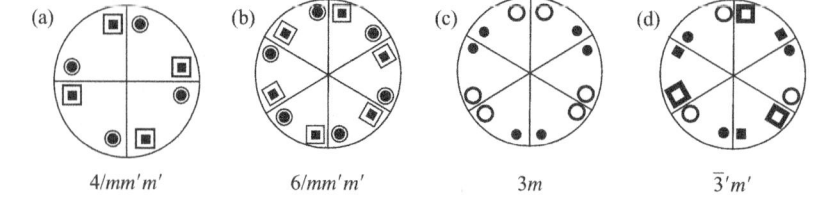

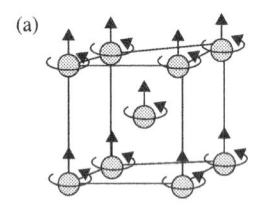

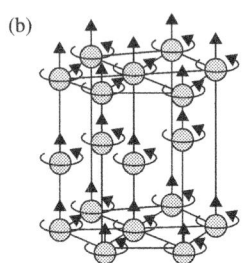

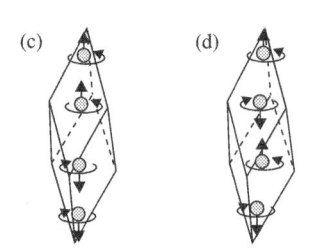

$$I'_i = |a|a_{ij}I_j \quad \text{and} \quad H'_i = |a|a_{ij}H_j.$$

Fig. 14.6 Magnetic structures of (a) α-Fe, (b) Co, (c) α-Fe$_2$O$_3$ ($T < 250$ K), and (d) Cr$_2$O$_3$.

Fig. 14.7 Magnetic point groups of (a) α-Fe, (b) Co, (c) α-Fe$_2$O$_3$ ($T < 250$ K), and (d) Cr$_2$O$_3$. The squares and circles are related by time reversal.

(a) $4/mm'm'$ (b) $6/mm'm'$ (c) $\bar{3}m$ (d) $\bar{3}'m'$

Table 14.2 Maxwell's Equations in vector and tensor notation along with the linear constitutive equations (ρ is the free charge density [C/m^3])

Vector form	Tensor form
$\vec{\nabla} \times \vec{E} = -\dfrac{\partial \vec{B}}{\partial t}$	$\varepsilon_{ijk} \dfrac{\partial E_k}{\partial Z_j} = -\dfrac{\partial B_i}{\partial t}$
$\vec{\nabla} \times \vec{H} = \vec{J} + \dfrac{\partial \vec{D}}{\partial t}$	$\varepsilon_{ijk} \dfrac{\partial H_k}{\partial Z_j} = J_i + \dfrac{\partial D_i}{\partial t}$
$\vec{\nabla} \cdot \vec{B} = 0$	$\dfrac{\partial B_i}{\partial Z_i} = 0$
$\vec{\nabla} \cdot \vec{D} = \rho$	$\dfrac{\partial D_i}{\partial Z_i} = \rho$

If time reversal is involved in the transformation, there will be an additional sign change. Reversing the sign of the flow in the current loop reverses the magnetic field vector. Therefore in applying Neumann's Principle to an axial tensor property there will be two possible sign changes, one for the handedness change $|a|$ and the other for time reversal. The nature of the transformation will become clearer when axial tensor properties such as pyromagnetism, magnetoelectricity, and piezomagnetism are considered.

14.5 Saturation magnetization and pyromagnetism

Ferromagnetic and ferrimagnetic materials are hysteretic. When a large magnetic field is applied and all the magnetic moments are aligned with the field, the magnetization saturates to I_s [Wb/m^2]. The magnetization process involves domain wall motion together with rotation of the magnetization vector into the field direction. Field reversal leads to hysteresis and magnetization saturation to $-I_s$ in the reverse direction. A schematic view of the experiment is shown in Fig. 14.8. The specimen is shaped as a toroid (or a picture frame for single crystals) with coils wrapped around opposite arms. The hysteresis loop is obtained under an AC drive using an oscilloscope. Extrapolating the saturation value back to the $H = 0$ vertical axis gives the spontaneous magnetization I_s. The variation of I_s with temperature is the pyromagnetic coefficient Q (in units of Wb/m^2 K):

$$\Delta I_s = Q \Delta T.$$

When written in tensor in tensor form, the defining transformation goes as follows

$$\Delta I_i' = \pm |a| a_{ij} \Delta I_j = \pm |a| a_{ij} Q_j \Delta T$$
$$= \pm |a| a_{ij} Q_j \Delta T' = Q_i' \Delta T'.$$

Therefore the pyromagnetic coefficients transform as an axial first rank tensor.

$$Q_i' = \pm |a| a_{ij} Q_j.$$

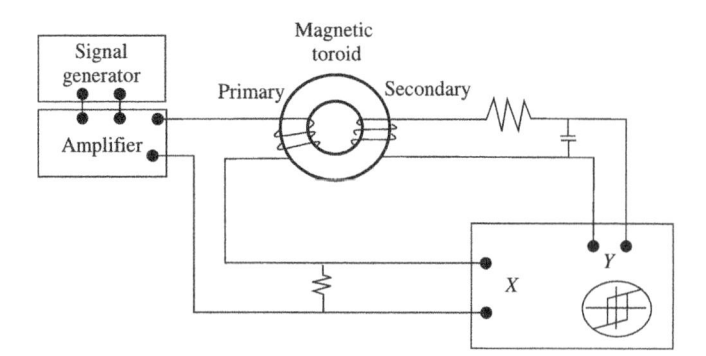

Fig. 14.8 Experimental determination of the saturation magnetization I_s under high drive conditions. The pyromagnetic coefficient is determined by measuring I_s as a function of temperature.

In matrix form it is a 3 × 1 column matrix.

$$\begin{matrix} 3\times 1 & 3\times 3 & 3\times 1 \\ (Q') = \pm|a| & (a) & (Q). \end{matrix}$$

To illustrate Neumann's Principle for the pyromagnetic effect we use ferromagnetic cobalt as an example. As shown in Fig. 14.6(b), Co has a hexagonal close-packed structure with the magnetic spins parallel to [001] = Z_3. The magnetic point group is $6/mm'm'$ and the three generating symmetry elements are $6 \parallel Z_3$, $m \perp Z_3$ and $m' \perp Z_1$.

For the mirror plane perpendicular to Z_3

$$(Q') = (+1)(-1)\begin{pmatrix} 1 & 0 & 0 \\ 0 & 1 & 0 \\ 0 & 0 & -1 \end{pmatrix} \quad (Q) = \begin{pmatrix} -Q_1 \\ -Q_2 \\ Q_3 \end{pmatrix}.$$

Equating this result to the starting matrix tell us that $Q_1 = -Q_1 = 0$, $Q_2 = -Q_2 = 0$, $Q_3 = Q_3$ is unchanged.

For the mirror accompanied by time reversal perpendicular to Z_1,

$$(Q') = (-1)(-1)\begin{pmatrix} -1 & 0 & 0 \\ 0 & 1 & 0 \\ 0 & 0 & 1 \end{pmatrix}\begin{pmatrix} 0 \\ 0 \\ Q_3 \end{pmatrix} = \begin{pmatrix} 0 \\ 0 \\ Q_3 \end{pmatrix}.$$

The coefficient Q_3 remains. Applying the sixfold axis leads to no further change. As expected from the structure, there is only one nonzero coefficient Q_3.

The 31 magnetic point groups exhibiting pyromagnetism are listed in Table 14.3. Generally only one measurement is required to specify the effect. Only low symmetry monoclinic and triclinic classes require two or three measurements.

The temperature dependence of the saturation magnetization in two of the magnetic garnets is illustrated in Fig. 14.9. The curve for YIG (=$Y_3Fe_5O_{12}$) is quite normal with I_s dropping to zero at the Curie temperature. The largest Q values occur just below T_c. Similar behavior is observed for most other pyromagnetic materials and for many pyroelectric materials.

Gadolinium iron garnet ($Gd_3Fe_5O_{12}$) has a more unusual behavior. The spontaneous magnetization begins with a very high value at low temperatures, then drops rapidly to zero near room temperature, and then changes sign before decreasing to zero at T_c. The pyromagnetic coefficient changes sign as well. This unusual behavior is caused by the competition between the magnetic

Table 14.3 The 31 pyromagnetic point groups

Magnetic point groups	Pyromagnetic matrix
$1, \bar{1}$	(Q_1, Q_2, Q_3)
$2', m', 2'/m'$	$(Q_1, 0, Q_3)$
$2, m, 2/m$	$(0, Q_2, 0)$
$2'2'2, m'm'2, m'2'm,$ $m'm'm, 3, \bar{3}, 32', 3m',$ $\bar{3}m', 4, \bar{4}, 4/m, 42'2',$ $4m'm', \bar{4}2'm', 4/mm'm', 6$ $\bar{6}, 6/m, 62'2', 6m'm', \bar{6}m'2',$ $6/mm'm', \infty, \infty/m, \infty 2', \infty m',$ ∞/mm'	$(0, 0, Q_3)$
All others	$(0, 0, 0)$

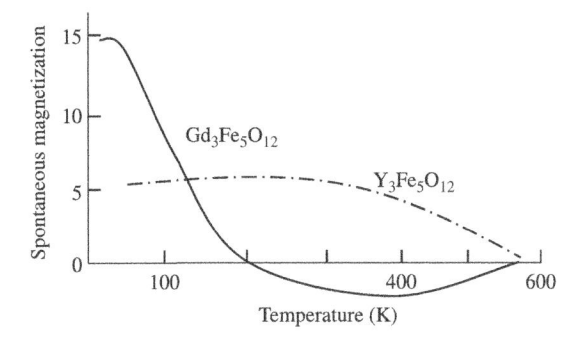

Fig. 14.9 Spontaneous magnetization I_s (in Bohr magnetons/molecule) of two garnet crystals plotted as a function of temperature.

dipoles associated with gadolinium and iron. At high temperature above T_c both moments are randomized by thermal fluctuations. On cooling through T_c the iron atoms align to create the spontaneous magnetization. Further cooling leads to alignment of the gadolinium moments in the opposite direction. As the large Gd moments align, the net magnetization decreases to zero and then reverses sign. Magnetism arises from the unpaired 3d electrons in Fe and the 4f electrons of Gd. The 4f electrons are buried further inside the atoms and do not interact with neighboring atoms as extensively as 3d electrons.

Pyroelectricity has a thermodynamically related converse effect in which an applied electric field produces a change in entropy. The magnetic analog to this electrocaloric effect is the magnetocaloric effect observed in Fe, Ni, and other ferromagnetic materials. In these materials the magnetocaloric effect is due to the change in the net number of aligned spins within one domain as the temperature changes. As expected the effect is largest near the Curie temperature where a small magnetic field can cause a large change in entropy. In Fe, temperature changes of 1 K are observed near 1000 K.

As discussed in Section 6.2, the thermodynamic relationship between the magnetocaloric and pyromagnetic effects is

$$\left(\frac{\partial S}{\partial H}\right)_T = \left(\frac{\partial I}{\partial T}\right)_H.$$

At $T = 0$ K, the entropy is zero irrespective of all parameters including the magnetic field H. Therefore $(\partial S/\partial H)_T = 0$ at $T = 0$, and the magnetocaloric and pyromagnetic effects disappear at very low temperatures. The same argument applies to the electrocaloric and pyroelectric effects.

14.6 Magnetic susceptibility and permeability

All materials respond to a magnetic field, producing a magnetization $I = \chi H$ and a magnetic flux density $B = \mu H$. The susceptibility χ and permeability μ are in H/m. As with the dielectric constant, it is often easier to work in dimensionless quantities by expressing χ and μ in units of μ_0: $\bar{\chi} = \chi/\mu_0$ and $\bar{\mu} = \mu/\mu_0 = \bar{\chi} + 1$. For strongly magnetic materials $\bar{\mu}$ and $\bar{\chi}$ can be used interchangeably. At room temperature the relative susceptibility $\bar{\chi}$ ranges from about 10^{-5} in weakly magnetic materials to 10^6 in strong magnets. At low temperatures, superconductors expel magnetic lines of force through the Meissner Effect. This makes $B = 0$ and $\bar{\chi} = -1$, a perfect diamagnet. In general, $\bar{\chi}$ may be linear or nonlinear, and positive or negative, and is often temperature-sensitive.

Magnetic susceptibility and magnetic permeability are both second rank polar tensors. This can be seen by examining the transformation between coordinate systems. Consider the susceptibility tensor χ_{ij} relating I_i to H_j:

$$I'_i = \pm|a|a_{ij}I_j = \pm|a|a_{ij}\chi_{jk}H_k$$
$$= \pm|a|a_{ij}\chi_{jk}(\pm|a|a_{lk}H'_l) = \chi'_{il}H'_l.$$

In this relation the magnetization I and the magnetic field H transform as axial first rank tensors (axial vectors). The same is true for the flux density B. Therefore the axial nature cancels out for χ and μ, making them polar second rank tensors:

$$\chi'_{ij} = a_{ik}a_{il}\chi_{kl}$$
$$\mu'_{ij} = a_{ik}a_{jl}\mu_{kl} = \chi'_{ij} + \mu_0.$$

In matrix form

$$\begin{matrix} 3\times 3 & 3\times 3 & 3\times 3 & 3\times 3 & & 3\times 3 & 3\times 3 & 3\times 3 & 3\times 3 \\ (\chi') & = & (a) & (\chi) & (a)_t & \text{and} & (\mu') & = & (a) & (\mu) & (a)_t. \end{matrix}$$

These expressions are identical to those of the electric permittivity and the dielectric constant (the relative permittivity).

Applying Neumann's Principle to χ and μ leads to the same result as the dielectric constant. As shown in Table 9.1, the matrices contain between 1 and 6 independent coefficients, depending on symmetry. Triclinic crystals require six measurements and cubic crystals only one. The procedure for finding the principal axes follows that for the dielectric constant (Section 9.4). It is important to point out, however, that $\bar{\chi}$ and $\bar{\mu}$ often depend strongly on temperature, pressure, frequency, and the size of the magnetic field.

The permeabilities of metals and ceramics with long-range magnetic order are strongly influenced by domain structure. Values of the relative permeability of ferromagnetic metals and ferrimagnetic oxides are comparable to the dielectric constants of ferroelectric oxides. Soft magnets like permalloy have permeabilities of 10^5–10^6 while values of 10–1000 are more common in hard magnets where the domain walls are fixed in position.

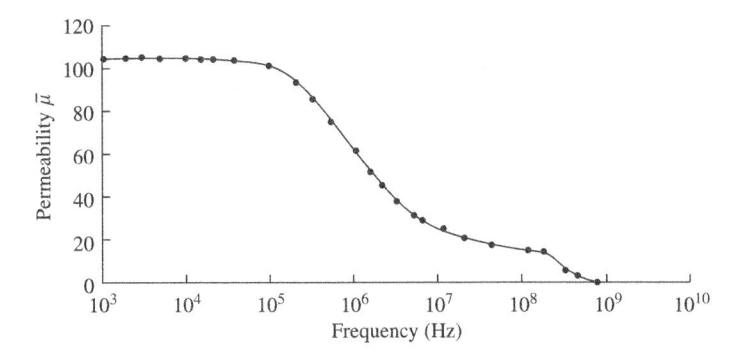

Fig. 14.10 The permeability spectrum of yttrium iron garnet measured in small fields.

The magnetic permeability spectrum of cubic yttrium iron garnet is shown in Fig. 14.10. The spectrum is characterized by two distinct regions of dispersion. Relaxation of the domain-wall contribution takes place between 10^5 and 10^7 Hz. The dispersion in the microwave range between 10^8 and 10^9 Hz is due to gyromagnetic resonance of the Fe^{3+} spin system caused by anisotropy and the internal demagnetization field.

Above a few GHz the permeability of magnetic oxides is almost zero. There is no magnetic refractive index analogous to the optical permittivity ($\bar{K} \approx n^2$), so that $\mu \cong \mu_0$ in the infrared and optical range.

Other causes of loss in ferromagnetic and ferrimagnetic materials include hysteresis loss caused by irreversible domain wall motion, and eddy current losses from conduction. These losses are minimized by keeping the measurement fields small and by using magnetic oxides like YIG with high electrical resistivity. The high frequency limit f for magnetic oxides is governed by *Snoek's Law* which states that $\bar{\mu} f \cong 0.56$ GHz. Soft ferrites with high permeability ($\bar{\mu}$) have lower limits than low permeability ferrites. The limit is controlled by the gyromagnetic resonance. Further discussion of magnetic resonance can be found in Section 31.3.

14.7 Diamagnetic and paramagnetic crystals

The magnetic susceptibility of a diamagnetic material is negative so that a diamagnet tends to move *out* of a magnetic field. Paramagnetic materials possess positive susceptibilities and move into a magnetic field.

Most compounds are weakly diamagnetic because of Lenz's Law. A change in magnetic flux passing through an electric circuit induces a current that opposes the flux change by creating a field in the opposite direction. In atoms or molecules, electrons moving about the nucleus create the "electric currents"; an applied field causes changes in the orbits, inducing a diamagnetic moment that disappears when the field is removed.

Diamagnetic anisotropy is generally small in most inorganic crystals. Quartz and calcite crystals are typical. Quartz (point group 32) the relative susceptibilities at room temperature are $\bar{\chi}_{11} = \bar{\chi}_{22} = -1.51 \times 10^{-5}$ and $\bar{\chi}_{33} = -1.52 \times 10^{-5}$. For calcite (point group $\bar{3}m$) the values are $\bar{\chi}_{11} = \bar{\chi}_{22} = -1.24 \times 10^{-5}$ and $\bar{\chi}_{33} = -1.38 \times 10^{-5}$.

Fig. 14.11 (a) Relative diamagnetic susceptibilities of four aromatic crystals (b) Susceptibilities referred to molecular axes. Dimensionless, all multiplied by 10^{-5}.

(a)

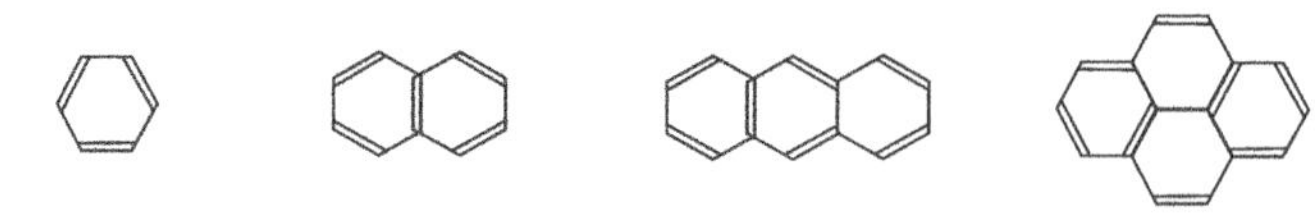

Benzene (C_6H_6) Naphthalene ($C_{10}H_8$) Anthracene ($C_{14}H_{10}$) Pyrene ($C_{16}H_{10}$)

(b)

	(a) Principal axes			(b) Molecular axes		
	χ_{11}	χ_{22}	χ_{33}	χ_{11}^M	χ_{22}^M	χ_{33}^M
Benzene	−6.5	−3.8	−6.2	−3.5	−3.5	−9.1
Naphthalene	−5.4	−15.0	−7.6	−5.5	−5.3	−17.4
Anthracene	−7.6	−21.2	−10.3	−7.6	−6.3	−25.2
Pyrene	−8.0	−17.8	−20.6	−8.1	−8.1	−30.3

Much larger anisotropy is observed in graphite and various aromatic crystals. Flat aromatic molecules show large diamagnetic susceptibilities when the molecules are oriented perpendicular to the applied magnetic field. This can be explained in terms of the π-orbitals associated with the benzene-like rings. The applied field induces electron currents that move around the molecule in wide orbits, and since the induced magnetization is proportional to the product of the current times the enclosed area, a large diamagnetic effect results. The susceptibility in graphite, measured perpendicular to the hexagonal layers, is about five times larger than in the perpendicular direction ($\bar{\chi}_{11} = \bar{\chi}_{22} = -5.0 \times 10^{-5}$ and $\bar{\chi}_{33} = -27.4 \times 10^{-5}$). Magnetic properties of other aromatic molecules are described in Fig. 14.11. When referred to molecular axes the susceptibilities are very anisotropic, approaching those of graphite.

In paramagnetic solids, the magnetic susceptibility is positive and usually somewhat larger in magnitude. Paramagnetism is common in dilute transition-metal salts in which the metal ions with unpaired spins interact with one another only weakly. In alums such as $KCr(SO_4)_2 \cdot 12H_2O$, the magnetic chromium ions are widely spaced because of the intervening sulfate groups and water molecules, and the interactions between spins are very small. At room temperature the spins are randomly oriented and align only slightly when a field is applied. Alignment becomes easier at very low temperatures where the susceptibility becomes large in accordance with *the Curie Law* ($\chi = C/T$). As pointed out in Section 7.3, alums and other dilute paramagnetic salts are used in cryogenic systems to achieve temperatures below 0.001 K by the process of adiabatic demagnetization.

Yet another type of paramagnetism is found in metals like titanium. Pauli paramagnetism is caused by small changes in band structure in the presence of a magnetic field. Conduction electrons with spins parallel to the field have slightly lower energy than those with antiparallel spins. The susceptibility of many nonmagnetic metals is nearly temperature independent.

Anisotropy also appears at low temperature in antiferromagnetic crystals. In manganese oxide the spins are parallel to $\langle 110 \rangle$ directions (Fig. 14.12). Above the Neel temperature T_N in the paramagnetic state, the magnetic susceptibility of cubic MnO is isotropic, following a Curie–Weiss Law. Below the transition

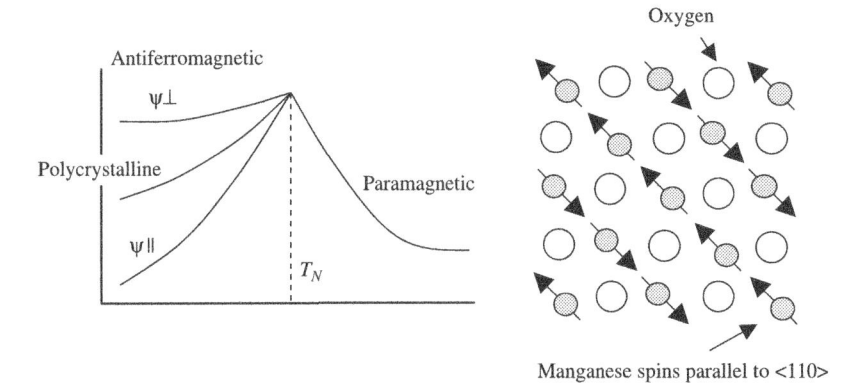

Fig. 14.12 Anisotropic magnetic susceptibility of antiferromagnetic MnO. The component parallel to the magnetic spins is much smaller than those measured in the perpendicular directions.

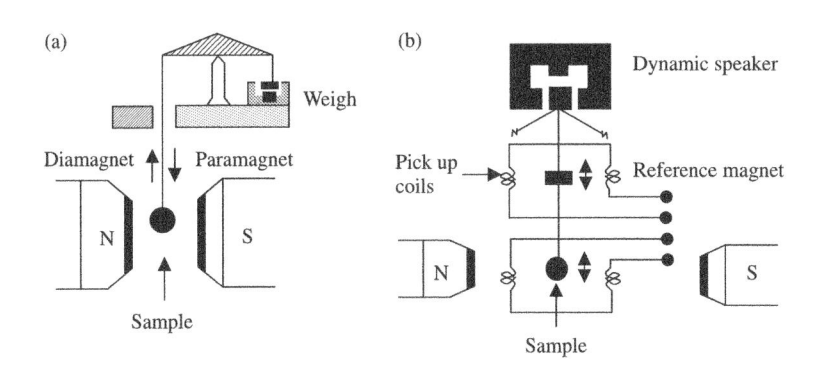

Fig. 14.13 Two methods of measuring magnetic susceptibility. The Guoy Balance (a) measures the weight change of a sample placed in a magnetic field gradient. The vibrating sample magnetometer (b) compares the voltage generated in pick-up coils positioned next to a vibrating reference magnet with that of the vibrating sample.

in the antiferromagnetic state, there is marked anisotropy with the susceptibility perpendicular to the spins much greater than the parallel component. The large susceptibility for $\chi_\perp$ comes from the reorientation of Mn spins. For polycrystalline MnO, the susceptibility averages in the same way as the dielectric constant: $\langle \chi \rangle = (2\chi_\perp + \chi_\parallel)/3$.

14.8 Susceptibility measurements

Two of the ways of measuring magnetic susceptibility in diamagnetic, paramagnetic and antiferromagnetic substances are illustrated in Fig. 14.13. The magnetic Guoy balance involves weighing the sample in a magnetic field gradient, dH/dZ. Diamagnetic samples move out of the field and weigh less when the field is switched on. Paramagnetic solids are attracted to the field and gain weight. The magnitude of the weight change is a measure of the induced magnetization.

The vibrating sample magnetometer (Fig. 14.13(b)) is another commonly-used method. The sample and a reference magnet are attached to a vibrating rod. The sample is positioned between the poles of a large magnet with pick-up coils nearby. Coils are also placed near the vibrating reference magnet. AC voltages are generated in both sets of coils as the magnetic lines of force intersect the coils. The induced magnetization in the sample is obtained by comparing the two voltages.

Fig. 14.14 The direct and converse magnetoelectric effects. In the electrically-induced effect (ME)$_E$, an electric voltage is applied to the sample and a magnetic response is measured with a pick-up coil. In the magnetically-induced magnetoelectric effect (ME)$_H$, a current is passed through a coil wrapped around the sample. The resulting electric polarization creates a voltage across electrodes attached to the sample.

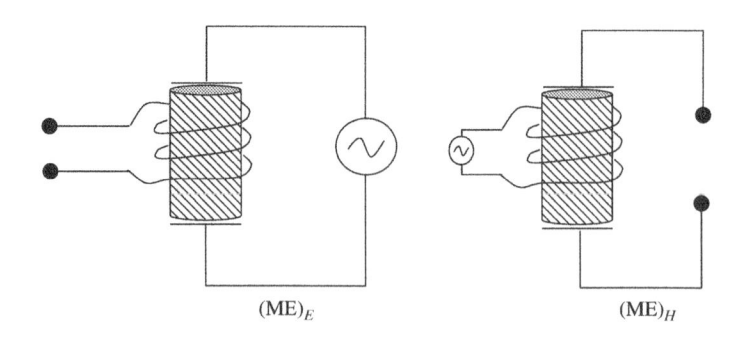

(ME)$_E$ (ME)$_H$

14.9 Magnetoelectricity

The magnetoelectric effect is especially interesting because it involves an unexpected coupling between electric and magnetic variables that does not normally appear in Maxwell's Equations. For Cr_2O_3 and other magnetoelectric materials, an applied magnetic field not only induces magnetization, but electric polarization as well (Fig. 14.14). The converse phenomenon, electrically-induced magnetization, is thermodynamically related to the direct effect.

The induced magnetization I_i [Wb/m^2] is linearly proportional to the applied electric field E_j [V/m] through the magnetoelectric coefficients Q_{ij} [Wb/V m]. Since E is a polar first rank tensor and I is an axial first rank tensor, the magnetoelectric effect is an axial second rank tensor which transforms as follows.

$$I'_i = \pm |a| a_{ij} I_j = \pm |a| a_{ij} Q_{jk} E_k$$
$$= \pm |a| a_{ij} Q_{jk} a_{lk} E'_l = Q'_{il} E'_l$$
$$Q'_{il} = \pm |a| a_{ij} a_{lk} Q_{jk}.$$

In matrix form, the linear magnetoelectric effect transforms in a similar way in going from the old (unprimed) system to the new (primed) system.

$$\begin{array}{cccccc} 3 \times 1 & & 3 \times 3 & 3 \times 1 & 3 \times 3 & 3 \times 3 & 3 \times 1 \\ (I') & = \pm |a| & (a) & (I) & = \pm |a| & (a) & (Q) & (E) \end{array}$$

$$\begin{array}{ccccccc} & & 3 \times 3 & 3 \times 3 & 3 \times 3 & 3 \times 1 & 3 \times 3 & 3 \times 1 \\ & = \pm |a| & (a) & (Q) & (a)_t & (E') & = (Q') & (E') \end{array}$$

$$\begin{array}{ccccc} 3 \times 3 & & 3 \times 3 & 3 \times 3 & 3 \times 3 \\ (Q') & = \pm |a| & (a) & (Q) & (a)_t. \end{array}$$

The magnetoelectric effect vanishes for all symmetry groups containing time reversal symmetry (1′).

$$(Q') = (-1)(+1)(+1)(Q)(+1) = (-Q) = (Q) = 0.$$

Therefore it only occurs in magnetic point groups. The effect also disappears for ordinary inversion ($\bar{1}$) operations:

$$(Q') = (+1)(-1)(-1)(Q)(-1) = (-Q) = (Q) = 0.$$

For space inversion accompanied by time inversion ($\bar{1}'$), the magnetoelectric effect is permitted.

$$(Q') = (-1)(-1)(-1)(Q)(-1) = (Q) = (Q).$$

The magnetoelectric effect was first observed experimentally in antiferromagnetic chromium oxide. Below $T_N = 307$ K, Cr_2O_3 belongs to magnetic point group $\bar{3}'m'$ (see Figs. 14.6 and 14.7). The generating elements are a mirror plane accompanied by time reversal (m') perpendicular to Z_1 and a threefold roto-inversion axis along Z_3, again accompanied by time reversal ($\bar{3}'$). Applying Neumann's Principle, first for $m' \perp Z_1$:

$$\begin{pmatrix} Q'_{11} & Q'_{12} & Q'_{13} \\ Q'_{21} & Q'_{22} & Q'_{23} \\ Q'_{31} & Q'_{32} & Q'_{33} \end{pmatrix} = (-1)(-1) \begin{pmatrix} -1 & 0 & 0 \\ 0 & 1 & 0 \\ 0 & 0 & 1 \end{pmatrix} \begin{pmatrix} Q_{11} & Q_{12} & Q_{13} \\ Q_{21} & Q_{22} & Q_{23} \\ Q_{31} & Q_{32} & Q_{33} \end{pmatrix}$$

$$\times \begin{pmatrix} -1 & 0 & 0 \\ 0 & 1 & 0 \\ 0 & 0 & 1 \end{pmatrix}$$

$$= \begin{pmatrix} Q_{11} & -Q_{12} & -Q_{13} \\ -Q_{21} & Q_{22} & Q_{23} \\ -Q_{31} & Q_{32} & Q_{33} \end{pmatrix} = \begin{pmatrix} Q_{11} & Q_{12} & Q_{13} \\ Q_{21} & Q_{22} & Q_{23} \\ Q_{31} & Q_{32} & Q_{33} \end{pmatrix}.$$

This equality can be satisfied if $Q_{12} = Q_{13} = Q_{21} = Q_{31} = 0$. Next the matrix is transformed under $\bar{3}' \parallel Z_3$.

$$\begin{pmatrix} Q'_{11} & Q'_{12} & Q'_{13} \\ Q'_{21} & Q'_{22} & Q'_{23} \\ Q'_{31} & Q'_{32} & Q'_{33} \end{pmatrix} = (-1)(-1) \begin{pmatrix} \frac{1}{2} & -\frac{\sqrt{3}}{2} & 0 \\ \frac{\sqrt{3}}{2} & \frac{1}{2} & 0 \\ 0 & 0 & -1 \end{pmatrix} \begin{pmatrix} Q_{11} & 0 & 0 \\ 0 & Q_{22} & Q_{23} \\ 0 & Q_{32} & Q_{33} \end{pmatrix}$$

$$\times \begin{pmatrix} \frac{1}{2} & \frac{\sqrt{3}}{2} & 0 \\ -\frac{\sqrt{3}}{2} & \frac{1}{2} & 0 \\ 0 & 0 & -1 \end{pmatrix}$$

$$= \begin{pmatrix} (\frac{1}{4}Q_{11} + \frac{3}{4}Q_{22}) & (-\frac{\sqrt{3}}{4}Q_{11} + \frac{\sqrt{3}}{4}Q_{22}) & (\frac{\sqrt{3}}{2}Q_{23}) \\ (\frac{\sqrt{3}}{4}Q_{11} - \frac{\sqrt{3}}{4}Q_{22}) & (\frac{3}{4}Q_{11} + \frac{1}{4}Q_{22}) & (\frac{-1}{2}Q_{23}) \\ (\frac{\sqrt{3}}{2}Q_{32}) & (-\frac{1}{2}Q_{32}) & (Q_{33}) \end{pmatrix}$$

$$= \begin{pmatrix} Q_{11} & 0 & 0 \\ 0 & Q_{22} & Q_{23} \\ 0 & Q_{32} & Q_{33} \end{pmatrix}$$

that can be satisfied only if $Q_{11} = Q_{22}$ and $Q_{23} = Q_{32} = 0$. Therefore the magnetoelectric matrix for point group $\bar{3}'m'$ is

$$\begin{pmatrix} Q_{11} & 0 & 0 \\ 0 & Q_{11} & 0 \\ 0 & 0 & Q_{33} \end{pmatrix}.$$

Magnetoelectric matrices for other magnetic point groups are listed in Table 14.4. Only 58 of the 90 magnetic point groups are magnetoelectric.

Magnetoelectric coefficients of chromium oxide crystals are shown in Fig. 14.15. Note that Q_{33} and Q_{11} go to zero at the Neel point where

Table 14.4 Magnetoelectric matrices for the 90 magnetic crystal classes and 14 magnetic Curie groups

$1, \bar{1}'$
$$\begin{pmatrix} Q_{11} & Q_{12} & Q_{13} \\ Q_{21} & Q_{22} & Q_{23} \\ Q_{31} & Q_{32} & Q_{33} \end{pmatrix}$$

$2, m', 2/m'$
$$\begin{pmatrix} Q_{11} & 0 & Q_{13} \\ 0 & Q_{22} & 0 \\ Q_{31} & 0 & Q_{33} \end{pmatrix}$$

$2', m, 2'/m$
$$\begin{pmatrix} 0 & Q_{12} & 0 \\ Q_{21} & 0 & Q_{23} \\ 0 & Q_{32} & 0 \end{pmatrix}$$

$222, m'm'2, m'm'm'$
$$\begin{pmatrix} Q_{11} & 0 & 0 \\ 0 & Q_{22} & 0 \\ 0 & 0 & Q_{33} \end{pmatrix}$$

$22'2', 2mm, m'm2', m'mm$
$$\begin{pmatrix} 0 & 0 & 0 \\ 0 & 0 & Q_{23} \\ 0 & Q_{32} & 0 \end{pmatrix}$$

$3, \bar{3}', 4, \bar{4}', 4/m', 6, \bar{6}', 6/m', \infty, \infty/m'$
$$\begin{pmatrix} Q_{11} & Q_{12} & 0 \\ -Q_{12} & Q_{11} & 0 \\ 0 & 0 & Q_{33} \end{pmatrix}$$

$4', \bar{4}, 4'/m'$
$$\begin{pmatrix} Q_{11} & Q_{12} & 0 \\ Q_{12} & -Q_{11} & 0 \\ 0 & 0 & 0 \end{pmatrix}$$

$32, 3m', \bar{3}'m', 422, 4m'm', \bar{4}'2m', 4/m'm'm', 622, 6m'm', \bar{6}'m'2,$
$6/m'm'm', \infty 2, \infty/m'm'm', \infty m'$
$$\begin{pmatrix} Q_{11} & 0 & 0 \\ 0 & Q_{11} & 0 \\ 0 & 0 & Q_{33} \end{pmatrix}$$

$4'22, 4'mm', \bar{4}2m, \bar{4}2'm', 4'/m'mm'$
$$\begin{pmatrix} Q_{11} & 0 & 0 \\ 0 & -Q_{11} & 0 \\ 0 & 0 & 0 \end{pmatrix}$$

$32', 3m, \bar{3}'m, 42'2', 4mm, \bar{4}'2'm, 4/m'mm, 62'2', 6mm, \bar{6}'m2',$
$6/m'mm, \infty 2', \infty/m'm$
$$\begin{pmatrix} 0 & Q_{12} & 0 \\ -Q_{12} & 0 & 0 \\ 0 & 0 & 0 \end{pmatrix}$$

$23, m'3, 432, \bar{4}'3m', m'3m', \infty\infty, \infty\infty m'$
$$\begin{pmatrix} Q_{11} & 0 & 0 \\ 0 & Q_{11} & 0 \\ 0 & 0 & Q_{11} \end{pmatrix}$$

Other magnetic groups
$$\begin{pmatrix} 0 & 0 & 0 \\ 0 & 0 & 0 \\ 0 & 0 & 0 \end{pmatrix}$$

long-range order in the magnetic structure disappears. Antiferromagnetic crystals like Cr_2O_3 contain antiferromagnetic domains, similar to ferromagnetic substances. It is therefore necessary to "pole" the crystals in parallel magnetic and electric fields before measuring the magnetoelectric coefficients of a single domain. This point will be discussed further in Section 16.8. The antiferromagnetic domains in Cr_2O_3 are illustrated in Fig. 14.15.

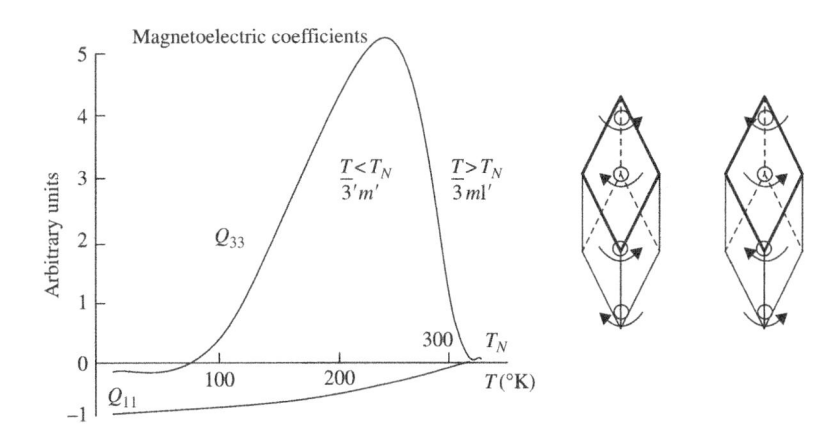

Fig. 14.15 Magnetoelectric effect in Cr_2O_3 single-domain single crystals. The two domain states are related by the time reversal operator $1'$. Annealing in parallel E and H fields are required to produce the single domain state.

The magnetoelectric poling process has also been carried out on ceramic specimens of Cr_2O_3 by applying electric and magnetic fields. The fields are applied to the ceramic as it cools through the Neel point. The poling fields are then removed and the magnetoelectric coefficients are evaluated using either the direct or the converse effect. If the poling fields are parallel the effective magnetic point group is $\infty m'$. If they are perpendicular, the point group is $2'mm'$. Experiments have verified these effects and their relationships to the single crystal value.

The atomistic origin of the magnetoelectric effect is illustrated in Fig. 14.16. Consider an electrically-induced magnetoelectric effect in an antiferromagnetic crystal. Initially, with no applied field, both the net polarization and the net magnetization are zero. When a field is applied parallel to the chain direction, the positive ions move in the field direction and negative ions in the opposite direction, creating electric polarization. For the configuration in Fig. 14.16, the cation and anion move closer together in one pair, and further apart in the other. The resulting increase or decrease in electron overlap will effect the electron orbital motion in the cations, changing their magnetic moments. Since "up" moments are differently affected than "down" moments, the net magnetization is no longer zero. Thus an applied electric field induces a small magnetization, the magnetoelectric effect.

Note that the model structure in Fig. 14.16 possesses a center of symmetry accompanied by time reversal accompanied by time reversal ($\bar{1}'$). As pointed out earlier, magnetoelectricity is forbidden in magnetic materials with ordinary spatial inversion ($\bar{1}$). This means that ferromagnetic materials like α-Fe and Co do not exhibit magnetoelectricity. The symmetry restriction is similar to that for piezoelectricity since all centric magnetic groups are nonmagnetoelectric.

Measurements have been reported for about thirty magnetoelectric materials. Coupling coefficients are small, limiting interest in device development, but composite magnetoelectrics offer some interesting possibilities.

Sintered diphasic ceramics containing grains of a poled ferroelectric in intimate contact with grains of a magnetized ferromagnet exhibit surprisingly large magnetoelectric coefficients. Under an electric field the ferroelectric grains change shape through the piezoelectric effect. This shape change distorts the neighboring magnetic grains causing a changed in magnetization though the converse magnetostriction effect. Measurements on $BaTiO_3$–$CoFe_2O_4$

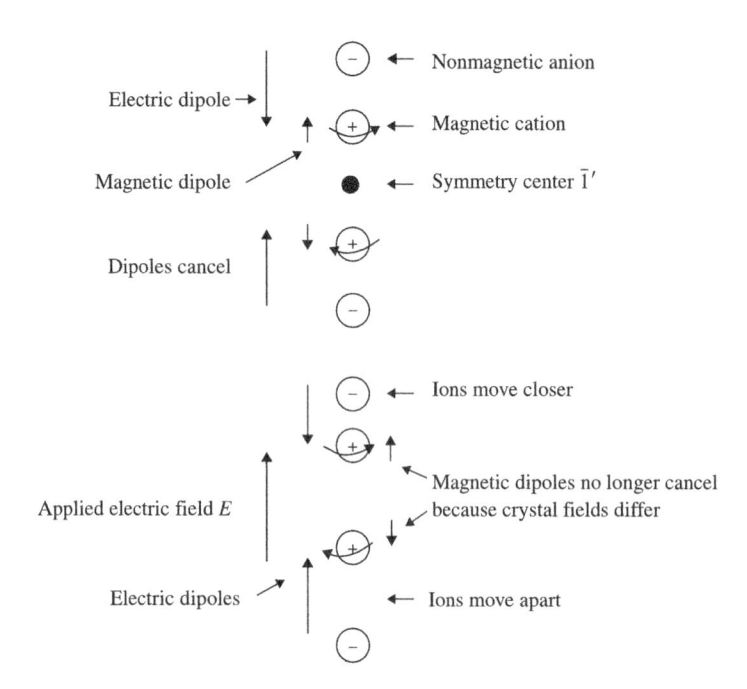

Fig. 14.16 One-dimensional model of the magnetoelectric effect in antiferromagnets.

composites gave larger magnetostriction coefficients than Cr_2O_3. A converse magnetoelectric effect takes place under a magnetic field.

Problem 14.3
Poled ferroelectric ceramics have the same symmetry as an electric vector. Magnetized ferrites possess the symmetry of a magnetic field. Using Curie's Principle of symmetry superposition, determine the symmetry group of the magnetoelectric composite (a) when the two poling fields are parallel, and (b) when they are perpendicular. What are the appropriate magnetoelectric matrices for these two geometries? Explain how the coefficients should be measured.

14.10 Piezomagnetism

Piezomagnetism is the magnetic analog to piezoelectricity. The direct piezomagnetic effect is a linear relation between magnetization I and mechanical stress X:
$$I_i \, [\text{Wb/m}^2] = Q_{ijk} \, [\text{Wb/N}] X_{jk} \, [\text{N/m}^2].$$
The converse effect relates strain x to magnetic field H.
$$x_{ij} \, [\,] = Q_{ijk} \, [\text{m/A}] H_k \, [\text{A/m}].$$
The piezomagnetic coefficients transform as a third rank axial tensor.
$$I'_i = \pm |a| a_{ij} I_j = \pm |a| a_{ij} Q_{jkl} X_{kl}$$
$$= \pm |a| a_{ij} a_{mk} a_{nl} Q_{jkl} X'_{mn} = Q'_{imn} X'_{mn}$$
$$Q'_{imn} = \pm |a| a_{ij} a_{mk} a_{nl} Q_{jkl}.$$

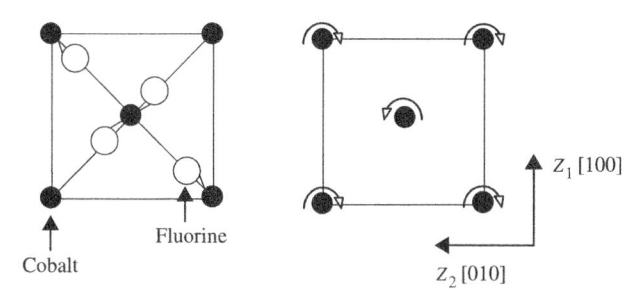

Fig. 14.17 Cobalt fluoride crystal structure and magnetic structure below $T_N = 38$ K.

In matrix form, the transformation involves the $6 \times 6 \ \alpha$ matrix from Table 10.1.

$$\underset{(I')}{3 \times 1} = \pm|a| \ \underset{(a)}{3 \times 3} \ \underset{(I)}{3 \times 1} = \pm|a| \ \underset{(a)}{3 \times 3} \ \underset{(Q)}{3 \times 6} \ \underset{(X)}{6 \times 1}$$

$$= \pm|a| \ \underset{(a)}{3 \times 3} \ \underset{(Q)}{3 \times 6} \ \underset{(\alpha)^{-1}}{6 \times 6} \ \underset{(X')}{6 \times 1} = \underset{(Q')}{3 \times 6} \ \underset{(X')}{6 \times 1}$$

$$\underset{(Q')}{3 \times 6} = \pm|a| \ \underset{(a)}{3 \times 3} \ \underset{(Q)}{3 \times 6} \ \underset{(\alpha)^{-1}}{6 \times 6}.$$

Problem 14.4
What is Q_{113} in matrix form? Are the two equal in magnitude? Describe how they might be measured experimentally.

The piezomagnetic 3×6 matrix is similar in form to the 3×6 piezoelectric matrix but the symmetry requirements are entirely different because of its axial nature. Like the magnetoelectric effect, the piezomagnetic effect disappears whenever the time reversal element is present. Therefore the normal crystallographic point groups do not show the effect. For time reversal ($1'$):

$$(Q') = (-1)(+1)(+1)(Q)(+1) = -(Q) = (Q) = 0.$$

Piezomagnetism has been demonstrated in cobalt fluoride (CoF_2). The crystal structure and magnetic structure are illustrated in Fig. 14.17. CoF_2 is antiferromagnetic and piezomagnetic below 38 K with Co spins aligned along the $[001] = Z_3$ axis. The crystal structure is isomorphous with rutile (TiO_2), tetragonal point group $4/mmm1'$. Below T_N the magnetic point group is $4'/mmm'$ as determined by neutron diffraction.

We now proceed to derive the piezomagnetic matrix for magnetic group $4'/mmm'$ using tensor notation and the direct inspection method. The three symmetry elements that generate the group are $4' \parallel Z_3$, $m \perp Z_3$, and $m \perp Z_1$. The $4'$ axis along Z_3 consists of fourfold rotation combined with time reversal $1'$. By direct inspection coefficient Q_{111} undergoes the following transformations.

$$111 \xrightarrow{4} 222 \xrightarrow{1'} -222 \xrightarrow{4} 111 \xrightarrow{1'} -111 \to 0 \quad \therefore Q_{111} = Q_{222} = 0$$

or in matrix terms, $Q_{11} = Q_{22} = 0$.

$$333 \xrightarrow{4} 333 \xrightarrow{1'} -333 \to 0 \quad \therefore Q_{333} = Q_{33} = 0.$$

$$122 \xrightarrow{4} 211 \xrightarrow{1'} -211 \xrightarrow{4} 122 \xrightarrow{1'} -122 \to 0 \quad \therefore Q_{122} = Q_{211} = 0,$$

$$Q_{12} = Q_{21} = 0.$$

In a similar way it can be shown that $Q_{13} = Q_{23} = 0$, $Q_{16} = Q_{26} = 0$, $Q_{14} = Q_{25}$, $Q_{31} = -Q_{32}$, and $Q_{15} = -Q_{24}$.

Therefore for point group $4'$,

$$(Q) = \begin{pmatrix} 0 & 0 & 0 & Q_{14} & Q_{15} & 0 \\ 0 & 0 & 0 & -Q_{15} & Q_{14} & 0 \\ Q_{31} & -Q_{31} & 0 & 0 & 0 & Q_{36} \end{pmatrix}.$$

The mirror plane m perpendicular to Z_3 does not change this matrix but the mirror perpendicular to $[100] = Z_1$ leads to further simplification. For a mirror operation the handedness changes so $|a| = -1$, and $1 \to 1$, $2 \to -2$, and $3 \to -3$. For the four remaining coefficients,

$$14 = 123 \to 123$$
$$15 = 113 \to -113 \quad \therefore Q_{15} = 0$$
$$31 = 311 \to -311 \quad \therefore Q_{31} = 0$$
$$36 = 312 \to 312.$$

The final result for point group $4'/mmm'$ is

$$(Q) = \begin{pmatrix} 0 & 0 & 0 & Q_{14} & 0 & 0 \\ 0 & 0 & 0 & 0 & Q_{14} & 0 \\ 0 & 0 & 0 & 0 & 0 & Q_{36} \end{pmatrix}.$$

Experiments on cobalt fluoride crystals at low temperatures gave

$$Q_{14} = 2.7 \times 10^{-11} \text{ Wb/N} \quad \text{and} \quad Q_{36} = 1.0 \times 10^{-11} \text{ Wb/N}.$$

Problem 14.5
A tensile stress is applied to a CoF_2 crystal in an arbitrary direction specified by spherical coordinates θ and ϕ. What are the resulting components of magnetization along [100] and [001]?

Piezomagnetic matrices for other magnetic point groups are listed in Table 14.5. Of the 90 magnetic groups 66 are piezomagnetic along with 5 of the 14 magnetic Curie groups. Thirty-five of the 66 piezomagnetic groups are antiferromagnetic, 31 are not.

Table 14.5 Piezomagnetic matrices for the various magnetic symmetry groups

$1, \bar{1}$	$\begin{pmatrix} Q_{11} & Q_{12} & Q_{13} & Q_{14} & Q_{15} & Q_{16} \\ Q_{21} & Q_{22} & Q_{23} & Q_{24} & Q_{25} & Q_{26} \\ Q_{31} & Q_{32} & Q_{33} & Q_{34} & Q_{35} & Q_{36} \end{pmatrix}$
$2, m, 2/m$	$\begin{pmatrix} 0 & 0 & 0 & Q_{14} & 0 & Q_{16} \\ Q_{21} & Q_{22} & Q_{23} & 0 & Q_{25} & 0 \\ 0 & 0 & 0 & Q_{34} & 0 & Q_{36} \end{pmatrix}$
$2', m', 2'/m'$	$\begin{pmatrix} Q_{11} & Q_{12} & Q_{13} & 0 & Q_{15} & 0 \\ 0 & 0 & 0 & Q_{24} & 0 & Q_{26} \\ Q_{31} & Q_{32} & Q_{33} & 0 & Q_{35} & 0 \end{pmatrix}$
$222, mm2, mmm$	$\begin{pmatrix} 0 & 0 & 0 & Q_{14} & 0 & 0 \\ 0 & 0 & 0 & 0 & Q_{25} & 0 \\ 0 & 0 & 0 & 0 & 0 & Q_{36} \end{pmatrix}$

Continued overleaf

Table 14.5 (*Continued*)

Magnetic groups	Matrix
$2'2'2, m'm'2, m'2'm, m'm'm$	$\begin{pmatrix} 0 & 0 & 0 & 0 & Q_{15} & 0 \\ 0 & 0 & 0 & Q_{24} & 0 & 0 \\ Q_{31} & Q_{32} & Q_{33} & 0 & 0 & 0 \end{pmatrix}$
$3, \bar{3}$	$\begin{pmatrix} Q_{11} & -Q_{11} & 0 & Q_{14} & Q_{15} & -2Q_{22} \\ -Q_{22} & Q_{22} & 0 & Q_{15} & -Q_{14} & -2Q_{11} \\ Q_{31} & Q_{31} & Q_{33} & 0 & 0 & 0 \end{pmatrix}$
$32, 3m, \bar{3}m$	$\begin{pmatrix} Q_{11} & -Q_{11} & 0 & Q_{14} & 0 & 0 \\ 0 & 0 & 0 & 0 & -Q_{14} & -2Q_{11} \\ 0 & 0 & 0 & 0 & 0 & 0 \end{pmatrix}$
$32', 3m', \bar{3}m'$	$\begin{pmatrix} 0 & 0 & 0 & 0 & Q_{15} & -2Q_{22} \\ -Q_{22} & Q_{22} & 0 & Q_{15} & 0 & 0 \\ Q_{31} & Q_{31} & Q_{33} & 0 & 0 & 0 \end{pmatrix}$
$4, \bar{4}, 4/m, 6, \bar{6}, 6/m, \infty, \infty/m$	$\begin{pmatrix} 0 & 0 & 0 & Q_{14} & Q_{15} & 0 \\ 0 & 0 & 0 & Q_{15} & -Q_{14} & 0 \\ Q_{31} & Q_{31} & Q_{33} & 0 & 0 & 0 \end{pmatrix}$
$4', \bar{4}', 4'/m$	$\begin{pmatrix} 0 & 0 & 0 & Q_{14} & Q_{15} & 0 \\ 0 & 0 & 0 & -Q_{15} & Q_{14} & 0 \\ Q_{31} & -Q_{31} & 0 & 0 & 0 & Q_{36} \end{pmatrix}$
$422, 4mm, \bar{4}2m, 4/mmm, 622, 6mm, \bar{6}m2,$ $6/mmm, \infty 2$	$\begin{pmatrix} 0 & 0 & 0 & Q_{14} & 0 & 0 \\ 0 & 0 & 0 & 0 & -Q_{14} & 0 \\ 0 & 0 & 0 & 0 & 0 & 0 \end{pmatrix}$
$4'22', 4'mm', \bar{4}'2m', \bar{4}'2'm, 4'/mmm'$	$\begin{pmatrix} 0 & 0 & 0 & Q_{14} & 0 & 0 \\ 0 & 0 & 0 & 0 & Q_{14} & 0 \\ 0 & 0 & 0 & 0 & 0 & Q_{36} \end{pmatrix}$
$42'2', 4m'm', \bar{4}2'm', 4/mm'm', 62'2',$ $6m'm', \bar{6}m'2', 6/mm'm', \infty 2', \infty/mm'$	$\begin{pmatrix} 0 & 0 & 0 & 0 & Q_{15} & 0 \\ 0 & 0 & 0 & Q_{15} & 0 & 0 \\ Q_{31} & Q_{31} & Q_{33} & 0 & 0 & 0 \end{pmatrix}$
$6', \bar{6}', 6'/m'$	$\begin{pmatrix} Q_{11} & -Q_{11} & 0 & 0 & 0 & -2Q_{22} \\ -Q_{22} & Q_{22} & 0 & 0 & 0 & -2Q_{11} \\ 0 & 0 & 0 & 0 & 0 & 0 \end{pmatrix}$
$6'22', 6'mm', \bar{6}'m'2, \bar{6}'m2', 6'/m'mm'$	$\begin{pmatrix} Q_{11} & -Q_{11} & 0 & 0 & 0 & 0 \\ 0 & 0 & 0 & 0 & 0 & -2Q_{11} \\ 0 & 0 & 0 & 0 & 0 & 0 \end{pmatrix}$
$23, m3, 4'32, \bar{4}'3m', m3m'$	$\begin{pmatrix} 0 & 0 & 0 & Q_{14} & 0 & 0 \\ 0 & 0 & 0 & 0 & Q_{14} & 0 \\ 0 & 0 & 0 & 0 & 0 & Q_{14} \end{pmatrix}$
All other magnetic groups	$\begin{pmatrix} 0 & 0 & 0 & 0 & 0 & 0 \\ 0 & 0 & 0 & 0 & 0 & 0 \\ 0 & 0 & 0 & 0 & 0 & 0 \end{pmatrix}$

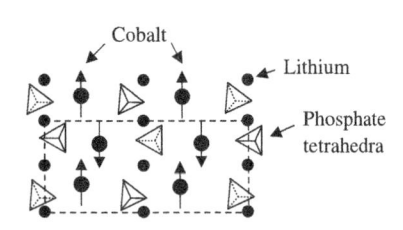

Fig. 14.18 Structure of antiferromagnetic LiCoPO$_4$. Cobalt spins are aligned along Z_3 below the Neel temperature of 25 K.

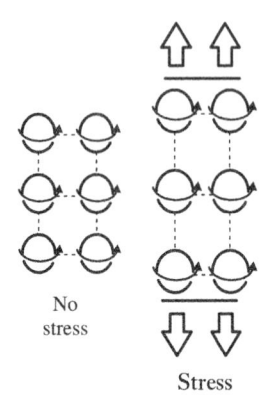

Fig. 14.19 A simple model representing the intrinsic piezomagnetic effect under mechanical stress, the changes in interatomic distances lead to changes in the local crystal fields and the magnetic exchange interactions. These electronic effects create a stress-induced change in magnetization. The converse effect is a field-induced mechanical strain. Piezomagnetic strains are generally small compared to magnetostrictive strains. Linear effects are usually larger than higher order effects but it is not true in this case.

Problem 14.6
Orthorhombic LiCoPO$_4$(Fig. 14.18) belongs to magnetic point group mmm'. What other symmetry elements are present in this point group? Using Neumann's Law and either tensor or matrix transformations, derive the pyromagnetic, magnetoelectric, piezomagnetic, and magnetic susceptibility matrices for this point group.

The piezomagnetic coefficients reported for CoF$_2$ are two orders of magnitude smaller than the effective piezomagnetic coefficients associated with magnetostriction. As explained in Section 15.5 the linearized portion of the magnetostriction effect has piezomagnetic coefficients in the 10^{-9} to 10^{-8} m/A (=Wb/N) range. Magnetostrictive strains are caused by rotation of the magnetization vector and by domain wall movements. They are much larger than the intrinsic piezomagnetic effect pictured in Fig. 14.19.

14.11 Summary

Four magnetic properties were discussed in this chapter: Linear relationships between magnetization and temperature (pyromagnetism), magnetization and magnetic field (magnetic susceptibility), magnetization and electric field (magnetoelectricity), and magnetization and mechanical stress (piezomagnetism). Pyromagnetism, magnetoelectricity, and piezomagnetism are axial tensor properties that occur only among the 90 magnetic point groups. All three effects are absent in the nonmagnetic crystallographic groups. Magnetic susceptibility is a polar second rank tensor property found in all crystallographic and magnetic point groups. It is interesting to compare the symmetry requirements for pyromagnetism and piezomagnetism with their electric analogs, pyroelectricity and piezoelectricity. Very similar group–subgroup relationships apply as shown in Fig. 14.20.

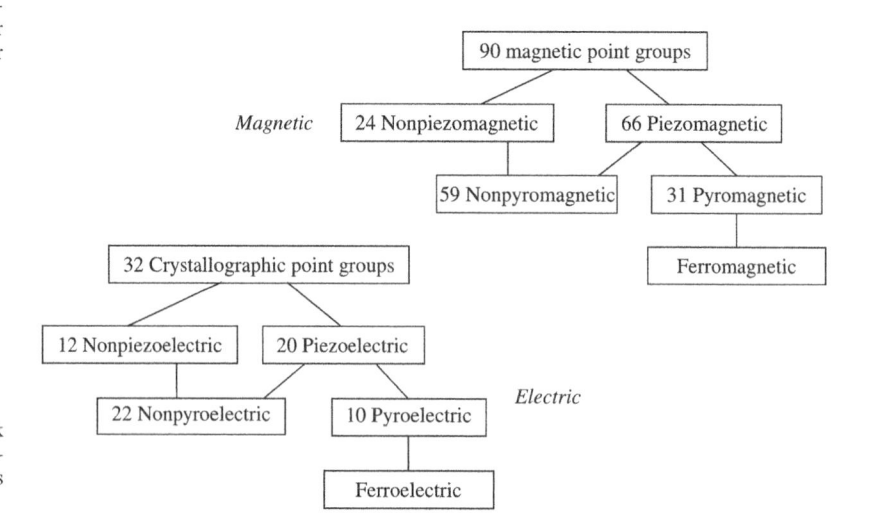

Fig. 14.20 Effect of symmetry on odd-rank polar tensors (pyroelectricity and piezoelectricity) and odd-rank axial tensor properties (pyromagnetism and piezomagnetism).

Nonlinear phenomena 15

The physical properties discussed thus far are linear relationships between two measured quantities. This is only an approximation to the truth, and often not a very good approximation, especially for materials near a phase transformation. A more accurate description can be obtained by introducing higher order coefficients.

To illustrate nonlinearity we discuss electrostriction, magnetostriction, and higher order elastic, and dielectric effects. These phenomena are described in terms of fourth and sixth rank tensors.

15.1	Nonlinear dielectric properties	147
15.2	Nonlinear elastic properties	148
15.3	Electrostriction	151
15.4	Magnetostriction	153
15.5	Modeling magnetostriction	154
15.6	Magnetostrictive actuators	159
15.7	Electromagnetostriction and pseudopiezoelectricity	160

15.1 Nonlinear dielectric properties

Many of the recent innovations in the field of electroceramics have exploited the nonlinearities of material properties with factors such as electric field, mechanical stress, temperature, or frequency. The nonlinear dielectric behavior of ferroelectric ceramics (Fig. 15.1), for example, has opened up new markets in electronics and communications. In these materials the electric polarization saturates under high fields. Electric displacement D_i varies with applied electric field E_j as

$$D_i = \varepsilon_{ij}E_j + \varepsilon_{ijk}E_jE_k + \varepsilon_{ijkl}E_jE_kE_l + \cdots,$$

where ε_{ij} is the dielectric permittivity and ε_{ijk} and ε_{ijkl} are higher order terms. The data in Fig. 15.1 were collected for a relaxor ferroelectric in its paraelectric state above T_c where the symmetry is centrosymmetric. Therefore the third rank

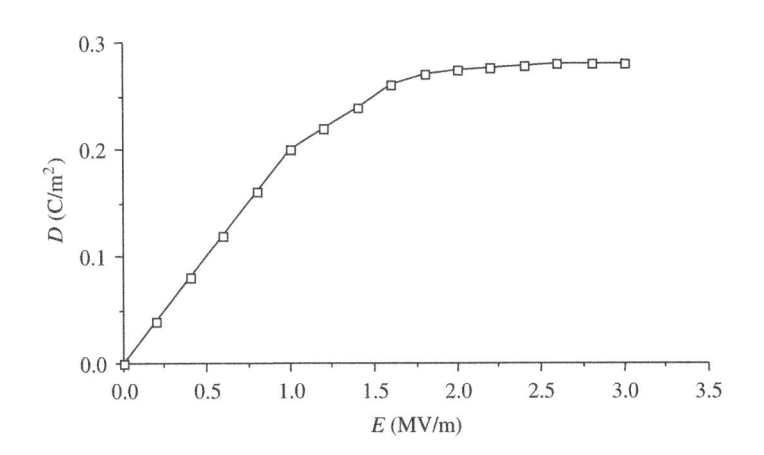

Fig. 15.1 Nonlinear dielectric behavior for the relaxor ferroelectric 0.9PMN–0.1PT [Pb(Mg$_{0.3}$Nb$_{0.6}$Ti$_{0.1}$)O$_3$] in its cubic paraelectric state. The electric permittivity is the slope of this curve. Under a bias field the dielectric constant drops to much lower values.

tensor ε_{ijk} is zero, and the shape of the curve is largely controlled by the first and third terms. For cubic crystals, the fourth rank tensor ε_{ijkl} is similar in form to the elastic constants discussed in Chapter 13.

Tunable microwave devices utilize nonlinear dielectrics in which the polarization saturates as in Fig. 15.1. By applying a DC bias the dielectric constant can be adjusted over a wide range.

Problem 15.1
Polycrystalline PMN ceramics contain randomly oriented grains conforming to Curie group $\infty\infty m$. Using the data in Fig. 15.1, derive approximate values for the linear and nonlinear dielectric constants.

15.2 Nonlinear elastic properties

In a similar way, the nonlinear properties of elastomers and other polymers have been exploited in composite structures. In natural vulcanized rubber (Fig. 15.2), the strain saturates as stress is increased. Linear stress–strain behavior conforming to Hooke's Law is observed only in regions of low stress. Nonlinear elastic behavior in general may be described by

$$X_{ij} = c_{ijkl} x_{kl} + c_{ijklmn} x_{kl} x_{mn} + \cdots,$$

where X_{ij} is stress, x_{kl} strain, c_{ijkl} the elastic stiffness, and c_{ijklmn} the higher order elastic constants.

c_{ijklmn} is a sixth rank tensor which transforms as the product of six direction cosines:

$$c'_{ijklmn} = a_{io} a_{jp} a_{kq} a_{lr} a_{ms} a_{nt} c_{opqrst}.$$

When written in matrix form the linear and nonlinear terms are

$$X_i = c_{ij} x_j + c_{ijk} x_j x_k + \cdots.$$

The c_{ij} and c_{ijk} coefficients have the same dimensions [N/m^2] and are often referred to as the second order and third order elastic constants. Typical values for third order elastic constants are given Table 15.1.

Note that the principal third order stiffnesses (c_{111}) are negative for inorganic crystals. The third order elastic constants of NaCl, Cu, and other inorganic crystals are negative while those of many polymers are positive. This is because

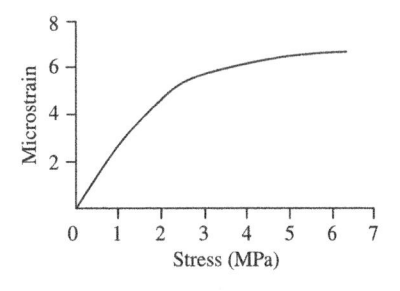

Fig. 15.2 Nonlinear stress–strain behavior of a typical vulcanized rubber. The strain is measured in parts per million.

Table 15.1 Third order elastic constants for rocksalt, germanium, copper, and magnesium (stiffnesses are given in units of 10^{11} N/m^2)

Cubic crystals ($m3m$)	c_{111}	c_{112}	c_{123}	c_{144}	c_{155}	c_{456}
NaCl	−8.43	−0.5	+0.46	+0.29	−0.6	+0.26
Ge	−7.20	−3.80	−0.3	−0.1	−3.05	−0.45
Cu	−13.50	−8.00	−1.2	+0.66	−7.20	−0.32

Hexagonal crystals ($6/mmm$)	c_{111}	c_{112}	c_{113}	c_{123}	c_{133}
Mg	−6.63	−1.78	+0.30	−0.76	−0.86

	c_{144}	c_{155}	c_{222}	c_{333}	c_{344}
	−0.30	−0.58	−8.64	−7.26	−1.93

the nonlinearity has different origins. Consider the crystals first. When a tensile stress (positive sign) is applied to the crystal, the atoms are pulled further apart and the interatomic force is weakened. This reduces the elastic stiffness and when a large compressive force is applied (negative stress) the atoms are pushed close together causing the electron clouds to overlap. This results in repulsive forces that increase the elastic stiffness. Therefore the third order elastic constant is *negative*.

In a polymer the situation is different. Here there are two kinds of forces between atoms: strong covalent bonds within the polymer chain and weak van der Waals bonds between the chains. When a strong tensile force (positive) is applied the randomly oriented chains are rotated into the force direction. This aligns the strong covalent bonds with the stress and leads to a stiffening of the material. Just the opposite occurs when a compressive force (negative) is applied. The chains are now rotated perpendicular to the stress so that now the force acts primarily on the weaker van der Waals bonds. This leads to a decrease in stiffness and a *positive* third order stiffness coefficients. The stiffening under stress is illustrated in Fig. 15.2.

The third order elastic stiffnesses in Table 15.1 are written in the three subscript matrix notation. Conversion from the six subscript tensor form to the three subscript matrix form goes as follows: $11 = 1$, $22 = 2$, $33 = 3$, 23 or $32 = 4$, 31 or $13 = 5$, 12 or $21 = 6$. Therefore $c_{111111} = c_{111}$, $c_{121111} = c_{611}$, $c_{122111} = c_{661}$, etc. As with the second order stiffnesses, there are no factors of 2 or 4 between the third-order tensor and matrix stiffness coefficients.

In general, there are $3^6 = 729$ tensor coefficients, but the coefficients are commutative within the index pairs ($c_{ijklmn} = c_{jiklmn} = \cdots$) and between the index pairs ($c_{ijklmn} = c_{klijmn} = \cdots$). In matrix notation this means that $c_{ijk} = c_{jik} = \cdots$, where i, j, and $k = 1\text{--}6$. Commutation reduces the number of tensor coefficients from 729 to 216, of which 56 are independent. Thus, 56 measurements are needed for a triclinic crystal.

As shown in Table 15.2, a large number of measurements are required even for the most symmetric point groups. A full listing of the matrix coefficients is given in the Landolt–Bornstein Tables.

The most important third order measurements are those carried out on earth materials. The pressure derivatives of the elastic coefficients of minerals control

Table 15.2 The number of independent second and third order elastic constants for the 32 crystallographic point groups

Symmetry	Second order	Third order
Triclinic $(1, \bar{1})$	21	56
Monoclinic $(2, m, 2/m)$	13	32
Orthorhombic $(222, mm2, mmm)$	9	20
Trigonal $(3, \bar{3})$	7	20
$(32, 3m, \bar{3}m)$	6	14
Tetragonal $(4, \bar{4}, 4/m)$	7	16
$(422, 4mm, \bar{4}2m, 4/mmm)$	6	12
Hexagonal $(6, \bar{6}, 6/m)$	5	12
$(\bar{6}m2, 622, 6mm, 6/mmm)$	5	10
Cubic $(23, m3)$	3	8
$(432, \bar{4}3m, m3m)$	3	6

Table 15.3 Comparison of elastic stiffness and their initial pressure derivatives for four oxide minerals (Adiabatic stiffness c_{ij} are expressed in megabars, and the pressure derivatives $\delta c_{ij}/\delta P$ are dimensionless. The point groups are given in parentheses.)

ij	Beryl ($6/mmm$) c_{ij}	$\delta c_{ij}/\delta p$	Quartz (32) c_{ij}	$\delta c_{ij}/\delta p$	Corundum ($\bar{3}m$) c_{ij}	$\delta c_{ij}/\delta p$	Forsterite (mmm) c_{ij}	$\delta c_{ij}/\delta p$
11	2.80	4.5	0.86	3.3	4.98	6.2	3.29	8.3
22	2.80	4.5	0.86	3.3	4.98	6.2	2.01	5.9
33	2.48	3.4	1.07	10.8	5.02	5.0	2.36	6.2
44	0.66	−0.2	0.59	2.7	1.47	2.2	0.67	2.1
55	0.66	−0.2	0.59	2.7	1.47	2.2	0.81	1.7
66	0.91	0.3	0.41	−2.7	1.68	1.5	0.81	2.3
12	0.99	3.9	0.051	8.7	1.63	3.3	0.66	4.3
13	0.67	3.3	0.105	6.0	1.17	3.7	0.68	4.2
23	0.67	3.3	0.105	6.0	1.17	3.7	0.73	3.5
14	0	0	0.18	1.9	−0.23	0.1	0	0

changes in seismic wave velocities deep within the earth, and are strong indicators of the onset of phase transformations and earthquakes. Elastic stiffness coefficients and their initial pressure derivatives for four oxides are listed in Table 15.3.

Three observations can be made: (1) The pressure derivatives are all about 1 to 10 megabars/megabar (dimensionless). (2) Large stiffnesses usually show greater pressure derivatives than small ones: if $c_{11} > c_{22}$, then $(\delta c_{11}/\delta p) > (\delta c_{22}/\delta p)$. (3) Pressure derivatives of the stiffnesses are positive in dense-packed structures but in open structures are occasionally negative. Quartz and beryl each have one negative derivative but the close-packed corundum and forsterite structures show none. When a close-packed structure is compressed, the atoms move closer together but this need not be true in an open structure where rotations can take place. The stiffness coefficients for bending are considerably smaller than for stretching. Thus, rotation can lead to negative pressure dependence of shear stiffness coefficients. Note that the pressure derivatives are positive numbers because pressure is negative.

The second observation, that the largest stiffness coefficients tend to have larger pressure derivatives, can be rationalized as follows. When subjected to very high hydrostatic pressures, the atoms are crowded together, closer and closer, until the electron shells begin to overlap and strong repulsive forces come into play. In this highly compressed state, the overlap is greatest in the directions of strong bonding which are directions of high stiffness. This leads to added stiffening of the strongly bonded directions under compressive forces, so that if $c_{11} > c_{22}$, then $\delta c_{11}/\delta p > \delta c_{22}/\delta p$.

Problem 15.2
Forsterite is very similar to olivine, one of the very common minerals in the upper mantle, where pressures range up to 10 GPa at depths near 400 km. Using the data in Table 15.3, make a plot of c_{11}, c_{22}, and c_{33} as a function of pressure in the earth's mantle.

At the earth's surface the density of forsterite is 3.22 g/cm^3. Estimate its density under pressures of 10 GPa. In the lower mantle forsterite converts to denser crystalline phases with the spinel and perovskite structures.

15.3 Electrostriction

The converse piezoelectric effect is a linear relation between strain and electric field. Electrostriction relates strain to the square of the electric field. Written in tensor form

$$x_{ij} = d_{ijk}E_k + M_{ijkl}E_kE_l.$$

Piezoelectricity is a third rank polar tensor that disappears in centrosymmetric materials, but electrostriction does not. Electrostriction is a fourth rank polar tensor that, like the elastic constants, is present in all point groups. The matrices are listed in Table 15.5.

Electrostriction can also be defined as a quadratic relationship between strain and electric polarization:

$$x_{ij} = Q_{ijkl}P_kP_l.$$

The need for defining a polarization-related electrostriction tensor arises from the fact that ferroelectrics often show nonlinear dielectric properties (see Fig. 15.1). Electrostrictive strain is not a quadratic function of field in relaxor ferroelectrics such as lead magnesium niobate (PMN). Both the polarization and the strain saturate at high field levels (Figs. 15.1 and 15.3(a)) When plotted as a function polarization, however, the electrostrictive strain is proportional to P^2 (Fig. 15.3(b)) showing that the Q coefficients are preferable to M. Electrostrictive strain arises from the field induced polarization. For linear dielectrics, the M and Q coefficients can be used interchangeably.

The electrostriction tensor can be written in matrix form using the following notation. For cubic crystals,

$$\begin{pmatrix} x_1 \\ x_2 \\ x_3 \\ x_4 \\ x_5 \\ x_6 \end{pmatrix} = \begin{pmatrix} Q_{11} & Q_{12} & Q_{12} & 0 & 0 & 0 \\ Q_{12} & Q_{11} & Q_{12} & 0 & 0 & 0 \\ Q_{12} & Q_{12} & Q_{11} & 0 & 0 & 0 \\ 0 & 0 & 0 & Q_{44} & 0 & 0 \\ 0 & 0 & 0 & 0 & Q_{44} & 0 \\ 0 & 0 & 0 & 0 & 0 & Q_{44} \end{pmatrix} \begin{pmatrix} P_1^2 \\ P_2^2 \\ P_3^2 \\ P_2P_3 \\ P_3P_1 \\ P_1P_2 \end{pmatrix}$$

while for ceramics and other isotropic materials, $2Q_{44} = Q_{11} - Q_{12}$ just as it does for the elastic stiffnesses.

Electrostrictive ceramics are used in active optic systems to eliminate vibrations and atmospheric turbulence. In adjusting the position of optical components, electrostrictive ceramics have an advantage over piezoelectric ceramics because there is little or no hysteresis from domain wall motions. The most widely used compositions are modified PMN ceramics, typically $Pb(Mg_{0.3}Nb_{0.6}Ti_{0.1})O_3$ relaxor ferroelectrics, operating in the pseudocubic state above T_c. Relaxor ferroelectrics consist of temperature-sensitive microdomains from the many different active ion linkages in the disordered octahedral framework.

Compared to piezoelectricity, which utilizes an acentric material, electrostrictive transducers make use of a cubic material poised on an instability with microregions fluctuating in polarization. On the average, the atoms reside in the ideal cubic sites but are continually shifting off these positions. An atomic view of the Q_{11}, Q_{12}, and Q_{44} motions in the cubic perovskite structure appear in Fig. 15.4. The underlying origin of these effects, as well as the large nonlinear polarization in Fig. 15.1, is a partial ordering of the octahedral cations.

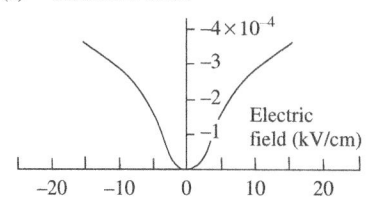

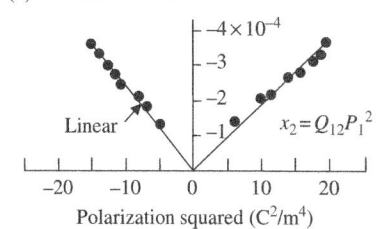

Fig. 15.3 (a) Field-induced transverse strain of PMN–PT $(Pb(Mg_{0.3}Nb_{0.6}Ti_{0.1})O_3)$ ceramic. For small fields, the strain is quadratic but the curve saturates under high fields. (b) The same data plotted as a function of the polarization squared.

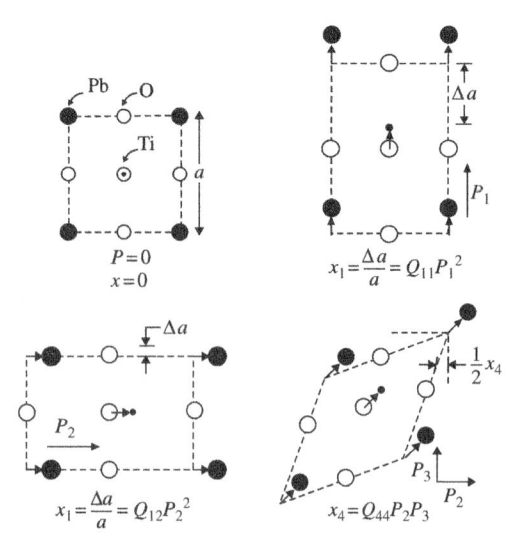

Fig. 15.4 Electrostriction in cubic perovskites, showing the physical basis of the electrostriction coefficients Q_{11}, Q_{12}, and Q_{44}.

Mg and Nb ions alternate in position but only over a very short range, typically 30–50 Å—just a few unit cells. Within these ordered islands there are fluctuating dipoles that produce large polarizations and large electrostrictive motions.

Relaxor ferroelectrics are very common among Pb-based perovskites, suggesting that Pb^{2+} and its "lone-pair" electrons play an important role in the microdomain process, possibly by adjusting the orientation of the lone pair and the associated electric dipole moment.

Electrostriction coefficients for various insulators range over about many orders of magnitude. Q_{11} values for soft elastomers are as high as 10^6 m^4/C^2 while those of relaxor ferroelectrics are only about 10^{-2} m^4/C^2. Other insulators are in between with the Q coefficients roughly proportional to s/ε, the elastic compliance divided by the dielectric permittivity. This can be rationalized in the following way. Since $Q = x/P^2$, and x is large when the material deforms easily (like an elastomer), then $Q \sim s$. The permittivity determines how a material interacts with an electric field. Since both the strain and the polarization are proportional to ε, the Q coefficient will be proportional to $1/\varepsilon$. Hence Q varies as s/ε, and the M coefficient as $s\varepsilon$. The large electrostrictive strain in polyurethane and other elastomers comes from the high elastic compliance, while that of the relaxor ferroelectrics is from the huge values of the dielectric constant and the induced polarization. Both are useful in electrostrictive transducers and actuators.

A variety of experimental methods have been employed to measure the wide range of electrostrictive effects in dielectric materials. The direct and converse electrostriction effects offer three independent ways of measuring the Q coefficients: (a) by measuring the strains induced in the dielectric in response to electrically-induced polarization, (b) measuring the change in permittivity (via the change in capacitance) under mechanical stress, and (c) the change in piezoelectric voltage coefficient resulting from a change in polarization. All three relations are derived from the Gibbs free energy function (Section 6.2).

Interferometric and compressometer measurements on four fluoride crystals are presented in Table 15.4. Elastic compliance coefficients are shown for comparison since both phenomena depend on how easily the crystals deform. The anisotropy plots in Fig. 15.5 show that fluorite crystals distort most easily

Table 15.4 Electrostriction and elastic compliance coefficients for cubic fluoride crystals (the Q values are in m^4/C^2 and the s coefficients in 10^{-11} N/m^2)

Material	Q_{11}	Q_{12}	Q_{44}	s_{11}	s_{12}	s_{44}
CaF$_2$	−0.49	−0.48	2.01	0.69	−0.15	2.99
SrF$_2$	−0.33	−0.39	1.90	0.99	−0.26	3.18
BaF$_2$	−0.31	−0.29	1.48	1.53	−0.47	4.06
KMnF$_3$	0.49	−0.10	1.15	1.21	−0.35	3.75

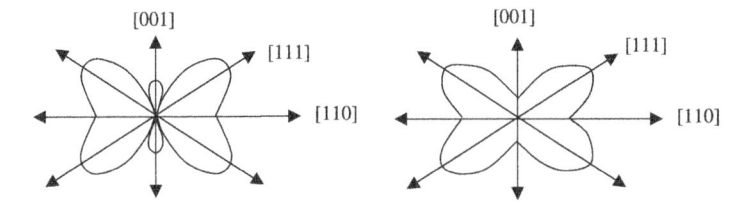

Fig. 15.5 Field- and stress-induced strain in cubic CaF$_2$. Elastic compliance and electrostrictive coefficients show similar anisotropy with the largest coefficients in body diagonal directions.

along ⟨111⟩ directions when driven by mechanical stress or electric field. In the fluorite structure the fluorine ions form continuous chains along ⟨100⟩ axes making the crystal stiffest along the cube axes. Note that the electrostrictive strain is negative in the [100] direction but the elastic strain is not. s_{11} coefficients are required to be positive numbers because the product of stress and strain is an energy density. This argument does not apply to electrostriction.

15.4 Magnetostriction

Magnetostriction is the magnetic analog to electrostriction and is governed by the same type of tensor transformation. The magnetostrictive coefficients N_{ijkl} relate strain x_{ij} to the square of the magnetization $I_k I_l$.

$$x_{ij} = N_{ijkl} I_k I_l.$$

As such, it is a higher order effect to piezomagnetism, but in practical magnetic materials the quadratic magnetostrictive effect is often far larger than the linear piezomagnetic effect. The tensor transformation goes as follows.

$$\begin{aligned} x'_{ij} &= a_{ik} a_{jl} x_{kl} = a_{ik} a_{jl} N_{klmn} I_m I_n \\ &= a_{ik} a_{jl} N_{klmn} (\pm |a| a_{om} I'_o)(\pm |a| a_{pn} I'_p) \\ &= N'_{ijop} I'_o I'_p \end{aligned}$$

$$N'_{ijop} = a_{ik} a_{jl} a_{om} a_{pn} N_{klmn}.$$

Therefore the axial nature of the magnetization cancels out and the magnetostriction coefficients transform as a polar fourth rank tensor. In the contracted matrix form the magnetostriction effect is

$$\begin{pmatrix} x_1 \\ x_2 \\ x_3 \\ x_4 \\ x_5 \\ x_6 \end{pmatrix} = \begin{pmatrix} N_{11} & N_{12} & N_{13} & N_{14} & N_{15} & N_{16} \\ N_{21} & N_{22} & N_{23} & N_{24} & N_{25} & N_{26} \\ N_{31} & N_{32} & N_{33} & N_{34} & N_{35} & N_{36} \\ N_{41} & N_{42} & N_{43} & N_{44} & N_{45} & N_{46} \\ N_{51} & N_{52} & N_{53} & N_{54} & N_{55} & N_{56} \\ N_{61} & N_{62} & N_{63} & N_{64} & N_{65} & N_{66} \end{pmatrix} \begin{pmatrix} I_1^2 \\ I_2^2 \\ I_3^2 \\ I_2 I_3 \\ I_3 I_1 \\ I_1 I_2 \end{pmatrix}$$

The tensor and matrix coefficients are related by $N_{mn} = N_{ijkl}$.

Magnetostriction matrices for the various symmetry groups are given in Table 15.5. The number of measurements required ranges from 36 in the triclinic classes (1 and $\bar{1}$) to 2 in isotropic materials ($\infty\infty$ and $\infty\infty m$). Note that the electrostriction and magnetostriction matrices are not required to be symmetric like the elastic constants. The energy argument does not apply here.

15.5 Modeling magnetostriction

Many properties of ferromagnetic and ferrimagnetic crystals are awkward to describe because the spontaneous magnetization is not tied strongly to the crystal lattice. Strong magnetic fields rotate I_s and eventually it aligns with the magnetic field (Fig. 15.6). This leads to an apparent change in symmetry as the magnetization saturates in the field direction.

The magnetostrictive strain saturates as well, and is characterized by its saturation value λ. As indicated in Fig. 15.7, the strain is mainly caused by rotation rather than domain wall motion. For most magnetic materials the λ values are small, typically 10^{-5}, but larger strains of 10^{-4}–10^{-3} are found in some ceramics and intermetallic compounds like $CoFe_2O_4$ and $TbFe_2$.

To model the rotation effects, we assign a set of direction cosines to the magnetization vector and another set to the direction in which the strain is measured (Fig. 15.8). The direction cosines for the components of the rotated magnetization are

$$I_1 = \alpha_1 I_s, \quad I_2 = \alpha_2 I_s, \quad \text{and} \quad I_3 = \alpha_3 I_s,$$

Table 15.5 Matrices for the electrostriction and magnetostriction coefficients in crystallographic and Curie groups

$1, \bar{1}$	$\begin{pmatrix} N_{11} & N_{12} & N_{13} & N_{14} & N_{15} & N_{16} \\ N_{21} & N_{22} & N_{23} & N_{24} & N_{25} & N_{26} \\ N_{31} & N_{32} & N_{33} & N_{34} & N_{35} & N_{36} \\ N_{41} & N_{42} & N_{43} & N_{44} & N_{45} & N_{46} \\ N_{51} & N_{52} & N_{53} & N_{54} & N_{55} & N_{56} \\ N_{61} & N_{62} & N_{63} & N_{64} & N_{65} & N_{66} \end{pmatrix}$
$2, m, 2/m$	$\begin{pmatrix} N_{11} & N_{12} & N_{13} & 0 & N_{15} & 0 \\ N_{21} & N_{22} & N_{23} & 0 & N_{25} & 0 \\ N_{31} & N_{32} & N_{33} & 0 & N_{35} & 0 \\ 0 & 0 & 0 & N_{44} & 0 & N_{46} \\ N_{51} & N_{52} & N_{53} & 0 & N_{55} & 0 \\ 0 & 0 & 0 & N_{64} & 0 & N_{66} \end{pmatrix}$
$222, mm2, mmm$	$\begin{pmatrix} N_{11} & N_{12} & N_{13} & 0 & 0 & 0 \\ N_{21} & N_{22} & N_{23} & 0 & 0 & 0 \\ N_{31} & N_{32} & N_{33} & 0 & 0 & 0 \\ 0 & 0 & 0 & N_{44} & 0 & 0 \\ 0 & 0 & 0 & 0 & N_{55} & 0 \\ 0 & 0 & 0 & 0 & 0 & N_{66} \end{pmatrix}$
$3, \bar{3}$ $N_{66} = \frac{1}{2}(N_{11} - N_{12})$	$\begin{pmatrix} N_{11} & N_{12} & N_{13} & N_{14} & -N_{25} & N_{16} \\ N_{12} & N_{11} & N_{13} & -N_{14} & N_{25} & -N_{16} \\ N_{31} & N_{31} & N_{33} & 0 & 0 & 0 \\ N_{41} & -N_{41} & 0 & N_{44} & N_{45} & N_{52} \\ -N_{52} & N_{52} & 0 & N_{45} & N_{44} & N_{41} \\ -N_{16} & N_{16} & 0 & N_{25} & N_{14} & N_{66} \end{pmatrix}$

Continued overleaf

15.5 Modeling magnetostriction

Table 15.5 (*Continued*)

$32, 3m, \bar{3}m$ $N_{66} = \frac{1}{2}(N_{11} - N_{12})$	$\begin{pmatrix} N_{11} & N_{12} & N_{13} & N_{14} & 0 & 0 \\ N_{12} & N_{11} & N_{13} & -N_{14} & 0 & 0 \\ N_{31} & N_{31} & N_{33} & 0 & 0 & 0 \\ N_{41} & -N_{41} & 0 & N_{44} & 0 & 0 \\ 0 & 0 & 0 & 0 & N_{44} & N_{41} \\ 0 & 0 & 0 & 0 & N_{14} & N_{66} \end{pmatrix}$
$4, \bar{4}, 4/m$	$\begin{pmatrix} N_{11} & N_{12} & N_{13} & 0 & 0 & N_{16} \\ N_{12} & N_{11} & N_{13} & 0 & 0 & -N_{16} \\ N_{31} & N_{31} & N_{33} & 0 & 0 & 0 \\ 0 & 0 & 0 & N_{44} & N_{45} & 0 \\ 0 & 0 & 0 & -N_{45} & N_{44} & 0 \\ N_{61} & -N_{61} & 0 & 0 & 0 & N_{66} \end{pmatrix}$
$422, 4mm, \bar{4}2m, 4/mmm$	$\begin{pmatrix} N_{11} & N_{12} & N_{13} & 0 & 0 & 0 \\ N_{12} & N_{11} & N_{13} & 0 & 0 & 0 \\ N_{31} & N_{31} & N_{33} & 0 & 0 & 0 \\ 0 & 0 & 0 & N_{44} & 0 & 0 \\ 0 & 0 & 0 & 0 & N_{44} & 0 \\ 0 & 0 & 0 & 0 & 0 & N_{66} \end{pmatrix}$
$6, \bar{6}, 6/m, \infty, \infty/m$ $N_{66} = \frac{1}{2}(N_{11} - N_{12})$	$\begin{pmatrix} N_{11} & N_{12} & N_{13} & 0 & 0 & N_{16} \\ N_{12} & N_{11} & N_{13} & 0 & 0 & -N_{16} \\ N_{31} & N_{31} & N_{33} & 0 & 0 & 0 \\ 0 & 0 & 0 & N_{44} & N_{45} & 0 \\ 0 & 0 & 0 & -N_{45} & N_{44} & 0 \\ -N_{16} & N_{16} & 0 & 0 & 0 & N_{66} \end{pmatrix}$
$622, 6mm, \bar{6}m2, 6/mmm,$ $\infty 2, \infty m, \infty/mm$ $N_{66} = \frac{1}{2}(N_{11} - N_{12})$	$\begin{pmatrix} N_{11} & N_{12} & N_{13} & 0 & 0 & 0 \\ N_{12} & N_{11} & N_{13} & 0 & 0 & 0 \\ N_{31} & N_{31} & N_{33} & 0 & 0 & 0 \\ 0 & 0 & 0 & N_{44} & 0 & 0 \\ 0 & 0 & 0 & 0 & N_{44} & 0 \\ 0 & 0 & 0 & 0 & 0 & N_{66} \end{pmatrix}$
$23, m3$	$\begin{pmatrix} N_{11} & N_{12} & N_{21} & 0 & 0 & 0 \\ N_{21} & N_{11} & N_{12} & 0 & 0 & 0 \\ N_{12} & N_{21} & N_{11} & 0 & 0 & 0 \\ 0 & 0 & 0 & N_{44} & 0 & 0 \\ 0 & 0 & 0 & 0 & N_{44} & 0 \\ 0 & 0 & 0 & 0 & 0 & N_{44} \end{pmatrix}$
$432, \bar{4}3m, m3m$	$\begin{pmatrix} N_{11} & N_{12} & N_{12} & 0 & 0 & 0 \\ N_{12} & N_{11} & N_{12} & 0 & 0 & 0 \\ N_{12} & N_{12} & N_{11} & 0 & 0 & 0 \\ 0 & 0 & 0 & N_{44} & 0 & 0 \\ 0 & 0 & 0 & 0 & N_{44} & 0 \\ 0 & 0 & 0 & 0 & 0 & N_{44} \end{pmatrix}$
$\infty\infty, \infty\infty m$ $N_{44} = \frac{1}{2}(N_{11} - N_{12})$	$\begin{pmatrix} N_{11} & N_{12} & N_{12} & 0 & 0 & 0 \\ N_{12} & N_{11} & N_{12} & 0 & 0 & 0 \\ N_{12} & N_{12} & N_{11} & 0 & 0 & 0 \\ 0 & 0 & 0 & N_{44} & 0 & 0 \\ 0 & 0 & 0 & 0 & N_{44} & 0 \\ 0 & 0 & 0 & 0 & 0 & N_{44} \end{pmatrix}$

156   Nonlinear phenomena

Fig. 15.6 Under strong magnetic fields the domain pattern changes and the magnetization rotates into the field direction. The usual sequence is pictured here.

Fig. 15.7 Magnetostriction strain $x'_3 = \Delta \ell/\ell$ saturates at a value λ under large magnetic fields. The dependence of strain on field can be approximated by an "effective" piezomagnetic coefficient Q_{33} given by the slope of the dotted line. For nickel, Q_{33} is about -3×10^{-9} m/A.

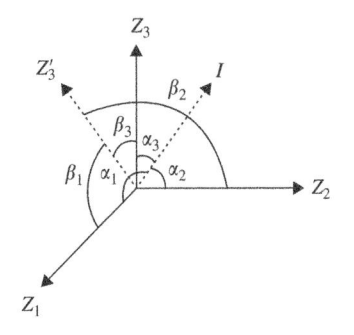

Fig. 15.8 Direction cosines $(\alpha_1\, \alpha_2\, \alpha_3)$ specify the direction of the magnetization relative to the principal axes. For the cubic crystals considered here, the principal axes Z_1, Z_2, Z_3 are along the cube edges [100], [010], and [001]. A second set of direction cosines $(\beta_1\, \beta_2\, \beta_3)$ specify the direction Z'_3 in which the strain x'_{33} is measured.

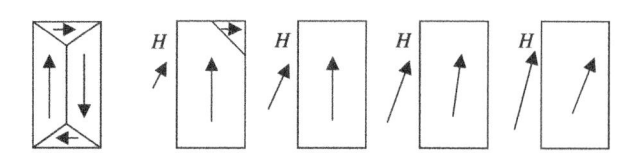

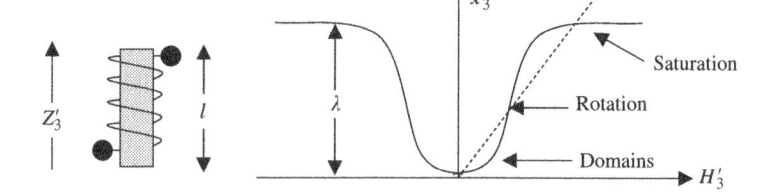

where I_s is the magnitude of the saturation magnetization. For cubic crystals the three components lie along [100], [010], and [001].

Z'_3 is the direction in which the strain is measured. Its orientation is specified by direction cosines β_1, β_2, and β_3. The longitudinal strain x'_{33} along Z'_3 is related to the strains along the crystal axes by

$$x'_{33} = \beta_i \beta_j x_{ij} = \beta_1^2 x_{11} + \beta_1 \beta_2 x_{12} + \cdots.$$

To simplify the mathematics the discussion is limited to cubic crystals (more properly these are pseudocubic magnetic crystals). Most of the magnetostrictive crystals of practical interest have high symmetry crystal structures belonging to point group $m3m$. The magnetostriction matrix for point group $m3m$ (Table 15.5) is

$$\begin{pmatrix} x_1 \\ x_2 \\ x_3 \\ x_4 \\ x_5 \\ x_6 \end{pmatrix} = \begin{pmatrix} N_{11} & N_{12} & N_{12} & 0 & 0 & 0 \\ N_{12} & N_{11} & N_{12} & 0 & 0 & 0 \\ N_{12} & N_{12} & N_{11} & 0 & 0 & 0 \\ 0 & 0 & 0 & N_{44} & 0 & 0 \\ 0 & 0 & 0 & 0 & N_{44} & 0 \\ 0 & 0 & 0 & 0 & 0 & N_{44} \end{pmatrix} \begin{pmatrix} \alpha_1^2 I_s^2 \\ \alpha_2^2 I_s^2 \\ \alpha_3^2 I_s^2 \\ \alpha_2 \alpha_3 I_s^2 \\ \alpha_3 \alpha_1 I_s^2 \\ \alpha_1 \alpha_2 I_s^2 \end{pmatrix}.$$

Expanding and remembering that $\alpha_1^2 + \alpha_2^2 + \alpha_3^2 = 1$,

$$x_1 = h_0 + h_1 \alpha_1^2$$
$$x_2 = h_0 + h_1 \alpha_2^2$$
$$x_3 = h_0 + h_1 \alpha_3^2$$
$$x_4 = h_2 \alpha_2 \alpha_3$$
$$x_5 = h_2 \alpha_3 \alpha_1$$
$$x_6 = h_2 \alpha_1 \alpha_2,$$

where the magnetostrictive coefficients are

$$h_0 = N_{12}I_s^2$$
$$h_1 = (N_{11} - N_{12})I_s^2$$
$$h_2 = N_{44}I_s^2.$$

Substituting the strain components into the expression for saturation strain measured along an arbitrary direction Z_3':

$$x_s = x_3' = h_0 + h_1(\alpha_1^2\beta_1^2 + \alpha_2^2\beta_2^2 + \alpha_3^2\beta_3^2)$$
$$+ 2h_2(\alpha_1\alpha_2\beta_1\beta_2 + \alpha_1\alpha_3\beta_1\beta_3 + \alpha_2\alpha_3\beta_2\beta_3).$$

In specifying strain, care must be taken to choose the correct reference state $x_s(0)$. Generally the measurements begin with the sample in a multidomain demagnetized state (Fig. 15.6).

For cubic magnetic materials the easy axes for magnetization are usually along the cube edges $\langle 100 \rangle$ or along body diagonal $\langle 111 \rangle$ directions. (These correspond to magnetic point groups $4/mm'm'$ or $\bar{3}m'$ for each domain state.) If the $\langle 111 \rangle$ directions are the easy axes there are eight domain states, with all equally abundant in the demagnetized reference state. The eight domains and their direction cosines are

$$[111] \quad \alpha_1 = \frac{1}{\sqrt{3}}, \quad \alpha_2 = \frac{1}{\sqrt{3}}, \quad \alpha_3 = \frac{1}{\sqrt{3}}$$

$$[\bar{1}11] \quad \alpha_1 = -\frac{1}{\sqrt{3}}, \quad \alpha_2 = \frac{1}{\sqrt{3}}, \quad \alpha_3 = \frac{1}{\sqrt{3}}$$

$$[1\bar{1}1] \quad \alpha_1 = \frac{1}{\sqrt{3}}, \quad \alpha_2 = -\frac{1}{\sqrt{3}}, \quad \alpha_3 = \frac{1}{\sqrt{3}}$$

plus five more. To obtain the strain of the reference state, $x_s(0)$, we assign equal weight (1/8) to each domain state and substitute the direction cosine values into the expression for x_s. The result is

$$x_s(0) = \frac{h_0 + h_1}{3}.$$

If the $\langle 100 \rangle$ axes are the easy directions, there are six domain states with the following direction cosines:

$$[100] \quad \alpha_1 = 1, \quad \alpha_2 = 0, \quad \alpha_3 = 0$$
$$[\bar{1}00] \quad \alpha_1 = -1, \quad \alpha_2 = 0, \quad \alpha_3 = 0$$
$$[010] \quad \alpha_1 = 0, \quad \alpha_2 = 1, \quad \alpha_3 = 0$$

plus three more. With all six domains of equal abundance, the resulting strain of the demagnetized state is

$$x_s(0) = \frac{h_0 + h_1}{3}.$$

Since this is the same result as the $\langle 111 \rangle$ case, the reference state strain is independent of the easy axis direction.

Therefore the net magnetostrictive strain relative to the demagnetized state is

$$\lambda = x_s - x_s(0) = h_1 \left(\alpha_1^2 \beta_1^2 + \alpha_2^2 \beta_2^2 + \alpha_3^2 \beta_3^2 - \frac{1}{3} \right)$$
$$+ h_2 (\alpha_1 \alpha_2 \beta_1 \beta_2 + \alpha_1 \alpha_3 \beta_1 \beta_3 + \alpha_2 \alpha_3 \beta_2 \beta_3).$$

The h_1 and h_2 coefficients are replaced with two measured coefficients λ_{100} and λ_{111}. If the magnetic field H is along [100], and the magnetostrictive strain is measured in the same direction,

$$\alpha_1 = \beta_1 = 1 \quad \text{and} \quad \alpha_2 = \beta_2 = \alpha_3 = \beta_3 = 0.$$

The saturation strain is

$$\lambda = h_1 \left(1 - \frac{1}{3} \right) = \frac{2h_1}{3} = \lambda_{100}.$$

If the field and strain are along [111],

$$\alpha_1 = \beta_1 = \alpha_2 = \beta_2 = \alpha_3 = \beta_3 = \frac{1}{\sqrt{3}}$$

and

$$\lambda = \frac{h_2}{3} = \lambda_{111}.$$

Using these two measured values of the saturation strain, the strains for any direction can be evaluated from the following expression.

$$\lambda = x_s(H) - x_s(0)$$
$$= \frac{3}{2} \lambda_{100} \left(\alpha_1^2 \beta_1^2 + \alpha_2^2 \beta_2^2 + \alpha_3^2 \beta_3^2 - \frac{1}{3} \right)$$
$$+ 3\lambda_{111} (\alpha_1 \alpha_2 \beta_1 \beta_2 + \alpha_1 \alpha_3 \beta_1 \beta_3 + \alpha_2 \alpha_3 \beta_2 \beta_3).$$

The magnetostriction coefficients of cubic crystals are usually measured with a single crystal disk cut parallel to the ($1\bar{1}0$) plane. The [001], [111], and [110] directions all lie in this plane. Piezoresistive strain gages can be attached to the plate parallel to these directions. The disk is then suspended between the pole pieces of an electromagnet and rotated about [$1\bar{1}0$] to magnetize the specimen along the various cube axes. The resulting strains are measured with the strain gages.

Magnetostriction constants for several metal and oxide single crystals are given in Table 15.6. Note the large coefficients in magnetic materials containing cobalt (see Section 16.4).

Table 15.6 Magnetostriction coefficients of three ferromagnetic metals and several ferrites (all coefficients are measured at 20°C and are multiplied by 10^{-6})

	λ_{100}	λ_{111}
Iron (Fe)	20.7	−21.2
Nickel (Ni)	−45.9	−24.3
$Co_{0.5}Fe_{0.5}$	119	41
Fe_3O_4	−20	78
$MnFe_2O_4$	−25	−33
$Co_{0.8}Fe_{2.2}O_4$	−590	120

Problem 15.3

Plot the saturation strain for magnetite (Fe_3O_4) using the coefficients in Table 15.6. Carry out the calculation in the $(1\bar{1}0)$ plane in two ways: First, with the strain measurement parallel to the magnetic field, and second with the strain measured perpendicular to the field.

15.6 Magnetostrictive actuators

Lead zirconate titanate (PZT) and lead magnesium niobate (PMN) ceramics are outstanding piezoelectric and electrostrictive actuator materials, but equally interesting developments are taking place in ferroelastic and ferromagnetic materials. As discussed in Chapter 16, all these ferroic materials have a domain structure in which the walls can be moved with electric fields, magnetic fields, or mechanical stresses.

A listing of some of the more useful polycrystalline magnetostrictors is given in Table 15.7. Magnetostrictive alloys like Terfenol-D ($Tb_{1-x} Dy_x Fe_2$) function well as both sensors and actuators. High-power actuators can deliver forces greater than 50 MPa with strains up to 0.6% while magnetostrictive sensor materials can provide hundreds of times the sensitivity of semiconductor strain gauges. Magnetoelastic materials also have tunable elastic moduli that can be controlled by magnetic fields.

The rare-earth atoms in Terfenol have large orbital moments that interact with magnetic fields to give large magnetostrictive strains. As pointed out earlier, rotation of the magnetization is largely responsible for the shape change. The field-induced strain (λ_s) in Terfenol-D is about a hundred times larger than those in iron and nickel (Table 15.7).

The iron in Terfenol produces the high Curie temperature. The rare-earth terbium and dysprosium atoms produce the large magnetostrictive effects. In combination these three elements produce the useful actuator alloys. A portion of the Tb–Dy–Fe magnetic phase diagram is shown in Fig. 15.9. $TbFe_2$ and $DyFe_2$ are cubic and paramagnetic at high temperatures and then undergo magnetic phase transformation to ferrimagnetic structures with

Table 15.7 Room temperature magnetostriction coefficients and Curie temperatures for several polycrystalline metals and ceramics

Material	$3/2 \lambda_s$ ($\times 10^{-6}$)	T_c (°C)
Ni	−50	360
Fe	−14	770
Co	−93	1130
Fe_3O_4	60	585
$CoFe_2O_4$	−165	520
$Y_3Fe_5O_{12}$	−3	275
$SmFe_2$	−2340	415
$GdFe_2$	59	525
$TbFe_2$	2630	431
$DyFe_2$	650	362
$HoFe_2$	120	332
$ErFe_2$	−449	320
$TmFe_2$	−185	287

160 *Nonlinear phenomena*

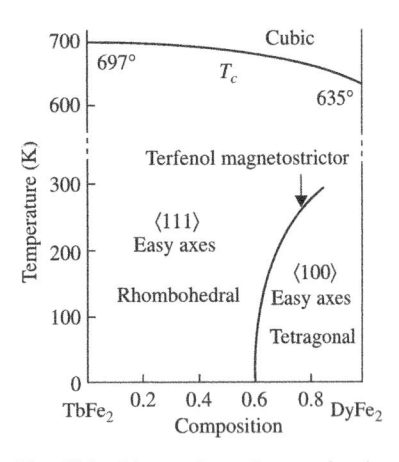

Fig. 15.9 Binary phase diagram for the Tb$_{1-x}$Dy$_x$Fe$_2$ system. Compositions near the magnetic spin boundary at $x = 0.7$ are used in magnetostrictive applications.

magnetic spins along either ⟨111⟩ or ⟨100⟩ directions. Compositions near Tb$_{0.3}$Dy$_{0.7}$Fe$_2$ are poised on an instability in spin orientation. Like PZT it is close to a rhombohedral–tetragonal phase boundary. The phase diagram of Terfenol is the magnetic equivalent to the morphotropic boundary of PZT (Fig. 12.9).

The figure of merit for magnetostrictive actuators is proportional to the saturation strain coefficient λ_s. But in addition to having a large shape change, the strain must be easy to move. Therefore the anisotropy coefficient K_1, which controls the rotation of the magnetization, also comes into the figure of merit, λ_s/K_1. Magnetic anisotropy is discussed in Section 16.4. TbFe$_2$ has a very large λ_s coefficient but also has a large K_1, which reduces the figure of merit. DyFe$_2$ has an anisotropy coefficient of the opposite sign. By tuning the composition close to the point where the anisotropy coefficient goes to zero, the strain is easy to re-orient. This maximizes the figure of merit.

Fig. 15.10 shows the two magnetic phase changes in Terfenol-D. At high temperature, it is paramagnetic with randomly oriented spins. Below T_c near 700 K, the spins align along ⟨111⟩ directions giving a rhombohedral distortion of the unit cell. The strong antiferromagnetic interactions between the iron and rare-earth spins make the alloy ferrimagnetic rather than ferromagnetic. Near room temperature there is spin reorientation to the tetragonal ⟨001⟩ directions of the cubic unit cell. The point group changes from $m3m1' \rightarrow \bar{3}m' \rightarrow 4/mm'm'$.

The large magnetostrictive strains in Terfenol-D are caused by the orbital motion of the rare-earth 4f electrons that impart a shape anisotropy to the atoms. Trivalent terbium and dysprosium atoms resemble flattened oblate ellipsoids, while other rare earths such as erbium, thulium, and samarium have elongated prolate shapes. Note the sign changes in λ_s in Table 15.7.

15.7 Electromagnetostriction and pseudopiezoelectricity

An unexpectedly large deformation, sometimes referred to as electromagnetostriction, has been observed in conducting MnZn ferrite crystals. When placed in an electric field an electric current passes through the crystal inducing a

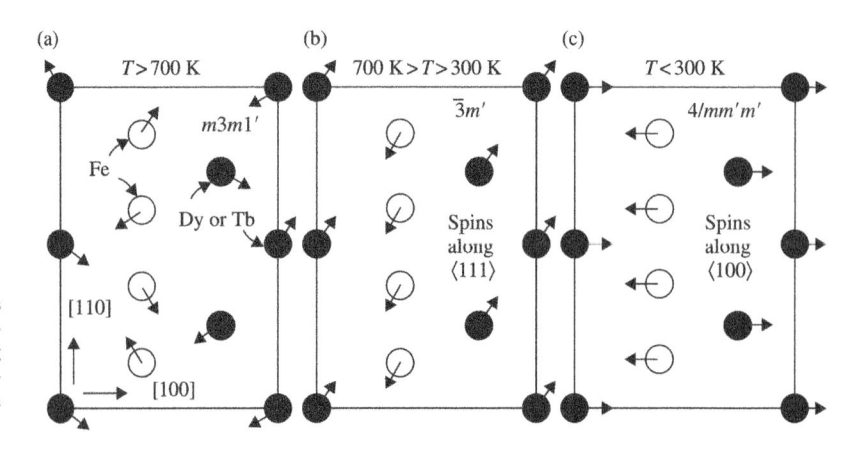

Fig. 15.10 Magnetic phase transformations in Terfenol-D (Tb$_{0.3}$Dy$_{0.7}$Fe$_2$) magnetostrictive actuators. At T_c the spins align along ⟨111⟩ and then to ⟨100⟩ near room temperature. The drawings show only a section through the complex C15 structure.

magnetic field around the current. If the deformation is regarded as electrostrictive, the M coefficient defined by $x = ME^2/2$ (x is strain, E is electric field) is about 10^{-10} m²/V². This is about 10^{11} times larger than the electrostriction coefficient of barium titanate. The M coefficient of BaTiO$_3$ is about 10^{-21} m²/V² in both the ferroelectric and paraelectric phases. More accurately, however, the deformation should be regarded as a magnetostrictive effect induced by an electric current within the ferrimagnetic crystal. The current induces a closed magnetic flux around the applied electric field. When evaluated as a magnetostrictive strain, $x = NH^2/2$ (x is strain, H is magnetic field), the coefficient N has a reasonable value of 10^{-8} m²/A².

When a static magnetic field is superimposed on the applied electric field, the resulting strain is linearly proportional to the electric field. The sign of the deformation changes when the magnetic field is reversed, and when mechanically stressed, electrical signals are observed. In other words, the ferrite crystal exhibits a magnetically-induced pseudopiezoelectric effect with both a direct and a converse effect. The pseudopiezoelectric coefficient is about 10^{-9} C/N, about the same size as a soft PZT ceramic.

16 Ferroic crystals

16.1 Free energy formulation 162
16.2 Ferroelasticity 165
16.3 Ferromagnetism 168
16.4 Magnetic anisotropy 170
16.5 Ferroelectricity 174
16.6 Secondary ferroics: Ferrobielectricity and ferrobimagnetism 177
16.7 Secondary ferroics: Ferrobielasticity and ferroelastoelectricity 179
16.8 Secondary ferroics: Ferromagnetoelectrics and ferromagnetoelastics 182
16.9 Order parameters 183

Twinned crystals are normally classified according to twin-laws and morphology, or according to their mode of origin, or according to a structural basis, but there is another classification that deserves wider acceptance, one that is based on the tensor properties of the orientation states. An advantage of such a classification is the logical relationship between free energy and twin structures, for it becomes immediately apparent which forces and fields will be effective in moving twin walls. The domain patterns in ferroelectric and ferromagnetic materials are strongly affected by external fields, but there are many other types of twinned crystals with movable twin walls and hysteresis. These materials are classified as ferroelastic, ferrobielastic, and various other ferroic species. As explained in the next section, each type of switching arises from a particular term in the free energy function.

Ferroic crystals possess two or more orientation states or domains, and under a suitably chosen driving force the domain walls move, switching the crystal from one domain state to another. Switching may be accomplished by mechanical stress (X), electric field (E), magnetic field (H), or some combination of the three. Ferroelectric, ferroelastic, and ferromagnetic materials are well known examples of primary ferroic crystals in which the orientation states differ in spontaneous polarization ($P_{(s)}$), spontaneous strain ($x_{(s)}$), and spontaneous magnetization ($I_{(s)}$), respectively. It is not necessary, however, that the orientation states differ in the primary quantities (strain, polarization, or magnetization) for the appropriate field to develop a driving force for domain walls. If, for example, the twinning rules between domains lead to a different orientation of the elastic compliance tensor, a suitably chosen stress can then produce different strains in the two domains. This same stress may act upon the difference in induced strain to produce wall motion and domain reorientation. Aizu suggested the term ferrobielastic to distinguish this type of response from ferroelasticity, and illustrated the effect with Dauphine twinning in quartz. Other types of secondary ferroic crystals are listed in Table 16.1, along with the difference between domain states, and the driving fields required to switch between states. The derivation of the various ferroic species from a free energy function is considered next.

16.1 Free energy formulation

The stability of an orientation state is governed by the free energy G. In differential form dG is comprised of thermal energy and various work terms:

$$dG = S dT - x_{ij} dX_{ij} - P_i dE_i - I_i dH_i,$$

Table 16.1 Classification of primary and secondary ferroics

Ferroic class	Orientation state differ in	Switching force	Example
Primary			
Ferroelectric	Spontaneous polarization	Electric field	BaTiO$_3$
Ferroelastic	Spontaneous strain	Mechanical stress	CaAl$_2$Si$_2$O$_8$
Ferromagnetic	Spontaneous magnetization	Magnetic field	Fe$_3$O$_4$
Secondary			
Ferrobielectric	Dielectric susceptibility	Electric field	SrTiO$_3$
Ferrobimagnetic	Magnetic susceptibility	Magnetic field	NiO
Ferrobielastic	Elastic compliance	Mechanical stress	SiO$_2$
Ferroelastoelectric	Piezoelectric coefficients	Electric field and mechanical stress	NH$_4$Cl
Ferromagnetoelastic	Piezomagnetic coefficients	Magnetic field and mechanical stress	FeCO$_3$
Ferromagnetoelectric	Magnetoelectric coefficients	Magnetic field and electric field	Cr$_2$O$_3$

S is entropy, T temperature, x_{ij} strain, X_{ij} stress, P_i electric polarization, E_i electric field, I_i magnetization, and H_i magnetic field. The directional subscripts refer to Cartesian coordinates: $i, j = 1, 2, 3$. The entropy term is neglected in what follows since we assume the experiments are performed under isothermal conditions.

Strain x is measured relative to an extrapolation of the prototype high temperature structure and can be written as a spontaneous strain $x_{(s)}$ plus an induced strain. Induced strain may arise from applied mechanical stress (elasticity), from applied electric fields (piezoelectricity) or from applied magnetic fields (piezomagnetism).

$$x_{ij} = x_{(s)ij} + s_{ijkl}X_{kl} + d_{kij}E_k + Q_{kij}H_k.$$

In this equation s_{ijkl} is a component of the fourth-rank elastic compliance tensor. The piezoelectric coefficients d_{kij} constitute a third rank tensor, as do the piezomagnetic coefficients Q_{kij}. Compliance and piezoelectricity are polar tensors whereas piezomagnetism is an axial tensor.

Electric polarization can be expanded in a manner similar to strain, with a spontaneous contribution $P_{(s)}$ and several induced effects.

$$P_i = P_{(s)i} + \psi_{ij}E_j + d_{ijk}X_{jk} + Q_{ij}H_j.$$

The second rank tensors ψ_{ij} and Q_{ij} represent the electric susceptibility and magnetoelectric coefficients, respectively. ψ_{ij} is a polar tensor and Q_{ij} is an axial tensor. Only the ten polar crystal classes in Table 8.1 possess spontaneous polarization $P_{(s)}$.

Magnetization can be expanded in terms of the spontaneous magnetization, and induced effects arising from electric and magnetic fields, and mechanical stress.

$$I_i = I_{(s)i} + \chi_{ij}H_j + Q_{ijk}X_{jk} + Q_{ji}E_j.$$

Only ferromagnetic and ferrimagnetic crystals have nonzero spontaneous magnetization. Table 14.3 lists the 31 magnetic symmetry groups with spontaneous magnetization.

Substituting the expressions for x_{ij}, P_i, and I_i into the differential form for free energy, combining terms, and integrating gives the thermodynamic potential G, which applies to all orientation states. Now consider a multidomain crystal together with a set of measurement axes. The domain states will be oriented differently relative to these measurement axes. Let ^1G represent the free energy for the first orientation state and ^2G for the second, with the tensor terms referred to the measurement axial system.

The driving potential for a state shift accompanied by domain wall motion is the difference in free energy $\Delta G = {}^1G - {}^2G$. In the absence of external fields and forces, the energies of all orientation states are equal, so that $\Delta G = 0$. Under external forces the difference in free energy for the two orientation states is

$$\Delta G = \Delta x_{(s)ij}X_{ij} + \Delta P_{(s)i}E_i + \Delta I_{(s)i}H_i + \tfrac{1}{2}\Delta s_{ijkl}X_{ij}X_{kl} + \tfrac{1}{2}\Delta \psi_{ij}E_iE_j$$
$$+ \tfrac{1}{2}\Delta \chi_{ij}H_iH_j + 2\Delta d_{ijk}E_iX_{jk} + 2\Delta Q_{ijk}H_iX_{jk} + 2\Delta Q_{ij}H_iE_j.$$

In this expression $\Delta x_{(s)ij}$ is $^2x_{(s)ij} - {}^1x_{(s)ij}$, the difference in a certain component of spontaneous strain for orientation states 1 and 2. $\Delta P_{(s)i}$ and $\Delta I_{(s)i}$ are the differences in the ith component of spontaneous polarization and spontaneous magnetization for the two domains. Differences in elastic compliance coefficients are represented by Δs_{ijkl}. The remaining five terms in ΔG arise from differences in electric and magnetic susceptibility, and from differences in piezoelectric, piezomagnetic, and magnetoelectric coefficients.

A wide variety of ferroic phenomena is possible, depending on which terms in ΔG are important. If $\Delta P_{(s)}$ is nonzero, the material is ferroelectric provided the coercive field does not exceed the electric breakdown limit. Materials with $\Delta x_{(s)} \neq 0$ are ferroelastic if the mechanical stress required to move domain walls does not result in fracture. Ferromagnetic domains—the third type of primary ferroic—possess finite differences in spontaneous magnetization.

In a ferroelectric the transition between domain states is said to be primary ferroic behavior because ΔG is proportional to E. If $\Delta P_{(s)} = 0$ and $\Delta \psi \neq 0$, then $\Delta G \sim E^2$ and the material is potentially ferrobielectric. This and other types of secondary ferroics are listed in Table 16.1. For a ferrobielastic $\Delta G \sim X^2$, and for a ferrobimagnetic $\Delta G \sim H^2$. Cross-coupled ferroics include ferroelastoelectrics ($\Delta G \sim EX$), ferromagnetoelastics ($\Delta G \sim HX$), and ferromagnetoelectrics ($\Delta G \sim EH$).

Because of these many coupling coefficients, the ferroic classes are not mutually exclusive. For example, in any dielectric material there may be a coupling between polarization and lattice strain through the piezoelectric or electrostriction coefficients; if a crystal spontaneously polarizes, it also spontaneously strains. If all the orientation states of a ferroelectric differ also in the orientation of the spontaneous strain tensor, then the material may be termed fully ferroelectric, fully ferroelastic. In most ferroelectrics, however, some but not all domain states differ in the orientation of the strain tensor. For example, in $BaTiO_3$, 90° domains differ in the orientation of the strain tensor, but 180° domains do not. Such systems may be described as fully ferroelectric, partially ferroelastic. Cobalt ferrite is fully ferromagnetic, partially ferroelastic, while nickel iodine boracite is fully ferromagnetic, fully ferroelectric, and fully ferroelastic. In this type of crystal, all domain walls can be driven by electric, magnetic, or mechanical stress fields.

16.2 Ferroelasticity

A crystal is ferroelastic if it has two or more orientation states differing in spontaneous strain, and can be transformed reversibly from one to another of these states by an external mechanical stress. It is a type of mechanical twinning in which the lattice reorients rapidly in response to an applied force. There is no diffusion or breaking of primary chemical bonds, only small rearrangements of the crystal structure with atomic displacements of the order of 0.1 Å. At elevated temperatures the domains disappear as the crystal transforms to a higher symmetry point group.

Twinning is often used in mineral identification and in elucidating the formation conditions of rocks. The distribution of mechanical twins in feldspars and other rock-forming minerals enables petrologists to analyze thermal environments and stress patterns.

Ferroelastic twinning is almost universal in the feldspar family that makes up more than half of the earth's outer crust. Two views of the stripe-like twins in the plagioclase ($NaAlSi_3O_8$-$CaAl_2Si_2O_8$) series and in microcline ($KAlSi_3O_8$) feldspars are shown in Fig. 16.1. The change in optical extinction angles across domain walls is one of the chief methods of identifying various feldspars. The two most important types of twins in feldspars are albite and pericline polysynthetic twins. Albite twin lamellae are parallel to (010) cleavage planes, and the twin structures are related by reflection across (010). This is a symmetry element of the prototype point group $2/m$ found in the high temperature feldspar structure. Microcline, low-temperature $KAlSi_3O_8$, belongs to triclinic point group $\bar{1}$, as do the plagioclase feldspars. The polysynthetic twins disappear when the triclinic structure transforms to monoclinic. In pericline twins, the two orientation states are related by rotation of 180° about [010], the twofold symmetry axis in the prototype point group. The pseudosymmetry in triclinic feldspars is caused by crumpling of the aluminosilicate framework about the alkali or alkaline earth ions (Fig. 16.2).

The spontaneous strain in the ferroelastic state is measured relative to the high temperature prototype structure. Strain is a symmetric second rank tensor with six components: three longitudinal strains x_{11}, x_{22}, x_{33}, and three shear components x_{23}, x_{31}, and x_{12}. For the feldspars the symmetry change giving rise to the ferroelastic twin states is from the prototype structure (point group $2/m$) to the twinned triclinic structure (point group $\bar{1}$). To determine which components of spontaneous strain appear at the phase transformation, we take

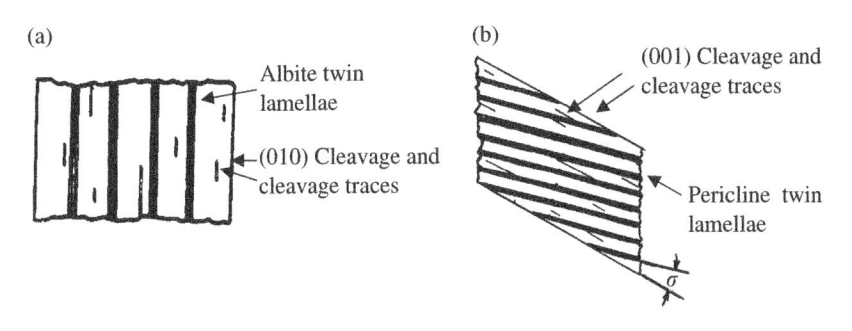

Fig. 16.1 Plagioclase feldspars exhibit perfect cleavage parallel to (001), and less perfect cleavage parallel to (010). Albite twin lamellae (a) are usually present on (001) cleavage surfaces albite twins are parallel to the straight edge formed by the intersecting (010) and (001) cleavage planes. Crystal platelets parallel to (010) cleavage surfaces (b) sometimes show pericline twins. Here the twin lamellae intersect the (010)–(001) edge at an angle σ, the so-called rhombic section.

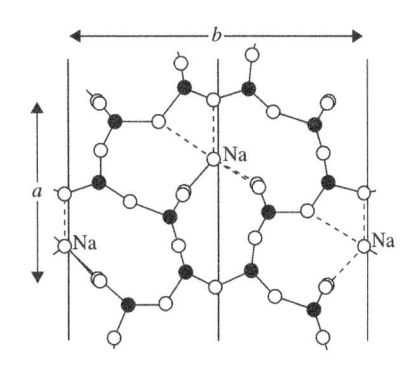

Fig. 16.2 Projection of the triclinic albite (NaAlSi$_3$O$_8$) structure along the c-axis. The tetrahedrally coordinated (Al, Si) ions (●) and oxygens (○) form an aluminosilicate framework with Na$^+$ ions in cavities. Sodium ions are not large enough to contact all the oxygen ions lining the cavity walls so the framework partially collapses, and the symmetry is lowered from monoclinic (point group $2/m$) to triclinic (point group $\bar{1}$). When heated, the sodium ions vibrate to fill the cavity and change the symmetry back to monoclinic. Pseudo-mirror planes in the triclinic unit cell are shown as vertical lines.

the difference between the strain matrices for $\bar{1}$ and $2/m$:

$$\begin{pmatrix} x_{11} & x_{12} & x_{13} \\ x_{12} & x_{22} & x_{23} \\ x_{13} & x_{23} & x_{33} \end{pmatrix} - \begin{pmatrix} x_{11} & 0 & x_{13} \\ 0 & x_{22} & 0 \\ x_{13} & 0 & x_{33} \end{pmatrix} = \begin{pmatrix} 0 & x_{12} & 0 \\ x_{12} & 0 & x_{23} \\ 0 & x_{23} & 0 \end{pmatrix}.$$

The two orientation states in the plagioclase feldspars are related by reflection across the (010) pseudo-mirror plane. Reflection reverses the sign of Z_2, leaving Z_1 and Z_3 unchanged. Tensor components with an odd number of two subscripts change sign under this operation. Thus the nonzero components of spontaneous strain are $x_{(s)23}$ and $x_{(s)12}$, shear strains about Z_1 and Z_3 respectively.

Domain wall orientations in ferroelastic crystals minimize strain mismatch. For the two domain states S_1 and S_2 the wall coordinates Z_i and Z_j satisfy the zero strain condition

$$(x_{(s)ij}(S_1) - x_{(s)ij}(S_2))Z_i Z_j = 0.$$

For the feldspars the two states have the following spontaneous strains.

$$x_{(s)ij}(S_1) = \begin{pmatrix} 0 & x_{12} & 0 \\ x_{12} & 0 & x_{23} \\ 0 & x_{23} & 0 \end{pmatrix},$$

$$x_{(s)ij}(S_2) = \begin{pmatrix} 0 & -x_{12} & 0 \\ -x_{12} & 0 & -x_{23} \\ 0 & -x_{23} & 0 \end{pmatrix}.$$

Substituting these values into the wall coordinate equation gives

$$2x_{12}Z_1 Z_2 + 2x_{12}Z_2 Z_1 + 2x_{23}Z_2 Z_3 + 2x_{23}Z_3 Z_2 = 0.$$

Simplifying, the zero strain condition is

$$x_{12}Z_1 Z_2 + x_{23}Z_2 Z_3 = 0.$$

The two solutions to this equation give the two wall orientations shown in Fig. 16.1. Albite twins parallel to (010) correspond to the solution $Z_2 = 0$. The second solution $Z_1/Z_3 = -x_{23}/x_{12}$ dictates the orientation of pericline twin walls. The rhombic angle σ is determined by the ratio of the two spontaneous strain components x_{23}/x_{12}.

Crystallographic data are often used to estimate the spontaneous strain components. For the feldspar varieties that exhibit mechanical twinning, the triclinic unit cell angles are $\alpha = 93.5 \pm 0.5°$, $\beta = 116.0 \pm 0.5°$, $\gamma = 90.5 \pm 0.5°$, compared to $\alpha = 90°$, $\beta = 116.0 \pm 0.5°$, $\gamma = 90°$ for the monoclinic prototype structure. Neglecting the small difference between the triclinic γ and $90°$, the two components of spontaneous strain are $x_{(s)12} \cong 0$ and $x_{(s)23} = (\pi/360°)(\alpha° - 90°) = 0.03$. Based on this analysis, the difference in free energy for the two orientation states takes the form $\Delta G \cong 4x_{(s)23}X_{23}$. The most effective stress in moving domain walls is $X_{23} = X_4$ a shearing stress about Z_1.

In mechanical twinning experiments, albite and pericline twin lamellae spaced by about 0.03 mm are introduced under uniaxial stresses of about 25×10^8 N/m^2. Similar observations have been made on cleavage flakes using

Table 16.2 Ferroelastic crystals with phase transition temperatures, spontaneous strains, and symmetry changes

	T_c (K)	$x_{(s)}$	Symmetry species
As_2O_5	578	0.023	$422F222$
HCN	170	0.10	$4mmFmm2$
$NdTaO_4$	1601	0.008	$\bar{4}2mF2/m$
$Pb_3(AsO_4)_2$	548	0.02	$\bar{3}mF2/m$
$YBa_2Cu_3O_7$	970	0.02	$4/mmmFmmm$

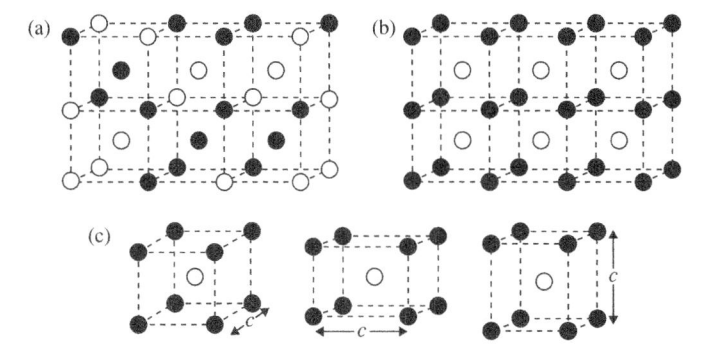

Fig. 16.3 Body-centered intermetallics often show a structural transformation from (a) a high-temperature disordered body-centered cubic phase (space group $Im3m$) to an annealed ordered austenite phase (b) with the CsCl structure (space group $Pm3m$). At lower temperatures, there is a second phase transition from austenite to a twinned martensite phase, pictured here as three variants of a body-centered tetragonal phase (space group $P4/mmm$) (c) with ferroelastic and antiphase domain walls.

a polarizing microscope. When pressed with the tip of a needle, twin lamellae appear and disappear.

Several other ferroelastic crystals are listed in Table 16.2. Typical strains are of the order of 1%.

Problem 16.1
Gadolinium molybdate has a phase transition at 160°C that changes its symmetry from tetragonal $\bar{4}2m$ at high temperature to orthorhombic $mm2$ at room temperature. Remembering that the symmetry of the high-temperature prototype structure must include the symmetry elements of the low-temperature ferroic structure, show that the polar axis of $mm2$ must lie along Z_3. Draw the stereographic projections of $\bar{4}2m$ and the two domain states of $mm2$. What symmetry elements are destroyed on cooling through the phase transformation?

Gadolinium molybdate is both ferroelastic and ferroelectric. Following the procedure used for the feldspars, determine the difference in free energy for the two domain states. What are the components of mechanical stress and electric field that will move domain walls? What are the orientations of strain-free walls?

Shape memory alloys are an important family of thermomechanical actuators utilizing ferroelasticity and stress-induced phase changes. NiTi intermetallic compounds, commonly known as Nitinol, exhibit martensitic phase transformations similar to those observed in the processing of steel. Two characteristics of martensitic phase changes are the absence of long-range diffusion and the appearance of a shape change. Ferroelastic phase transformations are also distortive and diffusionless, and have much in common with the martensites. Ferroelastic crystals exhibit mechanical hysteresis between stress and strain caused by the stress-induced movement of domain walls. Martensites are also internally twinned, but mechanical stress causes phase changes as well as domain wall motion.

Typically these shape memory alloys undergo a transformation from a disordered body-centered cubic structure to a partially ordered CsCl structure at high temperature (Fig. 16.3). In steel-making this is the so-called austenite phase. On further cooling the intermetallic compound undergoes a second phase change into a distorted multidomain martensite phase. Fig. 16.3 shows three martensite twins produced by a cubic to tetragonal phase change. Under stress the martensite easily deforms, and when reheated, goes back to the original morphology of the high temperature austenite structure. This is the so-called shape memory effect.

Some of the martensite structures are very complex (Fig. 16.4). The monoclinic structure of NiTi belongs to point group $2/m$ and has a β angle about $8°$ different from $90°$. This generates a large spontaneous strain accompanying the martensite phase change and is the shape change upon which applied mechanical stresses or thermally-induced stresses act. During the diffuse phase change, the high temperature austenite phase only partially transforms into the low-temperature martensite phase. Then under mechanical stress, two things happen: ferroelastic domain wall movement in the martensite phase is accompanied by conversion of the remaining metastable austenite phase into martensite. Both events make important contributions to the stress-induced shape change.

A magnetic shape memory (MSM) occurs in Ni_2MnGa alloys. This is a new class of actuator materials that combines the advantages of magnetostriction (magnetic drive) with conventional shape memory effects (large strains up to 10%). MSM alloys have the ability to develop large strokes under precise and rapid control. Reorienting martensite twins rather than rotating the magnetization away from the easy axis achieves the high strains. A large magnetic anisotropy is required (Section 16.4).

Ni_2MnGa has the cubic Heusler alloy structure at high temperature. Nickel atoms occupy the corners of a BCC unit cell with Mn and Ga atoms in alternate body-centered sites. On cooling through martensitic transition the unit cell becomes tetragonal with $c/a < 1$ similar to Fig. 16.3(c). Twin planes are parallel to $\{110\}$. At still lower temperatures, the alloy passes through a magnetic transition aligning the spins.

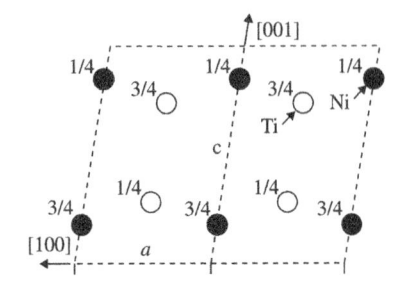

Fig. 16.4 Crystal structure of the martensitic phase of Nitinol, NiTi. The space group is $P2_1/m$ with lattice parameters $a = 2.884$, $b = 4.110$, $c = 4.665$ Å, $\beta = 98.10°$. Sometimes referred to as the low AuCd structure, there are 12 orientation states of this monoclinic phase with respect to the cubic austenite phase.

16.3 Ferromagnetism

The orientation states of a ferromagnet differ in spontaneous magnetization, and can be switched by a magnetic field. This broad definition encompasses ferrimagnets (magnetite, Fe_3O_4) and weak ferromagnets (hematite, α-Fe_2O_3), as well as ordinary ferromagnets (iron). It does not include antiferromagnetic, paramagnetic, and diamagnetic substances that have no spontaneous magnetization. Such materials are not ferromagnetic but may exhibit other types of ferroic behavior.

Magnetite is the best example of a magnetic mineral and is the forerunner of a large family of useful ferrite ceramics. Below $585°C$, it is ferrimagnetic with a magnetization of four Bohr magnetons per molecule corresponding to the four unpaired electron spins of Fe^{2+}. In the inverse spinel magnetic structure of magnetite, tetrahedral Fe^{3+} spins are directed antiparallel to octahedral Fe^{3+} and Fe^{2+} spins so that the Fe^{3+} moments cancel, leaving a spontaneous magnetization equivalent to one Fe^{2+} moment per molecule.

The direction of easy magnetization is $\langle 111 \rangle$, giving rise to eight orientation states for $I_{(s)}$. Magnetite belongs to magnetic point group $\bar{3}m'$ one of the 21 pyromagnetic classes (Table 14.1), although the trigonal distortion is too small to be seen by normal X-ray diffraction. Magnetic domains are difficult to see because magnetite is opaque in visible light, even in thin section.

In transparent ferromagnetic and ferrimagnetic crystals domains are visible in polarized light because of the Faraday Effect, a nonreciprocal rotation of the plane of polarization (Chapter 31). The angle of rotation ϕ is given by $\phi = \rho t \cos \theta$, where t is the specimen thickness, ρ the rotation per unit thickness, and θ the angle between the magnetization vector and the light path. Faraday rotation coefficients (ρ) for ferrites are typically 1000°/cm.

Hematite, α-Fe_2O_3, exhibits both antiferromagnetism and weak ferromagnetism. Above 950 K hematite is paramagnetic with the corundum (α-Al_2O_3) crystal structure. From 250 to 950 K the spins lie in the trigonal (001) plane and are nearly antiparallel, but with a small ferromagnetic component, also in (001). Crystallographers generally assign hematite to crystallographic point group $\bar{3}m$ (magnetic point group $\bar{3}m1'$) but the magnetic point group symmetry is $2/m$ at room temperature. Antiferromagnetic crystals often exhibit weak (parasitic) ferromagnetism when the ferromagnetic component does not violate the symmetry elements of the magnetic spin array. In hematite, weak spontaneous magnetization appears along the monoclinic twofold axis in $2/m$. Below room temperature at 250 K, the spin direction changes to the trigonal [001] direction and the weak ferromagnetic effect disappears. At temperatures below the spin flop transition, the magnetic point group is $\bar{3}m$ (see Fig. 14.5(c)).

Magnetic domains in hematite have been observed using the Faraday Effect. The white, gray, and black regions in Fig. 16.5 correspond to domains with three different magnetic axes; magnetic fields of only 10 A/m produce significant changes in the domain pattern. When cooled through the spin-flop transition at 250 K, the domains disappear, and then reappear in a different pattern on heating.

Barium ferrite is one of the most widely used hard magnetic materials. $BaFe_{12}O_{19}$ is a ferrimagnetic oxide with the magnetoplumbite structure (Fig. 16.6), a close-packed structure with a large hexagonal unit cell. The Ba^{2+} and O^{2-} ions from a close-packed array with Fe^{3+} cations distributed over octahedral, tetrahedral, and trigonal bipyramid interstices. The structure and saturation magnetization are similar to magnetite but the anisotropy is much larger because of the lower symmetry. Spins are locked tightly to the $[001] = Z_3$ direction giving a high coercive field. Hot pressing under a magnetic field is very effective in aligning the easy axes of fine-grained barium ferrite ceramics.

On cooling through T_c the symmetry changes from $6/mmm1'$ in the paramagnetic state to $6/mm'm'$ in the ferromagnetic state. Of the twelve Fe^{3+} ions in the unit cell, eight are spin up and four spin down. Only 180° domains are observed (Fig. 16.7). Domain wall orientations are not constrained because the spontaneous strain is the same for both domain states.

A listing of commonly used ferromagnetic and ferrimagnetic materials is given in Table 16.3.

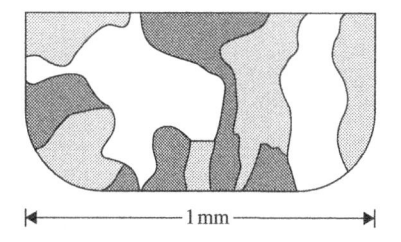

Fig. 16.5 Magnetic domains in a hematite crystal observed by means of the Faraday Effect. Three orientations of the parasitic ferromagnetism give rise to different light intensities when the platelet is tilted with respect to the light beam.

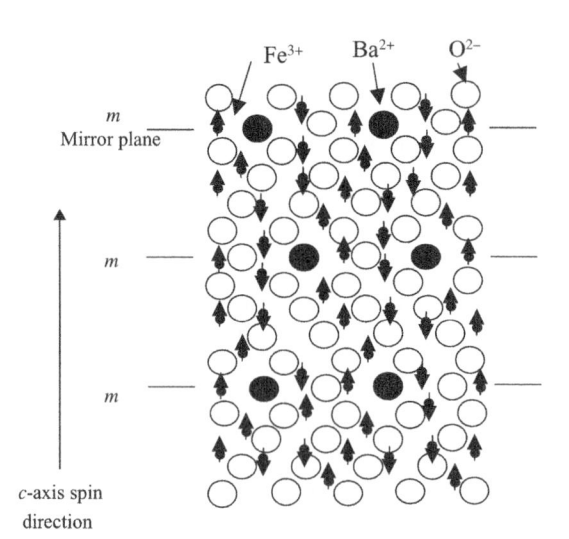

Fig. 16.6 Cross-section of the barium ferrite structure with the hexagonal c-axis vertical. Arrows denote the spin orientation of the iron atoms. Layers containing barium atoms are mirror planes.

Fig. 16.7 Barium ferrite 180° domains viewed along the hexagonal c-axis. Colloidal suspensions of magnetite particles make the domain walls visible.

16.4 Magnetic anisotropy

Experimentally it is found that spontaneous magnetization tends to lie along certain crystallographic axes, the so-called easy axes. In the magnetics literature, this effect is known as magnetic anisotropy or as crystalline anisotropy. The existence of magnetic anisotropy is demonstrated by the magnetization curves of ferromagnetic single crystals. Fig. 16.8 shows the magnetization curves for Fe, Co, and Ni.

When a magnetic field is applied to a ferromagnetic material the domain structure changes in such a way as to increase the magnetization parallel to the external field. This continues until the specimen is filled with favorably oriented domains. If the field is applied along an easy axis, the magnetization rapidly saturates. For other directions the magnetization process continues by rotating the magnetization vectors into the direction of the field until saturation is achieved. Thus the magnetization process generally involves domain wall motion followed by rotation. A third contribution—the induced magnetization coming from the intrinsic magnetic susceptibility—is generally small compared to the contributions from spontaneous magnetization.

Magnetic anisotropy energy is defined as the work required to align the magnetization in a certain direction compared to that required for the easy direction. In quantifying the energy it is customary to describe directions trigonometrically with a set of direction cosines. Let α_1, α_2, and α_3 be the direction cosines between the magnetization vector and the principal axes.

Since most of the important magnetic materials are cubic or hexagonal, we are primarily concerned with crystals belonging to point groups $m3m1'$ and $6/mmm1'$. Magnetic anisotropy energy is expressed as a power series in the direction cosines α_1, α_2, α_3. For centric crystals, odd power terms disappear, as do cross-coupled terms like $\alpha_1\alpha_2$. And for cubic crystals, the energy must be independent of interchange of any two coefficients. Moreover since $\alpha_1^2 + \alpha_2^2 + \alpha_3^2 = 1$ no anisotropy can arise from such a term. The two lowest power terms for the anisotropy energy of cubic crystals are

$$F = K_1(\alpha_1^2\alpha_2^2 + \alpha_2^2\alpha_3^2 + \alpha_3^2\alpha_1^2) + K_2\,\alpha_1^2\alpha_2^2\alpha_3^2.$$

16.4 *Magnetic anisotropy* 171

Table 16.3 Ferromagnetic and ferrimagnetic crystals with Curie temperature T_c (K), spontaneous magnetization (Bohr magnetons/formula unit) and symmetry changes

	T_c	$I_{(s)}$	Symmetry
Ferromagnetic metals			
Fe	1043	2.22	$m3m1'F4/mm'm'$
Co	1403	1.7	$6/mmm1'F6/mm'm'$
Ni	631	0.6	$m3m1'F\bar{3}m'$
Ferrimagnetic oxides			
MnFe$_2$O$_4$	573	5.0	$m3m1'F\bar{3}m'$
Fe$_3$O$_4$	858	4.2	$m3m1'F\bar{3}m'$
CoFe$_2$O$_4$	793	3.3	$m3m1'F4/mm'm'$
NiFe$_2$O$_4$	858	2.3	$m3m1'F\bar{3}m'$
Y$_3$Fe$_5$O$_{12}$	560	4.7	$m3m1'F\bar{3}m'$
Gd$_3$Fe$_5$O$_{12}$	564	15.2	$m3m1'F\bar{3}m'$
BaFe$_{12}$O$_{19}$	720	20	$6/mmm1'F6/mm'm'$

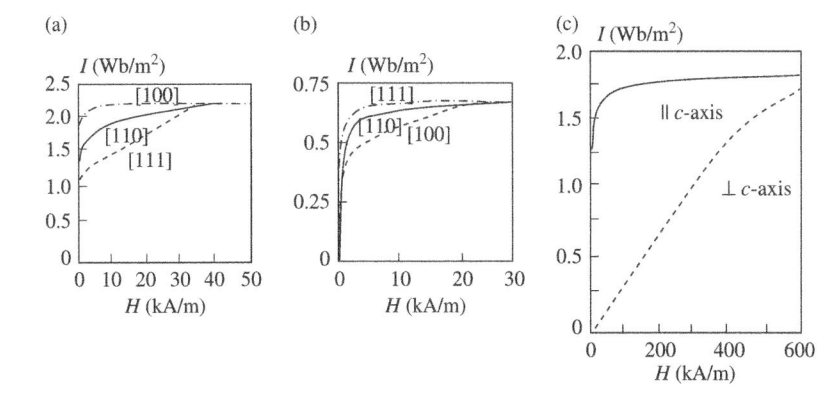

Fig. 16.8 Magnetization curves of single crystals of (a) iron, (b) nickel, and (c) cobalt. The direction of the applied field with respect to the crystallographic axes is indicated for each curve.

For hexagonal crystals the corresponding expression is

$$F = K_1 \sin^2 \theta + K_2 \sin^4 \theta,$$

where θ is the angle between the measurement direction and the hexagonal axis.

A listing of anisotropy coefficients for important magnetic materials is presented in Table 16.4. There are several different methods for measuring magnetic anisotropy, including the torque magnetometer in which the specimen is suspended between the pole pieces of rotatable electromagnet. The specimen is a single crystal cut in the form of a disk. When a strong magnetic field is applied, the disk rotates in such a way as to bring the easy axis into the field direction. The torque on the support wire can be measured optically from the angle of twist (Fig. 16.9). By rotating the magnet the torque can be measured for many different crystallographic orientations.

Problem 16.2
The anisotropy constants for cubic yttrium iron garnet (YIG = Y$_3$Fe$_5$O$_{12}$) are $K_1 = -0.60$ and $K_2 = -0.23$ J/m^3 at room temperature and $K_1 = -2.58$ and $K_2 = -0.17$ J/m^3 at 4 K. Plot the anisotropy energy F in the $(\bar{1}10)$ plane

172 *Ferroic crystals*

Table 16.4 Anisotropy constants of various cubic and hexagonal crystals in kJ/m³

	K_1	K_2	Easy axis
Cubic crystals			
Fe	46	15	[100]
Ni	−5	—	[111]
Fe_3O_4	−11	−28	[111]
$MnFe_2O_4$	−4	—	[111]
$NiFe_2O_4$	−7	−11	[111]
$Co_{0.8}Fe_{2.2}O_4$	290	440	[100]
$Y_3Fe_5O_{12}$	−0.6	−0.2	[111]
Hexagonal crystals			
Co	410	100	∥ [001]
$BaFe_{12}O_{19}$	330	—	∥ [001]
$BaFe_{18}O_{27}$	300	—	∥ [001]
$BaCo_2Fe_{16}O_{27}$	−400	—	⊥ [001]

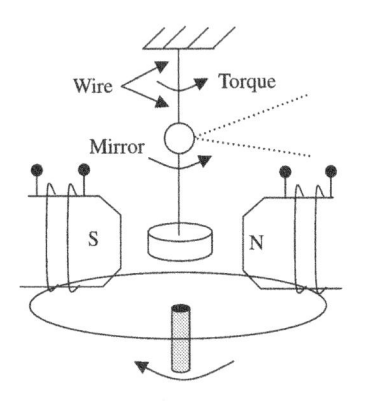

Fig. 16.9 Schematic diagram of a torque magnetometer used to measure magnetic anisotropy. When placed in a magnetic field, the single crystal disk tends to rotate to align the easy axes with the field. The resulting torque provides a measure of the anisotropy energy.

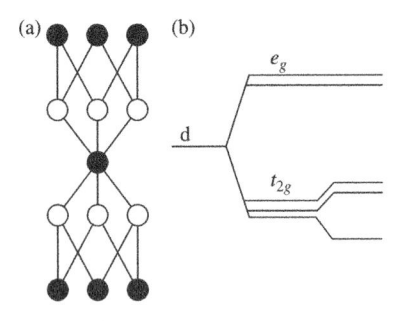

Fig. 16.10 (a) Environment of the octahedrally coordinated cation in the spinel structure. Metal ions and oxygen are denoted by solid and open circles, respectively. (b) d orbitals for an octahedrally coordinates transition metal ion in a slightly trigonal field.

showing the values along [001], [110], and [111] directions. What are the easy axes of magnetization? Discuss the temperature dependence.

Problem 16.3
The easy axis directions can be predicted from the K_1/K_2 ratio by setting $dF/d\theta = 0$. Determine the conditions for hexagonal crystals such as metallic cobalt.

As shown in Fig. 16.8, and Table 16.4 nickel magnetizes most easily along the ⟨111⟩ body diagonals (negative K_1). In the FCC structure of Ni the spins point perpendicular to the (111) close-packed layers, and away from the nearest neighbors. The same is true for α-Fe which has a BCC crystal structure and $K_1 > 0$. The spins point along ⟨100⟩ directions between four neighboring Fe atoms. Cobalt (HCP structure) is similar to nickel with the spins along [001] perpendicular to the close-packed layers. Note that the anisotropy coefficients of Co are much larger than those of Ni and Fe. The K_1 coefficient of cobalt decreases rapidly when heated causing the easy axis to shift away from [001] and eventually becoming perpendicular to [001]. The symmetry changes from $6/mm'm'$ to $2'/m'$ to $mm'm'$.

For most ferrites K_1 is small and negative—but in $CoFe_2O_4$ it is large and positive. Cobalt is commonly used as an additive to increase anisotropy and coercive field. The strong preference of Co^{2+} spins for certain crystal axes appears to be related to its orbital degeneracy. The two ions with the greatest orbital degeneracy in octahedral sites are Fe^{2+} and Co^{2+}. Since the orbital configuration can be changed without changing the energy, the degenerate state possesses appreciable orbital angular momentum.

In the spinel structure the octahedral sites are only approximately octahedral in symmetry. The true point symmetry of the so-called octahedral site is not $m3m$ but trigonal, crystal class $\bar{3}m$. Nearest and next-nearest neighbors are illustrated in Fig. 16.10. The next-nearest metal atoms superpose a trigonal field (with principal axis along ⟨111⟩) on the octahedral field created by the oxygens. As a result, the triply degenerate t_{2g} orbitals are split into a doublet and a singlet. The doublet wave functions lie principally in the plane perpendicular to [111], while the singlet extends parallel to [111]. Since the second nearest neighbors, six octahedral cations, are grouped close to [111], the singlet orbital is lower in energy because of the attraction between electrons and cations. Therefore, the energy level diagram is that shown schematically in Fig. 16.10.

In the case of Fe^{2+}, five electrons go in with spin up, one with spin down. The latter is in the singly degenerate ground state so that Fe^{2+} has no orbital angular momentum in the ground state. For Co^{2+}, which has one more electron, the extra electron occupies half the doubly degenerate state. Therefore Co^{2+} is free to change its state in the plane perpendicular to [111], giving the atom angular momentum parallel to [111]. The angular momentum is fixed in direction and, because of spin-orbit coupling, the spin magnetic moments are also strongly aligned along [111]. This explains why the anisotropy constant of $CoFe_2O_4$ is of opposite sign and larger than those of the other ferrites. This model has been used to explain magnetic anisotropy and magnetic annealing in ferrites. In Fe–Co ferrite ceramics annealed in a magnetic field, Co^{2+} ions diffuse to sites in which the trigonal axis is closest to the applied field direction. Upon cooling the unbalanced Co^{2+} distribution is retained giving rise to uniaxial anisotropy.

There are other ways to create magnetic anisotropy in addition to cobalt substitutions. One way is through poling, another is compositional gradients, and a third is through shape anisotropy. The iron oxide used in magnetic type, γ-Fe_2O_3, provides an example of the imaginative use of crystal chemistry in providing shape anisotropy through topotaxy.

Single-domain acicular particles of γ-Fe_2O_3 oriented parallel to the length of the tape are the magnetic constituent of most tape recorders. When prepared with particle lengths about 0.5μ and length-to-width ratios about five, γ-Fe_2O_3 particles behave like single domain arrays. But it is not easy to prepare cubic materials such as γ-Fe_2O_3 in fine-grained acicular (needlelike) morphology to develop the required shape anisotropy. The processing method involves pseudomorphism, the retention of crystal habit during conversion of one compound to another. Pseudomorphism is fairly common in mineralogy (limonite after pyrite is an example) and can be used to good advantage in materials science as well.

Beginning from a ferrous salt solution, precipitation conditions are carefully controlled to give α-FeOOH, an orthorhombic crystal of acicular habit. The ferric oxyhydroxide is gently oxidized to α-Fe_2O_3, then reduced to magnetite Fe_3O_4, and finally oxidized again to γ-Fe_2O_3, retaining the original habit. Firing temperatures do not exceed 250°C to 400°C. Magnetic γ-Fe_2O_3 is slightly superior to magnetite because the latter exhibits oxidation and magnetic accommodation—making it difficult to erase a recording.

Improvements in the magnetic anisotropy of polycrystalline materials have been achieved by controlling the directional alignment of the crystallites. There are several different ways of developing texture. Directional solidification is most suitable for brittle materials that cannot be mechanically deformed. By withdrawing heat in one direction from a molten metal, crystals grow from the cool side, following the direction of heat in the mold. Alnico and other permanent magnet materials are often cast by directional solidification.

Wire drawing, rolling and other deformation processes are used to produce texture in ductile materials. During deformation, the slip planes slide past each other, causing the crystals to assume a common directional alignment. The degree of control over the texture depends on the number of slip planes and the deformation techniques. By adjusting die size, or the spacing between rollers, the metal is reduced in stages, leading to different textures.

16.5 Ferroelectricity

Many capacitor formulations are based on $BaTiO_3$, one of a number of ferroelectric substances crystallizing with the perovskite structure. Barium atoms are located at the corners of the unit cell and oxygens at the face center positions. Both barium and oxygen have ionic radii of about 1.4 Å and together they make up a face centered cubic array having a lattice parameter near 4 Å. Octahedrally coordinated titanium ions located at the center of the perovskite unit cell are the active ions in promoting ferroelectricity. The low-lying d orbitals of Ti lead to displacive phase transformations and large electronic polarizability.

On cooling from high temperature, the crystal structure of $BaTiO_3$ undergoes three ferroelectric phase transitions. All three are displacive in nature with atomic movements of 0.1 Å or less. The point symmetry changes from cubic $m3m$ to tetragonal $4mm$ at the Curie point of 130°C. The tetragonal state with its spontaneous polarization along [001] persists down to 0°C where it transforms to orthorhombic symmetry (point group $mm2$) as $P_{(s)}$ shifts to a [110] direction. On further cooling, the orthorhombic state transforms to rhombohedral (point group $3m$) near −90°C. The structural changes are illustrated in Fig. 16.11. A peak in the dielectric constant occurs at each of the phase transformations. Note in Fig. 16.11 that the dielectric constants along the a and b axes are larger than along the polar c-axis. The instability of the structure makes it easy to tilt the spontaneous polarization vector and a transverse electric field. In regard to capacitor technology, it is extremely important that the dielectric constant is high over a wide temperature range. The presence of the two lower ferroelectric phase changes ensures that the permittivity remains high below T_c.

The two types of domains and domain walls are shown in Figs. 16.12–16.14. Barium titanate 180° domains with polarization along [001] and [00$\bar{1}$] are compared in Fig. 16.12. In this case the walls exhibit ferroelectric behavior but not ferroelasticity, and since there is no strain change across 180° walls, the wall orientations are not controlled by strain mismatch. There is a second criterion,

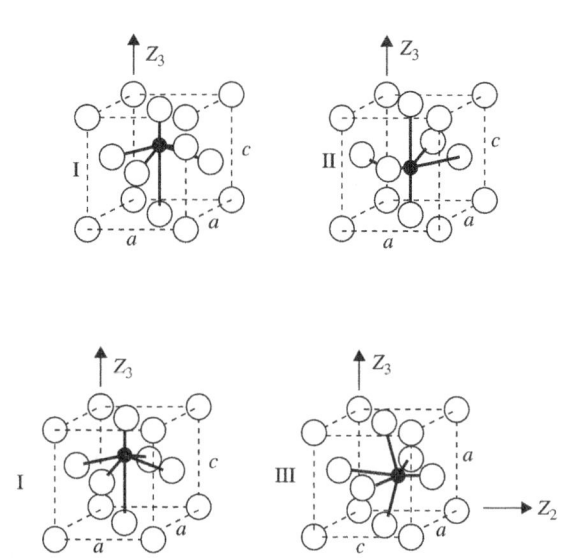

Fig. 16.11 Structural changes occurring at three ferroelectric phase transformations in $BaTiO_3$ result in large values of the dielectric constant over a wide temperature range.

Fig. 16.12 Tetragonal 180° domains in $BaTiO_3$ polarized along $\pm Z_3$. The difference in free energy shows that the walls can be driven electrically with fields along Z_3 but not with mechanical stress.

Fig. 16.13 Crystal structure drawings and free energy terms for two 90° domains polarized along Z_2 and Z_3. Mechanical stresses and electric fields are both effective in moving domain walls.

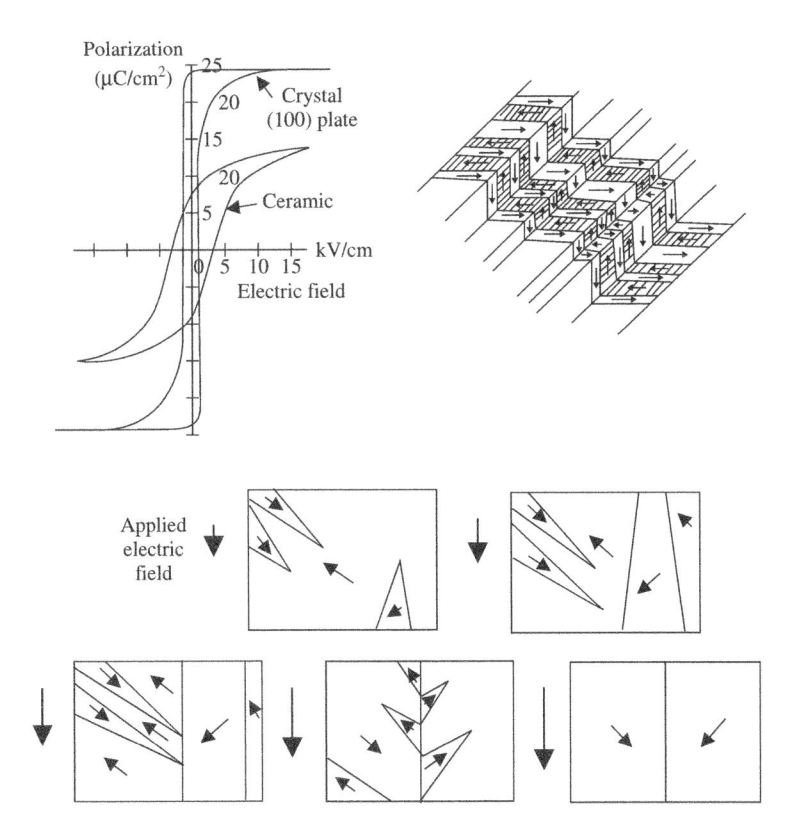

Fig. 16.14 Hysteresis loops for ferroelectric BaTiO$_3$ single crystals and ceramics. Near room temperature, the spontaneous polarization is 0.26 C/m^2. Typical domain structure and field-induced domain wall movements are shown below.

however, that involves the spontaneous polarization. As shown in Fig. 16.12 there is a substantial change in $P_{(s)}$ across 180° walls. Therefore the wall will be charged unless it is parallel to $P_{(s)}$ and to $-P_{(s)}$. For domains polarized along $\pm Z_3$, $(h\ k\ 0)$ planes satisfy this criterion. Similar rules apply for 180° domain walls separating domains polarized along $\pm Z_1$ or $\pm Z_2$.

$$x_{(s)}^{I} = \begin{pmatrix} -x_s/2 \\ -x_s/2 \\ +x_s \\ 0 \\ 0 \\ 0 \end{pmatrix}, \quad x_{(s)}^{II} = \begin{pmatrix} -x_s/2 \\ -x_s/2 \\ +x_s \\ 0 \\ 0 \\ 0 \end{pmatrix}, \quad \Delta x_{(s)} = \begin{pmatrix} 0 \\ 0 \\ 0 \\ 0 \\ 0 \\ 0 \end{pmatrix},$$

$$P_{(s)}^{I} = \begin{pmatrix} 0 \\ 0 \\ P_s \end{pmatrix}, \quad P_{(s)}^{II} = \begin{pmatrix} 0 \\ 0 \\ -P_s \end{pmatrix}, \quad \Delta P_{(s)} = \begin{pmatrix} 0 \\ 0 \\ 2P_s \end{pmatrix},$$

$$\Delta G = -2P_{(s)3}E_3 \Rightarrow \text{Driving force}.$$

The 90° domains in BaTiO$_3$ (Fig. 16.13) show both ferroelectric and ferroelastic behavior. Differences in spontaneous polarization and spontaneous strain for neighboring domains are both substantial. The spontaneous polarization is measured from hysteresis loops (Fig. 16.14) while the spontaneous strain can be estimated from lattice parameters. For a tetragonal crystal, $\Delta x_{(s)} = (c-a)/\sqrt[3]{a^2 c}$.

$$x^{I}_{(s)} = \begin{pmatrix} -x_s/2 \\ -x_s/2 \\ +x_s \\ 0 \\ 0 \\ 0 \end{pmatrix}, \quad x^{III}_{(s)} = \begin{pmatrix} -x_s/2 \\ +x_s \\ -x_s/2 \\ 0 \\ 0 \\ 0 \end{pmatrix}, \quad \Delta x_{(s)} = \begin{pmatrix} 0 \\ -3x_{(s)}/2 \\ +3x_{(s)}/2 \\ 0 \\ 0 \\ 0 \end{pmatrix},$$

$$P^{I}_{(s)} = \begin{pmatrix} 0 \\ 0 \\ P_s \end{pmatrix}, \quad P^{III}_{(s)} = \begin{pmatrix} 0 \\ P_s \\ 0 \end{pmatrix}, \quad \Delta P_{(s)} = \begin{pmatrix} 0 \\ -P_s \\ +P_s \end{pmatrix}$$

$$\Delta G = -P_s(E_3 - E_2) - \frac{3x_s}{2}(X_3 - X_2).$$

The 90° walls are parallel to {110} planes to minimize strain mismatch. Because of the c/a strain ratio, 90° domains are actually 90° 36′ apart resulting in slightly tapered domain wall configurations, and are optically distinguishable in a polarizing microscope.

There are many other ferroelectric oxides with the perovskite, tungsten bronze, pyrochlore, and bismuth titanate layer structure (Table 16.5). Compared with other oxides, all have high dielectric constants, high refractive indices, and large electromechanical coupling coefficients, and all contain corner-linked octahedral networks of Ti^{4+}, Nb^{5+}, or other d^0 ions. These transition-metal elements are the highly polarizable "active" ions promoting ferroelectricity, and the high permittivities and piezoelectric constants required for transducers and capacitors. With reference to the periodic system, there are two major groups of active ions, and both are near electronic "crossover" points where different types of atomic orbitals are comparable in energy and where hybrid bond formation is prevalent. The first group typified by Ti^{4+}, Nb^{5+}, Ta^{5+}, and W^{6+}, consists of d^0 ions octahedrally coordinated to oxygen. For Ti^{4+} the electronic crossover involves the 3d, 4s, and 4p orbitals, which combine

Table 16.5 Ferroelectric crystals, Curie temperatures (T_c), additional phase transitions (T_t), spontaneous polarization, and symmetry groups

Compound	Symmetry	T_c (K)	T_t (K)	$P_{(s)}$ (C/m^2)
Ba$_2$NaNb$_5$O$_{15}$	4/mmm F 4mm F mm2	858	573	
BaTiO$_3$	m3m F 4mm F mm2 F 3m	408	278,183	0.26
Bi$_4$Ti$_3$O$_{12}$	4/mmm F m	949		
(CH$_2$NHCOOH)$_3$H$_2$SO$_4$	2/m F 2	322		0.031
Gd$_2$(MoO$_4$)$_3$	$\bar{4}$2m F mm2	432		0.0017
KH$_2$PO$_4$	$\bar{4}$2m F mm2	122		0.048
KNa(C$_4$H$_4$O$_6$)·4H$_2$O	222 F 2 F 222	297	255	0.025
KNbO$_3$	m3m F 4mm F mm2 F 3m	708	498,263	0.30
LiNbO$_3$	$\bar{3}$m F 3m	1473		0.71
LiTaO$_3$	$\bar{3}$m F 3m	938		0.50
NaNO$_2$	mmm F mm2	437		0.085
Pb$_5$Ge$_3$O$_{11}$	$\bar{6}$ F 3	450		
Pb$_3$MgNb$_2$O$_9$	m3m F 3m	263		
PbNb$_2$O$_6$	4/mmm F mm2	843		
PbTiO$_3$	m3m F 4mm	763		0.57
SbSI	mmm F mm2	295		0.25
Sr$_2$Nb$_2$O$_7$	mmm F mm2	1623		

with the σ and π orbitals of its six O^{2-} neighbors to form a number of molecular orbitals for the $(TiO_6)^{8-}$ complex. The bond energy of the complex can be lowered by distorting the octahedron to a lower symmetry. This leads to molecular dipole moments and ferroelectric and ferroelastic hysteresis as the dipoles reorient under electric fields or mechanical stress. A second group of active elements contributing to polar distortions in oxide dielectrics are the lone-pair ions having two electrons outside a closed shell in an asymmetric hybrid orbital. Among oxides, the most important of these lone-pair ions are Pb^{2+} and Bi^{3+} that are found in a number of ferroelectrics ($PbTiO_3$, $PbNb_2O_6$, $Bi_4Ti_3O_{12}$) with high Curie temperatures. In many of these compounds, Pb^{2+} and Bi^{3+} are in pyramidal coordination with oxygen and therefore contribute to the spontaneous polarization.

Among the other ferroelectric compounds listed in Table 16.5, there are other molecular mechanisms for ferroelectricity. Hydrogen bonding plays a key role in Rochelle Salt, triglycine sulfate, and potassium dihydrogen phosphate. In many water-soluble ferroelectric crystals such as these, the transition from the paraelectric state above T_c to the low-temperature ferroelectric state is essentially an order—disorder phenomenon. Above the transition, protons are statistically distributed in double potential wells. Below T_c they freeze into ordered positions, lowering the symmetry and creating ferroelectric domains.

Problem 16.4
Bismuth titanate, $Bi_4Ti_3O_{12}$, is ferroelectric below the Curie temperature of 949 K where it changes from tetragonal (4/*mmm*) to monoclinic (*m*). At room temperature, the components of spontaneous polarization are $P_{(s)1} = 0.50$, $P_{(s)2} = 0$, and $P_{(s)3} = 0.05$ C/m² along the principal axes. Make plots of the spontaneous polarization as a function of direction for the eight domain states, remembering that Z_1 and Z_2 are equivalent to one another in the high temperature structure.

The spontaneous strains are $x_{(s)11} = -x_{(s)22} = 0.0068$. All other components are zero or near zero. Write out the equation for the difference in free energy between two domain states (ΔG) and determine the components of electric field and mechanical stress that will move the domain wall between them.

16.6 Secondary ferroics: Ferrobielectricity and ferrobimagnetism

Examples of the six types of secondary ferroic phenomena are listed in Table 16.1. Ferrobielectricity is a secondary ferroic effect arising from field-induced electric polarization, rather than spontaneous polarization as in a ferroelectric. Switching between orientation states occurs because of differences in the dielectric permittivity tensor. Permittivity is a secondary tensor like strain and magnetic susceptibility. Any orientation states differing in spontaneous strain will also differ in both electric and magnetic susceptibility. Therefore all ferroelastics are potentially ferrobielectric and ferrobimagnetic.

Ferrobielectricity can be expected in nonpolar crystals with mimetic twinning and substantial dielectric anisotropy. Antiferroelectric materials such as $NaNbO_3$ and $SrTiO_3$ are promising candidates because the dielectric permittivities are large enough to make a sizeable contribution to the induced polarization term in the free energy function. Below 110 K, SrTiO$_3$ is ferroelastic and possibly ferrobielectric. The phase transition involves a symmetry change from cubic (class $m3m$) to tetragonal ($4/mmm$) at low temperature. The TiO$_6$ octahedra of the ideal perovskite structure rotate about a fourfold axis. Alternate octahedra rotate clockwise and counterclockwise causing the structure to crumple about the Sr^{2+} ions. On cooling through the transition the tetragonal c-axis may develop along any of the three cubic edges, giving rise to 90° domains in SrTiO$_3$, the difference in free energy is proportional to $(K_{33} - K_{11}) E^2$. There is no apparent discontinuity in the dielectric constant or its slope at the cubic-tetragonal transition, but anisotropy in the permittivity develops at low temperatures. Dielectric constants as large as 25,000 have been reported below 50 K, where electric double hysteresis loops are observed, along with changes in weak-field permittivity under DC bias. Such behavior may be associated with ferrobielectric domain wall movement. The symmetry change at low temperature is from $m3m$ to $4/mmm$, the same as many of the shape memory alloys.

Multidomain crystals with anisotropic magnetic susceptibility can exhibit ferrobimagnetism. Magnetic susceptibility is a polar second-rank tensor like strain and electric permittivity, therefore ferrobimagnetism has the same symmetry requirements as ferroelasticity and ferrobielectricity. In materials with spontaneous magnetization, ferrobimagnetism will be masked by the larger ferromagnetic effect. The ferrobimagnetic effect is most likely to occur in antiferromagnetic crystals since χ_{ij} is relatively small and nearly isotropic in most paramagnetic and diamagnetic solids.

Antiferromagnetic nickel oxide is both ferroelastic and ferrobimagnetic. At temperatures above the Neel point of 523 K, NiO is paramagnetic with the cubic rocksalt structure. Below T_N, antiferromagnetic ordering of the Ni^{2+} spins results in a small rhombohedral distortion. The unit cell contracts slightly along one of the $\langle 111 \rangle$ body diagonals with the angle between cube axes changing from 90° to 90° 4'. Crystallographic twinning occurs because the contraction may take place along any of the four body diagonals. Each domain is optically uniaxial with the optical axis parallel to the contraction direction. The birefringence ($n_e - n_o = 0.003$ at 5900 Å) is large enough to make domains visible in polarized light.

In a well-annealed crystal, domain walls are easily displaced by a mechanical stress (ferroelasticity) or by a magnetic field (ferrobimagnetism). Elastic energy is lowest for domains with the contraction axis parallel to the applied stress. The walls can be moved distances of several mm and the movement observed with a polarizing microscope. Only small mechanical stresses ($<10^5$ N/m^2) are required to move domain walls. A multi-domain specimen can be converted to an untwinned state by pinching the crystal between thumb and index finger.

Untwinned NiO crystals possess anisotropic magnetic susceptibility. For domains contracted along [111], the magnetic susceptibility parallel to [111] exceeds those measured in the perpendicular directions. In such a domain, spins lie in the (111) plane perpendicular to the [111] contraction directions. As in most antiferromagnetic materials, the magnetic susceptibility is largest perpendicular to the spins.

Moderate magnetic field of 400 kA/m are sufficient to move domain walls in well-annealed crystals. Induced magnetic energy (and total free energy) is

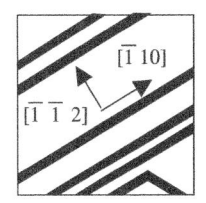

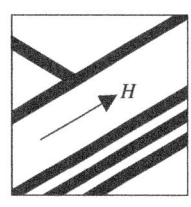

Fig. 16.15 Displacement of domain walls in antiferromagnetic NiO by a magnetic field. The crystal is a thin plate approximately 1 mm on edge with the major face parallel to (111). A magnetic field was first applied along [$\bar{1}\bar{1}2$] and then along [$\bar{1}10$] to produce ferrobimagnetic switching. Domains are visible in polarized light because of the spontaneous strain associated with antiferromagnetic ordering. Viewed between crossed polarizer and analyzer, the major portion of the crystal has domains contracted along [111]. Dark stripes sloping up to the left and up to the right correspond to domains contracted along [$\bar{1}11$] and [$11\bar{1}$], respectively.

minimized when the maximum magnetic susceptibility is parallel to the applied magnetic field. Antiferromagnetic domains with contraction direction parallel to H are favored over other orientations. The response to an applied field is highly erratic because the walls are easily pinned by crystal imperfections. Domain wall movement in ferrobimagnetic NiO is illustrated in Fig. 16.15.

16.7 Secondary ferroics: Ferrobielasticity and ferroelastoelectricity

Ferrobielastic crystals are a class of secondary ferroics in which orientation states differ in elastic compliance, a fourth-rank polar tensor. Ferrobielastic switching in α-quartz has been known for many years. Under applied stress the Dauphiné twins in quartz strain differently. This creates a difference in free energy favoring one domain over the other, causing domain walls to move. Ferrobielasticity is a second order effect in which the strain difference between orientation states is induced by applied stress. When the stress is removed, the induced strain and difference in free energy also disappear. Domain changes under stress can be observed optically because of differences in the photoelastic tensor for the two twin segments. Photoelasticity—the change in refractive indices with stress—is a fourth-rank tensor like elasticity. Orientation states differing in elastic constants will also differ in photoelastic coefficients (Chapter 27).

β-quartz is hexagonal, crystal class 622. On cooling through the phase transition at 573°C, the symmetry is lowered to 32. Transformation twins develop as β-quartz converts to α-quartz. The transformation twins, often called Dauphiné twins or electrical twins, consist of two orientation states related by 180° rotation about [001], the trigonal axis. Dauphiné twins combine two right-handed (or two left-handed) orientation states, often with irregular composition planes. Such twinning renders the crystals useless for piezoelectric applications because it reverses the direction of the Z_1-axis and the signs of the d_{11} piezoelectric coefficients. Because of the importance of piezoelectric quartz in communications applications, techniques for detwinning quartz were developed during the Second World War when untwinned mineral specimens were very scarce.

When referred to the same measurement axes, Dauphiné twin orientation states differ in elastic constants. Class 32 has six independent compliance coefficients: s_{1111}, s_{1122}, s_{1133}, s_{1123}, s_{3333}, and s_{2323}. The twins are related by

180° rotation about Z_3 which reverses the signs of Z_1 and Z_2. Polar tensor coefficients with an odd number of 1 or 2 subscripts change sign under such an operation. Therefore s_{1123} changes sign for the two orientation states but the other coefficients do not. Under an appropriate stress X the difference in free energy between Dauphiné states is proportional to $s_{1123}X^2$. As shown in Fig. 16.16, a uniaxial stress at 45° to Z_2, and Z_3 is effective in switching the ferrobielastic domains.

In matrix form, the applied stress X'_2 creates a difference in free energy of $\Delta G = s_{14}(X_1 X_4 - X_2 X_4 + 2 X_5 X_6) = -s_{14}(X'_2)^2/4$. The coercive stress required to introduce twins into quartz is about 5×10^8 N/m^2 at room temperature but drops to very small values above 250°C.

Atomic movements in the Dauphiné twin operation are small and do not involve the breaking of Si–O bonds. In shifting from one orientation state to other, silicon atoms are displaced by 0.3 Å, and oxygens by about twice amount. Across the composition plane there is a slight difference in bond angles. Dauphiné twinning disappears at the α–β transformation. The atomic structures of the twin states are illustrated schematically in Fig. 16.17.

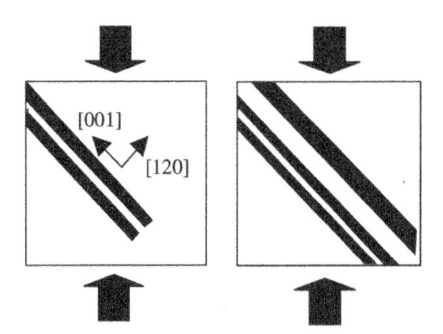

Fig. 16.16 Ferrobielastic switching of Dauphiné twins in quartz produced by uniaxial stress applied at 45° to Z_2 and Z_3. As the mechanical stress is increased slowly from 4.9 to 5.0×10^4 N/m^2, the striped twin pattern changes abruptly. Specimen dimensions are $5 \times 5 \times 3$ mm. Orthogonal property axes (Z_1, Z_2, Z_3) correspond to the [100], [120], [001] crystallographic axes, respectively. The crystal is viewed along Z_1 between crossed polarizer and analyzer. Domains are visible because of the photoelastic effect; the contrast in brightness disappears when the stress is removed.

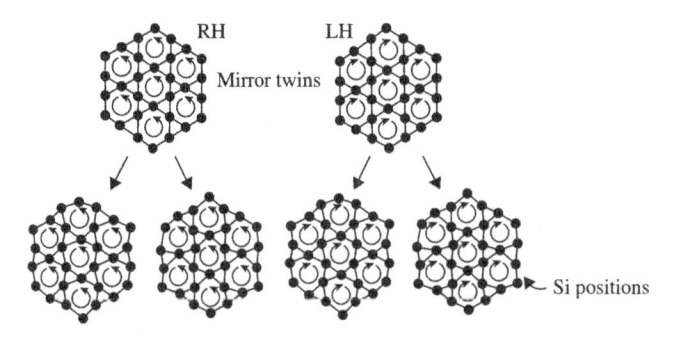

Fig. 16.17 Dauphiné and Brazil twins in quartz. Brazil twins are growth twins that differ in handedness. They are common in natural quartz but are not found in synthetic crystals where the handedness of seed crystals is carefully controlled. Twin walls in Brazil twins are immobile because motion would involve the breaking of Si–O chemical bonds. Dauphiné twins appear on cooling through the α–β phase transition where the symmetry changes from hexagonal (622) to trigonal (32). The slight puckering of the structure can be influenced by mechanical stresses applied in certain directions. Only the Si atoms are shown in the crystal structure drawings projected along Z_3 onto the {001} plane. The circular arrows indicate the direction of the helical spirals in the quartz structure.

16.7 Secondary ferroics: Ferrobielasticity and ferroelastoelectricity

When referred to a common set of axes, the domain states of a true ferroelastoelectric differ in the piezoelectric tensor coefficients. The crystal can be switched from one state to another when an electric field and a mechanical stress are applied simultaneously. A ferroelastoelectric is not simply a ferroelectric that is also ferroelastic. Such materials can be switched by either an electric or mechanical force. *Both* forces are required to switch a true ferroelastoelectric, for it is neither ferroelectric nor ferroelastic.

Since all polar classes are potentially ferroelectric, a likely source of ferroelastoelectrics are the ten nonpolar piezoelectric classes: 222, 32, $\bar{4}$, $\bar{4}2m$, 422, $\bar{6}$, $\bar{6}m2$, 622, 23, and $\bar{4}3m$. Quartz is a potential ferroelastoelectric since Dauphiné twins differ in piezoelectric constants as well as elastic constants, but ammonium chloride (sal ammoniac, NH_4Cl) is the only proven ferroelastoelectric. Ammonium chloride undergoes a near second-order transition at $-30°C$ accompanied by a λ-anomaly in the specific heat. The crystal structure is cubic, both above and below the transition, but the space group changes from $Pm3m$ at room temperature to $P\bar{4}3m$ at low temperatures. The NH_4Cl structure resembles CsCl with Cl at (0, 0, 0) and N at $(\frac{1}{2}, \frac{1}{2}, \frac{1}{2})$. Hydrogens lie along the body diagonals forming N–H–Cl hydrogen bonds. There are two possible orientations for tetrahedral NH_4 group with hydrogens at x, x, x; $x, \bar{x}, \bar{x}$; $\bar{x}, x, \bar{x}$; $\bar{x}, \bar{x}, x$ ($x = 0.347$), or at $\bar{x}, x, x$; $x, \bar{x}, x$; $x, x, \bar{x}$; $\bar{x}, \bar{x}, \bar{x}$. Neutron diffraction data recorded at room temperature indicate random disorder between the two orientations. Measurements below the transition at liquid air temperature have established an ordered model with only one set of positions occupied (Fig. 16.17). In the absence of external forces the two orientation states are equal in energy, giving rise to domains at low temperatures. Reflection across (100) brings the two states into coincidence. This is a very subtle type of twinning since the physical properties of the two orientation states are nearly identical. Only through third-rank tensor properties such as piezoelectricity and the linear electro-optic effect can the two states be distinguished.

Crystal class $\bar{4}3m$ has but one independent piezoelectric modulus d_{123} relating polarization along [100] to a shearing stress about [100]: $P_1 = d_{123}X_{23}$. For the two orientation states, d_{123} is equal in magnitude but opposite in sign. Reflection across (100) takes Z_1 to $-Z_1$, and leaves Z_2 and Z_3 unchanged. Therefore d_{123} transforms to $-d_{123}$ for two domains related by a mirror parallel to (100).

Ammonium chloride is a potential ferroelastoelectric because its two orientation states differ in piezoelectric coefficients (Fig. 16.18). Applying a uniaxial stress X along [011] together with an electric field E along [100] leads to a difference in free energy $\Delta G = 2d_{123}XE$. Domain switching will take place if the driving potential ΔG is large enough to overcome the resistance to domain wall motion. In matrix notation the full difference in free energy is

$$\Delta G = 4d_{14}(E_1 X_4 + E_2 X_5 + E_3 X_6),$$

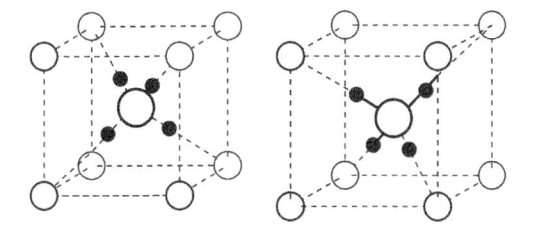

Fig. 16.18 Twin orientation states in ferroelastoelectric ammonium chloride at low temperatures. The twins are related by reflection across (100) which reverse the signs of the piezoelectric coefficients $d_{14} = d_{25} = d_{36}$.

where d_{14} is about 6 pC/N and the coercive stress-field is about 1 MN/m² × 1 MV/m at $-30°$C.

Problem 16.5
Dauphiné twin walls in quartz can be moved using the ferroelastoelectric effect. There are two independent piezoelectric coefficients in α-Quartz: d_{11} and d_{14}. Determine what components of electric field and mechanical stress are needed to move domain walls.

16.8 Secondary ferroics: Ferromagnetoelectrics and ferromagnetoelastics

The domains of a ferromagnetoelastic material differ in piezomagnetic coefficients. Siderite (FeCO$_3$) is antiferromagnetic below 30 K. The magnetic structure (Fig. 16.19) consists of antiparallel Fe^{2+} spins aligned along the hexagonal c-axis. Siderite belongs to crystal class $\bar{3}m1'$, and magnetic point group $\bar{3}m$. Symmetry elements $\bar{3}'$ and m', in which the spatial operation is accompanied by time reversal, are absent. Crystals with magnetic symmetry $\bar{3}m$ are potentially piezomagnetic (Table 14.5). There are two independent piezomagnetic coefficients, Q_{222} and Q_{123}. In matrix notation, these tensor coefficients correspond to Q_{22} and Q_{14}. Q_{222} relates a tensile stress along Z_2 to a magnetization in the same direction; Z_2 is the crystallographic [120] direction perpendicular to both the twofold (Z_1) and threefold (Z_3) symmetry axes. Piezomagnetic coefficient Q_{123} relates the magnetization component along Z_1 resulting from a shearing stress about Z_1.

The piezomagnetic effect has been studied in iron carbonate crystals at liquid hydrogen temperature using a magnetic torsion balance in which a press containing the specimen is suspended between the pole pieces of the magnet. Q_{123} was measured, but Q_{222} was below the limit of observation. The magnitude of Q_{123} is sensitive to bias during annealing. When cooled through the Neel point without stress bias, the effect was smaller, presumably because of antiferromagnetic domains. Domains in antiferromagnetic siderite are of the 180° type in which all spins are reversed (Fig. 16.19). The magnetic structures of neighboring domains are related by reflection across $(2\bar{1}0)$ accompanied by time reversal m' converting Q_{123} to $-Q_{123}$, so that the piezomagnetic coefficient is of opposite sign for the two domains. Siderite is therefore a potential ferromagnetoelastic crystal in which domains can be switched by applying mechanical

Fig. 16.19 Domain states in antiferromagnetic FeCO$_3$ at low temperatures. The two magnetic structures are related by the time reversal operator $1'$ which reverses all spin directions. Annealing in combining magnetic fields and mechanical stress is required before measuring the piezomagnetic effect. Carbonate groups are not shown.

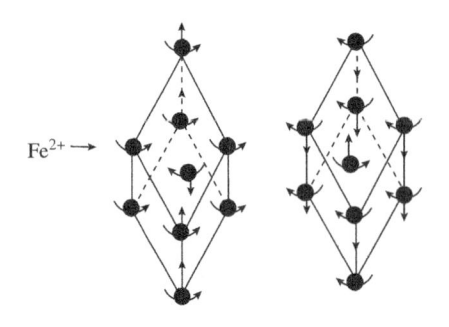

stress and magnetic field simultaneously. The field should be directed along Z_1 together with a shearing stress about Z_1.

The magnetoelectric coupling between electric and magnetic variables was described earlier in Section 14.9. In analytic form the electrically-induced magnetoelectric effects is $I_i = Q_{ij}E_j$, and the magnetically induced effect is $P_i = Q_{ij}H_j$. Magnetoelectric coefficients have been measured for about twenty materials, including Cr_2O_3, $LiFePO_4$ (triphylite) and $LiMnPO_4$ (lithiophilite). $LiMnPO_4$ undergoes a paramagnetic-antiferromagnetic phase transition at 50 K. Magnetic susceptibility data collected in the paramagnetic region show typical Curie–Weiss behavior with an effective atomic moment of 5.45 μB and an extrapolated Neel temperature of 88 K.

Triphyllite and lithiophyllite are isostructural with olivine; lattice parameters for the orthorhombic unit cell are $a = 10.31$, $b = 6.00$, $c = 4.69$ Å. The space group is $Pnma1'$ with four molecules per unit cell (Fig. 14.18). Divalent iron atoms occupy mirror plane positions (equipoint $4c$) with coordinates $\pm(0.28, 0.25, 0.98; 0.22, 0.75, 0.48)$. In triphyllite low temperature neutron diffraction studies gave a magnetic structure in which two of the Fe^{2+} spins are parallel to $+b$, the other two to $-b$. All four spins are reversed for the antiferromagnetic 180° domain.

The magnetic structure of triphylite conforms to magnetic point group mmm', one of the magnetoelectric groups. The symbol m' means that the mirror operation perpendicular to c includes a time reversal operator. Time reversal flips the spins by 180° since magnetic moments are associated with moving electric charge. The only nonzero magnetoelectric coefficients for mmm' are $Q_{12} = Q_{21}$.

Magnetoelectric measurements provide ample evidence for the existence of antiferromagnetic domains. The magnetoelectric coefficient Q_{12} is identical in magnitude for the two domains but opposite in sign. If the sample is raised above the Neel temperature and cooled through the transition, the sign of Q can be positive or negative. Rapid cooling produces both kinds of domains. Powder specimens exhibit no magnetoelectric effect unless annealed in bias fields to remove the degeneracy between two domains. The principle of the method is quite simple. In an electric field the induced magnetization for one domain is opposite to that of the other. If a magnetic field is then applied, the energies differ for the two time-reversed structures, making one more probable than the other. Poling works best just below the Neel point where coercive fields are usually smallest.

16.9 Order parameters

The macroscopic classification of ferroic crystals into primary and secondary ferroics is helpful in designing experiments, but provides very little understanding of the underlying atomistic phenomena involved in the phase transformation responsible for the twinning. A more useful approach is based on an order parameter Q and identifying its atomistic origin. The basic ideas of Landau theory are introduced in this section.

Ferroic phase transformations involve a change in symmetry from a high temperature prototype state to a low temperature ferroic state. The point group symmetry of the ferroic state is a subgroup of the prototype state. In the feldspar family, for instance, this symmetry change is represented by the symmetry species $2/m\ F\bar{1}$. The symmetry elements 2 and m are lost on cooling through the phase transformation but the center of symmetry $\bar{1}$ is retained as the crystal changes from monoclinic to triclinic. $\bar{1}$ is a subgroup of $2/m$ since both phases possess a center of symmetry.

The breaking of symmetry at the phase transition introduces a new variable called the order parameter. This order parameter Q is used to describe the difference in free energy ΔG between the ferroic state and the prototype phase. It is this difference in free energy that stabilizes the ferroic state. (Note that this ΔG is different than the ΔG in Section 16.1. There we were comparing the difference in free energy between two domain states. Here we are comparing the difference in free energy between the high and low temperature phases.)

This difference in free energy, sometimes called the excess free energy, depends on temperature through the order parameter Q. A typical relationship is

$$\Delta G = \left(\frac{A}{2}\right)(T - T_c)Q^2 + \left(\frac{B}{4}\right)Q^4 + \left(\frac{C}{6}\right)Q^6 + \cdots,$$

where A, B, and C are independent of temperature and T_c is the transition temperature. The exact form of the power series depends on the nature of the transition and the symmetries involved.

The structural meaning of the order parameter may be related to a macroscopic quantity such as polarization or strain, or it may come from the softening of an optic mode, an acoustic mode, or a spin mode (Fig. 16.20). Order–disorder phenomena involving hydrogen bonds or atom exchange are also potential order parameters. Some ferroics have more than one order parameter.

The displacive phase transition in the ferroelastic feldspars (Section 16.2) is strongly influenced by Al–Si ordering in the tetrahedral sites of the structure. Similar coupling between displacive transitions and cation ordering occurs in relaxor ferroelectrics like $Pb_3MgNb_2O_9$ and shape memory alloys such as NiTi. This often leads to broad, diffuse phase transitions.

Polarization is the order parameter in barium titanate and most other ferroelectric crystals. The excess free energy is expanded as a power series in P

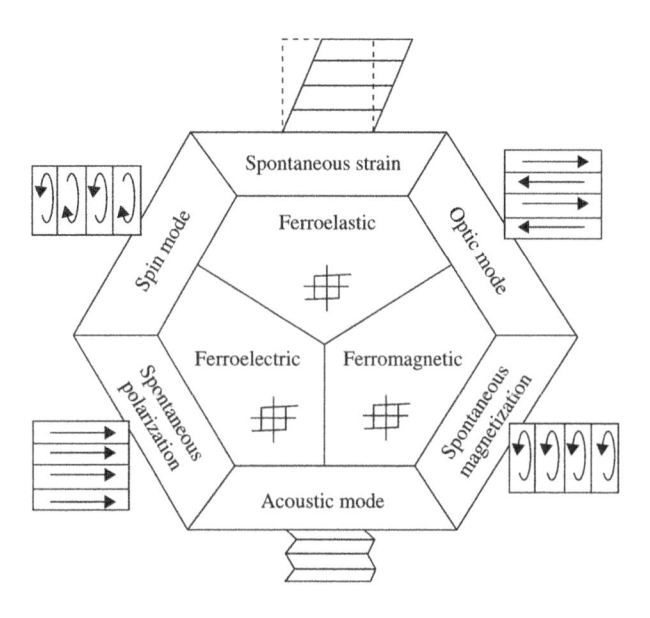

Fig. 16.20 Ferroic phase transformations have a number of different origins. In a *proper ferromagnet*, the order parameter is proportional to the spontaneous magnetization, but ferromagnetism can also arise from acoustic modes, spontaneous strain, and other improper causes. In weak ferromagnets the spontaneous magnetization is coupled to an antiferromagnetic spin mode. Polarization is the order parameter for proper ferroelectrics, and strain for proper ferroelastics.

and the various coefficients are determined by experiment using the measured values of spontaneous polarization, dielectric permittivity, pyroelectric effect, and spontaneous strain. *Proper ferroelectrics* like BaTiO$_3$ and KH$_2$PO$_4$ are characterized by a dielectric instability in which the dielectric constant follows a Curie–Weiss law

$$K = \frac{C}{(T - T_o)},$$

where C is the Curie constant and T_o is approximately equal to T_c, the transition temperature.

Proper ferroelectrics can be divided into two main groups. Displacive ferroelectrics like BaTiO$_3$, PbTiO$_3$, and KNbO$_3$ were discussed in Section 16.5. Perovskite ferroelectrics and related oxides have Curie constants of about 10^5 K. The second group of hydrogen-bonded ferroelectrics have much smaller Curie constants near 10^3 K. Triglycine sulfate, potassium dihydrogen phosphate, Rochelle salt, and other members of this family are driven by an order–disorder transformation in which protons oscillate in a double potential well. Below T_c they freeze into an ordered configuration creating the spontaneous polarization.

Improper ferroelectrics behave very differently. The classic example is gadolinium molybdate, Gd$_2$(MoO$_4$)$_3$, that has a dielectric constant of about 10 that hardly changes at T_c. The order parameter is a soft acoustic mode that couples to the spontaneous polarization through a piezoelectric coefficient. The spontaneous polarization in gadolinium molybdate is a hundred times smaller than barium titanate.

Proper ferroelastics, in which the order parameter is linearly coupled to the spontaneous strain, are relatively rare. Arsenic pentoxide (As$_2$O$_5$) and disordered feldspars are the best examples. Most ferroelastics are *improper ferroelastics* in which strain is not the order parameter.

The excess free energy and the order parameter describe the underlying causes of ferroic behavior. Lead zirconate titanate (PZT = PbZr$_{1-x}$Ti$_x$O$_3$) is a proper ferroelectric that is widely used as a piezoelectric transducer. Compositions near the morphotropic phase boundary at $x = 0.5$ have very large electromechanical coupling coefficients. The properties and phase diagram were described in Section 12.7. Above T_c the perovskite structure is cubic, point group $m3m$. Below T_c the ferroic phase is rhombohedral ($3m$) for $x < 0.5$ and tetragonal ($4mm$) for $x > 0.5$. For these symmetries, the leading terms in the excess free energy function is

$$\Delta G = \frac{A(T - T_c)(P_1^2 + P_2^2 + P_3^2)}{2} + \frac{B_\mathrm{I}(P_1^4 + P_2^4 + P_3^4)}{4} + \frac{B_\mathrm{II}(P_1^2 P_2^2 + P_3^2 P_1^2 + P_2^2 P_3^2)}{2} + \cdots.$$

In the cubic paraelectric state above T_c, the spontaneous polarization disappears and $\Delta G = 0$ when $P_1 = P_2 = P_3 = 0$. Just below T_c for Ti-rich compositions ($x > 0.5$), the excess free energy is maximized for the tetragonal polar state in which $P_1 = P_2 = 0$ and $P_3 = P_s$. In this case $\Delta G_T = B_\mathrm{I} P_s^4/4$. For the Zr-rich compositions at $x < 0.5$, the rhombohedral phase is stabilized with $P_1 = P_2 = P_3 = P_s/3^{1/2}$. The excess free energy for the rhombohedral state is $\Delta G_R = B_\mathrm{I} P_s^4/12 + B_\mathrm{II} P_s^4/6$. Comparing ΔG_R with ΔG_T, we see that

$\Delta G_R = \Delta G_T$ when $B_\mathrm{I} = B_\mathrm{II}$. This corresponds to the morphotropic phase boundary near $x = 0.5$. When $B_\mathrm{II} > B_\mathrm{I}$ the tetragonal state is stable, and when $B_\mathrm{I} > B_\mathrm{II}$ PZT is rhombohedral.

The PZT properties are especially useful near the tetragonal–rhombohedral phase boundary where $B_\mathrm{I} = B_\mathrm{II} = B$. Substituting this condition into ΔG gives

$$\Delta G = \frac{A(T - T_c)(P_1^2 + P_2^2 + P_3^2)}{2} + \frac{B(P_1^2 + P_2^2 + P_3^2)^2}{4} + \cdots .$$

The dielectric, elastic, and piezoelectric properties near the boundary can be derived using the thermodynamic relationships in Chapter 6. When this is done, it can be shown that for compositions on the tetragonal side of the boundary $K_{11} = K_{22} \sim 1/(B_\mathrm{II} - B_\mathrm{I}) \to \infty$ and $c_{44} = c_{55} \sim B_\mathrm{I} - B_\mathrm{II} \to 0$. At the phase boundary PZT is transversely unstable with large dielectric constants and large elastic shear compliance perpendicular to the polarization direction. It is this transverse instability that is responsible for the remarkable piezoelectric properties near the morphotropic phase boundary.

Over the past several decades there has been a growing interest in the fundamental causes of ferroic phase transitions, focusing on soft modes, order parameters, and improper transitions. Cross-coupled improper domain phenomena caused by electric, elastic, and magnetic interactions result in unusual ferroic behavior. Lithium ammonium tartrate, for instance, is a type of *elastoferroelectric* in which mechanical strain is the primary order parameter at the 98 K phase transition. Strain gives rise to ferroelectricity through piezoelectric coupling with the polarization. The opposite effect occurs in sodium potassium tartrate (Rochelle salt) where electric polarization is the primary order parameter. Domains can be switched with mechanical stress as well as with electric fields because of the small spontaneous strain resulting from piezoelectric coupling to the polarization. Since the ferroelastic effect has its origin in an electric instability, we refer to it as *electroferroelasticity*. An even more subtle type of instability takes place in lithium thallium tartrate in which the order parameter resides in the electromechanical coupling coefficient.

Other interesting cross-coupling effects arise when magnetic phenomena are included. There are a number of examples of *elastoferromagnetism* and *magnetoferroelasticity* and at least one good example of *electroferromagnetism*: nickel iodine boracite. At room temperature, NiI boracite is cubic, point group $\bar{4}3m$. Below room temperature at 120 K, it undergoes a transition to an antiferromagnetic state as the Ni^{2+} moments align; at this stage, the material is an antiferromagnetic piezoelectric, but it is neither ferromagnetic nor ferroelectric. On further cooling, a second phase transition to an orthorhombic ferroelectric state takes place at 64 K. As the crystal structure develops a spontaneous polarization, the magnetic structure is also altered, destroying the balance of spins in the antiferromagnetic state and producing a weak ferromagnetism. The ferromagnetic effect is of electric origin and can be referred to an *electroferromagnet*.

The opposite effect is *magnetoferroelectricity* in which a reversible spontaneous polarization is developed on passing through a magnetic phase transition. Chromium chrysoberyl (Cr_2BeO_4) exhibits such an effect at low temperature. Cr_2BeO_4 is orthorhombic and isostructural with the minerals chrysoberyl (Al_2BeO_4) and forsterite (Mg_2SiO_4). The crystal structure consists of a close-packed lattice of oxygen ions with Cr^{3+} in octahedral sites and

Be^{2+} in tetrahedral positions. At room temperature, Cr_2BeO_4 is paramagnetic and centrosymmetric (point group $mmm1'$) and, therefore, nonpiezoelectric, nonpyroelectric, and nonferroelectric.

Low-temperature magnetic susceptibility measurements show a cusp-shaped peak at 28 K characteristic of a paramagnetic–antiferromagnetic transition. The magnetic structure determined by neutron diffraction at 4.5 K is a cycloidal spiral with a periodicity of about 65 A, roughly 12 unit-cell lengths. Antiferromagnetic resonance experiments confirmed the spiral structure and Neel temperature. The symmetry of the magnetic structure is triclinic, point group 1, making chromium chrysoberyl potentially magnetoferroelectric below 28 K.

When electrically poled, chromium chrysoberyl shows a weak pyroelectric effect in its antiferromagnetic state. The sign of the pyroelectric coefficient changes when the bias field is reversed. Like other ferroelectrics, the pyroelectric effect disappears above the transition. The reversible spontaneous polarization in the magnetoferroelectric state is approximately a million times smaller than that of $BaTiO_3$.

Magnetoferroelectrics are a type of improper ferroelectric, like gadolinium molybdate, in which polarization is not the order parameter driving the transformation. Because of the weakness in coupling between magnetic and electric effects, magnetoferroelectrics might be termed the ultimate impropriety.

17 Electrical resistivity

17.1 Tensor and matrix relations 188
17.2 Resistivity measurements 189
17.3 Electrode metals 191
17.4 Anisotropic conductors 193
17.5 Semiconductors and insulators 194
17.6 Band gap and mobility 196
17.7 Nonlinear behavior: Varistors and thermistors 199
17.8 Quasicrystals 202

The next six chapters describe the transport phenomena associated with the flow of charge, heat, and matter. In each case there is a vector *flux* that is governed by a vector *field*. Linear relationships between flux and field include electrical resistivity (Chapter 17), thermal conductivity (Chapter 18), diffusion (Chapter 19), and thermoelectricity (Chapter 21). All are represented by second rank tensors similar to electric permittivity (Chapter 9), but the underlying physics is somewhat different. Transport properties are nonequilibrium phenomena governed by statistical mechanics and the concept of microscopic reversibility, rather than the second law of thermodynamics that applies to equilibrium properties such as specific heat, permittivity, and elasticity.

Higher order tensors appear when the transport experiments are carried out in the presence of magnetic fields or mechanical stresses. Galvanomagnetic, thermomagnetic (Chapter 20), and piezoresistance effects (Chapter 22) require third- and fourth-rank tensors.

17.1 Tensor and matrix relations

When an electric field is applied to a conductor, an electric current flows through the sample. The field E_i (in V/m) is related to the current density J_j (in A/m^2) through Ohm's Law, where ρ_{ij} is the electrical resistivity (in Ω m). In tensor form,

$$E_i = \rho_{ij} J_j.$$

E_i and J_j are polar vectors (first rank polar tensors) and ρ_{ij} is a second rank polar tensor property which follows Neumann's law in the usual way. Sometimes it is more convenient to use the reciprocal relation involving the electrical conductivity σ_{ij}:

$$J_i = \sigma_{ij} E_j.$$

Conductivity is expressed in Ω^{-1} m^{-1} or S/m.

In matrix form

$$\begin{array}{cccccccc} 3\times 1 & & 3\times 3 & 3\times 1 & & 3\times 1 & & 3\times 3 & 3\times 1 \\ (E) & = & (\rho) & (J) & \text{and} & (J) & = & (\sigma) & (E). \end{array}$$

The reciprocal relation between conductivity and resistivity is derived as follows

$$(E) = (\rho)(J) = (\rho)(\sigma)(E).$$

Therefore $(\rho)(\sigma) = 1$ or $(\rho) = (\sigma)^{-1}$.

Individual components are related through the relation

$$\rho_{ij} = \frac{(-1)^{i+j}\Delta^{\sigma}_{ij}}{\Delta^{\sigma}},$$

where Δ^{σ} is the determinant of the conductivity matrix and Δ^{σ}_{ij} is the cofactor of conductivity component σ_{ij}.

As an illustration, find the resistivity matrix (ρ) corresponding to the conductivity matrix

$$(\sigma) = \begin{pmatrix} 3 & 1 & 0 \\ 1 & 4 & 0 \\ 0 & 0 & 2 \end{pmatrix} \text{ S/m}$$

$$\rho_{11} = \frac{(-1)^2(8-0)}{24-2} = \frac{4}{11}$$

$$\rho_{12} = \frac{(-1)^3(2-0)}{24-2} = -\frac{1}{11} = \rho_{21}$$

$$\rho_{22} = \frac{(-1)^4(6-0)}{24-2} = \frac{3}{11} \quad \text{etc.}$$

$$(\rho) = \frac{1}{11}\begin{pmatrix} 4 & -1 & 0 \\ -1 & 3 & 0 \\ 0 & 0 & 5 \end{pmatrix} \Omega \text{ m}.$$

Note that in general $\rho_{ij} \neq 1/\sigma_{ij}$.

17.2 Resistivity measurements

Resistivity and conductivity range over many orders of magnitude (Fig. 17.1) For superconductors $\rho = 0$, and for common metals $\rho \sim 10^{-6}$–10^{-8} Ω m. For good insulators $\rho \sim 10^{12}$–10^{16} Ω m. Semiconductors like Si and Ge lie in between, typically in the 10^{-2}–10^2 Ω m range.

Electrical resistivity is defined in terms of Fig. 17.2, where the resistivity ρ is given by

$$\rho = \frac{VA}{IL}$$

and is measured in ohm m, provided the current I is measured in amperes, the voltage drop V in volts, and the dimensions in meters. The simplest technique involves measuring the voltage drop across the sample and the current through the sample. If a voltmeter and ammeter are used, it is called the I–V measurement (Fig. 17.2(a)).

When the contacts at the ends of the sample have appreciable resistance, as in many experiments on metals and semiconductors, the simple I–V method is subject to serious errors. Fig. 17.2(b) illustrates the potential-probe method which two extra electrodes are used to eliminate errors associated with contact resistance. In this technique, the voltage drop is measured across the probes, and the probe separation D replaces the sample length L. The potential drops used are the average obtained for both directions of current

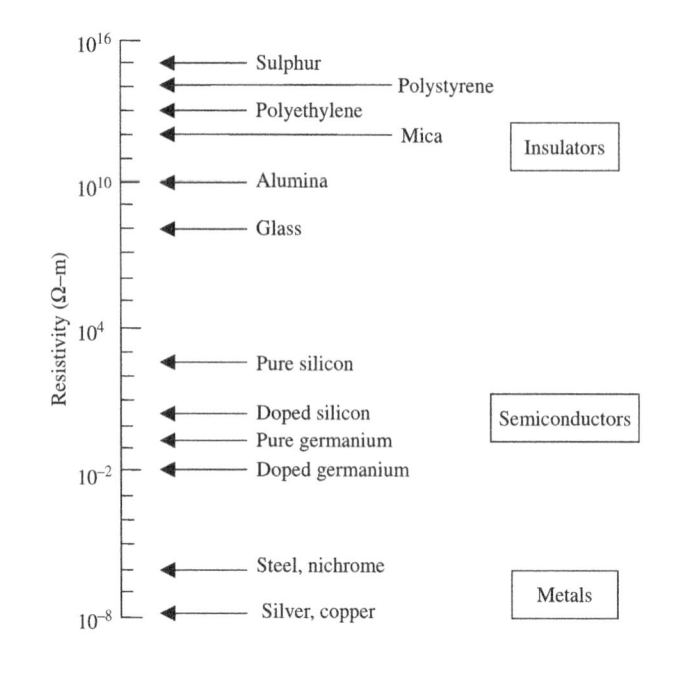

Fig. 17.1 Electrical resistivities vary over many orders of magnitude for metals, semiconductors, and insulators.

Fig. 17.2 Two ways of measuring electrical resistivity. (a) The *I–V* method using a voltmeter and ammeter. (b) the potential-probe method using two extra contacts to measure the voltage drop.

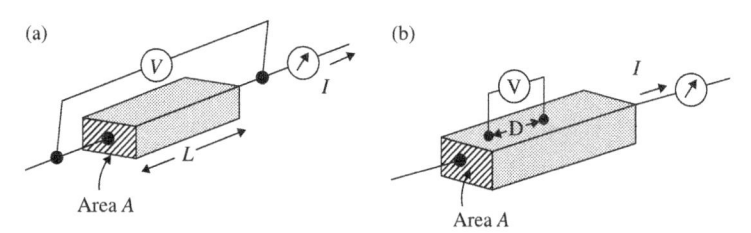

flow. This procedure eliminates some of the errors caused by thermoelectric effects or rectifying boundaries. Further improvement is achieved by using a potentiometer that draws very little current. With this method the resistivity is

$$\rho = \frac{V_D A}{DI},$$

where V_D is the potential drop across the probes and D is the distance between probes. The potential-probe method can be used for DC and AC measurements. A long thin sample shape defines the current flow direction. This is especially important in anisotropic materials where current and field directions often diverge. To simplify the field patterns, electrical measurements are taken along principal axes whenever possible.

Problem 17.1
A metal alloy belonging to point group $2/m$ crystallizes as very thin plates oriented perpendicular to the twofold symmetry axis. Describe how the resistivity coefficients can be measured, specifying the shape and orientation of the samples. How are the principal axes located? How will they change with temperature?

17.3 Electrode metals

Nearly all of the commonly used electrode metals are cubic, and therefore electrically isotropic. Four important properties are the electrical resistivity, melting point, thermal conductivity, and thermal expansion (Table 17.1).

In the processing of capacitors, resistors, and other electrical components, the electrodes are often sintered with the component. Melting points and thermal expansion coefficients are therefore important. Copper, silver, and gold, all excellent conductors, are limited to 1000°C, and are obviously unsuitable for co-firing with alumina substrates at 1500°C. Tungsten and molybdenum are a better match, despite their higher resistivity.

For many years silver–palladium alloys have been used as electrode metals for $BaTiO_3$ multilayer capacitors. Precious metals like Pd are expensive but they raise the melting point allowing silver to be cofired with barium titanate. The Ag–Pd phase diagram (Fig. 17.3) shows that the two metals form a complete solid solution with a continuous increase in melting point. Many of the electrode metals have face-centered cubic crystal structures (point group $m3m$) with lattice parameters near 4 Å. The phase diagrams and resistivity curves are generally similar to those of the Ag–Pd system. For solid solutions such as these the resistance of the alloy is greater than that of the end members. The random distribution of Ag and Pd atoms reduces the electron mobility and increases the electrical resistance. This is part of what is often referred to as *Matthiessen's Rule*. The electrical resistivity of dilute metal alloys $\rho(c, T)$ is partly dependent

Table 17.1 Physical properties of various electrode metals: melting point (MP, °C), electrical resistivity (ρ, 10^{-8} Ω m), thermal expansion (α, 10^{-6}/K), and thermal conductivity (k, W/m K)

Metal	MP (°C)	ρ ($\times 10^{-8}$ Ω m)	α ($\times 10^{-6}$/K)	k (W/m K)
Silver (Ag)	960	1.6	19.7	418
Gold (Au)	1063	2.2	14.2	297
Copper (Cu)	1083	1.7	17.0	393
Lead (Pb)	327	19.2	29.0	35
Palladium (Pd)	1552	10.8	11.0	71
Platinum (Pt)	1774	10.6	9.0	71
Nickel (Ni)	1455	6.8	13.3	92
Chromium (Cr)	1900	20.1	6.3	67
Molybdenum (Mo)	2625	5.2	5.0	146
Tungsten (W)	3415	5.5	4.5	201
Aluminum (Al)	660	2.5	23.8	238

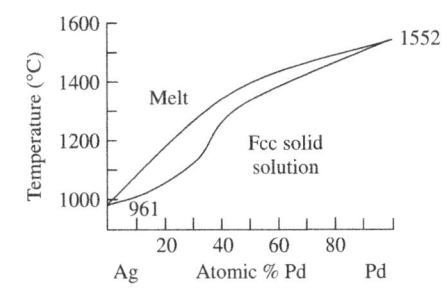

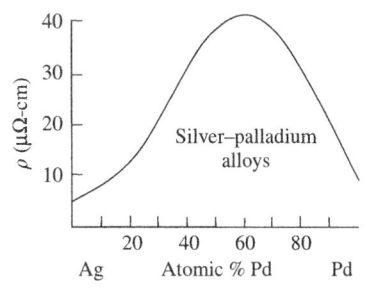

Fig. 17.3 The phase diagram and room temperature resistivity of Ag–Pd alloys used as electrodes in electroceramics. Melting points increase smoothly from Ag to Pd but the resistivity is greatest at intermediate compositions. Similar behavior is observed with many other electrode alloys including the Pd–Cu, Pd–Au, Pd–Ni, Pd–Pt, Ag–Au, and Ag–Pt systems.

on composition (c) and partly on temperature (T):

$$\rho(c, T) = \rho(c) + \rho(T).$$

As the equation indicates, the composition dependence and the temperature dependence are additive. Electrons are scattered by solute atoms and by thermal vibrations, giving rise to the total resistance.

Fig. 17.4 illustrates the composition dependence for Ag–Al alloys. For dilute alloys the resistivity increases linearly with composition, with the same increase occuring at room temperature and at liquid helium temperature. The $\rho(c)$ term is linearly proportional to c and independent of T.

Linde's Law further refines the composition dependence. Electrical resistivity increases more rapidly for some solutes than other. The resistivity of silver doped with elements of different valence is shown in Fig. 17.5. For a given host element, the increase in resistivity is proportional to the product of the solute concentration (c) times the square of the difference in valence between the host and solute element, $(\Delta Z)^2$:

$$\Delta \rho \sim c(\Delta Z)^2.$$

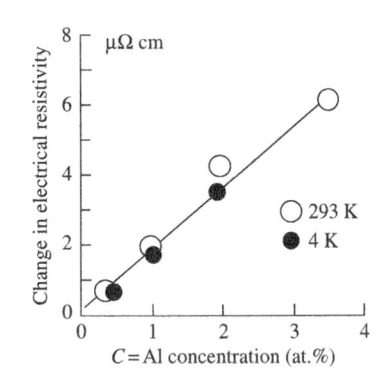

Fig. 17.4 The electrical resistivity of Ag–Al concentration. Data taken at high- and low-temperature show the same linear dependence. For dilute alloys the increase in resistivity is temperature independent.

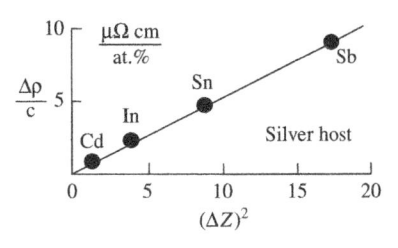

Fig. 17.5 Linde's Law for silver-based alloys doped with 1% of several different metals. The increase in resistivity depends on the type of dopant. For elements in different columns in the periodic system the increase is proportional to concentration c and also to $(\Delta Z)^2$, where ΔZ is the difference in valence.

In the case of silver, the increase in resistivity for Cd dopant is much less than that for In, Sn, and Sb. Silver is in Column I of the periodic system, Cd in II, In in III, Sn in IV, and Sb in V. Therefore the electron scattering power caused by the charge difference is greater for Sb than for Cd. Linde's Law is used to predict which solute elements increase the resistivity most rapidly. It works satisfactorily for alloys based on Ag, Au, Cu, Cd, Al, and Zn.

The effect of temperature on the resistivity of metals is remarkably constant across the periodic system. Fractional changes in ρ resistivity with temperature are listed in Table 17.2. With a few exceptions most of the values fall in the range 0.004 to 0.006/K. This is the second term $\rho(T)$ that appears in Matthiessen's Rule. Electrons are scattered by phonons leading to a linear increase in resistivity with temperature (Fig. 17.6).

Phase transitions can have a pronounced effect on electrical resistance. Resistivity values for a number of conducting oxides are compared in Fig. 17.7. Most contain transition metal elements with overlapping d orbitals. Some, like ReO_3, behave like electrode metals with a slight increase in resistivity with temperature. Others such as VO_2 show a large anomaly indicating a phase transition from a semiconducting state at low temperature to a metallic state at room temperature.

Thick film resistive glazes are widely used in hybrid microelectronic systems. Many of the successful resistor formulations are based on highly conducting oxides, with low temperature coefficients. Beginning with the PdO–Ag glaze introduced many years ago, the oxides of platinum-family metals have found wide application in the thick film industry.

Barium ruthenate is typical of the family. Ceramic specimens of $BaRuO_3$ are highly conducting ($\rho \sim 10^{-5}$ Ω m) with a positive temperature coefficient indicative of metallic-type conductivity. RuO_6 octahedra share faces in the $BaRuO_3$ structure bringing ruthenium ions into very close contact. In fact, the Ru–Ru distance for such pairs is 2.55 Å, which is shorter than the separation in ruthenium metal (2.65 Å). This suggests overlap of the d orbitals, leading to collective behavior and metallic conductivity.

Table 17.2 Electrical resistivity of metals increases with temperature. The resistivity ρ is in units of $\mu\Omega$ cm

	ρ	$(1/\rho)(d\rho/dT)$
Electrode metals		
Copper	1.54	0.0046/K
Silver	1.47	0.0040
Gold	2.03	0.0040
Aluminum	2.43	0.0046
Alkali metals		
Lithium	8.50	0.0044
Sodium	4.29	0.0048
Potassium	6.20	0.0050
Rubidium	11.24	0.0051
Cesium	18.0	0.0048
Alkaline earths		
Beryllium	2.71	0.0090
Magnesium	4.10	0.0042
Calcium	3.08	0.0040
Strontium	11.0	0.0040
Barium	29.8	0.0049
Transition metals		
Iron	8.70	0.0058
Cobalt	5.49	0.0056
Nickel	6.24	0.0059
Molybdenum	4.88	0.0047
Palladium	9.82	0.0040
Platinum	9.74	0.0039

17.4 Anisotropic conductors

Electrical resistivity and electric conductivity are symmetric second rank tensors like the dielectric constant. Cubic crystals and amorphous materials have the same resistivity in all directions so that ρ and σ are scalar properties. Tetragonal, hexagonal, and trigonal crystals have two independent coefficients measured parallel and perpendicular to the major symmetry axis, $Z_3 = [001] = c$. Lower symmetry crystals have three independent coefficients. Table 17.3 lists the room temperature resistivities of several anisotropic metals.

There is usually a change in anisotropy with temperature. Zinc and cadmium become more isotropic with increasing temperature.

The copper oxide superconductor compounds exhibit highly anisotropic electrical resistivity above T_c. Yttrium barium cuprate, $YBa_2Cu_3O_7$, has planar copper oxide layers separated by barium and yttrium ions (Fig. 17.8). Electrical resistivity is much lower parallel to the conducting Cu–O bonds than in the perpendicular direction.

The electrical anisotropy of nonstoichiometric rutile, TiO_{2-x}, can also be very large. In these crystals the titanium sites are occupied by a mixture of Ti^{4+} and Ti^{3+} ions. The Ti ions are octahedrally coordinated to oxygen with shared edges along the c-axis, and corner-sharing in the perpendicular directions. Extensive overlap of the d-orbitals along c leads to high conductivity values of 0.1 S/m near room temperature. In perpendicular directions where there is no overlap, the conductivity of these nonstoichiometric crystals is about three orders of magnitude smaller.

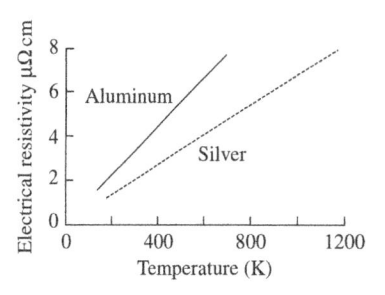

Fig. 17.6 Resistivity of two electrode metals showing the linear dependence on temperature.

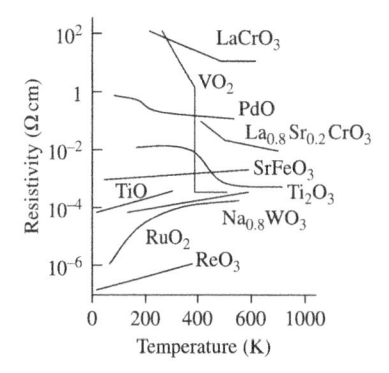

Fig. 17.7 Some conducting oxides show resistivities comparable to electrode metals.

Table 17.3 Principal resistivities of anisotropic metals in units of 10^{-8} Ω m

Metal	Point group	ρ_{11}	ρ_{22}	ρ_{33}
Antimony	$\bar{3}m$	36.0	36.0	26.3
Beryllium	$6/mmm$	3.13	3.13	3.57
Bismuth	$\bar{3}m$	109.0	109.0	138.3
Cadmium	$6/mmm$	6.8	6.8	8.3
Gallium	mmm	55.5	17.3	7.87
Indium	$4/mmm$	8.33	8.33	7.94
Magnesium	$6/mmm$	4.22	4.22	3.50
Tin	$4/mmm$	9.9	9.9	14.3
Zinc	$6/mmm$	5.91	5.91	6.13

Highly anisotropic conductivity also occurs in intermetallic compounds intercolated with organic molecules. Crystals of $TaS_2C_5H_5N$ consist of TaS_2 layers alternating with pyridine molecules giving a periodicity of 1.2 nm. A chemical bond is formed from the lone pair electrons of nitrogen and the half-filled conduction band of the TaS_2 layers. Conductivity in such two-dimensional materials is primarily parallel to the planes. Resistivity experiments require nearly perfect single crystals since dislocations and other defects tend to "short-out" the electrical anisotropy, which can be as large as $10^5 : 1$.

One-dimensional conductivity is observed in TCNQ salts. Organic crystals such as these have electrical conductivities comparable to metals yet contain no metal atoms. TTF–TCNQ crystals contain molecules of tetrathiofulrane and tetracyanoquinodimethane stacked face-to-face forming linear chains. The highly polarizable TTF molecule donates an electron to the TCNQ molecule, resulting in a half-filled conduction band with easy electron transfer parallel to the chain.

In anisotropic crystals, the current flow is generally not parallel to the electric field vector. Only for principal axes is the current parallel to E. Therefore care must be taken in resistivity measurement to define directions accurately. This is done by using either flat plate specimens or long thin rods as explained in Chapter 18. Flat plate specimens are used for conductivity measurements, and long thin bars for resistivity measurements.

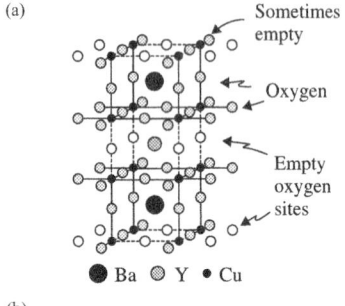

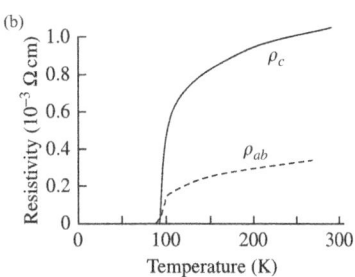

Fig. 17.8 (a) Crystal structure of $YBa_2Cu_3O_7$ showing the conducting Cu–O layers perpendicular to c. (b) Resistivity of twinned single crystals with highly anisotropic conduction. The true symmetry is orthorhombic (mmm) but the ferroelastic twinning in the (001) plane makes the resistivities along the a and b axes equal.

Problem 17.2

In a conducting crystal of symmetry $\bar{3}m$, determine

    a. The angle Δ between J and E for a given direction of E.
    b. The direction of E that maximizes Δ.
    c. The maximum value of Δ.
    d. Evaluate the numbers for antimony.

17.5 Semiconductors and insulators

Materials with ionic, covalent, or mixed ionic–covalent bonding are normally either insulators or semiconductors. Most useful semiconductors, including Si, Ge, and GaAs, have predominantly covalent bonding. An important distinction between metals and nonmetals is that the conductivity of a metal decreases with increasing temperature while the conductivity of semiconductors and insulators increases with increasing temperature.

Solid-state electronics is centered about the classical elemental semiconductor silicon that crystallizes in the diamond structure (Fig. 17.9). Each atom forms electron-pair covalent bonds to four nearest neighbors. Because of its importance in electronics, Si has been prepared at purity levels never achieved with other synthetic materials. Crystal perfection is correspondingly high, with large dislocation-free crystals a practical reality.

Compound semiconductors have also been of immense interest despite the formidable materials problems. Elements positioned symmetrically about Group IV are generally used in compound semiconductors. GaAs (III–V), ZnS (II–VI), and SiC (IV–IV) are representative of a class of materials finding applications at higher speeds, higher power, higher frequencies, and higher temperatures than Si or Ge. Gallium arsenide varactor diodes and oscillators operate well in the microwave region used for automobile radar. Narrow band gap semiconductors such as indium antimonide are used in Hall Effect devices and as infrared detectors. Yet compared with silicon, compound semiconductors have made little impact in solid-state electronics. In addition to the usual materials problems of purity and crystal perfection, the equally difficult problem of stoichiometry must be surmounted.

Structure–property relations in semiconductors are of great interest because of their importance to the electronics industry. Two of the questions one might ask are these: Given the chemical composition and structure, is it possible to predict whether the material is a semiconductor or not? And second, how do the important properties, such as band gap and mobility, depend on crystal chemistry?

Mooser and Pearson have given a formula that can be used to predict semiconductors. Semiconductor compounds satisfy the relation

$$\frac{n_e}{n_a} + N_a - N_c = 8,$$

where n_e is the number of valence electrons per formula unit, n_a is the number of the anions per formula unit, N_a is the average number of anion–anion bonds, per anion, and N_c is average number of cation–cation bonds formed by each cation. n_e and n_a are obtained from the chemical composition, and N_c and N_a determined from the structure. A number of illustrations are given in Table 17.4.

Some of the examples listed require further comment. The elements Ge, As, and Se point up the similarity between the 8-N rule and the Mooser–Pearson relation. The 8-N rule is used to predict the number of valence bonds formed by an element in the Nth column of the periodic system. For example, selenium in Column VI forms $8 - 6 = 2$ valence bonds, as observed in both the ring and helical forms. According to the Mooser–Pearson relation, selenium is a semiconductor because it forms electron-pair bonds with two near neighbors, and in this way every atom acquires a filled shell, and every electron participates in bonding. High electrical conductivity (metallic behavior) occurs in compounds in which not all of the valence electrons are involved in bonding.

The importance of cation–cation and anion–anion bonds is apparent in GaTe, FeS$_2$, and CdSb (Table 17.4). Gallium telluride forms a layer-like structure with one short Ga–Ga bond per gallium. Gallium would normally contribute three

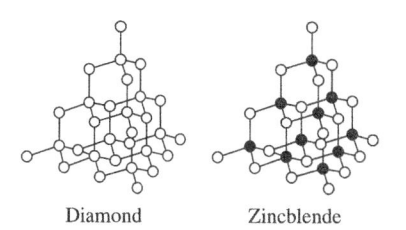

Fig. 17.9 In the diamond structure every carbon atom is tetrahedrally bonded to four neighboring carbon atoms. The zincblende structure is derived from that of diamond by replacing alternate atoms with zinc and sulfur. Most semiconductors of commercial importance are isomorphous with diamond or zincblende.

Table 17.4 Representative semiconductors satisfying the Mooser–Pearson relation

Compound	n_e	n_a	N_a	N_c	$(n_e/n_a) + N_a - N_c$
Ge	4	1	4	0	8
As	5	1	3	0	8
Se	6	1	2	0	8
SiC	8	2	4	0	8
GaAs	8	1	0	0	8
CdTe	8	1	0	0	8
AgInTe$_2$	16	2	0	0	8
PbS	8	1	0	0	8
Mg$_2$Sn	8	1	0	0	8
LiMgSb	8	1	0	0	8
Li$_3$Bi	8	1	0	0	8
Mg$_3$Sb$_2$	16	2	0	0	8
Bi$_2$Te$_3$	24	3	0	0	8
Fe$_2$O$_3$	24	3	0	0	8
BaTiO$_3$	24	3	0	0	8
FeS$_2$	14	2	1	0	8
CdSb	7	1	1	0	8
GaTe	9	1	0	1	8

valence electrons to the anion, but one electron is involved in an electron-pair bond with another cation, leaving two for each tellurium atom. Thus the anion electron shell is filled with eight electrons and GaTe is a semiconductor. Anion–anion bonds occur in FeS$_2$ and CdSb. Cadmium antimonide has a very deformed diamond arrangement in which Sb is coordinated to three Cd and one Sb, all near 2.8 Å. From the structure it is therefore reasonable to postulate that an Sb–Sb bond is formed, allowing the anions to satisfy their valence requirements. Pyrite is a better example of anion–anion bonds. The structure of FeS$_2$ contains Fe^{2+} ions and S$_2^{2-}$ dimers.

Additional semiconducting compounds can be derived from known examples by a process of cross-substitution, replacing one element by pairs from other columns of the periodic system while keeping the valence-electron to atom radio constant. The III–V and II–VI compounds based on the column IV semiconductors are familiar examples. The substitution process can be carried several steps further. The semiconductor AgInTe$_2$ is derived from CdTe by substituting equal amounts of monovalent silver and trivalent indium for divalent cadmium. Successive substitutions in the cation sublattice leads to Ag$_2$CdSnTe$_4$ and Ag$_5$InSn$_2$Te$_8$. Cation vacancies occur in semiconducting In$_2$Te$_3$ and HgIn$_2$Te$_4$. Substitutions need not be confined to the cation sublattice; cross-substitution involving the anion sublattice, or both sublattices are also possible. Symmetric substitutions in the anion sublattice occur in Al$_2$CO, a derivative of AlN. Double substitution takes place in LiMgSb, a semiconductor derived from Mg$_2$Sn by replacing half of the cations and all the anions. Similar derivative structures occur in nature. The minerals zincblende (ZnS), chalcopyrite (CuFeS$_2$), and stannite (Cu$_2$FeSnS$_4$) are three minerals in such a series.

17.6 Band gap and mobility

A number of relations between band gap and structure have been proposed, many of them empirical. Band gaps for Column IV elements are compared

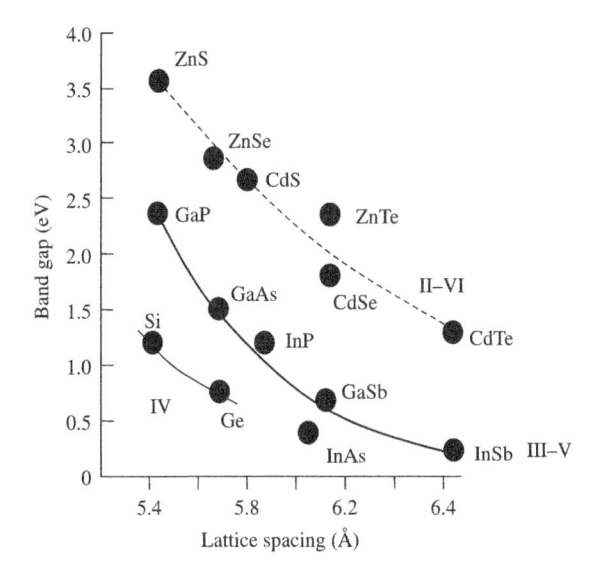

Fig. 17.10 Semiconductor band gaps plotted as a function of unit cell lattice parameters.

with III–V and II–VI compounds in Fig. 17.10. All have the tetrahedrally-bonded diamond or zincblende structures (Fig. 17.9). Within each family there is an inverse relationships between band gap (E_g) and bond length. In general, longer bonds are weaker bonds with more easily liberated electrons.

Band gaps of III–V compounds are larger than the corresponding Column IV elements: AlP has a larger gap than Si, GaAs exceeds Ge, and InSb is greater than Sn. From examples such as these, relations between electronegativity differences and E_g have been established. Ionic compounds have greater E_g values than most covalent compounds. To illustrate how the band gap can be adjusted, consider the two examples mentioned earlier. Comparing CdTe and its derivative AgInTe$_2$, we note that electronegativity differences to the three bonds are in the order Ag–Te > Cd–Te > In–Te, with the latter being the weakest bond. Energy gap is determined by the weakest bond since it has the smallest interband separation. Therefore this type of cross-substitution usually lowers E_g. The energy gap is 0.96 eV for AgInTe$_2$ and 1.5 eV for CdTe. In the second example, LiMgSb is derived from Mg$_2$Sn, increasing E_g, because electronegativity differences are in the order Li–Sb > Mg–Sb > Mg–Sn.

The electronegativity scale not only provides a measure of the bonding type but is also useful in predicting physical properties. An empirical relation between electronic band gap and electronegativity is shown in Fig. 17.11. Band gaps range from a fraction of an eV in semiconductors to several eV in good insulators.

The band gap E_g of a semiconductor is important in determining its electrical conductivity σ,

$$\sigma \sim \mu e^{-E_g/kT},$$

where the mobility μ depends chiefly on imperfections and temperature. When the band gap is comparable to thermal energies ($kT \sim 1/40$ eV at room temperature), large numbers of electrons are promoted from the valence to the conduction band, greatly increasing the electrical conductivity. Among compound semiconductors E_g increases with ionicity as shown in Fig. 17.11. Ionic compounds show a large band gap and optical transparency extending well into

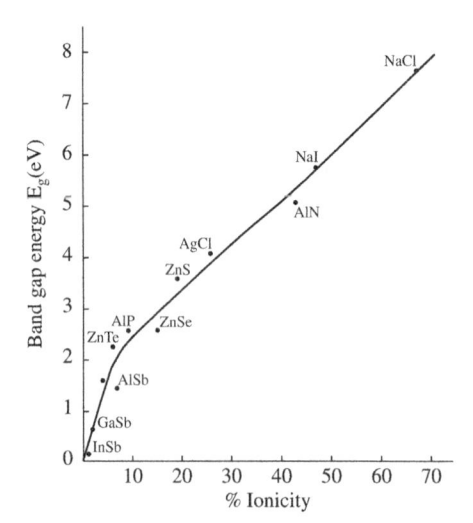

Fig. 17.11 Band gap generally increases with ionicity. The more ionic compounds with the rocksalt structure usually have wider band gaps than covalently bonded materials with the zincblende structure.

the ultraviolet region, whereas most covalent semiconductors are either opaque or colored because of their narrow band gaps. As ionicity increases, electrons are more tightly bound to the cores with a greater degree of localization of charge carriers. Fluorine has a very large electronegativity resulting in large band gaps in fluoride crystals. Calcium fluoride (CaF_2, fluorite) is very transparent in the near ultraviolet region where the processing of integrated circuits is carried out.

Mobility is very sensitive to imperfections so that most experimental measurements are more indicative of crystal growth technique than intrinsic limitations. Experimental mobility values for undoped Si and Ge have increased over the years as better crystals have been grown.

In pure materials free from defects, mobility is determined by the effective mass interaction with lattice vibrations. Heavy masses reduce the latter, decreasing the scattering by lattice vibrations and contributing to the large mobilities (>10,000 cm^2/V s) found in HgTe and InSb. Compounds with highly ionic bonding have extremely low mobilities, both for electrons and holes. When the electronegativity difference is greater than 1 unit, mobilities greater than 1000 cm^2/V s are unlikely.

Generally mobilities increase with molecular weight and decrease with electronegativity difference (Fig. 17.12). The explanation lies in the polarization effect of mobile electrons or holes on the surrounding atoms. The motion of charge carriers is accompanied by an adjustment in position of the ions that is intrinsically slow. Coulomb coupling between charge carriers and ions occurs in ionic compounds.

Chemical impurities are used to control the conductivity σ of semiconductors, $\sigma = n\mu_e + p\mu_h$. The density of conduction-band electrons and valence-band holes are n and p, respectively, with mobilities μ_e and μ_h. For wide band gap materials near 1 eV or larger, the room-temperature carrier concentrations are controlled by doping. Donor atoms contribute electrons to the conduction band, and acceptor atoms remove electrons from the valence band, leaving holes behind. Donor (n) and acceptor (p) doped resistivities for silicon, germanium, and gallium arsenide (Fig. 17.13) extend over many orders of magnitude. Silicon and germanium have four outer electrons forming tetrahedral sp^3 hybrid

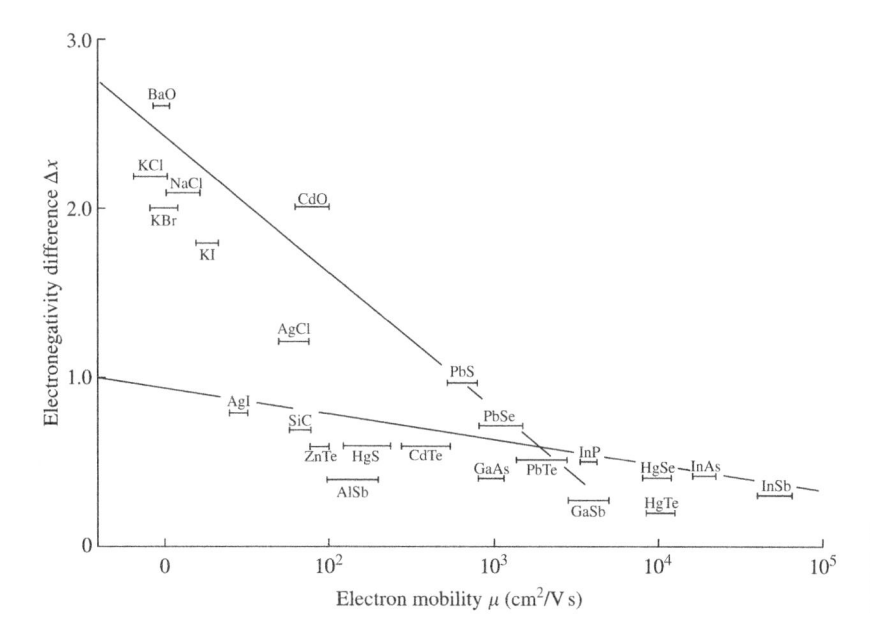

Fig. 17.12 Correlation between the electron mobility and the electronegativity difference of some zincblende and rocksalt type compounds.

bonds to four neighboring atoms in the crystal. Substitutional impurities with more than four outer electrons tend to be donors because only four electrons are needed for bonding. The remaining electrons can be ionized away to the conduction band if the temperature is high enough.

Column V and VI elements are donors whereas Column III, Column II, and transition-metal elements tend to be acceptors. Gallium, a Column III element, requires one additional outer electron to form four covalent bonds. An electron is thermally-excited to this acceptor state from a filled state of lower energy, leaving a hole behind. Zinc needs two electrons and copper three, therefore Ga, Zn, and Cu create 1-, 2-, and 3-acceptor states, respectively.

Many elements do not substitute for Si or Ge because of size mismatch. Small atoms such as hydrogen enter interstitial sites and are not electrically active.

Impurity atoms also affect the mobility. The coulomb fields associated with ionized impurities exert forces on charge carriers, tending to scatter them and reduce the mobility. Defects and disorder also reduce mobility.

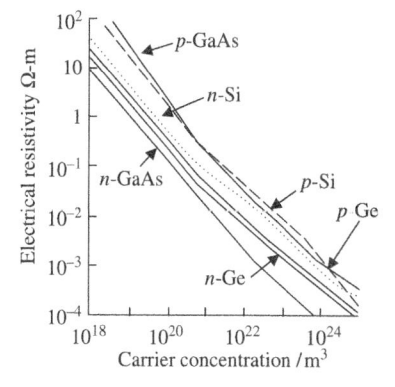

Fig. 17.13 The room-temperature resistivities of high quality semiconductor crystals are largely controlled by dopant concentrations.

17.7 Nonlinear behavior: Varistors and thermistors

Not all conductors are ohmic. *Varistors* are ceramic semiconductors with a highly nonlinear current–voltage relationship (Fig. 17.14). They are used to protect circuit elements against inductive surges that often damage contacts, relays, and rectifiers. By connecting the varistor in parallel with the circuit element, any voltage spikes greater than the reversible breakdown voltage cause currents to flow through the varistor rather than the circuit elements. Ceramic varistors have proved especially useful as lightning arrestors. Most varistors are made from doped zinc oxide with a bismuth-rich phase in the

200 *Electrical resistivity*

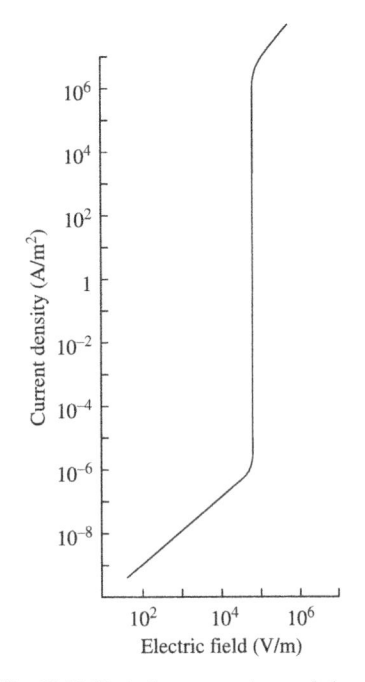

Fig. 17.14 Typical current–voltage relationship in a zinc oxide varistor. Ohmic behavior is observed at low and high fields with a reversible breakdown effect between. Quantum mechanical tunneling through Schottky barriers formed at the grain boundaries are responsible for this behavior.

grain boundaries. At low field levels, the varistor behaves like a semiconductor with small temperature-dependent currents. At a certain critical voltage, however, the resistance diminishes suddenly and currents increase dramatically. The phenomenon differs from normal electric breakdown in that the J–E characteristic is reversible and controllable by the ceramic composition and microstructure.

Ohm's Law breaks down for varistors. Higher order terms such as

$$E_i = \rho_{ij}J_j + \rho_{ijkl}J_jJ_kJ_l + \rho_{ijklmn}J_jJ_kJ_lJ_mJ_n + \cdots$$

are required to fit the experimental data. Odd-rank terms such as ρ_{ijk} are zero because the J–E relation is the same when the current flow is reversed.

This is not true for samples exhibiting diode behavior. In this case both odd- and even-rank tensor coefficients are needed to model rectification. Poled $BaTiO_3$ thermistors and tourmaline crystals are also capable of exhibiting asymmetric J–E characteristics.

In most metals the electrical resistivity increases linearly with temperature. Data for silver, copper, and other electrode metals were presented in Table 17.2. Typical values of $(1/\rho)(d\rho/dT)$ are about 0.5% per degree. Thermistors have much larger temperature coefficients, and often the effects are highly nonlinear. Three types of ceramic thermistors are in widespread use: NTC thermistors, PTC thermistors, and critical temperature thermistors. Typical resistance changes with temperature are illustrated in Fig. 17.15.

Vanadium dioxide is often used in critical temperature thermistors. Below 80°C, VO_2 is a semiconductor with a negative temperature coefficient of resistance. Above 80°C it shows metallic behavior with a great increase in conductivity (typically two orders of magnitude) and very little change with temperature. The critical temperature of 80°C can be modified by changes in chemical composition.

The V^{4+} ion in VO_2 has a peculiar electron configuration with one 3d electron outside a closed shell. In the low temperature state, adjacent V^{4+} ions form electron-pair bonds giving rise to a band gap and semiconductor behavior. A phase transition takes place near 80°C in which the 3d electrons are liberated from the pair bonds and are free to conduct electricity (see Fig. 17.7).

Electrical resistance decreases exponentially with increasing temperature in an NTC thermistor. Unlike critical temperature thermistors, there is no phase transition involved. Most NTC thermistors are composed of doped transition-metal oxides. Typical of these controlled valency semiconductors are Fe_2O_3 : Ti and NiO : Li.

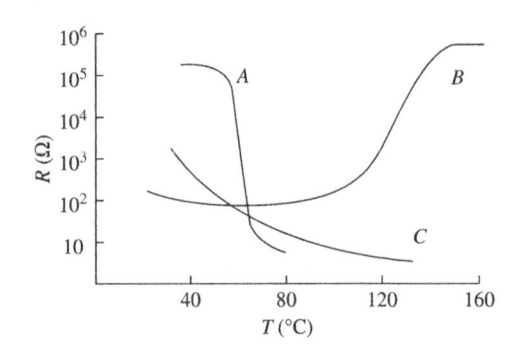

Fig. 17.15 Typically the electrical resistance of thermistors changes by several orders of magnitude with temperature. In NTC thermistors the resistance decreases steadily with increasing temperature, but sudden changes at phase transitions are involved in PTC and critical temperature thermistors. A, critical temperature thermistor; B, PTC thermistor; C, NTC thermistor.

Doped nickel oxide has the rocksalt structure with lithium partially replacing nickel in the cation sites. Ionic radii for Ni^{2+} (0.84 A), Ni^{3+} (0.74 A), and Li^+ (0.88 A) all favor octahedral coordination with oxygen. For semiconducting compositions near $Ni_{0.95}Li_{0.05}O$, the band gap is about 0.15 eV. The physical origin of this band gap is attributed to the attractive forces between Li^+ dopant ions and the compensating Ni^{3+} ions.

NTC thermistors are used in flowmeters in which the velocity is measured by monitoring the temperature difference between two thermistors. A heater positioned between the two thermistors provides the temperature difference. Thermistors are also used as inrush limiters to protect diodes, fuses, switches, and light bulbs. The sudden surge of current which occurs when a light bulb is turned on often ruptures the bulb filament. With an NTC thermistor in series with the bulb, the energy of the initial surge is dissipated as heat in the thermistor. The temperature coefficient $(1/\rho)(d\rho/dT)$ for a commercial NTC thermistor is about $-4\%/K$ near room temperature.

PTC thermistors differ from NTC thermistors in several important respects. The resistance of a PTC thermistor increases with temperature, but only over a limited temperature range near a phase transition. The resistance change is very large at this temperature because of grain boundary effects.

Barium titanate ceramics are widely used in PTC thermistors. When doped with donor ions such as La^{3+} or Ce^{3+} (for Ba^{2+}) or Nb^{5+} (for Ti^{4+}), the resistivity material shows a pronounced PTC effect. Explanation of the PTC effect rests upon understanding the defect structure. When sintered at high temperature, lanthanum-doped $BaTiO_3$ becomes an n-type semiconductor with conduction taking place via transfer of electrons between titanium ions, Ti^{4+} and Ti^{3+}. Thus the barium titanate grains in the ceramic are conducting, and remain conducting on cooling to room temperature.

But the grain boundary region changes during cooling. Oxygen is adsorbed on the surface of the ceramic and diffuses to grain boundary sites, altering the defect structure along the grain boundaries. The added oxygen ions attract electrons from nearby Ti^{3+} ions, thereby creating an insulating barrier between grains.

To explain the PTC effect it is necessary to consider the ferroelectric phase transition in $BaTiO_3$ and its effect on the insulating barriers between grains. Barium titanate is cubic and paraelectric above 130°C, the Curie temperature. Below this temperature the perovskite structure distorts to a tetragonal ferroelectric state in which a large spontaneous polarization P_s develops at the grain boundaries. Near room temperature the resistance of a PTC thermistor is low because the electron charge trapped in grain boundary regions is partially neutralized by spontaneous polarization. Wherever the domain structure is advantageously positioned, positive polarization charge will cancel the negatively charged barriers between conductive grains, thereby establishing low resistance paths across the ceramic. Above T_c the spontaneous polarization disappears and the resistivity increases, giving rise to the PTC effect.

PTC thermistors are used as protection against overvoltage and short circuits. When connected in series with the load, a PTC thermistor limits the current to safe levels. Large currents cause the temperature of the thermistor to rise into the PTC range, thereby raising the resistance and lowering the current.

Regarding the symmetry of ceramic thermistors, the appropriate point groups for randomly oriented grains and grain boundaries would be Curie group $\infty\infty m$, but textured ceramics have lower symmetry. Grain orientation would give cylindrical symmetry (∞/mm) and poled BaTiO$_3$ thermistors would have conical symmetry (∞m). Under these circumstances the resistivity is anisotropic. The coefficients relating resistivity to temperature over the operating range ΔT of the thermistors can be described by a power series in ΔT, $(\Delta T)^2$, $(\Delta T)^3$, etc. All the coefficients are second rank tensors since ΔT is a scalar.

17.8 Quasicrystals

Quasicrystals are an interesting form of solid matter with symmetries not found in normal crystals. They lack the three-dimensional translational periodicity characteristic of crystals, but often possess fivefold symmetry along one or more directions. Two commonly observed point groups in intermetallic quasicrystals are icosahedral ($m\bar{3}5$) and decagonal ($\bar{10}/mmm$).

The existence of quasicrystalline matter has been explained in terms of Penrose Tiling in which two or more cells pack together to fill space. An example of aperiodic two-dimensional tiling is shown in Fig. 17.16. Parallelograms with 72° and 36° angles pack together to form aligned star-like patterns with fivefold rotational symmetry. Certain Al–Mn alloys show quasicrystalline structures like this which are periodic along the fivefold axis but not in the perpendicular directions.

Compared to other alloys, intermetallic quasicrystals exhibit unusually high electrical resistance. Quasicrystals have higher resistance than amorphous alloys of the same composition. The temperature dependence of the electrical resistivity of an Al$_7$Pd$_2$Re alloy is shown in Fig. 17.17. These icosahedral quasicrystals have a very large negative temperature coefficient, especially at low temperatures. Most metals have a positive temperature coefficient, as pointed out in Section 17.3. The absence of periodicity promotes electron scattering and increases the resistance.

Decagonal quasicrystals are periodic along the $\bar{10}$-fold symmetry axis and aperiodic in the perpendicular directions. As a result, the resistivity is about 10 times smaller parallel to the axis than in the perpendicular directions.

Problem 17.3
Using Neumann's Principle, determine the number of independent tensor coefficients for quasicrystals with a fivefold symmetry axis parallel to Z_3. Carry out the analysis for polar tensors of rank 1, 2, 3, and 4. Compare the results with those obtained for point groups 4 and 6 in Chapters 8, 9, 12, and 13.

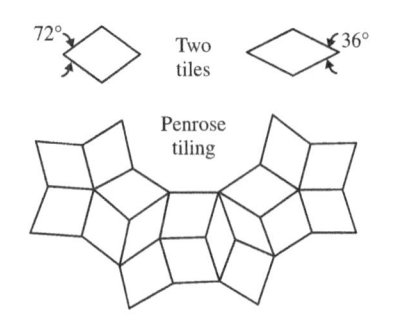

Fig. 17.16 Aperiodic tiling of two different parallelograms can produce quasicrystalline structures with fivefold symmetry.

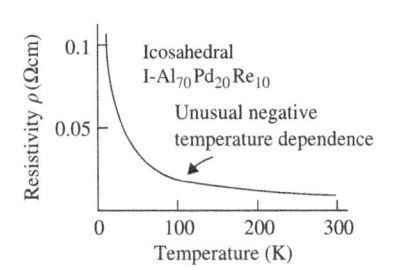

Fig. 17.17 Quasicrystalline icosahedral alloys are electrically isotropic with high resistivity.

Thermal conductivity

18

When different portions of a solid are at different temperatures, thermal energy is transported from the warmer to the cooler regions. The thermal conductivity coefficient provides a quantitative measure of the rate at which thermal energy is transported along the thermal gradient.

Thermal conductivity coefficients k relate the heat flux h [W/m^2] to temperature gradient dT/dZ. In tensor form,

$$h_i = -k_{ij}\left(\frac{dT}{dZ_j}\right).$$

The minus sign appears because heat flows from hot to cold. Thermal conductivity is measured in units of W/m K.

Four contributions to thermal conductivity are illustrated in Fig. 18.1. The two principal mechanisms are from conduction electrons and from lattice vibration phonons. In transparent solids, especially at high temperature, photon transport can also be important. In porous media, convection currents from gas or liquid molecules can contribute to the thermal conductivity.

18.1	Tensor nature and experiments	203
18.2	Structure–property relationships	206
18.3	Temperature dependence	208
18.4	Field dependence	210

18.1 Tensor nature and experiments

Thermal conductivity is a polar second rank tensor like electric permittivity, magnetic susceptibility, and electrical resistivity but there is a basic question

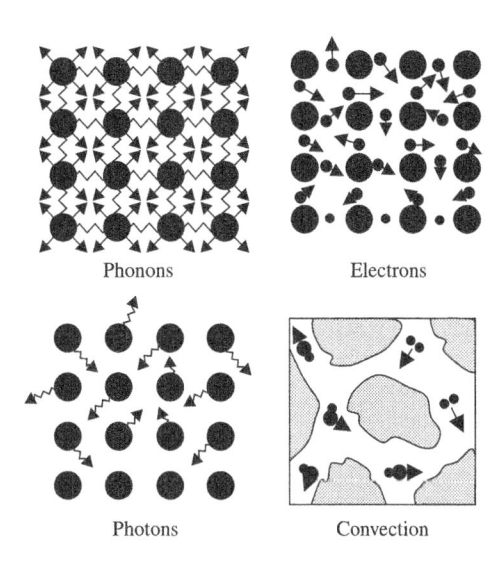

Phonons   Electrons

Photons   Convection

Fig. 18.1 Thermal conduction mechanisms in solids. Near room temperature electron transport dominates in metals, and phonon transport in ceramics. Photon transport contributes to thermal conductivity in glasses, and convection currents in gas or liquid filled porous materials.

regarding the symmetry of transport properties such as electrical and thermal conductivity. The symmetry of tensors is partly dictated by geometrical considerations through Neumann's Principle, and partly through thermodynamic arguments. For triclinic crystals there are nine nonzero conductivity coefficients k_{ij}. If the tensor is symmetric then $k_{ij} = k_{ji}$, and there are only six independent coefficients to be determined. For the dielectric constant it was shown that $K_{ij} = K_{ji}$, based on thermostatic energy arguments (Section 9.2). This argument does not hold for transport properties, but there is another principle based on irreversible thermodynamics. Onsager's Theorem states that for transport properties, involving the flow of charge, heat, or atomic species, then

$$k_{ij} = k_{ji}.$$

The proof of Onsager's Theorem depends on statistical mechanics and is beyond the scope of this book.

From a practical point of view, Onsager's Theorem is not very important because most transport experiments are performed on high symmetry metals, semiconductors, and insulators. Neumann's Principle controls the conductivity matrices for $m3m$, $\bar{4}3m$, $m3$, 432, 23, $6/mmm$, $\bar{6}m2$, 622, $4/mmm$, $\bar{4}2m$, $4mm$, 422, $\bar{3}m$, 32, $3m$, mmm, $mm2$, and 222. For triclinic crystals (point groups 1 and $\bar{1}$) Onsager's Theorem ensures that $k_{12} = k_{21}$, $k_{23} = k_{32}$, and $k_{13} = k_{31}$, and for monoclinic crystals (point groups 2, m, and $2/m$) it shows that $k_{13} = k_{31}$. For the remaining point groups (3, $\bar{3}$, 4, $\bar{4}$, $4/m$, 6, $\bar{6}$, and $6/m$) $k_{12} = -k_{21}$ according to symmetry arguments, but the Onsager Theorem together with Neumann's Principle combine to give $k_{12} = k_{21} = -k_{12} = 0$. Crystals belonging to these point groups provide a testing ground for Onsager's Theorem. If Onsager's Theorem is wrong, $k_{12} = -k_{21} \neq 0$, then there should be spiral heat flow in disk-shaped specimens oriented perpendicular to the principal symmetry axis. A careful search for spiral heat flow led to negative results, supporting the Onsager Theorem.

Therefore the electrical resistivity, thermal conductivity, and diffusion tensors are generally assumed to be symmetric. The tensor and matrix forms are identical to those of the dielectric constant.

For anisotropic materials the heat flow h is generally not parallel to the temperature gradient dT/dZ. In such a situation it is important to define the measured values carefully. By far the most common geometry is that of rod- or disk-shaped specimens with axial heat flow. If L is the distance between isothermals T_1 and T_2, and A is the cross-sectional area, then the heat flow is

$$h = \frac{k(T_1 - T_2)A}{L}.$$

As pointed out earlier, the shape of the sample is very important. A long thin rod forces the heat flow to be parallel to the axial direction, while a flat plate specimen defines the direction of the temperature gradient (Figs. 18.2 and 18.3).

The flat plate experiment is used to determine the thermal conductivity of good insulators. A thin layer of the insulator is positioned between metal blocks of known thermal conductivity k_M. The heat flow h is the same in the metal and the insulators, so that

$$h = k_M A \left(\frac{dT}{dZ}\right)_M = k_I A \left(\frac{dT}{dZ}\right)_I,$$

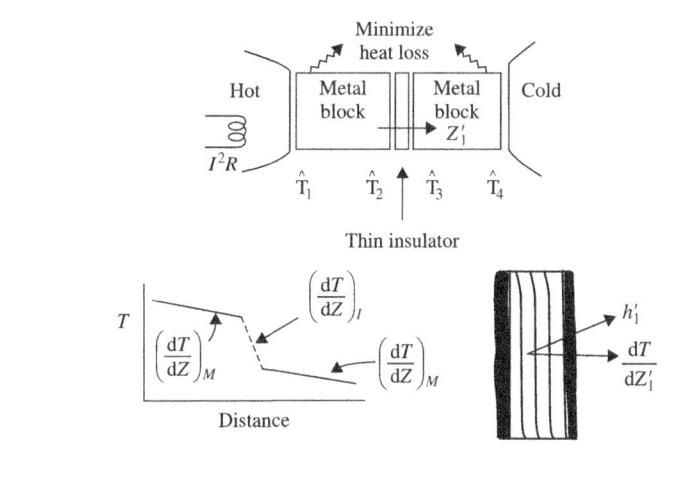

Fig. 18.2 Thermal conductivity measurement for thin plate insulators. Four thermocouples are used to determine the temperature gradients across the sample and the two supporting metal blocks. By making the sample very thin, the temperature gradient is fixed perpendicular to the axial direction.

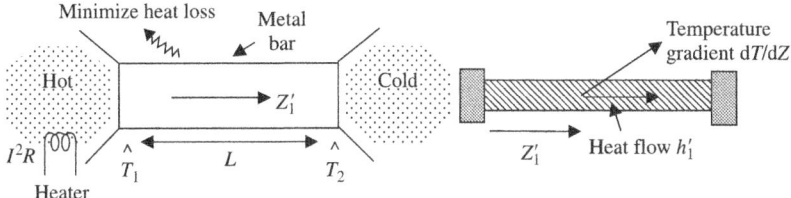

Fig. 18.3 Thermal resistivity measurement using long slender rods. The temperature difference $T_1 - T_2$ is monitored by two thermocouples. When heat losses to the surroundings are small, the heat flow is parallel to the axial direction Z_1'.

where A is the cross-sectional area, and k_I the thermal conductivity of the insulator. As shown in Fig. 18.2, four thermocouples are used to determine the temperature gradients $(dT/dZ)_M$ and $(dT/dZ)_I$ across the metal block and the insulator. The thermal conductivity of the insulator is equal to

$$k_I = \frac{k_M (dT/dZ)_M}{(dT/dZ)_I}.$$

As with all thermal measurements, great care must be taken to minimize heat loss to the surroundings.

In regard to anisotropic materials, where the heat flow is not parallel to the temperature gradient, the flat plate geometry ensures that the temperature gradient is parallel to the axial direction Z_1'. Therefore the radial gradients are zero and $dT/dZ_2' = dT/dZ_3' = 0$. The thermal conductivity component measured in this experiment is $k_{11}' = h_1'/(dT/dZ_1')$. For the general case of a triclinic crystal, six measurements of this type would be required with plates cut in six different orientations. Fewer plates would be needed for crystals of higher symmetry. The general procedure is the same as that for the dielectric constant (Section 9.4).

A thin rod geometry is preferred for metals and other good conductors. This is a thermal resistivity measurement rather than a thermal conductivity measurement. For a thin rod, heat flow is confined to the axial direction Z_1'. Therefore $h_2' = h_3' = 0$ and the temperature gradient along the rod is

$$\left(\frac{dT}{dZ_1'}\right) = r_{11}' h_1',$$

where r_{11}' is the thermal resistivity measured along Z_1'. Knowing the length of the rod L and its cross-sectional area A, the value of r_{11}' can be determined from the heat input I^2R and the temperature drop across the specimen, $T_1 - T_2$.

$$r'_{11} = \frac{A(T_1 - T_2)}{LI^2R}.$$

As before, the experiment must be carefully insulated from the environment to ensure that all the heat flows down the metal rod.

Triclinic crystals would require six such measurements to specify the three principal coefficients and the three orientation angles. Thermal resistivity coefficients are related to thermal conductivity coefficients by the matrix inversion procedure described in Section 17.1. Fig. 18.3 illustrates the thermal resistivity experiment for metals and other good conductors.

Problem 18.1
The hexagonal lattice parameters of graphite are $a = 2.46$ and $c = 6.70$ Å. Using the k_{11} and k_{33} values in Table 18.1, evaluate the thermal conductivities in the [110], [111], and [123] directions.

Problem 18.2
A 1 mm thick plate of silicon is in good thermal contact with a 2 mm thick layer of germanium. The outer temperatures are 30°C and 100°C. What is the heat flow per unit area? What is the temperature at the junction between the two crystals?

18.2 Structure–property relationships

Room-temperature thermal conductivity coefficients for solid materials range over about four orders of magnitude (Table 18.1). The best thermal conductors are either metals like copper or covalent crystals like diamond with short strong chemical bonds. For metals, conduction involves free electrons while in diamond phonons are very easily transmitted by the tightly bonded carbon atoms. At the other extreme are some superinsulators made from supercritically-dried porous silica aerogels. The thermal conductivity of these light-weight tiles is around 0.01 W/m K, which is even lower than air. Molecular crystals like Rochelle salt also have low thermal conductivity. There are no free electrons

Table 18.1 Principal thermal conductivity coefficients measured near room temperature (W/m K)

	k_{11}	k_{22}	k_{33}
Diamond $m3m$	550	550	550
Silicon $m3m$	175	175	175
Germanium $m3m$	65	65	65
Copper $m3m$	400	400	400
Graphite $\bar{3}m$	355	355	89
Quartz 32	6.5	6.5	11.3
Corundum $\bar{3}m$	22.5	22.5	25.1
Rutile $4/mmm$	9.3	9.3	12.9
Lithium fluoride $m3m$	16	16	16
Sodium chloride $m3m$	6.5	6.5	6.5
Rochelle salt 222	0.50	0.61	0.60

to transmit heat and the organic molecules are only loosely bonded together by van der Waals forces or hydrogen bonds. Weak intermolecular bonding makes polymers and other molecular materials good thermal insulators.

Metals contain many free electrons so that the electronic thermal conductivity predominates over the lattice contribution. The electric conductivity σ and thermal conductivity k are proportional to one another since both are controlled by the free electron concentration. The Wiedemann–Franz law ($k/\sigma T = L$) works well for many metals (Fig. 18.4). The Lorenz Ratio L is usually near its theoretical value $(\pi^2/3)(k_B/e)^2 = 2.443 \times 10^{-8}$ W Ω/K^2 as predicted by the free electron model for metals.

For glasses and cubic crystals thermal conductivity is scalar, but k depends on direction for lower symmetries. Among anisotropic materials, chain and layer structures show greater conductivity in the directions of closest bonding. Tellurium crystallizes in helices parallel to the trigonal c-axis and the ratio of conductivities parallel and perpendicular to the chains is about 1.5. In the layer structure of graphite, the conductivity of single crystals is four times greater within the layer than perpendicular to it. Graphite conducts heat best parallel to the carbon layers where phonons are transmitted easily along the strong covalent bonds. Thermal conductivity is much lower perpendicular to the layers where the bonds are weak (Fig. 18.5). Because of this, graphite and boron nitride have been used as heat shields in space vehicles. Below room temperature both k_{11} and k_{33} increase to a peak below 100 K.

Immense anisotropy is observed in thermal conductivity measurements on drawn polyethylene fibers. For draw ratios of 350 : 1 the thermal conductivity parallel to the draw axis is about three times that of steel. Normally one thinks of polymers as thermal insulators and metals as thermal conductors, but it is not true in this case. The thermal conductivity measured perpendicular to the draw axis is 0.3 W/m K at room temperature, giving a k_{33}/k_{11} anisotropy of more than two orders of magnitude. C–C covalent bonds control k_{33} while van der Waal forces are operative in the transverse directions. The temperature dependence

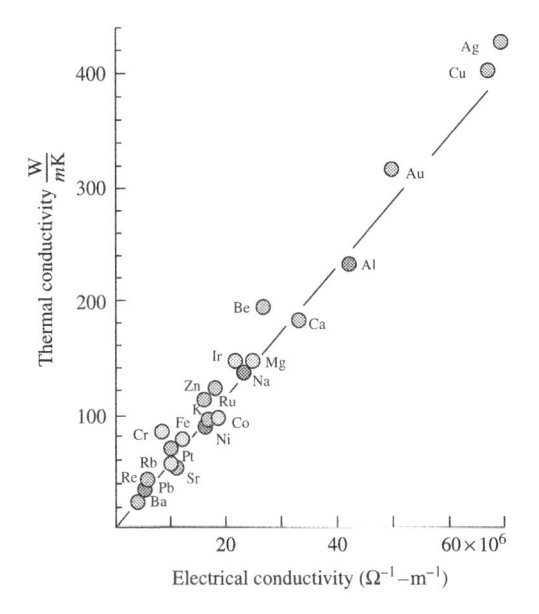

Fig. 18.4 Room temperature thermal and electrical conductivity of metals demonstrating the Wiedemann–Franz relationship.

Fig. 18.5 Crystal structures of (a) graphite and (b) boron nitride. Both are layer structures with highly anisotropic physical properties. The stacking of the layers is slightly different in the two structures.

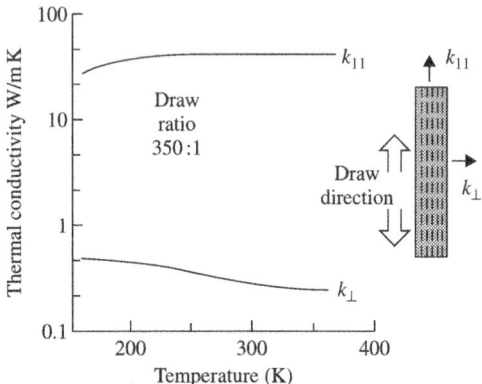

Fig. 18.6 Highly anisotropic thermal conductivity is observed in drawn polymer fibers. For polyethylene fibers the k values parallel to the fiber axis are about 100 times larger than those measured perpendicular to the draw direction (Choy).

is shown in Fig. 18.6. In ultradrawn polyethylene the protofibrils $(CH_2)_n$ begin to approximate a one-dimensional diamond structure.

18.3 Temperature dependence

The thermal conductivity of a solid can be expressed as the sum of lattice and electronic components: $k = k_L + k_e$. For nonmetals, the electronic components are negligible, and $k \equiv Cv\ell/3$, where C is the contribution of the lattice waves to the specific heat, v is the wave velocity and ℓ the mean free path. At high temperatures, k_L varies inversely with T because the mean free path is shortened by thermal vibrations. As the temperature is lowered, thermal conductivity goes through a maximum and then approaches zero as T goes to 0 K. The low temperature behavior is governed by imperfections and by sample dimensions.

The thermal conductivity of rutile (TiO_2) between room temperature and 800°C is shown in Fig. 18.7. Rutile is tetragonal (point group $4/mmm$) with chains of edge-sharing TiO_6 octahedra extending along the c-axis. The bonding is somewhat weaker in the directions perpendicular to c where the octahedra share only corners. As a result, the conductivity is higher along c. At room temperature the thermal conductivity of polycrystalline rutile ceramic lies between k_{11} and k_{33} as would be expected. Ideally the average conductivity for randomly oriented grains would be $(2k_{11} + k_{33})/3$ but at high temperature both single crystal values are far larger than that of the ceramic. The reason becomes apparent when one realizes that rutile single crystals are transparent but the ceramic is opaque. Photons make a strong contribution to thermal

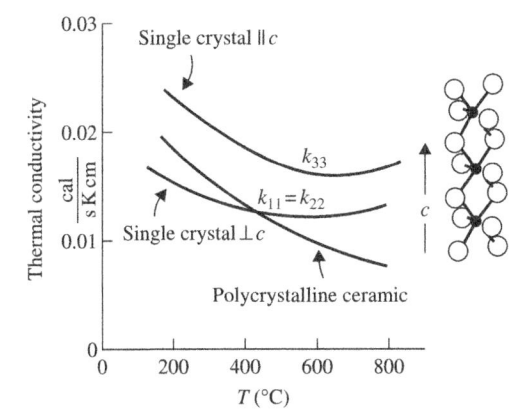

Fig. 18.7 Thermal conductivity of rutile crystals and rutile ceramic.

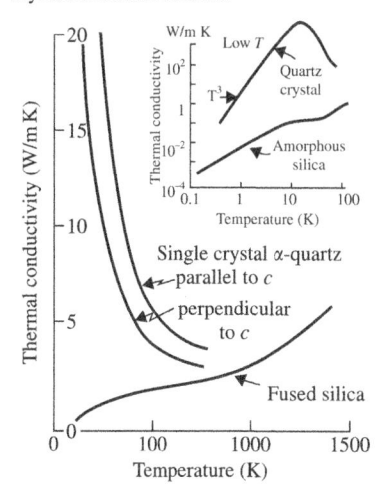

conductivity at elevated temperature. Normally k decreases above room temperature because of phonon–phonon scattering but the photons cause an increase in k in transparent media.

Similar behavior is observed in crystalline and amorphous silica (Fig. 18.8). The crystal structure of α-quartz consists of corner-linked SiO_4 tetrahedra arranged in a spiral around the trigonal c-axis. Quartz belongs to point group 32 so there are two independent thermal conductivity coefficients, $k_{11} = k_{22}$ and k_{33}. The k_{33} coefficient is significantly larger than k_{11} because of the spiral chains along c. Both components increase rapidly as the temperature is cooled below room temperature. At about 10 K the conductivities pass through a peak and then drop to very low values near 0 K. At very low temperatures there are very few phonons to transport heat, but as temperature increases, the number of activated vibrations increases raising the thermal conductivity. It reaches a peak when the phonons begin to interact strongly causing thermal resistance. As the phonon scattering increases, k continues to decrease until photons begin to make a contribution at high temperatures.

In silica glass (fused quartz) the situation is somewhat different. Again the structure consists of corner-linked SiO_4 tetrahedra, but there is no long range periodicity characteristic of crystalline materials. Amorphous solids like silica glass have smaller conductivities than crystals because the phonon waves are scattered more often in an aperiodic structure. Gemologists distinguish glass imitations from crystalline gems by touching the stone with the tongue. Glass feels warm compared to crystal because of its smaller thermal conductivity. The thermal conductivity of glass is usually smaller than that of a crystal of similar composition because of phonon scattering by the irregular structure. At ordinary temperatures the mean free path is of the order of 10 Å, about the size of silicate rings in the disordered structure. The thermal conductivity of glass increases sharply at high temperature because of its transparency to photon transmission. Silica glass does not show the large low-temperature peak in thermal conductivity characteristic of crystalline materials.

Radiation damage has a similar effect. When quartz is exposed to neutrons, the thermal conductivity is suppressed and approaches the values observed for silica glass. Chemical impurities and isotopes also promote phonon scattering.

Solid solutions have lower thermal conductivities than do the end member compositions. The solid solution of Si and Ge (Fig. 18.9) illustrate the reduction in thermal conductivity caused by phonon scattering. The random occupation of

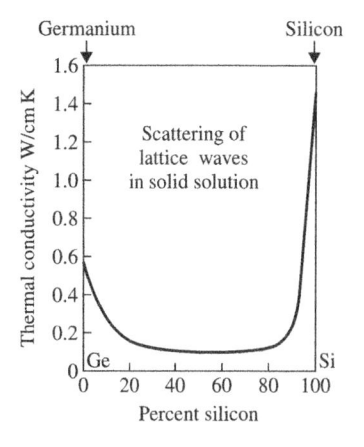

Fig. 18.8 Thermal conductivity of crystalline α-quartz and vitreous fused silica.

Fig. 18.9 Thermal conductivity of the Ge–Si solid solution. The scattering of lattice waves leads to a marked decrease in conductivity.

lattice sites by Si and Ge atoms reduces the mean free path of phonons. This is a thermal analog to Matthiessen's Rule for electrical resistance (Section 17.3).

In metals, thermal conduction is mainly due to free electrons. The lattice contribution can generally be ignored except for heavily alloyed specimens. The electronic thermal conductivity is nearly constant at high temperatures. For pure, well-annealed crystals there is a large peak in k at very low temperatures, similar to that observed in quartz crystals. Measurements on high quality gold crystals show a peak of 2800 W/m K at 20 K, almost an order of magnitude larger than the room temperature value.

18.4 Field dependence

Magnetic and electric fields have an influence on thermal conductivity. In the presence of a magnetic field, the thermal conductivity of metals and ceramics decrease noticeably. These thermomagnetic phenomena arise from the Lorentz force and are discussed in Section 20.5.

Interacting electric fields and temperature gradients lead to a number of interesting thermoelectric effects in metals and semiconductors. These are discussed in Chapter 21. For insulators, the changes in thermal conductivity with electric field are normally rather small, except near phase transformations at low temperatures. Low frequency phonons dominate the heat transport and storage capacity of materials at very low temperatures. Appreciable control of thermal conductivity by external electric fields is possible in soft-mode materials. Strontium titanate ($SrTiO_3$) exhibits soft-mode behavior at low temperatures where TiO_6 octahedra undergo rotational motions about [001]. At 5 K an electric field of 23 kV/cm induces a fivefold increase in the thermal conductivity of $SrTiO_3$. Rapid thermal switches have been constructed from soft-mode materials with field-tunable thermal conductivity.

In tensor form these *dielectrothermal effects* are given by

$$h_i = -k_{ij}(\partial T/\partial Z_j) - k_{ijk}(\partial T/\partial Z_j)E_k - k_{ijkl}(\partial T/\partial Z_j)E_k E_l - \cdots.$$

The heat flow h_i created by a temperature gradient dT/dZ_j is altered by electric fields components E_k and E_l through third- and fourth-rank polar tensors k_{ijk} and k_{ijkl}. For centrosymmetric crystals such as $SrTiO_3$, third rank tensors are zero. Therefore the change in thermal conductivity is proportional to E^2.

Diffusion and ionic conductivity

19

The phenomenon of atomic and ionic migration in crystals is called solid-state diffusion, and its study has shed light on many problems of technological and scientific importance. Diffusion is intimately connected to the strength of metals at high temperature, to metallurgical processes used to control alloy properties, and to many of the effects of radiation on nuclear reactor materials. Diffusion studies are important in understanding the ionic conductivity of the materials used in fuel cells, the fabrication of semiconductor integrated circuits, the corrosion of metals, and the sintering of ceramics.

19.1	Definition and tensor formulation	211
19.2	Structure–property relationships	212
19.3	Ionic conductivity	217
19.4	Superionic conductors	219
19.5	Cross-coupled diffusion	220

19.1 Definition and tensor formulation

When two miscible materials are in contact across an interface, the quantity of diffusing material which passes through the interface is proportional to the concentration gradient. The atomic flux J is given by

$$J = -D\frac{dc}{dZ},$$

where J is measured per unit time and per unit area, c is the concentration of the diffusing material per unit volume, and Z is the gradient direction. The proportionality factor D, the diffusion coefficient, is measured in units of m^2/s. This equation is sometimes referred to as *Fick's First Law*. It describes atomic transport in a form that is analogous to electrical resistivity (Ohm's Law) or thermal conductivity.

There are several objections to Fick's Law, as discussed in Section 19.5. Strictly speaking, it is valid only for self-diffusion coefficients measured in small concentration gradients.

Since J and Z are both vectors, the diffusion coefficient D is a second rank tensor.

$$J_i = -D_{ij}\frac{dc}{dZ_j}.$$

As with other symmetric second rank tensors, between one and six measurements are required to specify D_{ij}, depending on symmetry.

The relationship between structure and anisotropy is more apparent in PbI$_2$. Lead iodide is isostructural with CdI$_2$ in trigonal point group $\bar{3}m$ (Fig. 19.1). The self-diffusion of Pb is much easier parallel to the layers where the Pb atoms are in close proximity to one another. Diffusion is more difficult along $Z_3 = [001]$ because Pb atoms have a very long jump distance in this direction.

The mineral olivine, (Mg, Fe)$_2$SiO$_4$, is an important constituent of the deeper parts of the earth's crust. Geochemists have studied diffusion in olivine single

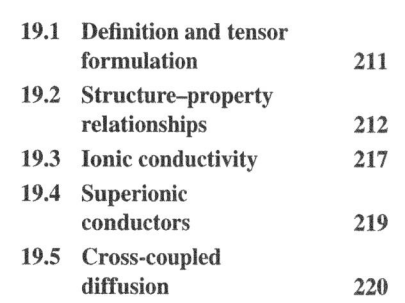

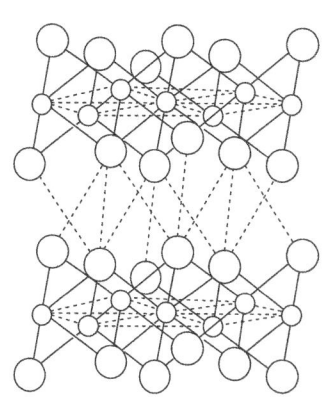

Fig. 19.1 Two layers of the PbI$_2$ structure. Small circles are divalent lead atoms which diffuse most easily parallel to the layers.

crystals. The crystal structure is orthorhombic, point group *mmm*, with three independent diffusion coefficients. Experiments on the diffusion of Ni in olivine at high temperature show that D_{33} is much larger than D_{11} and D_{22}. The origin of this large anisotropy can be explained by the crystal structure. In the olivine structure, nickel substitutes for magnesium and iron which occupy octahedral positions. Edge-sharing chains of octahedra lie along $Z_3 = [001]$ making diffusion especially easy in this direction. Only corner sharing is found in the Z_1 and Z_2 directions so the cation–cation spacing is much longer.

Experimental methods for measuring diffusion coefficients are relatively simple conceptually but require a good deal of skill and patience. Typically two metal bars are brought into contact and placed in a furnace. After being held at a carefully controlled temperature for a specified time, the bicrystal is removed from the furnace and sectioned into thin slices parallel to the interface. Each slice is then analyzed to determine its chemical composition. From these results, the amount of metal A that has diffused into B, and the amount of B that has diffused into A can be determined as a function of the distance from the interface. The diffusion coefficient D is determined from this information. For anisotropic materials one uses either long thin bars which define the direction of J, or flat plates which define the concentration gradient dc/dZ (Section 18.2). Long bars are generally easier to section into thin slices. The atomic distribution can be determined by chemical analysis or by using radioactive tracers in the diffusion specimens. With radioactive tracers it is possible to study self-diffusion (e.g. oxygen in MgO) which is of great significance in studying the densification process.

Diffusion coefficients depend exponentially on temperature, and are generally written in the form

$$D = D_0 \exp\left(\frac{-Q}{RT}\right),$$

where Q is the activation energy, R the universal gas constant, T the absolute temperature, and D_0 the diffusion pre-exponential factor which is generally assumed to be independent of temperature. This equation is often referred to as the Arrhenius relationship for the diffusion coefficient.

Converting the equation to logarithmic form gives

$$\ln D = \frac{-Q}{RT} + \ln D_0$$

and a plot of $\ln D$ against $1/T$ for the experimental values over a range of temperatures gives a straight line. The heat of activation Q is calculated from the slope of the line, and D_0 from the intercept. Experimental results are summarized by giving Q and D_0 values (Table 19.1).

Problem 19.1
Using the coefficients in Table 19.1, draw a graph of the diffusion coefficient (D) of tungsten from room temperature to the melting point.

19.2 Structure–property relationships

When normalized to the melting point, the self-diffusion coefficients are surprisingly similar for many materials. This can be seen in several ways. Near the

Table 19.1 Self-diffusion in cubic crystals. Melting points T_m (K), pre-exponential constants D_0 (m²/s), activation energies Q (kJ/mole) and diffusion coefficients near the melting point, $D(T_m)$ in m²/s (the ratio Q/T_m is remarkably constant)

	T_m	D_0	Q	$D(T_m)$	Q/T_m
Alkali metals (BCC)					
Li	454	2.3×10^{-5}	55.3	9.9×10^{-12}	0.122
Na	371	2.4×10^{-5}	43.8	1.6×10^{-11}	0.118
K	337	3.1×10^{-5}	40.8	1.5×10^{-11}	0.121
Rb	312	2.3×10^{-5}	39.4	5.8×10^{-12}	0.126
Transition metals (BCC)					
Ti	1933	1.1×10^{-4}	251	1.8×10^{-11}	0.130
V	2163	2.8×10^{-5}	309	9.7×10^{-13}	0.143
Cr	2130	2.0×10^{-5}	309	5.4×10^{-13}	0.145
Zr	2125	1.3×10^{-4}	274	2.5×10^{-11}	0.129
Nb	2741	1.2×10^{-3}	440	5.2×10^{-12}	0.160
Mo	2890	1.8×10^{-4}	461	8.4×10^{-13}	0.159
Ta	3269	1.2×10^{-4}	413	3.1×10^{-11}	0.126
W	3683	4.3×10^{-3}	641	3.4×10^{-12}	0.174
Electrode metals (FCC)					
Al	933	1.7×10^{-4}	142	1.9×10^{-12}	0.152
Ni	1726	1.9×10^{-4}	280	6.5×10^{-13}	0.162
Cu	1356	3.1×10^{-5}	200	5.9×10^{-13}	0.147
Pd	1825	2.1×10^{-5}	266	4.9×10^{-13}	0.146
Ag	1234	4.0×10^{-5}	185	6.1×10^{-13}	0.150
Pt	2046	2.2×10^{-5}	278	1.7×10^{-13}	0.136
Au	1336	1.1×10^{-5}	177	1.3×10^{-12}	0.133
Alkali halides (Rocksalt structure)					
LiF	1115	6.4×10^{-3}	212	7.5×10^{-13}	0.190
NaCl	1074	6.2×10^{-3}	206	5.8×10^{-13}	0.192
NaBr	1028	5.0×10^{-3}	195	6.5×10^{-13}	0.190
KCl	1049	3.6×10^{-3}	202	3.0×10^{-13}	0.193
RbCl	988	3.3×10^{-3}	192	2.4×10^{-13}	0.194

melting point the diffusion coefficients are generally within an order of magnitude of 10^{-12} m²/s. Second, the ratios of the activation energy to the melting point are in the range 0.12–0.19, and are even more tightly grouped within the various structure types. Most engineering metals are close to 0.15. The activation energy for diffusion is also closely related to the heat of fusion L_m. The ratio Q/L_m is about 15 for many metals.

Tammann's Rules are useful in choosing annealing temperatures for sintering or for microstructural homogenization. The rules state that diffusion in metals becomes significant at 33% of the melting point in K. Diffusion is generally more difficult in ionic crystals and only becomes significant at about 57% of the melting point. The bonding is even stronger in covalent materials where annealing is often done at temperatures above 90% of the melting temperature.

Somewhat higher temperatures are required to promote plastic flow and deformation. For metals, the minimum hot-working temperature is about 50% of the melting point in K, with corresponding increases in other materials.

The relationship between melting point and diffusion is quite apparent in many solid solutions. Fig. 19.2 shows the phase diagrams and diffusion coefficients for the Pd–Cu and Cu–Au systems. The diffusion coefficients were measured at temperatures slightly below the lowest melting point in each system. Note that in both cases the diffusion coefficients are highest for compositions close to their melting point.

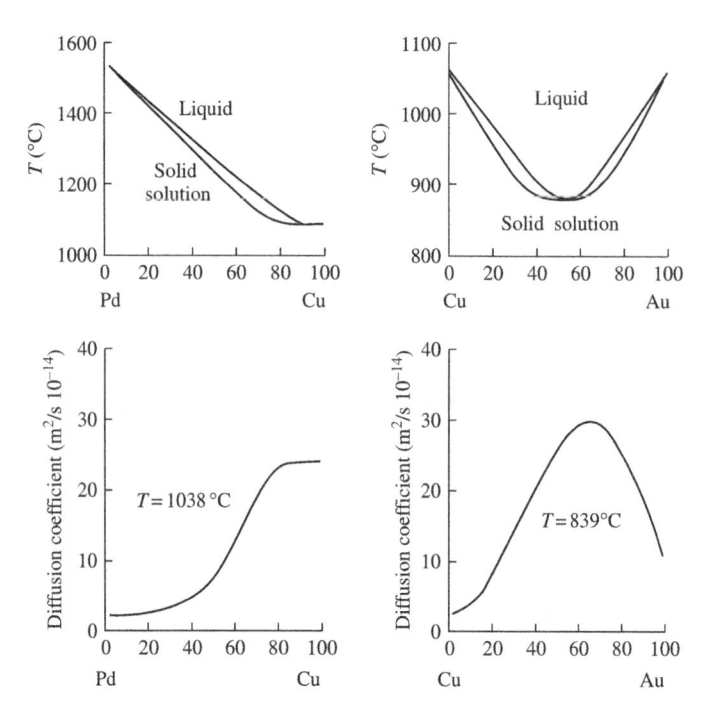

Fig. 19.2 Binary phase equilibrium diagrams and diffusion coefficients for Pd–Cu and Cu–Au alloys. Note that compositions with low melting points have large diffusion coefficients.

Table 19.2 Comparison of self-diffusion coefficients for four HCP metals. Note the correlation between and D_{11}/D_{33} ratio and the crystallographic c/a ratio

	$(D_0)_{11}$ ($\times 10^{-4}$ m²/s)	$(D_0)_{33}$ ($\times 10^{-4}$ m²/s)	Q_{11}	Q_{33}	D_{11}/D_{33}	c/a
Cd	0.1	0.05	82	76	1.21	1.89
Mg	1.5	1.0	136	134	0.82	1.63
Tl	0.4	0.4	94	96	0.73	1.60
Zn	0.58	0.13	102	91	1.67	1.86

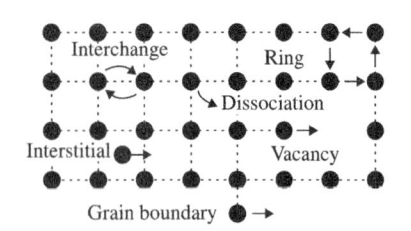

Fig. 19.3 Six diffusion mechanisms involving defects in crystals.

Anisotropy in self-diffusion can be correlated with crystal structure. Magnesium single crystals (point group $6/mmm$) have an almost ideal hexagonal close-packed structure with twelve nearly equidistant neighbors surrounding each Mg atom and a c/a ratio of 1.63. The structure is more distorted for other HCP metals. Diffusion coefficients for four HCP metals (Table 19.2) show that the D_{11}/D_{33} ratio scales with c/a. In zinc and cadmium diffusion takes place best in the (001) plane within the tightly bonded close-packed layers.

D values are extremely temperature-sensitive since most diffusion processes are thermally activated. At elevated temperatures, thermal vibration increases in amplitude leading to atom transport and diffusion, usually by one of the processes illustrated in Fig. 19.3. The *ring* and *interchange* mechanisms are more important in metals than in ionic solids where an exchange of neighboring cations and anions requires a great deal of energy. *Vacancy* and *interstitial* processes require less energy than interchange because only one atom is displaced, rather than two or more. Diffusion via vacancies requires a neighboring lattice vacancy and therefore tends to be less rapid than the interstitial mechanism. Atom sizes are very important. In many metals small atoms like hydrogen and carbon diffuse rapidly via interstitial sites. Thus D depends on the nature of

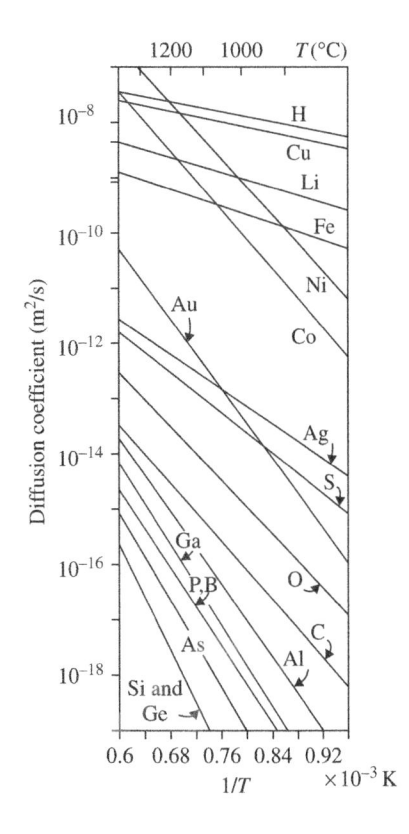

Fig. 19.4 Diffusion coefficients for various dopants in silicon. Self-diffusion is far more difficult than for H and other small atoms which diffuse interstitially.

the diffusing atom and the host lattice. For rapid diffusion, the host lattice must provide suitable interstitial sites and the "necking down" between sites must not be excessive. Zeolites are ideal from this point of view, with many cavities connected by open channels, making them useful in ion-exchange applications.

The importance of defects can be illustrated with the diffusion rates of various dopants in single crystal silicon (Fig. 19.4). Because of its importance in the processing of integrated circuits, the diffusion of impurities in silicon has been studied extensively. Small monovalent atoms such as lithium are thought to occupy interstitial sites in the diamond-like structure of silicon. The geometry of the jump process involved in interstitial diffusion is pictured in Fig. 19.5(a). In order to move from one interstitial site to another, the lithium atom must pass through a puckered hexagon of six silicon atoms. The activation energy for this process is quite low, leading to rapid diffusion through the crystal.

Self-diffusion in silicon is much more difficult and probably involves a vacancy mechanism (Fig. 19.5(b)). Four strong covalent bonds must be broken to create a vacancy in the diamond structure. This leads to a high activation energy and much lower diffusion coefficients as shown in Fig. 19.4. This also applies to substitutional dopants such as As, P, B, Ga and Al. These n- and p-type dopants from columns III and V of the periodic system are less well bonded than tetravalent Si or Ge atoms so their diffusion coefficients are somewhat larger. In general, poorly bonded atoms, which differ in size and valence from the host element, diffuse more rapidly than those which are similar in size and valence.

Vacancy diffusion is also important in most alloys which have close-packed crystal structures and where only very small foreign atoms can be

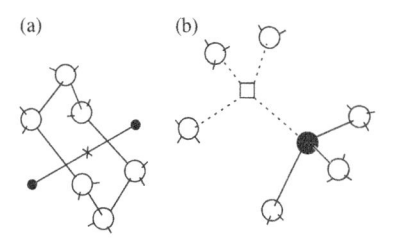

Fig. 19.5 (a) Interstitial migration in silicon involves passing through a hexagonal ring of Si atoms which is fairly easy for small atoms like Li and H. (b) The vacancy mechanism for self-diffusion and for substitutional dopants is much less favorable energetically (Girifalco).

accommodated in interstitial sites. Fig. 19.6 shows the activation energies in silver and copper alloys doped with neighboring element of higher or lower atomic number (Z). Note that diffusion is easier for dopants like Cd and As ($\Delta Z > 0$) and more difficult for elements like Ru and Co ($\Delta Z < 0$). This effect has been attributed to Coulomb interactions between the dopants and vacancies in Ag or Cu lattice. A missing Cu^+ or Ag^+ core in the metal alloy has a negative charge and is attracted to the positively charged cores of the Cd^{2+} or As^{5+} dopants. Thus Cd and As diffuse more easily because of the presence of a nearby vacancy. Elements with $\Delta Z < 0$ tend to repulse vacancies and diffuse more slowly.

In compounds, different ion species often diffuse at vastly different rates, although electrical neutrality requires that the diffusion rates be coupled. Diffusion coefficients for several oxides are shown in Fig. 19.7. Thermal vibration assists diffusion, causing the increase in D with temperature. Multiple-valence transition metals have high diffusion rates because of the importance of defects. The diffusion coefficients of the large oxygen ions are usually small except for structures like calcium-stabilized zirconia which contains anion vacancies.

Uraninite, one of the principal ores of uranium, has an interesting diffusion mechanism involving interstitial oxygens (Fig. 19.8). Despite its large ionic radius, oxygen diffuses faster than uranium as it migrates in and out of the interstitial sites. Ideally, uraninite has the chemical formula UO_2 with tetravalent uranium coordinated to eight oxygens in the fluorite structure. In practice, however, the chemical formula is

$$U^{4+}_{1-x}U^{6+}_{x}O_{2+x} \approx U_3O_8.$$

The excess oxygen occupies interstitial sites at the center of the unit cell and diffuses relatively easily through other empty interstitial sites.

Crystal chemistry is helpful in controlling diffusion coefficients. The rapid sintering of ceramics calls for large diffusion coefficients, but it is important to realize that all types of atoms must be transported. Thus the most important diffusion coefficient is the smallest diffusion coefficient, the rate-limiting coefficient. In oxides and similar materials, the anion usually has the smallest coefficient because it is usually larger than the cations. Anion diffusion can be

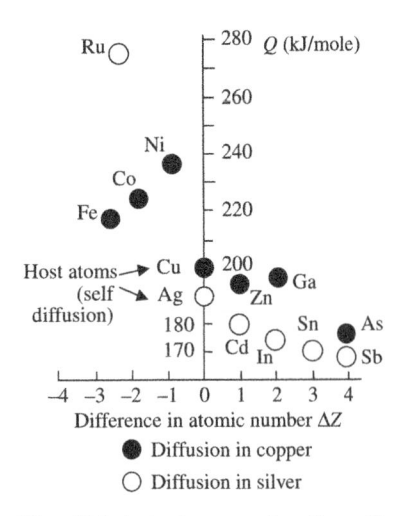

Fig. 19.6 Activation energies for self-diffusion in copper and silver together with activation energies for the diffusion of dopants from neighboring columns of the periodic system. Elements with $\Delta Z > 0$ diffuse more easily than those with $\Delta Z < 0$.

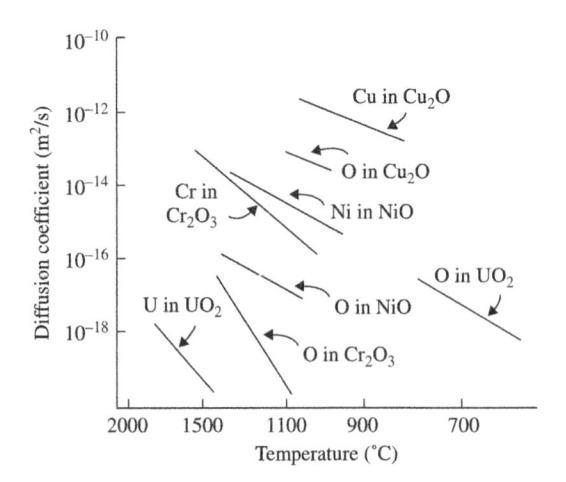

Fig. 19.7 Diffusion coefficients for cations and anions in several oxides. Generally the cations diffuse faster than oxygen, although there are exceptions such as uraninite.

increased by introducing anion vacancies. There are several ways of doing this. Doping the material with cations of smaller valence (Li$_2$O in MgO) or anions of higher valence (oxygen in a fluoride) are two chemical methods. Aluminum oxide can be sintered to theoretical density by adding about 0.1% MgO in solid solution. Another related technique is to heat the material in a reducing atmosphere, thereby creating oxygen vacancies.

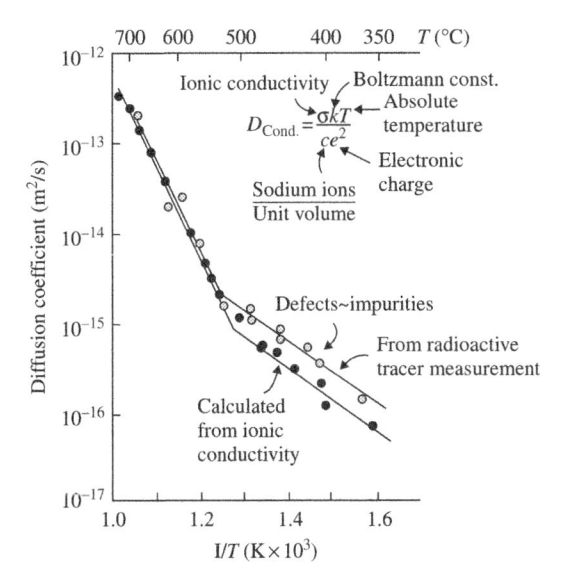

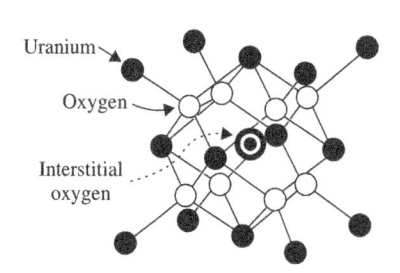

Fig. 19.8 Uraninite has the fluorite structure with excess oxygen occupying interstitial sites.

19.3 Ionic conductivity

Ionic conductivity σ is a second rank tensor like electronic conductivity

$$J_i = \sigma_{ij} E_j,$$

but is closely related to diffusion through the *Nernst–Einstein equation*

$$\sigma = \frac{Dnq^2}{kT}.$$

Here n is the number of charge carriers per unit volume, q the charge per ion, k is Boltzmann's Constant, and T the absolute temperature. The diffusion coefficient (D) generally refers to the more rapidly moving species, as pointed out earlier. Diffusion generally occurs by the movement of ions to neighboring vacancies. For stoichiometric compounds, the vacancy concentration and ionic conductivity are very small, although suitable doping will increase both.

Impurities and other defects often play a decisive role in electrolytic conduction. In a salt crystal, the movable charges may be interstitial ions (*Frenkel defects*) or vacancies (*Schottky defects*). The position of a vacancy changes when a neighboring ion moves in to fill it. Schottky defects predominate in KCl where both cation and anion vacancies occur. In AgCl, some Ag$^+$ ions occupy interstitial sites, producing positive Frenkel defects and negative Schottky defects.

The close relationship between ionic conductivity and diffusion is illustrated in Fig. 19.9 where the diffusion coefficient of NaCl is measured in two ways.

Fig. 19.9 Diffusion and ionic conduction in rocksalt (NaCl).

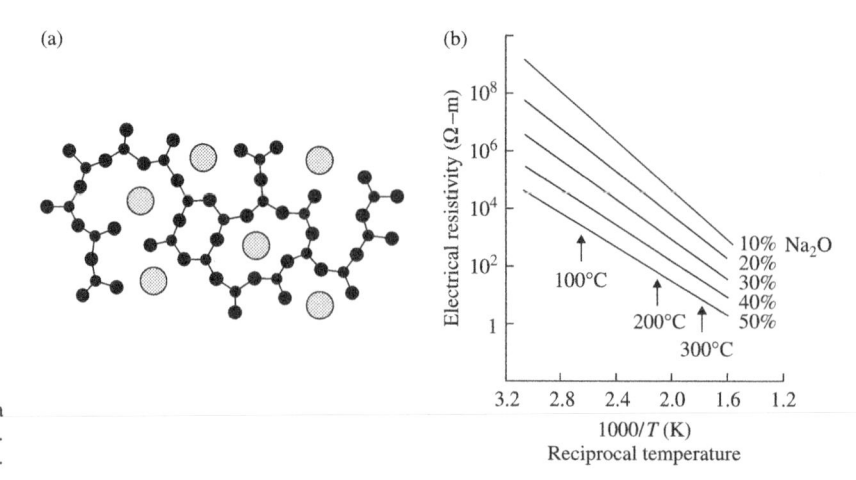

Fig. 19.10 (a) Mobile Na$^+$ ions in a framework of corner-linked SiO$_4$ tetrahedra. (b) Electrical resistivity of soda–silica glasses.

First, using a concentration gradient and the radioactive tracer method, and second from the ionic conductivity and the Nernst–Einstein relation. At high temperature, where thermally induced vacancies predominate, the two values for D are in very close agreement. The agreement is not so good at lower temperatures where various types of crystal defects control the movement of ions.

Transport numbers are often used to characterize conduction processes. In most solids, charge is carried by cations (t_c), anions (t_a), electrons (t_e) and holes (t_h). (Transport numbers (t_j) are defined as the fraction of the total current carried by the jth particle.) The sum of the transport numbers is one: $t_c + t_a + t_e + t_h + \cdots = 1$.

For NaCl and Na$_2$O–SiO$_2$ glasses the charge is carried by Na$^+$ ions ($t_c \approx 1$). In glass (Fig. 19.10) the Na$^+$ ions move from one cavity to another within the stationary silicate network. As expected, the electrical resistivity decreases with increasing sodium content and with increasing temperature.

Ionic conduction also controls the resistivity of crystalline silica. Quartz has a relatively open structure with channels along the trigonal c-axis, causing anisotropy in the electrical conductivity. In directions normal to the c-axis, quartz is a good insulator, but parallel to c, conduction of an electrolytic nature occurs readily. Ionic conduction may involve impurity ions already in the structure, or deliberately-introduced foreign ions. At 250°C, Na$^+$ ions are easily transported through the crystal from a sodium amalgam anode, but K$^+$ ions from a potassium amalgam anode pass through with much less facility. Activation energies for diffusion increase steadily with ionic radius, rising from 75 kJ/mole for Li$^+$ (0.6 Å radius) to 138 kJ/mole for Cs$^+$ (1.7 Å). Passage through the c-axis channels becomes increasingly difficult for ions exceeding the channel diameter in size. The behavior of Ag$^+$ ions in quartz is especially interesting because of the visible silver deposits left behind. Thread-like formations parallel to c, and platelets perpendicular to c are observed.

In BaCl$_2$ and most other ionic compounds with the fluorite structure, charge is carried by anions ($t_a \sim 1$). The fluorite structure favors anion motion because the anions have less charge and are closer together than the cations. The possibility of using interstitial sites, as in uraninite, also promotes Cl$^-$ motion in BaCl$_2$.

19.4 Superionic conductors

There are a few solids that have ionic conductivities comparable to liquids. At room temperature, aqueous electrolytes have conductivities near 100 S/m, while non-aqueous electrolytes are generally somewhat lower. Two of the best solid ionic conductors are RbAg$_4$I$_5$, and NaAl$_{11}$O$_{17}$, which have rather open structures in which cations diffuse rapidly. The room temperature conductivities of these compounds lie between 10 and 100 S/m. Anions will also diffuse but with greater difficulty because of their larger size. At 1000°C, rapidly moving oxygen ions increase the conductivity of defect solid solutions in the ZrO$_2$–Y$_2$O$_3$ system to as high as 10 S/m (Fig. 19.11).

Superionic conductors exhibit very high ionic conductivity with negligibly small electronic conductivity. Three main groups of superionic conductors are listed in Table 19.3. The silver halide group is characterized by cation disorder, as are the beta-alumina structures. In both these groups the univalent cations Ag$^+$, Na$^+$, Cu$^+$ are especially mobile. Oxygens are the conducting species in the defect-stabilized oxide group. In calcia-stabilized zirconia, calcium introduces a large number of anion vacancies into zirconia as indicated by the formula Ca$_x^{2+}$Zr$_{1-x}^{4+}$O$_{2-x}^{2-}$. The oxygen sites are only 2.5 Å apart in the zirconia structure.

The ionic conductivity of RbAg$_4$I$_5$ near room temperature is about 17 orders of magnitude larger than that of NaCl. The activation energies are small, since the large conductivities are attained well below the melting point of the superionic conductor. Such materials contain a large number of mobile charge carriers which is independent of temperature. Schottky or Frenkel defects are responsible for ionic conductivity in ordinary ionic compounds, so that the number of charge carriers is small and temperature dependent. Not only is the number of carriers large in a superionic conductor, but the mean free paths may be large as well. In the stabilized zirconia group, mobile oxygen ions are transported over many interatomic distances by a cooperative mechanism. In conventional ionic conductors, the mean free path is one "hop" between neighboring sites.

Several silver halides are excellent solid electrolytes with conductivities approaching those of liquid electrolytes. The silver ions occupy interconnected passageways formed by face-sharing anion polyhedra. The number of polyhedra exceeds the number of mobile cations. In α-AgI and other electrolytes based on a body-centered cubic arrangement of iodine atoms, the anions form passageways of face-sharing tetrahedra. The α-Ag$_2$HgI$_4$ structure consists of

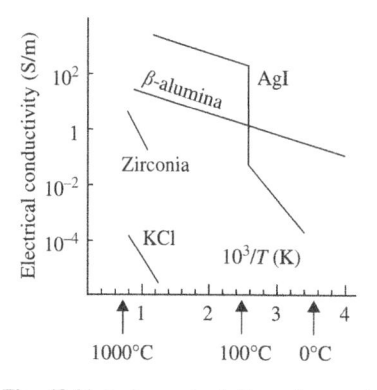

Fig. 19.11 Ionic conductivities of several solid electrolytes compared with KCl, a typical ionic solid.

Table 19.3 Superionic conductors

Halides and chalcogenides

α-AgI	Na$_2$S	KAg$_4$I$_5$
α-CuI	Ag$_3$SBr	CsAg$_4$I$_5$
α-Ag$_2$Te	α-Ag$_3$SI	NH$_4$Ag$_4$I$_5$
α-Ag$_2$Se	α-Ag$_3$HgI$_4$	RbAg$_4$I$_5$
α-Cu$_2$Se	Ag$_4$HgSe$_2$I$_2$	

Beta-alumina $A_2O \cdot nM_2O_3$
A = Na, Rb, Ag, K, Li, Tl
M = Al, Ga, Fe^{3+}
n = 5–11 (integer)

Defect-stabilized ceramic oxides
CaO·AO_2 A = Zr, Hf, Th, Ce
M_2O_3·ZrO$_2$ M = La, Sm, Y, Yb, Sc

a face-centered cubic arrangement of iodine atoms with silver ions moving through passageways formed by face-sharing octahedra and tetrahedra. In the more complex RbAg₄I₅ structure, the passageways are made up of face-sharing tetrahedra.

Solid electrolytes with anomalously high ionic conductivity are of interest for solid state batteries and fuel cells. Thermodynamic data can be helpful in selecting new superionic conductors. In good conductors like AgI there is a phase transition to the high conductivity states (Fig. 19.11). The entropy change (ΔS_t) accompanying this transition is anomalously large compared to other solid state transitions, and is comparable to the entropy of melting (ΔS_m). This is consistent with an highly disordered atomistic model in which Ag⁺ ions flow through a body-centered cubic lattice of I⁻ anions. Other superionic conductors such as Cu₂S and Ag₂S also have large ΔS_t values.

Anisotropic conduction is observed in β-alumina (NaAl₁₁O₁₇) where ionic motion take place in the (001) plane perpendicular to the hexagonal c-axis. The structure consists of spinel-type layers separated by Na–O layers. The oxygen functions as spacers between spinel blocks as sodium migrates with ease (Fig. 19.12).

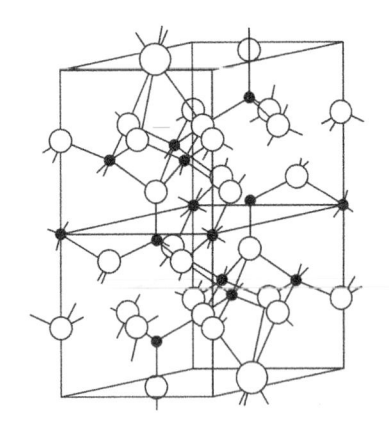

Fig. 19.12 A perspective drawing of half of the unit cell of β-alumina. The large circles represent Na, the small ones oxygen, and the black dots aluminum atoms.

19.5 Cross-coupled diffusion

19.5.1 Binary diffusion equations

Most inorganic, organic, and metal alloys contain two or more different elements and complex crystal structures. In β-brass, for example, the Cu and Zn atoms undergo an order–disorder phase transformation similar to the shape memory alloys (Section 16.2). Because of their different sizes and surroundings the Cu and Zn atoms are subject to different internal forces, and have different diffusion rates. The situation is similar to the different transport numbers noted in NaCl and BaCl₂ where one species has a much higher diffusion coefficient than the other. A generalized binary diffusion equation can be used to describe experiments where two (or more) elements contribute to atomic transport. For atoms A and B, the tensor relations involve three second rank tensor diffusion coefficients and two composition gradients:

$$J_i^A = -D_{ij}^{AA} \frac{dc^A}{dZ_j} - D_{ij}^{AB} \frac{dc^B}{dZ_j}$$

$$J_i^B = -D_{ij}^{AB} \frac{dc^A}{dZ_j} - D_{ij}^{BB} \frac{dc^B}{dZ_j}.$$

These equations point out the fact that the atomic flux in element A depends on the gradient in element B as well as the concentration gradient in A. This is one of the major objections to Fick's First Law (Section 19.1). It is for this reason that many authors prefer to describe diffusion and ionic conduction in terms of chemical potentials rather than concentration gradients. Chemical potential μ^A of element A is defined by

$$\mu^A = \frac{\partial G}{\partial n^A},$$

where G is the Gibbs free energy and n^A is the number of A atoms per unit volume. For binary systems $n^A + n^B = N$, the total number of lattice sites.

A term in the free energy arising from mixing entropy is required to establish the relationships between diffusion coefficients and chemical potential gradients.

Binary diffusion experiments are often carried out with two metals in contact with one another. Metals A and B are brought together and held at high temperature to promote diffusion across the A–B interface. As diffusion proceeds, the interface spreads out as A diffuses into B and B into A. The diffusion of A into B will, in general, differ from that of B into A. This gives rise to a movement of the interface, the so-called *Kirkendall Effect*.

19.5.2 Thermal diffusion equations

Since diffusion coefficients are extremely sensitive to temperature, great care must be taken in ordinary diffusion experiments to maintain isothermal conditions. Thermal diffusion experiments involve the transport of atoms in a controlled temperature gradient dT/dZ. As with the case of binary diffusion, there are two gradients and three second rank tensor coefficients. The atom flow is represented by

$$J_i = -D_{ij}\frac{dc}{dZ_j} - \beta_{ij}\frac{dT}{dZ_j}$$

and the heat flow by

$$h_i = -\beta_{ij}\frac{dc}{dZ_j} - k_{ij}\frac{dT}{dZ_j},$$

where D_{ij} is the diffusion tensor [m²/s], k_{ij} the thermal conductivity [W/mK], and β_{ij} the so-called *Soret coefficient* [m^{-1}s^{-1}K^{-1}] describing atom motion in a temperature gradient.

An important example of the Soret effect is the migration of carbon atoms in steel. A temperature gradient is established across an iron bar containing a small amount of uniformly distributed carbon. One end is hot, the other cool. When this condition is maintained for a period of time, and various parts of the bar are then analyzed for carbon content, it is found that carbon has migrated from the cold to the hot regions.

In the Soret experiment, a temperature gradient induces mass flow; sometimes the flow is from hot to cold regions, sometimes from cold to hot. In α-Fe, for example, interstitial H moves from hot to cold, and interstitial C and N in the reverse direction. Similar effects occur in other metals. Soret experiments have also been performed on zinc doped with In, Tl, or Ag. In this case In and Tl move in the direction of the temperature gradient, while silver migrates in the opposite direction.

19.5.3 Electrolysis and electron wind

Ionic conductivity involves the movement of atoms under applied electric fields, but what about metals and heavily doped semiconductors with large electronic conductivities? Both electron flow J^e and atom flow J^a take place under the combined influence of concentration gradients and electric fields. In tensor form

$$J_i^a = -D_{ij}\frac{dc}{dZ_j} - \gamma_{ij}E_j$$

$$J_i^e = -\gamma_{ij}\frac{dc}{dZ_j} - \sigma_{ij}E_j.$$

When a large electric current is passed through a hot metal wire it is found that solute atoms in the alloy migrate toward an electrode.

Originally, the solute motion was attributed to electrolysis, similar to the ionic conductivity in alkali halides, but two observations argued against this interpretation. First, the solute motion was always toward the anode, regardless of the charge difference between the solute and the solvent. It did not matter whether the effective charge was positive or negative, the motion was always toward the positive electrodes. The second observation concerns the field-induced diffusion in pure metal wires. Again mass transport is toward the anode, confirming the result on alloy systems.

Electron wind is one of the proposed explanations. In a metal, an applied voltage causes large numbers of electrons to move toward the anode. In a hot wire, the negative electrons that constitute the electric current collide with migrating atoms and give them a push toward the positive electrode. The collisions between electrons and atoms constitute an electron wind that controls the migration direction of diffusing metal atoms.

Problem 19.2
Thermal diffusion experiments are to be carried out on an orthorhombic crystal (point group *mmm*) in the presence of an electric field. The electric field, temperature gradient, and concentration gradient are in perpendicular directions along the orthorhombic axes. Write out the defining equations for heat flow, electric current, and diffusion. Which coefficients must be measured?

19.5.4 Effects of mechanical stress

Hydrostatic pressure generally slows down atomic migration in crystals. The underlying cause of this decrease in diffusion coefficients has to do with the decrease in free volume. Defects such as vacancies and interstitials are critical diffusion mechanisms, and both types of defects lead to increased volume. Hydrostatic pressure inhibits the formation of these defects, and further inhibits the additional volume increases accompanying atom motion from one crystallographic site to another. If V^* is the activation volume required for diffusion, an energy PV^* is needed to work against the external pressure P. The resulting effect on diffusion appears as an activation energy. Under pressure, the diffusion coefficient is given by

$$D(P) = D(0)\exp\left(\frac{-PV^*}{RT}\right).$$

Tracer measurements have been carried out on several metals at pressures of a thousand atmospheres or more. In zinc, sodium, and most other metals the diffusion coefficient decreases exponentially with increasing pressure.

Galvanomagnetic and thermomagnetic phenomena

20

The Lorentz force that a magnetic field exerts on a moving charge carrier is perpendicular to the direction of motion and to the magnetic field. Since both electric and thermal currents are carried by mobile electrons and ions, a wide range of galvanomagnetic and thermomagnetic effects result. The effects that occur in an isotropic polycrystalline metal are illustrated in Fig. 20.1. As to be expected, many more cross-coupled effects occur in less symmetric solids.

The galvanomagnetic experiments involve electric field, electric current, and magnetic field as variables. The Hall Effect, transverse magnetoresistance, and longitudinal magnetoresistance all describe the effects of magnetic fields on electrical resistance.

Analogous experiments on thermal conductivity are referred to as thermomagnetic effects. In this case the variables are heat flow, temperature gradient, and magnetic field. The Righi–Leduc Effect is the thermal Hall Effect in which

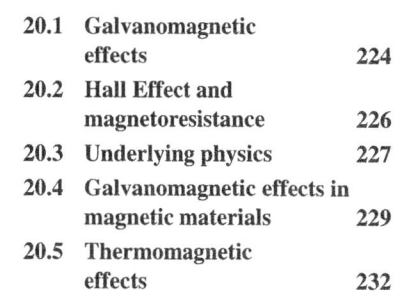

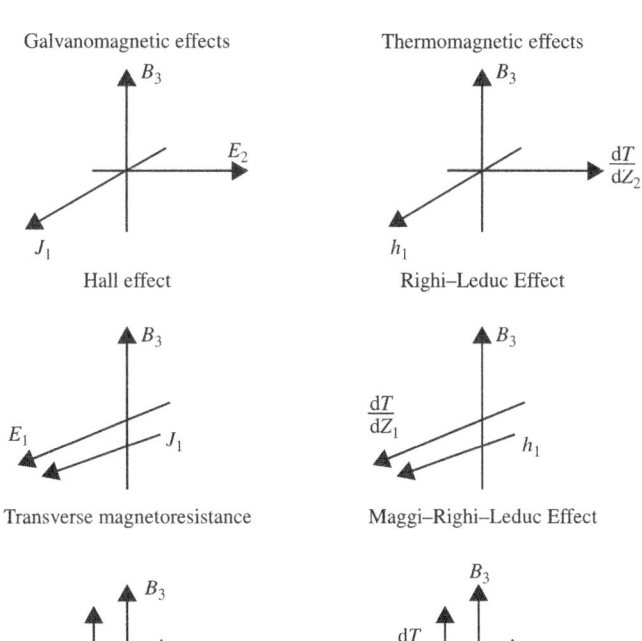

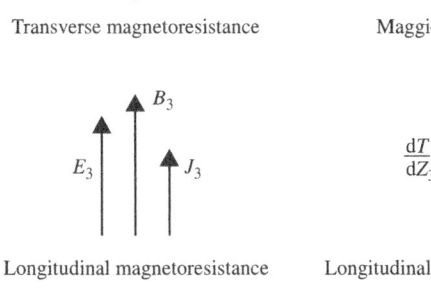

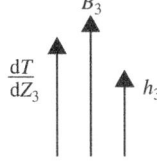

20.1	Galvanomagnetic effects	224
20.2	Hall Effect and magnetoresistance	226
20.3	Underlying physics	227
20.4	Galvanomagnetic effects in magnetic materials	229
20.5	Thermomagnetic effects	232

Fig. 20.1 Galvanomagnetic effects describe the influence of magnetic fields on electrical resistivity measurements. Thermomagnetic effects describe the effect of a magnetic field on thermal conduction. For an isotropic solid there are six such phenomena.

magnetic fields deflect heat flow rather than electric current. The transverse thermal magnetoresistance (the Maggi–Righi–Leduc Effect) and the longitudinal thermal magnetoresistance are analogous to the two galvanomagnetic magnetoresistance effects.

Additional interaction phenomena related to the thermoelectric and piezoresistance effects will be discussed in the next two chapters.

20.1 Galvanomagnetic effects

In tensor form Ohm's Law is

$$E_i = \rho_{ij} J_j,$$

where E_i is electrical field, J_j electric current density, and ρ_{ij} the electrical resistivity in Ω m. In describing the effect of magnetic field on electrical resistance, we expand the resistivity in a power series in magnetic flux density B. B is used rather than the magnetic field H because the Lorentz force acting on the charge carriers depends on B not H. In tensor form the Lorentz force F_i acting on a moving electron is represented by the vector product

$$F_i = -e\varepsilon_{ijk} v_j B_k,$$

where e is the charge on the electron, v_j its velocity, B_k the flux density and ε_{ijk} the rotation tensor used to represent cross products (Section 14.4).

The power series for the electric resistivity relation is

$$E_i = \rho_{ij} J_j + \rho_{ijk} J_j B_k + \rho_{ijkl} J_j B_k B_l + \cdots.$$

To determine the types of tensors that represent the higher order terms ρ_{ijk} and ρ_{ijkl}, we examine the way in which they transform between coordinate systems. The electric field in the new system is

$$E'_i = a_{im} E_m = a_{im} \rho_{mn} J_n + a_{im} \rho_{mnp} J_n B_p + a_{im} \rho_{mnpq} J_n B_p B_q + \cdots.$$

Electric field and electric current density J are first rank polar tensors and B is a first rank axial tensor. In going from new to old coordinates,

$$J_n = a_{jn} J'_j, \quad B_p = |a| a_{kp} B'_k \quad \text{and} \quad B_q = |a| a_{lq} B'_l.$$

Substituting into the expression for E'_i:

$$E'_i = a_{im} a_{jn} \rho_{mn} J'_j + |a| a_{im} a_{jn} a_{kp} \rho_{mnp} J'_j B'_k$$
$$+ |a||a| a_{im} a_{jn} a_{kp} a_{lq} \rho_{mnpq} J'_j B'_k B'_l + \cdots$$
$$= \rho'_{ij} J'_j + \rho'_{ijk} J'_j B'_k + \rho'_{ijkl} J'_j B'_k B'_l + \cdots.$$

Identifying, term by term, it is apparent that the tensor transformations are

$$\rho'_{ij} = a_{im} a_{jn} \rho_{mn}$$
$$\rho'_{ijk} = |a| a_{im} a_{jn} a_{kp} \rho_{mnp}$$
$$\rho'_{ijkl} = a_{im} a_{jn} a_{kp} a_{lq} \rho_{mnpq}.$$

In this expansion, even-rank tensors like ρ_{mn} and ρ_{mnpq} are polar, while odd-rank tensors such as ρ_{mnp} are axial.

Neumann's Principle applies to galvanomagnetic phenomena in the usual way. Most experimental studies have been carried out on semiconductors with the diamond, zincblende or wurtzite structures, or metals with the BCC, FCC, or HCP structures. The point groups of interest are cubic $m3m$ and $\bar{4}3m$, or hexagonal point groups $6/mmm$ and $6mm$.

For point group $m3m$ and $\bar{4}3m$, second rank polar properties like resistivity are isotropic with $\rho_{11} = \rho_{22} = \rho_{33}$. For fourth rank polar tensors there are three independent coefficients:

$$\rho_{1111} = \rho_{2222} = \rho_{3333}$$
$$\rho_{1122} = \rho_{2211} = \rho_{3311} = \rho_{1133} = \rho_{2233} = \rho_{3322}$$
$$\rho_{1212} = \rho_{1221} = \rho_{2112} = \rho_{2121} = \rho_{3131} = \rho_{1313} =$$
$$\rho_{1331} = \rho_{3113} = \rho_{2323} = \rho_{2332} = \rho_{3223} = \rho_{3232}.$$

Axial third rank tensors for point group $m3m$ have one independent coefficient, ρ_{123}. The proof goes as follows

$$\rho'_{ijk} = |a| a_{im} a_{jn} a_{kp} \rho_{mnp}.$$

For the mirror perpendicular to Z_1:

$$|a| = -1, 1 \to -1, 2 \to 2, 3 \to 3.$$

Therefore

$$111 \to -(-1)(-1)(-1) \to 111$$
$$222 \to -(2)(2)(2) \to -222 \to 0$$
$$123 \to -(-1)(2)(3) \to 123$$
$$112 \to -(-1)(-1)(2) \to -112 \to 0$$
$$\vdots$$

For the threefold rotation axes parallel to [111],

$$|a| = +1, 1 \to 2, 3 \to 1, 2 \to 3$$
$$111 \to 222 \to 0$$
$$333 \to 111 \to 0$$
$$123 \to 231 \to 312 \to 123$$
$$213 \to 321 \to 132 \to 213.$$

All coefficients with two identical subscripts are zero.

For the mirror perpendicular to [110]

$$|a| = -1, 1 \to -2, 2 \to -1, 3 \to 3$$
$$123 \to (-1)(-2)(-1)(3) \to -213$$
$$\vdots$$

Therefore $\rho_{123} = \rho_{231} = \rho_{312} = -\rho_{213} = -\rho_{321} = -\rho_{132}$ are the only nonzero coefficients.

Thus expanding out to second order in magnetic field leads to five independent coefficients, for point group $m3m$. An identical result is achieved for $\bar{4}3m$. The five coefficients (ρ_{11}, ρ_{123}, ρ_{1111}, ρ_{1122}, ρ_{1212}) correspond to five experiments: The resistivity in zero field plus four galvanomagnetic effects.

20.2 Hall Effect and magnetoresistance

Consider a series of experiments in which a current J is flowing along the Z_1 (=[100]) axis of a cubic crystal. The resulting electric field components are given by
$$E_m = \rho_{m1}J_1 + \rho_{m1p}J_1B_p + \rho_{m1pq}J_1B_pB_q.$$
For point group $m3m$ the resulting voltage components are
$$E_1 = \rho_{11}J_1 + \rho_{1111}J_1B_1^2 + \rho_{1122}J_1B_2^2 + \rho_{1133}J_1B_3^2$$
$$E_2 = \rho_{213}J_1B_3 + (\rho_{2112} + \rho_{2121})J_1B_1B_2$$
$$E_3 = \rho_{312}J_1B_2 + (\rho_{3113} + \rho_{3131})J_1B_1B_3.$$

If no magnetic field is present, then the only remaining term is $E_1 = \rho_{11}J_1$, which is the standard electrical resistivity measurement.

The first two of the galvanomagnetic coefficients are obtained with the magnetic induction aligned along Z_3 (=[001]) such that $B_1 = B_2 = 0 \neq B_3$. The electric field components are
$$E_1 = \rho_{11}J_1 + \rho_{1133}J_1B_3^2$$
$$E_2 = \rho_{213}J_1B_3$$
$$E_3 = 0.$$

From the measured value of E_2 we get the *Hall Effect*, $R = \rho_{213}$, which is linearly proportional to the applied field.

The second galvanomagnetic effect comes from the measurement of E_1. After applying the magnetic field, the change in resistivity is
$$\Delta\rho_{11} = \frac{E_1 - \rho_{11}J_1}{J_1} = \rho_{1133}B_3^2.$$

The fractional change in resistivity $\Delta\rho_{11}/\rho_{11}$ is called the *transverse magnetoresistance*. The *longitudinal magnetoresistance*, the third of the galvanomagnetic effects, is obtained by measuring E_1 with the current J_1 and magnetic induction B_1 in the same direction.
$$E_1 = \rho_{11}J_1 + \rho_{1111}J_1B_1^2.$$
The change in resistivity is
$$\Delta\rho_{11} = \frac{E_1 - \rho_{11}J_1}{J_1} = \rho_{1111}B_1^2$$
and the fractional change in resistivity, $\Delta\rho_{11}/\rho_{11}$, is the longitudinal magnetoresistance.

The fourth galvanomagnetic effect is obtained from the remaining nonlinear coefficient ρ_{1212}. With current J_1 flowing along [100] and magnetic induction directed along [110], the field E_2 is measured along [010]. Under these conditions $J_1 \neq 0 = J_2 = J_3$ and $B_1 = B_2 \neq 0 = B_3$, giving
$$E_2 = (\rho_{2112} + \rho_{2121})J_1B_1B_2 = \rho_{2112}J_1B_{110}^2,$$
where B_{110} is the magnetic induction in the [110] direction. This is called the *Planar Hall Effect*.

A typical sample shape for measuring the resistivity and galvanomagnetic effects is shown in Fig. 20.2. All measurements are carried out under isothermal conditions to keep temperature gradients small.

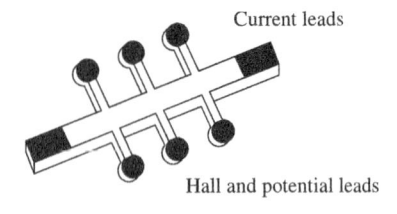

Fig. 20.2 Resistivity Hall Effect, and magnetoresistance coefficients are measured on samples with several contact probes. Silicon and other brittle materials are prepared with an ultrasonic cutter.

Table 20.1 The number of second, third and fourth rank galvanomagnetic coefficients for the 32 crystal classes (the number of independent coefficients is given in parentheses)

Point groups	ρ_{ij}	ρ_{ijk}	ρ_{ijkl}
Triclinic			
$1, \bar{1}$	9(6)	18(9)	54(36)
Monoclinic			
$2, m, 2/m$	5(4)	10(5)	28(20)
Orthorhombic			
$222, mm2, mmm$	3(3)	6(3)	15(12)
Trigonal			
$3, \bar{3}$	3(2)	10(3)	47(11)
$32, 3m, \bar{3}m$	3(2)	6(2)	25(7)
Tetragonal			
$4, \bar{4}, 4/m$	3(2)	10(3)	25(10)
$422, 4mm, \bar{4}2m, 4/mmm$	3(2)	6(2)	15(7)
Hexagonal			
$6, \bar{6}, 6/m$	3(2)	10(3)	25(7)
$622, 6mm, \bar{6}m2, 6/mmm$	3(2)	6(2)	15(6)
Cubic			
$23, m3$	3(1)	6(1)	15(4)
$432, \bar{4}3m, m3m$	3(1)	6(1)	15(3)

Relatively few galvanomagnetic measurements have been made on noncubic crystals. One of the reasons is the large number of property coefficients (Table 20.1) for low symmetry crystals. Even in fairly symmetric structures such as hexagonal close-packed metals (point group $6/mmm$) and wurtzite family semiconductors (point group $6mm$) there are twice as many effects as in cubic crystals.

Problem 20.1
Magnesium and other hexagonal close-packed metals belong to point group $6/mmm$. Wurtzite family semiconductors are in $6mm$. Work out the Hall Effect and magnetoresistance tensors for these point groups. Describe the experimental arrangements required to measure each of the eight independent coefficients.

When plotted as a function of direction, galvanomagnetic properties reflect the symmetry of the crystal. Fig. 20.3 shows the transverse magnetoresistance effect for three metal crystals. Tin is tetragonal (point group $4/mmm$), lead is cubic ($m3m$), and thallium is hexagonal ($6/mmm$). The experimental results shown in Fig. 20.3 are for resistivity measurements made along symmetry directions in the three crystals. The plots of $\Delta\rho(B)/\rho(0)$ for various orientation of the magnetic field clearly show the symmetry of the direction in which the resistivity is measured.

20.3 Underlying physics

Two interesting effects occur in a conductor when a magnetic field is applied at an angle to the current. The first is the *Hall Effect* which results in a voltage at right angles to the current and is caused by the Lorentz force. A second effect is the change in current density across the sample, leading to an increase in

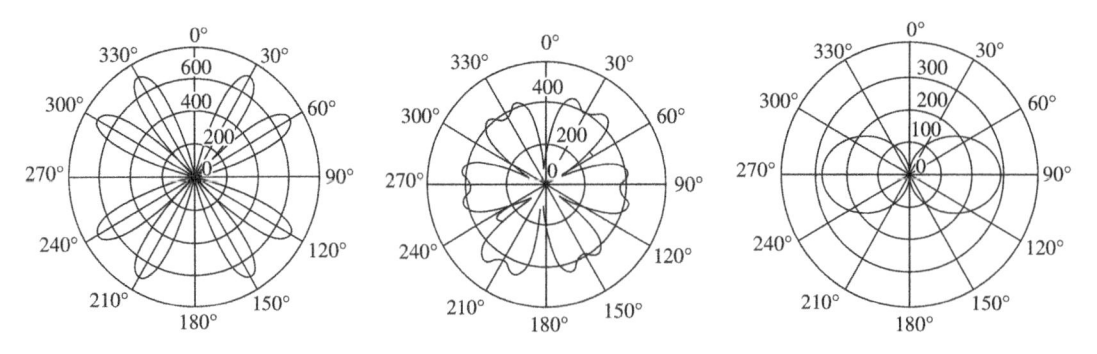

Fig. 20.3 Polar diagrams of the resistance change $\Delta\rho/\rho$ in a constant transverse magnetic field as a function of the angle, ϕ, between the crystallographic axes of the specimen and the direction of H. $T = 4.2$ K. (a) Sn, current in [001] direction. (b) Pb, current in [111] direction. (c) Tl, in [110] direction.

electric resistivity. The *magnetoresistance effect* is also caused by the Lorentz force. In the presence of the magnetic field the current has a longer path and the charge carriers are crowded together on one side of sample. The crowding phenomenon is called the magnetoconcentration or *Suhl Effect*.

Electrical conductivity σ is proportional to the product of the number of charge carriers per unit volume (N) times the mobility μ. Mobility is the drift velocity divided by the electric field. For a metal or an n-type semiconductor,

$$\sigma = N_e e \mu_e,$$

where e is the electronic charge. In p-type semiconductors where holes are the charge carriers,

$$\sigma = N_h e \mu_h.$$

The Hall coefficient R is inversely proportional to the number of charge carriers. $R = 1/N_e e$ for n-type materials and for p-type, $R = 1/N_h e$. Combining the Hall coefficient with the conductivity gives the mobility: $\mu = |R|\sigma$.

The resistivity of a metal usually increases when a magnetic field is applied. In low fields the increase in resistance is proportional to B^2, but in some metals $\Delta\rho/\rho$ saturates in very large fields. *Kohler's Rule* states that the fractional change in resistivity in a magnetic field is independent of temperature and resistivity. It is quite well obeyed experimentally.

For a typical metal like copper the charge carrier density is very large, usually on the order of the number of atoms per unit volume. Therefore R is small (Table 20.2). The mobility μ is also small compared to semiconductors. Structure–property relationships between mobility and crystal chemistry were discussed in Section 17.5. Typically for a metal μ is about 10^{-3} m^2/V s, approximately a hundred times smaller than the semiconductors listed in Table 20.3. Silicon and indium antimonide are the preferred materials for the Hall Effect and the magnetoresistive devices used for measuring magnetic flux, or as microphones and variable resistors.

Although most metals have small galvanomagnetic properties, magnetic conductors sometimes have large permeabilities μ_m which amplify the magnetic flux, $B = \mu_m H$. Thus we can distinguish three major classes of magnetic sensor materials (Fig. 20.4).

Magnetic field sensors sometimes make use of the high permeability of soft magnets to concentrate the flux for increased sensitivity. Enhanced sensitivity

Table 20.2 Hall constant R and mobility μ for several metals. R is in V m^3/A Wb = m^3/C. Calculated values obtained by assuming one conduction electron per atom in Cu, Li, and Na, and three per atom in Al (observed μ values are in m^2/V s)

Metal	R_{obs}	R_{calc}	μ
Cu	-0.55×10^{-10}	-0.74×10^{-10}	0.0032
Li	-1.70	-1.35	0.0018
Na	-2.50	-2.46	0.0053
Al	-0.30	-0.35	0.0012

Table 20.3 Carrier mobilities (in m^2/V s) and energy gaps for semiconductors with large galvanomagnetic coefficients (Mason)

Material	μ_e	μ_h	E_g
Si	0.12	0.025	1.11 eV
Ge	0.36	0.17	0.80
InSb	7.5	0.11	0.25
InAs	2.3	0.024	0.45
GaAs	0.85	0.040	1.35

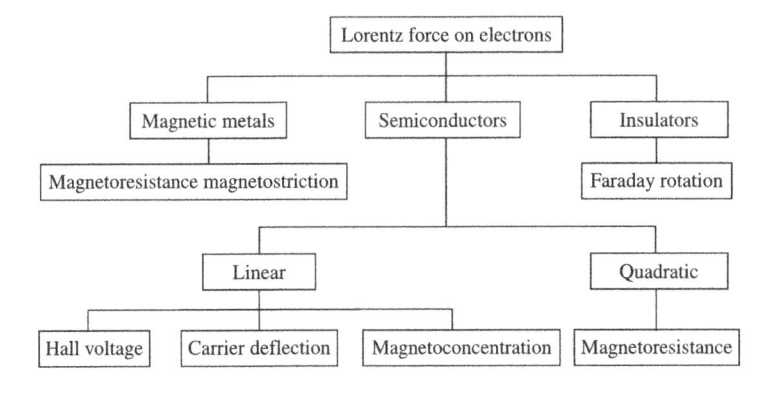

Fig. 20.4 Family tree of magnetic phenomena used to sense magnetic fields.

to magnetic fields is achieved in magnetoresistors made of permalloy (Ni$_{0.81}$Fe$_{0.19}$), in magneto-optic sensors made of ferromagnetic garnets, and in optical fibers clad with magnetostrictive coatings.

20.4 Galvanomagnetic effects in magnetic materials

The effects of magnetic phase transformations on electrical resistivity are apparent even in ordinary resistivity measurements. Departures from the normal linear temperature dependence are clearly visible with enhanced resistance at the transition temperatures (Fig. 20.5). Antiferromagnetic metals like europium show a peak similar to the ferromagnetic alloys.

Anisotropy becomes apparent in single-domain single crystals. Fig. 20.6 shows the fractional change in resistivity of nickel plotted as a function of

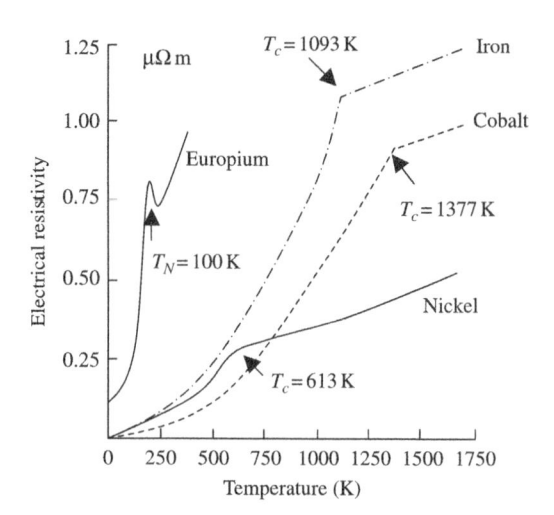

Fig. 20.5 Internal magnetoresistive effects lead to resistivity anomalies at magnetic phase transitions.

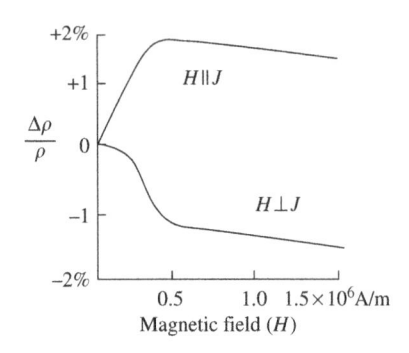

Fig. 20.6 The field-induced change in electrical resistivity of ferromagnetic nickel at room temperature. The rapid changes at low field are caused by domain wall movement. The linear decrease at higher fields takes place in the single domain state.

magnetic field. Nickel has a face-centered cubic crystal structure above its Curie temperature (613 K) but is tetragonal in its ferromagnetic state below T_c. At room temperature the average symmetry of the unmagnetized multidomain state is $m3m1'$, changing to $4/mm'm'$ for the single domain state. This change in symmetry takes place in a magnetic field applied parallel to one of the $\langle 100 \rangle$ easy axes.

The resulting change in resistivity is shown in Fig. 20.6. At zero field the nickel crystal is unmagnetized and is electrically isotropic. When the magnetic field is switched on, domain walls begin to move and anisotropy develops. The resistivity component measured parallel to the magnetic field is several percent larger than the resistivity perpendicular to H. At about 300,000 A/m, a single domain state is achieved. The symmetry then changes to magnetic point group $4/mm'm'$ and a linear decrease in $\Delta\rho/\rho$ is observed for both components. This linear magnetoresistance effect comes from the third rank axial components ρ_{333} and ρ_{113}. Component ρ_{333} corresponds to the experiment in which the resistivity is measured parallel to the magnetic field, and ρ_{113} from the experiment where the resistivity is measured perpendicular to the magnetic field.

As shown in Section 20.1, ρ_{ijk} is a third rank axial tensor. Components ρ_{113} and ρ_{333} are zero in nonmagnetic cubic crystals but not in magnetic group $4/mm'm'$. To prove this, refer to Section 14.10. The linear magnetoresistance effect is analogous to piezomagnetism that is also a third rank axial tensor. The nonzero coefficients for the magnetic symmetry groups are listed in Table 14.5. Piezomagnetic matrix coefficient Q_{31} corresponds to linear magnetoresistance coefficient ρ_{113} and Q_{33} to ρ_{333}. Both components are permitted in $4/mm'm'$.

There is a third contribution to the magnetoresistivity of magnetic materials, one that is closely related to magnetostriction (Chapter 15). Under strong magnetic fields, the magnetization vector will rotate away from the easy axis into the direction of the field. For cubic crystals, this results in mechanical deformations leading to a saturation strain λ given by the equation in Section 15.5:

$$\lambda = x_s(H) - x_s(0) = \frac{3}{2}\lambda_{100}\left(\alpha_1^2\beta_1^2 + \alpha_2^2\beta_2^2 + \alpha_3^2\beta_3^2 - \frac{1}{3}\right) + 3\lambda_{111}$$
$$\times (\alpha_1\alpha_2\beta_1\beta_2 + \alpha_1\alpha_3\beta_1\beta_3 + \alpha_2\alpha_3\beta_2\beta_3),$$

where α_1, α_2, and α_3 are the direction cosines specifying the direction in which the strain is measured, and β_1, β_2, and β_3 are direction cosines for the magnetic field vector. λ_{100} is the measured value of λ when the strain and field are both along [100]. λ_{111} is the measured value when both are oriented along the body diagonal [111].

Other properties, including the electrical resistivity are also affected by the rotation of the magnetization. Strain and electric resistivity are both polar second rank tensors so the mathematical expression for magnetostrictive strain can be adapted to the change in resistivity.

For cubic crystals, the saturation value is

$$\Delta\rho = \frac{3}{2}\Delta\rho_{100}\left(\alpha_1^2\beta_1^2 + \alpha_2^2\beta_2^2 + \alpha_3^2\beta_3^2 - \frac{1}{3}\right)$$
$$+ 3\Delta\rho_{111}(\alpha_1\alpha_2\beta_1\beta_2 + \alpha_1\alpha_3\beta_1\beta_3 + \alpha_2\alpha_3\beta_2\beta_3).$$

In this expression, the direction cosines α and β refer to the orientation of the resistivity measurement and the orientation of the magnetic field. $\Delta\rho_{100}$ and $\Delta\rho_{111}$ are experimental values along the cube edge and body diagonal. For nickel, $\Delta\rho_{100}/\rho = 4.3\%$ and $\Delta\rho_{111}/\rho = 1.9\%$. The corresponding values for iron are somewhat smaller: $\Delta\rho_{100}/\rho = 0.10\%$ and $\Delta\rho_{111}/\rho = 0.40\%$. Using these numbers, the change in resistivity caused by rotation of the magnetization can be readily estimated.

To summarize, there are at least three important factors governing the galvanomagnetic properties of magnetic crystals: domain wall motion, rotation of the magnetization, and intrinsic single domain effects.

In recent years two types of very large magnetoresistance have been reported. *Colossal magnetoresistance* has been observed in transition-metal oxides in which a magnetic transition is accompanied by a pronounced change in electrical conductivity. One of the best examples is the perovskite phase (La$_2$Ca)Mn$_3$O$_9$ (Fig. 20.7). At the Curie temperature near 270 K, the oxide changes from ferromagnetic-metallic behavior to a paramagnetic insulator. Large negative magnetoresistance coefficients are observed with resistance values substantially reduced under a magnetic field. All three effects, the magnetic and electric transitions together with the colossal magnetostriction, are caused by the 3d electrons of the octahedrally-coordinated Mn^{2+} and Mn^{3+} ions. Trivalent manganese is a well-known Jahn–Teller ion that promotes displacive phase transformations in ceramics and minerals such as Hausmannite, Mn$_3$O$_4$.

Giant Magnetoresistance (GMR) is a very different phenomenon observed in layered magnetic thin films. GMR is quantum mechanical effect involving spin-polarized electron transport in thin-film structures composed of alternating layers of ferromagnetic and nonmagnetic metals. When the magnetic moments of the ferromagnetic layers are parallel (Fig. 20.8), the spin-dependent scattering of the conduction electrons is minimized, and the material has its lowest resistance. When the magnetization vectors are antiparallel, the spin-dependent scattering is maximized, and the material has its highest resistance. Magnetization and magnetoresistance coefficients are manipulated with applied magnetic fields. The first major market for GMR devices are the "read" heads for magnetic hard disk drives.

Spin-polarized transport is emerging into a new family of anisotropic transport properties and devices. The field of *magnetoelectronics* is based on the

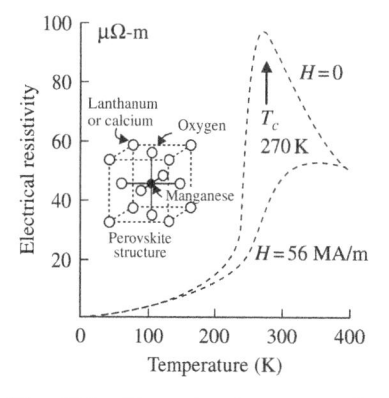

Fig. 20.7 The colossal magnetoresistance effect in manganese oxides with the perovskite structure. A large decrease in resistance occurs in strong magnetic fields.

Fig. 20.8 (a) Schematic representations of electron transport that is parallel to the plane of a layered magnetic sandwich structure for aligned (low resistance) and antialigned (high resistance) orientations of GMR devices. (b) The density of electronic states available to electrons in a normal metal and in a ferromagnetic metal whose majority spin states are completely filled. E, the electron energy; E_F, the Fermi level; $N(E)$, density of states.

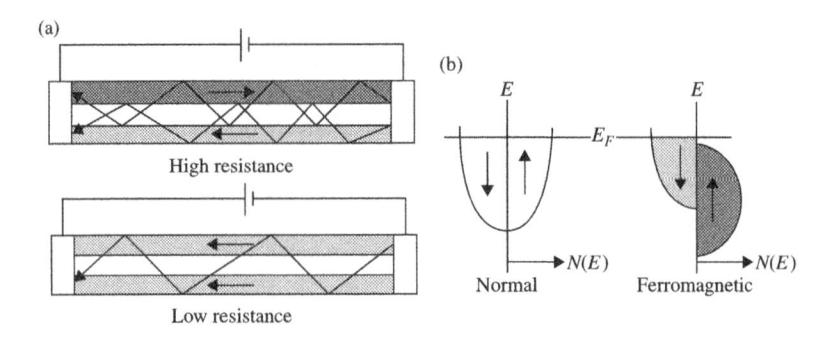

transport of up or down spins of the electrons rather than electrons or holes, as in traditional semiconductor electronics. Spin-polarized transport occurs naturally in ferromagnetic metals because the density of states available to spin-up and spin-down electrons are shifted in energy with respect to each other (Fig. 20.8). This shift causes the number of spin-up and spin-down charge carriers to be different in number and in mobility. When sandwiched with another conductor, spin-polarized carriers can be injected into a semiconductor, superconductor, or normal metal, and can also be used to tunnel through an insulating barrier. The most dramatic effects are generally seen for highly polarized currents with only one occupied spin band at the Fermi level. At present only the partially polarized conduction in the alloys of Fe, Co, and Ni are available. The best values of $(n\uparrow - n\downarrow)/(n\uparrow + n\downarrow)$ are typically 0.4 to 0.5.

The exploitation of spin polarized charge carriers represents a new field of magnetic transport phenomena with great promise for future magnetoelectronic systems.

20.5 Thermomagnetic effects

Under an applied magnetic field, the thermal resistivity of metals increases just as the electrical resistivity does. This is to be expected since electrons contribute to both the thermal conductivity and electrical conductivity. The tensor formulation for the various thermomagnetic effects follows the same procedure used for galvanomagnetic phenomena in Section 20.1. Writing out the heat flow h as a power series in magnetic induction gives

$$h_m = -k_{mn}\frac{dT}{dZ_n} - k_{mnp}\frac{dT}{dZ_n}B_P - k_{mnpq}\frac{dT}{dZ_n}B_P B_q,$$

where k_{mn} is the thermal conductivity and k_{mnp} and k_{mnpq} are third and fourth rank tensors describing the dependence of the thermal conductivity on the magnetic flux density. dT/dZ and B are the temperature gradient and flux density, respectively.

Again confining attention to point groups $m3m$ and $\bar{4}3m$, the symmetry groups of many important semiconductors and metals, there are just four independent thermomagnetic phenomena. Three are pictured in Fig. 20.1: The thermal Hall Effect and the longitudinal and transverse thermal magnetoresistance. The fourth phenomenon is the planar thermal Hall Effect.

If a temperature gradient dT/dZ is established along the [100] axis with a magnetic field along [001], the result is a heat flow and a thermal

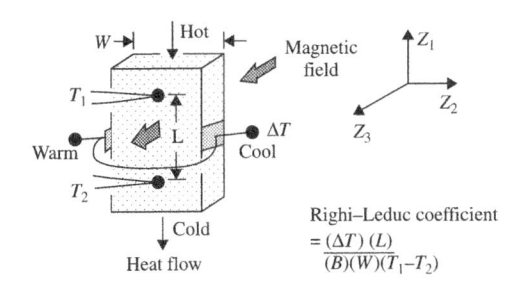

Fig. 20.9 In the Righi–Leduc experiment a temperature gradient is measured perpendicular to the heat flow direction and the applied magnetic field.

Righi–Leduc coefficient
$$= \frac{(\Delta T)(L)}{(B)(W)(T_1 - T_2)}$$

gradient along [010].

$$h_2 = -k_{213}\frac{dT}{dZ_1}B_3 = -k_{11}\frac{dT}{dZ_2}.$$

As shown in Fig. 20.9, the experiment involves measuring the thermal gradient in a direction perpendicular to the heat flow and the magnetic field. In terms of the tensor coefficients, $dT/dZ_2 = (k_{213}/k_{11})(dT/dZ_1)B_3$. This is the so-called *Righi–Leduc Effect* or the *Thermal Hall Effect*.

The second of the thermomagnetic phenomena, the transverse thermal magnetoresistance, is measured under similar conditions with heat flowing along $Z_1 = [100]$ and a magnetic field parallel to $Z_3 = [001]$. In this experiment, the change in thermal conductivity is measured under an applied magnetic field. The change in conductivity Δk_{11} is proportional to B_3^2:

$$\Delta k_{11} = k_{1133}B_3^2.$$

The fractional change in thermal conductivity $\Delta k_{11}/k_{11} = k_{1133}/k_{11}$ per unit flux density is the Maggi–Righi–Leduc coefficient.

The remaining two effects are the thermal analogs to the electrical longitudinal magnetoresistance and the electrical planar Hall Effect. The tensor coefficient for the longitudinal effect is k_{1111} relating the change in thermal conductivity $\Delta k_{11} = k_{1111}B_1^2$ in a magnetic field parallel to the heat transport.

The thermal resistivity of metals generally increases in magnetic fields just as the electrical resistivity does. Surprisingly large effects have been observed in cadmium at low temperatures. At 4 K a magnetic field of 3000 gauss (0.21 MA/m), doubles the longitudinal thermal resistance while the transverse effect is increased by a factor of four.

In the experiment called the thermal planar Hall Effect, the temperature gradient is applied parallel to [100] with the magnetic flux directed along [110]. The resulting heat flow is along [010]. The size of the effect is governed by tensor coefficient k_{2121}.

For lower symmetry crystals the number of thermomagnetic coefficients is identical to the galvanomagnetic effects in Table 20.1. Thermomagnetic effects are further complicated by accompanying thermoelectric effects that appear when temperature gradients are present. Thermoelectricity is discussed in the next chapter.

Problem 20.2
In this chapter the galvanomagnetic and thermomagnetic coefficients are written in tensor notation. Show how these coefficients can be rewritten in matrix notation. Define the symbols carefully and show how the matrix and tensor coefficients are related to one another.

21 Thermoelectricity

21.1 Seebeck Effect 234
21.2 Peltier Effect 235
21.3 Thomson Effect 235
21.4 Kelvin Relations and absolute thermopower 236
21.5 Practical thermoelectric materials 238
21.6 Tensor relationships 239
21.7 Magnetic field dependence 240

When two different metals are connected together in a circuit (Fig. 21.1(a)) and the two junctions are held at different temperatures, five physical phenomena take place simultaneously. Thermal and electric currents flow in the circuit, giving rise to *Joule heating* and *thermal conduction*. The driving forces for these currents are three interrelated thermoelectric phenomena: the *Seebeck Effect*, the *Peltier Effect*, and the *Thomson Effect*.

21.1 Seebeck Effect

For commonly used thermocouples (Fig. 21.1(b)), a voltage is developed when the junctions are held at different temperatures. In practice, one junction is held at a constant temperature (often the melting point of ice), and the open circuit voltage is measured as a function of the temperature of the second junction. If the reference temperature is 0°C, then the thermocouple voltage can be expressed as a power series.

$$V = \alpha T + \beta T^2 + \gamma T^3 + \cdots,$$

where T is the temperature in °C and the coefficients depend on the choice of metals. Data for Cu–Ni and Cu–Fe thermocouples are presented in Fig. 21.2 along with the governing equations. If the cold junction is at a temperature other than 0°C, it is only necessary to add a constant term. From this it follows that dV/dT at one junction is independent of the temperature of the second junction. The *Seebeck coefficient* α is defined as

$$\alpha = \lim_{\Delta T \to 0} \frac{\Delta V}{\Delta T} = \frac{dV}{dT},$$

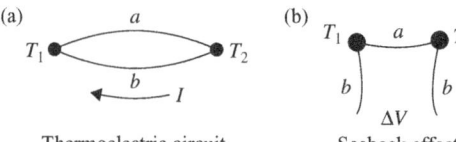

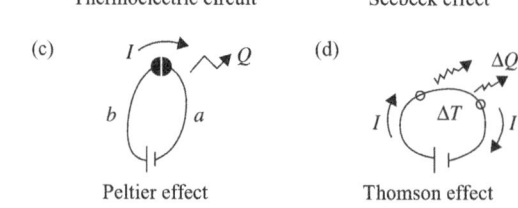

Fig. 21.1 Thermoelectric phenomena. (a) Simple circuit of two metals. (b) Seebeck Effect. (c) Peltier Effect. (d) Thomson Effect.

where ΔT is the difference in temperature between the two junctions of metals a and b. ΔV is the resulting open circuit voltage. The *absolute Seebeck coefficient* or *thermoelectric power* for a single metal will be introduced later after deriving the Kelvin Relations.

21.2 Peltier Effect

Consider a thermocouple with a battery as part of the circuit (Fig. 21.1(c)). A current I flows through the circuit causing Joule heating (I^2R), but there is an additional thermal effect caused by the Peltier Effect. Heat is either lost or gained, depending on the direction of the current through the junction between metals a and b. The *Peltier coefficient* π is defined by the relation

$$Q = \pi I,$$

where Q is the heat generated (or withdrawn) per second from the junction, and I is the current flowing through the junction.

The Peltier coefficient is measured by passing a known current through a junction and measuring the change in temperature with time. Knowing the rate of change of temperature and the heat capacity of the junction gives the rate at which heat is exchanged with the surroundings. The Peltier coefficient is obtained after correcting for I^2R loss.

Based on these measurements it is found that the Peltier heat is linearly proportional to the current I, and is reversed in sign when I is reversed in direction. The magnitude of the Peltier coefficient depends on the materials and on temperature. Peltier coefficients for a Cu–Ni thermocouple are listed in Table 21.1.

Cu–Ni $V = +20.4 \times 10^{-6}T + 0.023 \times 10^{-6}T^2$
Cu–Fe $V = -13.4 \times 10^{-6}T + 0.014 \times 10^{-6}T^2$

Fig. 21.2 The Seebeck Effect is used in thermocouples to measure temperature. Equations and graphs are given for Cu–Ni and Cu–Fe thermocouples showing the voltages and their temperature dependence.

Problem 21.1
Using the data for the Cu–Fe thermocouple given in Fig. 21.2, calculate the Peltier heat transferred at a junction at a temperature T by a current of 1 μA in an hour. Carry out the calculations for temperatures between 100°C and 500°C. At what temperature is the Peltier heat zero?

21.3 Thomson Effect

Heat is conducted along both wires of a thermocouple when the junctions are at different temperatures. When a battery is connected and current flows through

Table 21.1 Peltier coefficients for Cu–Ni junctions at temperatures between 0°C and 100°C. The data for the Peltier coefficient and the temperature dependence of the Seebeck voltage provide a test of the first Kelvin Relation, $\pi/T = \alpha$

T (K)	π (mV)	π/T (μV/K)	α (μV/K)
273	5.08	18.6	20.4
302	6.73	22.3	21.7
373	9.10	24.4	24.9

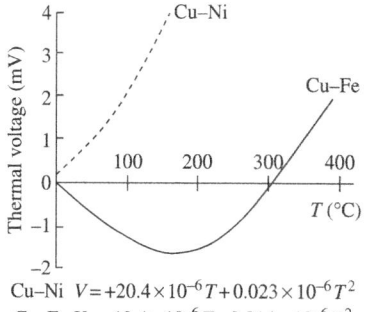

Table 21.2 Thomson coefficients for Cu and Fe. The difference between the coefficients is related to the Seebeck and Peltier coefficients through the second Kelvin Relation $(\gamma_{Cu} - \gamma_{Fe})/T = -d\alpha/dT$

	273 K	373 K
γ_{Cu}	1.6 μV/K	2.0 μV/K
γ_{Fe}	7.2 μV/K	16.4 μV/K
$(\gamma_{Cu} - \gamma_{Fe})/T$	−0.020 μV/K²	−0.039 μV/K²
$d\alpha/dT$	+0.028 μV/K²	+0.033 μV/K²

the wires (Fig. 21.1(d)), heat must be added to keep the temperature gradient constant. Allowing for Joule heating, heat must be added or extracted all along the wires to restore the original temperature distribution.

To measure the Thomson heat in a portion of the wire, one allows a known current to pass along a known temperature gradient. The rate at which Thomson heat is transferred between the wire and the environment is equal to the rate at which energy is dissipated in the wire minus the rate at which it is conducted away. After correcting for Joule heating and thermal conductivity, the Thompson heat is obtained from the remainder.

In a differential element of the wire, the Thomson heat term is

$$dQ = \gamma I \, dT,$$

where Q is the heat, T the absolute temperature, I the current, and γ the Thomson coefficient. The magnitude of the Thomson coefficient depends on the metal and the mean temperature of the differential element. The sign of the Thomson and Peltier coefficients reverse with current. Thomson coefficients for copper and iron are given in Table 21.2.

21.4 Kelvin Relations and absolute thermopower

Thermodynamic relationships between the Seebeck, Peltier, and Thomson coefficients are expressed through the two Kelvin Relations. For the thermocouple depicted in Fig. 21.1(a), operating under steady state conditions with a small current I, the Joule heating is negligible. From energy conservation and the First Law of Thermodynamics,

$$\underbrace{\{(\pi_{ab})_2 - (\pi_{ab})_1\}I}_{\text{Peltier Heat}} + \underbrace{\int_1^2 (\gamma_a - \gamma_b) I \, dT}_{\text{Thomson Effect}} = \underbrace{\int_1^2 \alpha_{ab} I \, dT}_{\text{Seebeck Effect}}$$

Differentiating,

$$\left(\frac{d\pi_{ab}}{dT}\right) + (\gamma_a - \gamma_b) = \alpha_{ab}.$$

Under these closed-system conditions there is no overall change in entropy. From the Second Law of Thermodynamics,

$$\int I \, d\left(\frac{\pi_{ab}}{T}\right) + \int \frac{\gamma_a - \gamma_b}{T} I \, dT = 0.$$

Again differentiating,

$$\left(\frac{d\pi_{ab}}{dT}\right) - \left(\frac{\pi_{ab}}{T}\right) + (\gamma_a - \gamma_b) = 0.$$

Combining the two differential forms gives the two Kelvin Relations:

First Kelvin Relation $$\alpha_{ab} = \frac{\pi_{ab}}{T}$$

Second Kelvin Relation $$\left(\frac{d\alpha_{ab}}{dT}\right) = \frac{\gamma_a - \gamma_b}{T}.$$

Typical experimental verification of the two relations are given in Tables 21.1 and 21.2.

The values of the Seebeck and Peltier effects are obviously dependent on the nature of *both* metals in the thermocouple. From the second Kelvin Relation we can define an absolute Seebeck coefficient (or Thermopower) for a single metal: To define the Seebeck coefficient α_a for metal a, let

$$\frac{d\alpha_a}{dT} = \frac{\gamma_a}{T},$$

where γ_a is the Thomson coefficient for metal a. From the Third Law of Thermodynamics it is known that $\alpha \to 0$ as $T \to 0$. Therefore, for any metal,

$$\int_0^\alpha d\alpha = \int_0^T \gamma \frac{dT}{T} = \alpha(T) = \frac{\pi(T)}{T}.$$

To evaluate the absolute coefficients, it is necessary to measure one coefficient that can be used as a standard. This has been done for lead by measuring the Thomson coefficient γ_{Pb} as a function of temperature, and then integrating the expression

$$\alpha_{Pb}(T) = \int_0^T \frac{\gamma_{Pb}(T)}{T} dT.$$

The thermopower of other elements can then be determined by measuring the Seebeck coefficients of thermocouples with Pb as one of the metals (Fig. 21.1(b)). For copper,

$$\alpha_{Cu} = \alpha_{CuPb} - \alpha_{Pb}.$$

The absolute Seebeck coefficients (thermopower) for ten metals are listed Table 21.3, along with their pressure dependence. The temperature dependence of the thermopower of aluminum and silver is shown in Fig. 21.3. α can be either positive or negative and generally changes linearly with temperature. At low temperatures below 100 K, the situation is more complicated, just as it is for thermal conductivity (Fig. 18.8). The low temperature peak is caused by an effect known as *phonon drag*. The electron transport caused by a temperature gradient leads to changes in the phonon–electron interactions. As phonons flow from the hot to the cold side, they drag electrons along, increasing the thermopower.

Table 21.3 Thermopower coefficients of several metals measured at 273 K. The pressure dependence $d\alpha/dp$ is determined by compressing the central wire in a thermocouple (Fig. 21.1(b)) and measuring the voltage in a temperature gradient. α is expressed in μV/K and the pressure p in kbar

Metal	α	$d\alpha/dp$	Metal	α	$d\alpha/dp$
Ag	1.4	8	Mg	−1.4	−8
Al	−1.85	−1.5	Ni	−19	5.6
Au	1.9	4.0	Pd	−9.0	22
Co	−40	−14	Pt	−4.6	14
Cu	1.7	3.0	W	0.1	11

Fig. 21.3 Absolute Seebeck Effect for Ag and Al. Near room temperature α changes slowly with temperature but phonon effects increase α rapidly near 0 K.

21.5 Practical thermoelectric materials

Thermoelectric refrigerators and generators are heat engines that employ the Peltier or Seebeck Effects. Except for measuring temperature with thermocouples, the early work on thermoelectric devices made from metals was disappointing because of the small thermoelectric coefficients and large thermal conductivities. The situation changed with the discovery of much larger Seebeck coefficients in semiconductors.

To understand the reasons, consider the origin of electrical conduction and thermoelectric phenomena in metals and semiconductors. In metals each atom contributes at least one electron that is able move freely. In semiconductors, the number of charge carriers is hundreds or thousands of times smaller, which accounts for the higher resistivity of these materials.

When one end of the conductor is hotter than the other, electrons leave the hot end more often than the cold end. Electrons tend to flow toward the cold end, and since they are negatively charged, the cold end becomes negative. (For p-type semiconductors, holes migrate to the cold end causing it to become positively charged). As the charge builds up at the cold end, it begins to repulse additional charge carriers, and an equilibrium is reached in which the net flow is zero. Charges no longer accumulate but the cold end retains a negative charge. The fewer the number of electrons available for the return flow to the hot end, the higher will be the voltage attained at the cold end before equilibrium is reached. Since the number of charge carriers is smaller in a semiconductor, a temperature gradient creates a larger voltage than in a metal.

The efficiency for energy conversion in a thermoelectric power source or refrigerator is proportional to the product

$$\left(\frac{T_H - T_C}{T_H}\right)\left(\frac{\alpha^2}{\rho k}\right).$$

The first term is the Carnot efficiency which involves the absolute temperatures of the hot (T_H) and cold (T_C) junctions. Greater efficiency is attained with a large difference in temperature. The second term is the material figure of merit where α is the Seebeck coefficient, ρ the electrical resistivity and k the thermal conductivity. Schematically, when plotted as a function of the density of charge carriers, α^2 and ρ decrease while k increases. The material figure of merit reaches a maximum at about 10^{25} charge carriers per cubic meter.

The thermoelectric circuit in most Seebeck and Peltier devices use p-type and n-type semiconductors joined at the hot end (Fig. 21.4). Between the cold ends may be an electric load such as a lamp, radio, or an electric motor. The current produced in the n-type semiconductor flows from the hot to the cold end, while in the p-type semiconductor it flows from cold to hot. Thus the current flows around

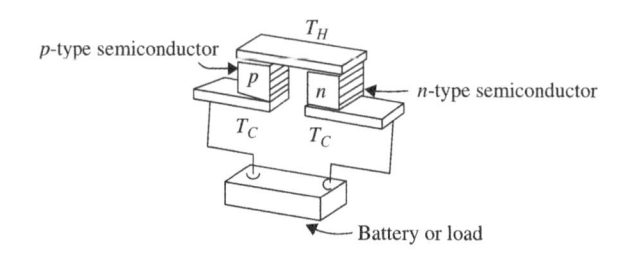

Fig. 21.4 Thermoelectric generators and refrigerators operate between hot (T_H) and cold (T_C) temperatures using doped semiconductors. Typical efficiencies are in the 10–20% range.

the entire circuit, including the load. It is important that the electrical resistivity ρ in the thermoelectric material be low to minimize Joule heat dissipation. It is also important that the thermal conductivity be small. Otherwise most of the heat supplied to the hot side flows directly to the cold end without doing useful work. This is why ρ and k appear in the figure of merit $\alpha^2/\rho k$.

Near room temperature bismuth telluride (Fig. 21.5) is one of the most efficient thermoelectrics. The Seebeck coefficients are about ± 200 μV/K for p- and n-type Bi_2Te_3, depending on doping level. Undoped material has mixed electron–hole conduction, and therefore much smaller α coefficients. Like most heavy-element compounds it also has low thermal conductivity which further enhances the figure of merit. The thermopower of Bi_2Te_3, PbTe, and other semiconductors are about two orders of magnitude larger than metals.

As pointed out earlier, the Carnot efficiency $(T_H - T_C)/T_H$ is also important in thermoelectric devices. Semiconducting ceramics made of silicon carbide are capable of operating over a wide temperature range, leading to much higher efficiencies. Porous SiC has large α coefficients (Fig. 21.6), high electrical conductivity, low thermal conductivity, and can operate at temperatures exceeding 1000°C.

The use of functionally graded materials (FGM) is a second approach to improving Carnot efficiency. Since the figure of merit $(\alpha^2/\rho k)$ material of each thermoelectric material is rather sharply peaked with respect to temperature (Fig. 21.7), a graded composite of two or three thermoelectrics can extend the working range substantially. Semiconductors with increasingly wide band gaps, are used in such a composite. The band gaps of Bi_2Te_3, PbTe, and $Si_{0.7}Ge_{0.3}$ are 0.13, 0.27, and 0.95 eV, respectively. Functionally graded materials such as this generally belong to Curie group ∞m.

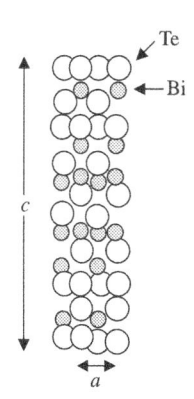

Fig. 21.5 The crystal structure of Bi_2Te_3 consists of cubic close-packed layers of tellurium and bismuth atoms. Easy cleavage takes place between adjacent tellurium layers. The point group is trigonal, $\bar{3}m$. Thermal and electrical conductivities are largest in the (001) planes. Similar anisotropy in the thermopower has been observed in p-type bismuth telluride.

21.6 Tensor relationships

For anisotropic solids, the absolute Seebeck coefficient is a second rank tensor α_{mn} which relates electric field E_m to a temperature gradient dT/dZ_n:

$$E_m = \alpha_{mn} \frac{dT}{dZ_n}.$$

The corresponding relationship for the absolute Peltier coefficient π_{mn} $(=T\alpha_{mn})$ is

$$h_m = \pi_{mn} J_n,$$

where h_m is the heat flow vector, and J_n is the electric current density. In describing thermoelectric phenomena these terms are combined with the second rank tensors ρ_{mn} and k_{mn} representing electrical resistivity and thermal conductivity.

$$E_m = \rho_{mn} J_n + \alpha_{mn} \frac{dT}{dZ_n}$$

$$h_m = \pi_{mn} J_n - k_{mn} \frac{dT}{dZ_n}.$$

There is, however, an important difference between the transport tensors and the thermoelectric tensors.

Onsager's Principle requires that the electric resistivity and thermal conductivity tensors be symmetric, but this does not hold for the Seebeck and Peltier

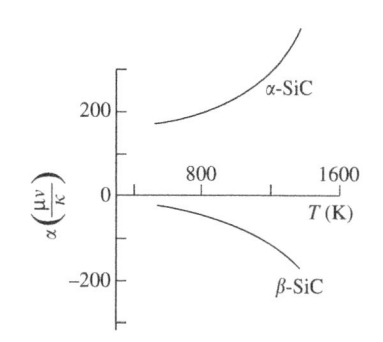

Fig. 21.6 Thermopower of silicon carbide ceramics. p-type α-SiC is sintered in an argon atmosphere while n-type β-SiC is prepared by sintering in N_2.

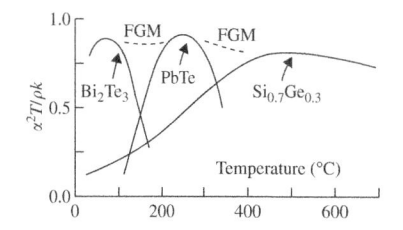

Fig. 21.7 Functionally graded thermoelectric devices utilize several different semiconductors to provide a wider temperature range.

Table 21.4 Seebeck and Peltier coefficients for the 32 crystallographic point groups and the seven Curie groups for textured solids

Point groups	Tensor terms
1, $\bar{1}$	$\begin{pmatrix} \alpha_{11} & \alpha_{12} & \alpha_{13} \\ \alpha_{21} & \alpha_{22} & \alpha_{23} \\ \alpha_{31} & \alpha_{32} & \alpha_{33} \end{pmatrix}$
2, m, 2/m	$\begin{pmatrix} \alpha_{11} & 0 & \alpha_{13} \\ 0 & \alpha_{22} & 0 \\ \alpha_{31} & 0 & \alpha_{33} \end{pmatrix}$
222, $mm2$, mmm	$\begin{pmatrix} \alpha_{11} & 0 & 0 \\ 0 & \alpha_{22} & 0 \\ 0 & 0 & \alpha_{33} \end{pmatrix}$
3, $\bar{3}$, 4, $\bar{4}$, 4/m, 6, $\bar{6}$, 6/m, ∞, ∞/m	$\begin{pmatrix} \alpha_{11} & \alpha_{12} & 0 \\ -\alpha_{12} & \alpha_{11} & 0 \\ 0 & 0 & \alpha_{33} \end{pmatrix}$
32, 3m, $\bar{3}m$, 422, 4mm, $\bar{4}2m$, 4/mmm 622, 6mm, $\bar{6}m2$, 6/mmm, ∞m, $\infty 2$, ∞/mm	$\begin{pmatrix} \alpha_{11} & 0 & 0 \\ 0 & \alpha_{11} & 0 \\ 0 & 0 & \alpha_{33} \end{pmatrix}$
23, $m3$, 432, $\bar{4}3m$, $m3m$, $\infty\infty m$, $\infty\infty$	$\begin{pmatrix} \alpha_{11} & 0 & 0 \\ 0 & \alpha_{11} & 0 \\ 0 & 0 & \alpha_{11} \end{pmatrix}$

coefficients which relate two different flows. In the most general (triclinic) case $\alpha_{mn} \neq \alpha_{nm}$ and $\pi_{mn} \neq \pi_{nm}$. Thus there are nine coefficients to be determined rather than six.

The thermoelectric tensor coefficients for other symmetry groups are given in Table 21.4. Neumann's Principle dictates that the Seebeck and Peltier coefficients are symmetric for all the important thermoelectric materials in high symmetry groups.

21.7 Magnetic field dependence

When a magnetic field is applied to a thermoelectric device there can be substantially improved performance. Bi–Sb and other semimetal alloys show very large changes in the figure of merit $\alpha^2/\rho k$. In the presence of a transverse field, the thermopower and the electrical resistivity increase while the thermal conductivity decreases. The increase in resistivity is caused by the transverse magnetoresistance effect (Section 20.2) and the decrease in thermal conductivity by the Maggi–Righi–Leduc Effect (Section 20.3). The net result of these changes (Fig. 21.8) is an optimum value for the magnetic field.

To describe the effect of magnetic field on the absolute Seebeck and Peltier coefficients, expand α in a power series in the magnetic flux density B. The derivation follows the same procedure used earlier for the galvanomagnetic effects (Section 20.1). The electromotive force E_m coming from a temperature gradient dT/dZ_n in the presence of a magnetic field is

$$E_m = \alpha_{mn}\left(\frac{dT}{dZ_n}\right) + \alpha_{mnp}\left(\frac{dT}{dZ_n}\right)B_p + \alpha_{mnpq}\left(\frac{dT}{dZ_n}\right)B_p B_q + \cdots .$$

α_{mn} is a polar second rank tensor, α_{mnp} an axial third rank tensor, and α_{mnpq} a polar fourth rank tensor.

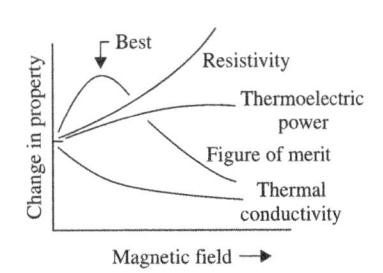

Fig. 21.8 Magnetic fields affect the thermopower, resistivity and thermal conductivity. The result can be an improvement in the thermoelectric figure of merit.

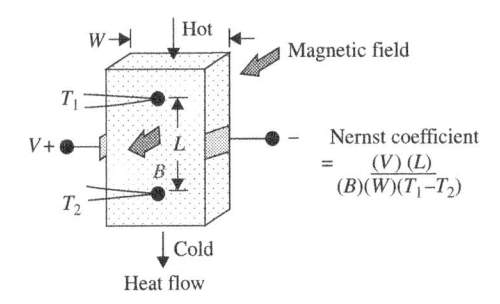

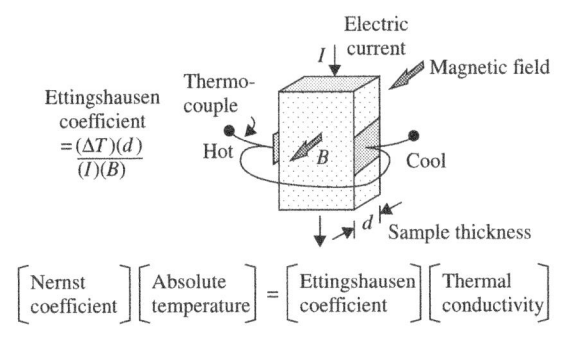

Fig. 21.9 Nernst and Ettingshausen coefficients are related through the thermal conductivity.

For cubic crystals and transverse magnetic fields, two important cross-coupled phenomena are the Nernst and Ettingshausen Effects (Fig. 21.9). With the magnetic field along $Z_3 = [001]$, the temperature gradient along $Z_1 = [100]$, and the voltage measured along $Z_2 = [010]$,

$$E_2 = \alpha_{213}\left(\frac{dT}{dZ_1}\right) B_3.$$

This is the *Nernst Effect* that describes a voltage gradient perpendicular to the heat flow. The *Ettingshausen Effect* relates a transverse thermal gradient transverse to an electric current. Both are third rank tensor coefficients analogous to the galvanomagnetic Hall Effect.

The *magneto-Seebeck Effect* controls the changes in thermopower with magnetic field. With a transverse magnetic flux B_3, the change in Seebeck coefficient is

$$\Delta\alpha_{11} = \frac{E_1}{dT/dZ_1} - \alpha_{11}(0) = \alpha_{1133} B_3^2,$$

where $\alpha_{11}(0)$ is the Seebeck coefficient measured in zero magnetic field. This effect is analogous to the transverse magnetoresistance (Section 20.2). Several other higher order effects are also involved in the coupling between the thermoelectric effects and magnetic fields.

For lower symmetry crystals the number of cross-coupled effects proliferates rapidly. The Onsager relations do not apply so there are far more second-, third-, and fourth-rank coefficients than for the galvanomagnetic effects (Table 21.5).

Ferromagnetic crystals develop a thermoelectric effect that depends of the direction of the saturation magnetization (specified by direction cosines $\alpha_1, \alpha_2, \alpha_3$). It also depends on the direction of the temperature gradient along which the thermoelectric voltage is measured (direction cosines $\beta_1, \beta_2, \beta_3$). For cubic ferromagnets the dependence of the thermoelectric voltage V

Table 21.5 The number of second-, third-, and fourth-rank coefficients for the Seebeck Effect and its magnetic field dependence (the total number of nonzero coefficients is followed by the number of independent coefficients in parentheses)

Point groups	α_{mn}	α_{mnp}	α_{mnpq}
Triclinic $1, \bar{1}$	9(9)	27(27)	54(54)
Monoclinic $2, m, 2/m$	5(5)	13(13)	28(28)
Orthorhombic $222, mm2, mmm$	3(3)	6(6)	15(15)
Trigonal $3, \bar{3}$ $32, 3m, \bar{3}m$	5(3) 3(2)	21(9) 10(4)	47(18) 25(10)
Tetragonal $4, \bar{4}, 4/m$ $422, 4mm, \bar{4}2m, 4/mmm$	5(3) 3(2)	13(7) 6(3)	27(14) 15(8)
Hexagonal $6, \bar{6}, 6/m$ $622, 6mm, \bar{6}m2, 6/mmm$	5(3) 3(2)	13(7) 6(3)	27(13) 15(7)
Cubic $23, m3$ $432, \bar{4}3m, m3m$	3(1) 3(1)	6(3) 6(1)	15(15) 15(3)

on the direction of I_s is given by

$$V = \frac{3}{2} V_{100} \left(\alpha_1^2 \beta_1^2 + \alpha_2^2 \beta_2^2 + \alpha_3^2 \beta_3^2 - \frac{1}{3} \right)$$
$$+ 3 V_{111} (\alpha_1 \alpha_2 \beta_1 \beta_2 + \alpha_1 \alpha_3 \beta_1 \beta_3 + \alpha_2 \alpha_3 \beta_2 \beta_3).$$

The coefficients V_{100} and V_{111} refer to measurements carried out along the [100] and [111] crystallographic axes with the temperature gradient and magnetization parallel to the voltage measurement.

The derivation follows that of the magnetostrictive effect in Section 15.5. Strain and thermopower are both second rank polar tensors so the mathematics is similar.

Measurements on single crystals of iron and nickel gave the following results

Fe $V_{100} = 0.70 \, \mu\text{V/K}$ $V_{111} = -0.13 \, \mu\text{V/K}$
Ni $V_{100} = 0.57 \, \mu\text{V/K}$ $V_{111} = 0.69 \, \mu\text{V/K}$

Problem 21.2
Calculate the thermoelectric voltage V for Fe and Ni crystals fully magnetized along the [110] direction. Plot V as a function of direction in the (110) plane perpendicular to I_s.

Piezoresistance

22

Piezoresistivity, the change in electrical resistivity with mechanical stress, is commonly used to monitor static or slowly varying stresses and strains. The sensitivity of piezoresistive elements are often compared by means of a strain gage factor: $G = \Delta R/Rx$. In this expression, $\Delta R/R$ is the fractional change in resistance associated with a strain x. G is a dimensionless quantity which has a value of 2–4 for strain gages made from metal wire, and about 200 for lightly-doped silicon.

22.1	Tensor description	243
22.2	Matrix form	244
22.3	Longitudinal and transverse gages	245
22.4	Structure–property relations	247

22.1 Tensor description

In the discussion of galvanomagnetic effects, a power series expansion was used to examine the dependence of resistivity on magnetic field (Section 20.1). Other series expansions work equally well. Piezoresistivity involves the linear and nonlinear relationships between electric field E_i, electric current density J_j, and mechanical stress X_{kl}. Here the change in electric field dE_i with current and stress is expanded in a McLaurin Series.

$$dE_i = \left(\frac{\partial E_i}{\partial J_j}\right) dJ_j + \left(\frac{\partial E_i}{\partial X_{kl}}\right) dX_{kl} + \frac{1}{2}\left(\frac{\partial^2 E_i}{\partial J_j \partial J_m}\right) dJ_j \, dJ_m$$
$$+ \frac{1}{2}\left(\frac{\partial^2 E_i}{\partial X_{kl} \partial X_{no}}\right) dX_{kl} \, dX_{no} + \left(\frac{\partial^2 E_i}{\partial J_j \partial X_{kl}}\right) dJ_j \, dX_{kl} + \cdots.$$

Each term in this series corresponds to a different physical property. $(\partial E_i/\partial J_j)$ is the electrical resistivity ρ_{ij}, a second rank polar tensor. $(\partial E_i/\partial X_{kl})$ is the piezoelectric voltage coefficient, g_{ikl}, a third rank polar tensor. $(\partial^2 E_i/\partial J_j \partial J_m)$ is the change in electrical resistivity with current level. It is a third rank polar tensor which describes deviations from Ohm's Law. $(\partial^2 E_i/\partial X_{kl} \partial X_{no})$ is a fifth rank polar tensor representing the stress dependence of the piezoelectric voltage coefficient.

The fifth term in the expansion, $(\partial^2 E_i/\partial J_j \partial X_{kl})$, is the property of interest in this chapter. It is a fourth rank polar tensor describing the dependence of electrical resistivity on mechanical stress. The symbol π_{ijkl} is used to represent piezoresistance in tensor form, and π_{ij} in matrix form.

Most strain gages are made from silicon and germanium crystals (point group $m3m$). This is a centrosymmetric point group for which all odd rank polar tensors disappear. Therefore the McLaurin series reduces to

$$dE_i = \rho_{ij} dJ_j + \pi_{ijkl} dJ_j \, dX_{kl}.$$

Integrating, this becomes

$$E_i = \rho_{ij}J_j + \pi_{ijkl}J_jX_{kl}.$$

The change in resistivity under stress is

$$\Delta\rho_{ij} = \frac{E_i - \rho_{ij}J_j}{J_j} = \pi_{ijkl}X_{kl}.$$

In cubic crystals where ρ_{ij} is a scalar, $\Delta\rho_{ij}$ is often written as a fractional change in resistivity. For point group $m3m$ there are three independent tensor coefficients:

$$\pi_{1111} = \pi_{2222} = \pi_{3333}$$

$$\pi_{1122} = \pi_{1133} = \pi_{2233} = \pi_{3322} = \pi_{2211} = \pi_{3311}$$

$$\pi_{1212} = \pi_{1221} = \pi_{2112} = \pi_{2121} = \pi_{1313} = \pi_{1331}$$

$$= \pi_{3113} = \pi_{3131} = \pi_{2323} = \pi_{2332} = \pi_{3223} = \pi_{3232}.$$

All other tensor coefficients are zero.

The fourth rank piezoresistive tensor is similar, but not identical, to the fourth rank elastic compliance tensor (Section 13.1). Comparing π_{ijkl} with s_{ijkl}, k and l can be interchanged because the stress tensor X_{kl} is symmetric. Subscripts i and j can also be interchanged because the conductivity σ_{ij} and strain x_{ij} tensors are also symmetric. But for π_{ijkl}, i and j cannot be interchanged with k and l because the energy argument in Section 13.1 does not apply.

Therefore the piezoresistivity tensor is not quite as symmetric as the compliance tensor in that $s_{ijkl} = s_{klij}$ but $\pi_{ijkl} \neq \pi_{klij}$. For silicon and other highly symmetric crystals this does not make any difference because Neumann's Law applies in the usual way, and the matrices are identical.

22.2 Matrix form

Piezoresistive components are generally presented in the shortened matrix form. For triclinic crystals,

$$\begin{pmatrix}\Delta\rho_1\\\Delta\rho_2\\\Delta\rho_3\\\Delta\rho_4\\\Delta\rho_5\\\Delta\rho_6\end{pmatrix} = \begin{pmatrix}\Delta\rho_{11}\\\Delta\rho_{22}\\\Delta\rho_{33}\\\Delta\rho_{23}\\\Delta\rho_{13}\\\Delta\rho_{12}\end{pmatrix} = \begin{pmatrix}\pi_{11} & \pi_{12} & \pi_{13} & \pi_{14} & \pi_{15} & \pi_{16}\\\pi_{21} & \pi_{22} & \pi_{23} & \pi_{24} & \pi_{25} & \pi_{26}\\\pi_{31} & \pi_{32} & \pi_{33} & \pi_{34} & \pi_{35} & \pi_{36}\\\pi_{41} & \pi_{42} & \pi_{43} & \pi_{44} & \pi_{45} & \pi_{46}\\\pi_{51} & \pi_{52} & \pi_{53} & \pi_{54} & \pi_{55} & \pi_{56}\\\pi_{61} & \pi_{62} & \pi_{63} & \pi_{64} & \pi_{65} & \pi_{66}\end{pmatrix}\begin{pmatrix}X_1\\X_2\\X_3\\X_4\\X_5\\X_6\end{pmatrix}.$$

Thirty-six measurements would be required to specify piezoresistivity in a triclinic crystal but symmetry simplifies the matrix for other point groups. For silicon and other crystals belonging to $m3m$, $\bar{4}3m$, or 432, there are just three independent coefficients, π_{11}, π_{12}, and π_{44}. For isotropic polycrystalline materials there are only two independent coefficients since $\pi_{44} = \pi_{11} - \pi_{12}$.

$$\begin{pmatrix} \Delta\rho_1 \\ \Delta\rho_2 \\ \Delta\rho_3 \\ \Delta\rho_4 \\ \Delta\rho_5 \\ \Delta\rho_6 \end{pmatrix} = \begin{pmatrix} \pi_{11} & \pi_{12} & \pi_{12} & 0 & 0 & 0 \\ \pi_{12} & \pi_{11} & \pi_{12} & 0 & 0 & 0 \\ \pi_{12} & \pi_{12} & \pi_{11} & 0 & 0 & 0 \\ 0 & 0 & 0 & \pi_{44} & 0 & 0 \\ 0 & 0 & 0 & 0 & \pi_{44} & 0 \\ 0 & 0 & 0 & 0 & 0 & \pi_{44} \end{pmatrix} \begin{pmatrix} X_1 \\ X_2 \\ X_3 \\ X_4 \\ X_5 \\ X_6 \end{pmatrix}.$$

Matrices for other crystallographic point groups are nearly identical to the magnetostrictive matrices given in Table 14.3. The only differences are in the expressions for N_{44} and N_{66} in those point groups where these elements are equal to $\frac{1}{2}(N_{11} - N_{12})$. The factor of $\frac{1}{2}$ is not needed for the piezoresistivity coefficients. Otherwise the matrices are the same for the π and N coefficients.

The relationship between the matrix and tensor coefficients involves factors of 2 whenever π_{ij} has $i = 1$–6, $j = 4$–6. Therefore, $\pi_{11} = \pi_{1111}$, $\pi_{12} = \pi_{1122}$, $\pi_{16} = 2\pi_{1112}$, $\pi_{66} = 2\pi_{1212}$, etc.

The three most important experimental configurations are for the longitudinal and transverse stresses, and for hydrostatic pressures. Two of the three piezoresistance coefficients for cubic crystals can be determined from these measurements. Using a long slender crystal oriented along $Z_1 = [100]$, with current, voltage and tensile stress in the same direction, the fractional change in resistivity is

$$\frac{\Delta\rho_{11}}{\rho(0)} = \frac{\Delta\rho_1}{\rho(0)} = \frac{\pi_{1111}X_{11}}{\rho(0)} = \frac{\pi_{11}X_1}{\rho(0)},$$

where $\rho(0)$ is the resistivity measured at zero stress.

Coefficient π_{12} is measured with current and voltage along $[100] = Z_1$ with stress applied in the transverse $[010] = Z_2$ direction.

$$\frac{\rho_{11}}{\rho(0)} = \frac{\rho_1}{\rho(0)} = \frac{\pi_{1122}X_{22}}{\rho(0)} = \frac{\pi_{12}X_2}{\rho(0)}.$$

For hydrostatic conditions the resistivity is monitored under pressure p. In this case $X_1 = X_2 = X_3 = -p$ and $X_4 = X_5 = X_6 = 0$.

$$\frac{\Delta\rho_1}{\rho(0)} = -\frac{p(\pi_{11} + 2\pi_{12})}{\rho(0)}.$$

The third independent coefficient, π_{44}, can be determined from measurements in other directions.

22.3 Longitudinal and transverse gages

Piezoresistive stress and strain gages are generally made of doped silicon or germanium crystals. The crystals are mounted on the test specimen with stress applied either in the longitudinal or transverse direction (Fig. 22.1).

To determine the most sensitive orientation for the longitudinal gage, coefficient π'_{11} is evaluated as a function of direction. For the longitudinal gage, the stress is applied along Z'_1, the same direction as the resistivity measurement. The direction cosines between the arbitrary direction Z'_1 and the cube

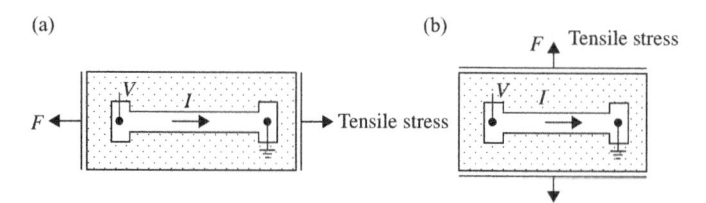

Fig. 22.1 Longitudinal (a) and transverse (b) piezoresistive strain gages.

Table 22.1 Piezoresistive coefficients for selected directions in cubic crystals. All directions are equally sensitive if the anisotropy factor $\pi_{11} - \pi_{12} - \pi_{44} = 0$. Isotropy has been observed in the Si–Ge solid solution series

Longitudinal direction	π_ℓ	Transverse direction	π_t
[100]	π_{11}	[010]	π_{12}
[001]	π_{11}	[110]	π_{12}
[111]	$\frac{1}{3}(\pi_{11} + 2\pi_{12} + 2\pi_{44})$	[1$\bar{1}$0]	$\frac{1}{3}(\pi_{11} + 2\pi_{12} - \pi_{44})$
[1$\bar{1}$0]	$\frac{1}{2}(\pi_{11} + \pi_{12} + \pi_{44})$	[111]	$\frac{1}{3}(\pi_{11} + 2\pi_{12} - \pi_{44})$
[$\bar{1}$10]	$\frac{1}{2}(\pi_{11} + \pi_{12} + \pi_{44})$	[001]	π_{12}
[110]	$\frac{1}{2}(\pi_{11} + \pi_{12} + \pi_{44})$	[1$\bar{1}$0]	$\frac{1}{2}(\pi_{11} + \pi_{12} - \pi_{44})$

axes are a_{11}, a_{12}, and a_{13}. For a cubic crystal,

$$\pi'_{11} = \pi'_{1111} = a_{1i}a_{1j}a_{1k}a_{1l}\pi_{ijkl}$$
$$= (a_{11}^4 + a_{12}^4 + a_{13}^4)\pi_{1111} + (a_{11}^2 a_{12}^2 + a_{11}^2 a_{13}^2 + a_{12}^2 a_{13}^2)(2\pi_{1122} + 4\pi_{1212}).$$

Substituting the matrix coefficients, and remembering that $a_{11}^2 + a_{12}^2 + a_{13}^2 = 1$, the expression for π'_{11} can be further simplified using the identity

$$(a_{11}^2 + a_{12}^2 + a_{13}^2)^2 = 1 = (a_{11}^4 + a_{12}^4 + a_{13}^4) + 2(a_{11}^2 a_{12}^2 + a_{11}^2 a_{13}^2 + a_{12}^2 a_{13}^2).$$

The longitudinal piezoresistive coefficient $\pi_\ell = \pi'_{11}$ becomes

$$\pi_\ell = \pi_{11} - 2(\pi_{11} - \pi_{12} - \pi_{44})(a_{11}^2 a_{12}^2 + a_{11}^2 a_{13}^2 + a_{12}^2 a_{13}^2).$$

This expression is evaluated for various directions in Table 22.1.

For the transverse gage in Fig. 22.1, the governing coefficient is $\pi'_{12} = \pi_t$. A tensile stress $X'_2 = X'_{22}$ is applied perpendicular to Z'_1, the direction of the resistivity measurement. For the cubic system,

$$\pi_t = \pi'_{12} = \pi'_{1122} = a_{1i}a_{1j}a_{2k}a_{2l}\pi_{ijkl}$$
$$= \pi_{1111}(a_{11}^2 a_{21}^2 + a_{12}^2 a_{22}^2 + a_{13}^2 a_{23}^2)$$
$$+ \pi_{1122}(a_{11}^2 a_{22}^2 + a_{12}^2 a_{21}^2 + a_{13}^2 a_{21}^2 + a_{13}^2 a_{22}^2 + a_{12}^2 a_{23}^2 + a_{11}^2 a_{23}^2)$$
$$+ \pi_{1212}(a_{11}a_{12}a_{21}a_{22} + a_{11}a_{13}a_{21}a_{23} + a_{12}a_{13}a_{22}a_{23}).$$

Converting to matrix coefficients and making use of the identities

$$(a_{11}a_{21} + a_{12}a_{22} + a_{13}a_{23})^2 = 0$$

or

$$(a_{11}^2 a_{21}^2 + a_{12}^2 a_{22}^2 + a_{13}^2 a_{23}^2) + 2(a_{11}a_{21}a_{12}a_{22} + a_{11}a_{21}a_{13}a_{23} + a_{12}a_{22}a_{13}a_{23})$$
$$= 0$$

and
$$(a_{11}^2 + a_{12}^2 + a_{13}^2)(a_{21}^2 + a_{22}^2 + a_{23}^2) = 1,$$
we find that
$$\pi_t = \pi_{12} + (\pi_{11} - \pi_{12} - \pi_{44})(a_{11}^2 a_{21}^2 + a_{12}^2 a_{22}^2 + a_{13}^2 a_{23}^2).$$

π_t is evaluated for several cubic directions in Table 22.1. These expressions are similar to those obtained for the elastic constants of cubic crystals. The maxima and minima for the π'_{11} surface will lie along [100] and [111] or vice versa.

When optimizing a longitudinal gage it is important to *maximize* the sensitivity to the longitudinal stress, but it is equally important to *minimize* the sensitivity to transverse stress. The reverse is true for a transverse piezoresistive gage in which the transverse sensitivity is *maximized* and the longitudinal sensitivity is *minimized*.

22.4 Structure–property relations

Numerical values of the piezoresistance coefficients in four semiconductor crystals are listed in Table 22.2. Note the very large anisotropy in the π coefficients. π_{11} is extremely large in *n*-type silicon while π_{44} is largest in *n*-Ge, *p*-Ge, and *p*-Si. The preferred orientation of longitudinal stress gages cut from *n*-Si has the stress along [1$\bar{1}$0] with [111] in the transverse direction. For the other three crystals stress is along [110], with [001] is the transverse direction. These choices ensure that the gages are not affected by transverse stresses and shear stresses.

The causes of the large anisotropy in piezoresistance coefficients have been explained from the energy band structure in momentum space (Fig. 22.2).

Table 22.2 Fractional changes in piezoresistivity for doped silicon and germanium crystals

Material-$\rho(0)$	$\pi_{11}/\rho(0)$ ($\times 10^{-11}$ m²/N)	$\pi_{12}/\rho(0)$ ($\times 10^{-11}$ m²/N)	$\pi_{44}/\rho(0)$ ($\times 10^{-11}$ m²/N)
n-Ge 150 Ω m	−2.3	−3.2	−138.1
p-Ge 110 Ω m	−3.7	+3.2	+96.7
n-Si 1170 Ω m	−102.2	+53.4	−13.6
p-Si 780 Ω m	+6.6	−1.1	+138.3

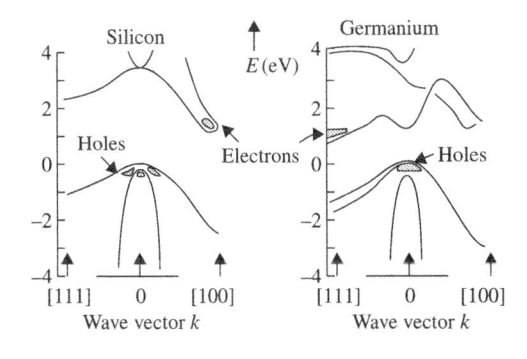

Fig. 22.2 $E(k)$ curves for doped silicon and germanium crystals.

For n-Si the energy valleys in k-space extend along $\langle 100 \rangle$ directions, but in the other three crystals (p-Ge, n-Ge, and p-Si) $\langle 111 \rangle$ directions are favored.

When a mechanical stress is applied to the crystal the atom spacings are changed and the energy levels shifted. As a result conduction electrons (or holes) shift between valleys, and the electrical resistivity changes. π_{11} is large in n-Si because the shifts are large with valleys along [100]. The [111] body diagonal valleys cause large π_{44} coefficients in the other crystals.

The piezoresistivity coefficients of the semiconductor sensors depend on temperature and dopant concentration. At room temperature and above, the coefficients are inversely proportional to temperature ($\pi \sim 1/T$) because of intervalley scattering. For p-Si, the inverse relation holds over the range from -100 to $80°C$.

The piezoresistance coefficients are also inversely proportional to dopant level when measured at 300 K. The π_{11} coefficient of n-Si drops from 150 ($\times 10^{-11}$ m^2/N) to 40 as the dopant concentrations is increased from 10^{16} to 20^{20} impurities/cm^3.

Problem 22.1
Describe how piezoresistance coefficient π_{44} can be measured using only tensile force experiments.

Problem 22.2
Most piezoresistive sensors use p-type silicon because of orientation limitations encountered during anisotropic etching (Section 32.6). Make plots of the longitudinal (π_ℓ) and transverse (π_t) piezoresistance coefficients in the ($1\bar{1}0$) plane where the [111], [110], and [001] directions are found. Numerical values for p-Si are listed in Table 22.2.

Problem 22.3
Elastoresistance refers to the linear relation between electrical resistivity and mechanical strain, rather than mechanical stress as in piezoresistance. Show how the two effects are related through the elastic constants. Set up the defining equations in both tensor and matrix form, and work out the relationships between the elastoresistive and piezoresistive coefficients of cubic crystals.

Acoustic waves I

23

In this chapter we treat plane waves (Fig. 23.1) specified by a wave normal $\vec{N}$ and a particle motion vector $\vec{U}$. Two types of waves, longitudinal waves and shear waves, are observed in solids. For low symmetry directions, there are generally three different waves with the same wave normal, a longitudinal wave and two shear waves. The particle motions in the three waves are perpendicular to one another. Only longitudinal waves are present in liquids because of their inability to support shear stresses. The transverse waves are strongly absorbed.

Acoustic wave velocities (v) are controlled by elastic constants (c) and density (ρ).

$$v = \upsilon\lambda = \sqrt{\frac{c}{\rho}}.$$

For a stiff ceramic ($c \sim 5 \times 10^{11}$ N/m^2) and density ($\rho \sim 5$ g/cm^3 = 5000 kg/m^3), the wave velocity is about 10^4 m/s. For low frequency vibrations near 1 kHz the wavelength λ is about 10 m. The shortest wavelengths are around 1 nm and correspond to infrared vibrations of 10^{13} Hz.

Acoustic wave velocities for polycrystalline alkali metals are plotted in Fig. 23.2. Longitudinal waves travel at about twice the speed of transverse shear waves since $c_{11} > c_{44}$. Sound is transmitted faster in light metals like Li which have shorter, stronger bonds and lower density than heavy alkali atoms like Cs.

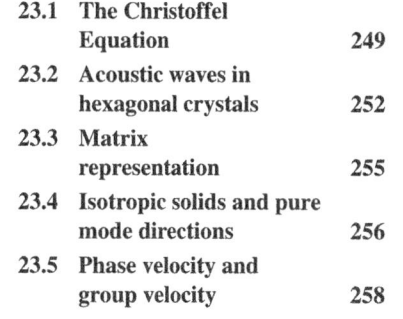

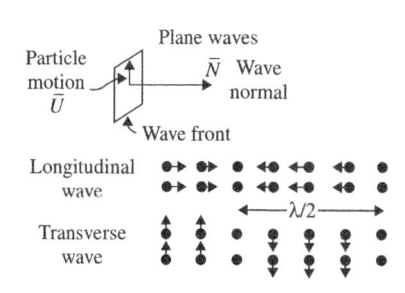

23.1	The Christoffel Equation	249
23.2	Acoustic waves in hexagonal crystals	252
23.3	Matrix representation	255
23.4	Isotropic solids and pure mode directions	256
23.5	Phase velocity and group velocity	258

Fig. 23.1 Acoustic waves are described by a wave normal $\vec{N}$ and a particle motion vector $\vec{U}$ which is parallel to $\vec{N}$ (or nearly so) for longitudinal waves. For shear waves the particle displacement is in the perpendicular direction.

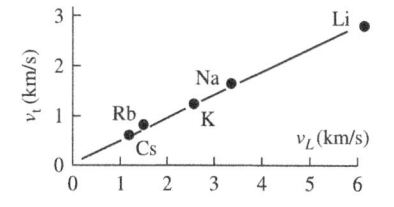

Fig. 23.2 Longitudinal waves in alkali metals travel at about twice the speed of transverse sound waves. Measurements were carried out on polycrystalline specimens at room temperature.

23.1 The Christoffel Equation

The tensor relation between velocity and elastic constants is derived using Newton's Laws and the differential volume element shown in Fig. 23.3(a). The volume is equal to $(\delta Z_1)(\delta Z_2)(\delta Z_3)$.

Acoustic waves are characterized by regions of compression and rarefaction because of the periodic particle displacements associated with the wave. These displacements are caused by the inhomogeneous stresses emanating from the source of the sound. In tensor form the components of the stress gradient are $\partial X_{ij}/\partial Z_k$ and will include both tensile stress gradients and shear stress gradients, as pictured in Fig. 23.3(b).

The force F acting on the volume element is calculated by multiplying the stress components by the area of the faces on which the force acts.

250 Acoustic waves I

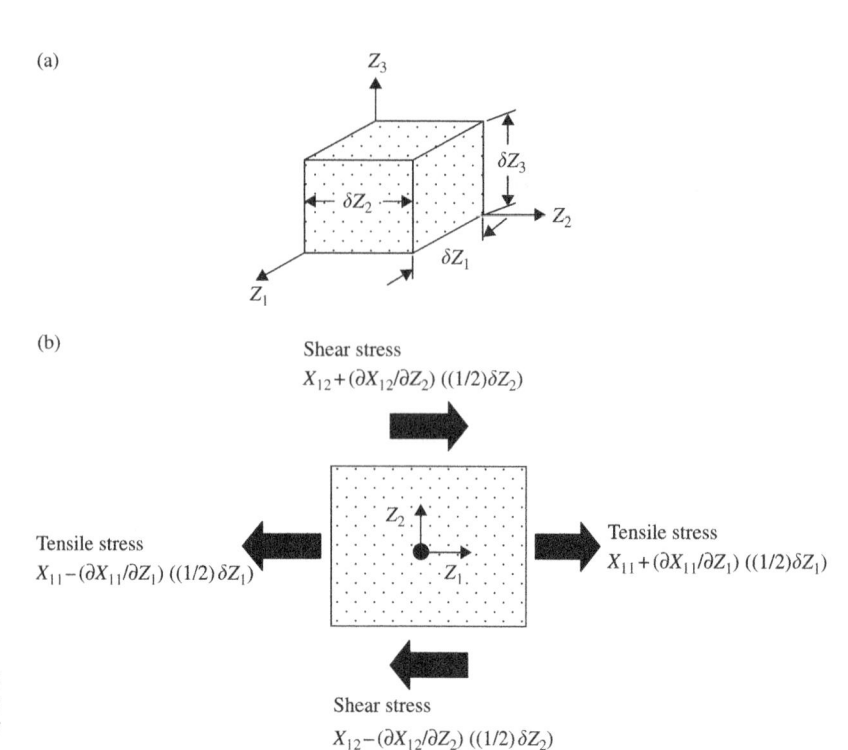

Fig. 23.3 A differential volume element (a) and a section perpendicular to Z_3 showing stress gradients for both tensile and shear forces (b).

F_1, the component of force along Z_1, is given by

$$F_1 = \frac{\partial X_{11}}{\partial Z_1}\left(\frac{1}{2}\delta Z_1 + \frac{1}{2}\delta Z_1\right)(\delta Z_2 \delta Z_3) + \frac{\partial X_{12}}{\partial Z_2}\left(\frac{1}{2}\delta Z_2 + \frac{1}{2}\delta Z_2\right)(\delta Z_1 \delta Z_3)$$

$$+ \frac{\partial X_{13}}{\partial Z_1}\left(\frac{1}{2}\delta Z_3 + \frac{1}{2}\delta Z_3\right)(\delta Z_1 \delta Z_2)$$

$$= \frac{\partial X_{1j}}{\partial Z_j}(\delta Z_1 \delta Z_2 \delta Z_3).$$

Generalizing this result to all three force components,

$$F_i = \frac{\partial X_{ij}}{\partial Z_j}\delta V = \frac{\partial X_{ij}}{\partial Z_j}\frac{\delta m}{\rho},$$

where δV is the volume of the differential element, δm is its mass, and ρ the density.

From Newton's Second Law, the force is equal to the product of the mass and the acceleration $\ddot{u}_i$.

$$F_i = (\delta m)\ddot{u}_i = \frac{\partial X_{ij}}{\partial Z_j}\frac{(\delta m)}{\rho}.$$

The equation of motion is

$$\rho \ddot{u}_i = \frac{\partial X_{ij}}{\partial Z_j},$$

where u_i is the displacement of the volume element in the ith direction.

Elastic constants c_{ijkl} are introduced into the equation of motion through Hooke's Law

$$X_{ij} = c_{ijkl} x_{kl}.$$

Referring back to the defining relation for strain (Section 10.3),

$$x_{kl} = \frac{\partial u_k}{\partial Z_l}.$$

Substituting these two relationships into the equation of motion gives

$$\rho \ddot{u}_i = c_{ijkl} \frac{\partial}{\partial Z_j} \left(\frac{\partial u_k}{\partial Z_\ell} \right).$$

The Christoffel Equation is obtained from the plane wave solution to the equation of motion. A plane wave is represented by

$$u_k = A_k \exp i(\omega t - \vec{k} \cdot \vec{Z}) = A_k \exp i(\omega t - k_i Z_i)$$

in which u_k is the kth component of the displacement of the volume element from its origin at rest. A_k is the wave amplitude, i is $\sqrt{-1}$, ω is the angular frequency, $\vec{k}$ is the wave vector parallel to the wave normal, and $\vec{Z}$ the coordinate vector. The magnitude of $|\vec{k}|$ is $2\pi/\lambda$, where λ is the wavelength, and the scalar product

$$\vec{k} \cdot \vec{Z} = k_1 Z_1 + k_2 Z_2 + k_3 Z_3 = k_i Z_i.$$

Taking the derivatives of u with respect to time and space gives the acceleration

$$\ddot{u}_i = \frac{\partial^2 u_i}{\partial t^2} = -\omega^2 u_i,$$

strain

$$x_{kl} = \frac{\partial u_k}{\partial Z_\ell} = -i k_\ell u_k$$

and strain gradient,

$$\frac{\partial x_{kl}}{\partial Z_j} = \frac{\partial^2 u_k}{\partial Z_\ell \partial Z_j} = -k_\ell k_j u_k.$$

The equation of motion now becomes

$$\rho \ddot{u}_i = -\rho \omega^2 u_i = c_{ijkl} \frac{\partial^2 u_k}{\partial Z_j \partial Z_\ell} = -c_{ijkl} k_\ell k_j u_k.$$

To simplify further, let $U_k = A_k/A$ where $\vec{U}$ is a unit vector denoting the orientation of the particle motion relative to the coordinate axes. U_k is the direction cosine of $\vec{U}$ with the kth axis. A is the amplitude of the wave and A_k is its projection on the kth axis.

In a similar way, $\vec{N}$ is defined as a unit vector parallel to the wave normal. Its projection on the reference axis Z_j is given by

$$N_j = \frac{k_j}{k},$$

where k ($=2\pi/\lambda$) is the wave vector and k_j its jth component.

$\vec{U}$ and $\vec{N}$ are illustrated in Fig. 23.1, and U_k and N_j are the direction cosines of the particle motion and wave normal.

Substituting these concepts into the equation of motion,

$$\rho \ddot{u}_i = c_{ijkl} k_\ell k_j u_k$$

$$\rho \omega^2 A_i \exp i(\omega t - k_i Z_i) = c_{ijkl} k_\ell k_j A_k \exp i(\omega t - k_i Z_i)$$

$$\rho \omega^2 A U_i = c_{ijkl} N_\ell N_j k^2 U_k A.$$

The wave velocity v is introduced through the relation

$$v^2 = \frac{(2\pi v)^2}{(2\pi/\lambda)^2} = \frac{\omega^2}{k^2}$$

giving the Christoffel Equation

$$\rho v^2 U_i = c_{ijkl} N_j U_k N_\ell.$$

The Christoffel Equation relates the ultrasonic wave velocities, wave normals, polarization directions, and elastic constants.

23.2 Acoustic waves in hexagonal crystals

As an illustration of the Christoffel Equation, we examine sound waves in hexagonal crystals. All seven hexagonal classes have the same elastic stiffness matrix.

$$\begin{pmatrix} c_{11} & c_{12} & c_{13} & 0 & 0 & 0 \\ c_{12} & c_{11} & c_{13} & 0 & 0 & 0 \\ c_{13} & c_{13} & c_{33} & 0 & 0 & 0 \\ 0 & 0 & 0 & c_{44} & 0 & 0 \\ 0 & 0 & 0 & 0 & c_{44} & 0 \\ 0 & 0 & 0 & 0 & 0 & c_{66} \end{pmatrix}$$

There are five independent elastic constants $c_{11} = c_{1111}$, $c_{12} = c_{1122}$, $c_{13} = c_{1133}$, $c_{33} = c_{3333}$, and $c_{44} = c_{2323}$. Coefficient $c_{66} = \frac{1}{2}(c_{11} - c_{12})$.

For waves traveling along the $Z_1 = [100]$ axis, the direction cosines of the wave normal are $N_1 = 1$, $N_2 = N_3 = 0$. Therefore $j = l = 1$ in the Christoffel Equation:

$$\rho v^2 U_i = c_{i1k1} U_k.$$

There are three solutions ($i = 1, 2, 3$) to the equation leading to three different sound waves. For $i = 1$,

$$\rho v^2 U_1 = c_{11k1} U_k$$
$$= c_{1111} U_1 + c_{1121} U_2 + c_{1131} U_3$$
$$= c_{11} U_1 + c_{16} U_2 + c_{15} U_3$$
$$= c_{11} U_1.$$

For hexagonal crystals since $c_{15} = c_{16} = 0$, the particle motion for this wave is parallel to the wave normal Z_1 making it a longitudinal wave. The velocity of this wave is $v = \sqrt{c_{11}/\rho}$.

The second solution with $i = 2$ is

$$\rho v^2 U_2 = c_{21k1} U_k$$
$$= c_{2111} U_1 + c_{2121} U_2 + c_{2131} U_3$$
$$= c_{61} U_1 + c_{66} U_2 + c_{65} U_3$$
$$= c_{66} U_2 = \tfrac{1}{2}(c_{11} - c_{12}) U_2$$

for hexagonal crystals. This is a transverse wave polarized along $Z_2 = [120]$, perpendicular to the wave normal along $Z_1 = [100]$. The speed is $v = \sqrt{(c_{11} - c_{12})/2\rho}$.

The third wave ($i = 3$) gives a Christoffel relation

$$\rho v^2 U_3 = c_{3111} U_1 + c_{3121} U_2 + c_{3131} U_3 = c_{44} U_3.$$

For the hexagonal case, the third wave is a transverse shear wave polarized along $Z_3 = [001]$ with velocity $v = \sqrt{c_{44}/\rho}$.

In summary, there are three acoustic waves moving along [100]: a longitudinal wave polarized parallel to [100], a shear wave polarized parallel to [120], and another polarized parallel to [001]. The longitudinal wave travels faster since c_{11} is generally larger than c_{44} and c_{66}.

With the wave normal along $Z_2 = [120]$ ($N_1 = N_3 = 0$ and $N_2 = 1$), the Christoffel Equation becomes

$$\rho v^2 U_i = c_{i2k2} U_k.$$

For hexagonal crystals, this leads to three similar waves with analogous polarization directions and the same velocities as the three waves traveling along $Z_1 = [100]$.

Waves along $Z_3 = [001]$ ($N_1 = N_2 = 0$, $N_3 = 1$) are described by the equation

$$\rho v^2 U_i = c_{i3k3} U_k.$$

For hexagonal crystals the three solutions are

$$i = 1 \quad \rho v^2 U_1 = c_{44} U_1$$
$$i = 2 \quad \rho v^2 U_2 = c_{44} U_2$$
$$i = 3 \quad \rho v^2 U_3 = c_{33} U_3.$$

The first and second waves are shear waves transversely polarized along $Z_1 = [100]$ and $Z_2 = [120]$. They are referred to as degenerate modes since they travel with the same velocity, $v = \sqrt{c_{44}/\rho}$.

The third wave is a longitudinally polarized wave with particle motion along $Z_3 = [001]$. Its velocity is $v = \sqrt{c_{33}/\rho}$.

As mentioned earlier (Section 13.5) elastic constants are often measured ultrasonically. Quartz transducers are used to generate longitudinal and transverse sound waves. AC-cuts are used for shear waves and the thickness mode of X-cuts for longitudinal waves. The crystals are operated in a pulsed mode and used to launch the wave and receive the reflected wave from the opposite face. Wave velocities are measured by timing the waves electronically.

Of the five independent elastic constants for hexagonal crystals, four can be measured from acoustic waves traveling along Z_1 and Z_3. Only stiffness coefficient c_{13} remains undetermined. This can be done from an inclined direction such as $N_1 = 1/\sqrt{2}$, $N_2 = 0$, $N_3 = 1/\sqrt{2}$.

The Christoffel Equation becomes $2\rho v^2 U_i = c_{i1k1} U_k + c_{i1k3} U_k + c_{i3k1} U_k + c_{i3k3} U_k$. Again there will be three waves corresponding to $i = 1, 2, 3$.

For $i = 1$, this simplifies to

$$2\rho v^2 U_1 = (c_{11} + c_{44}) U_1 + (c_{13} + c_{44}) U_3.$$

For $i = 2$

$$2\rho v^2 U_2 = (c_{66} + c_{44}) U_2 = \left(\frac{1}{2} c_{11} - \frac{1}{2} c_{12} + c_{44}\right) U_2.$$

For $i = 3$

$$2\rho v^2 U_3 = (c_{44} + c_{13}) U_1 + (c_{44} + c_{33}) U_3.$$

The second root is a pure shear wave polarized parallel to $Z_2 = [120]$. Coefficient c_{13} can be determined from the other two roots corresponding to $i = 1$ and $i = 3$. Multiplying the two equations together gives $(2\rho v^2 - c_{11} - c_{44}) \times (2\rho v^2 - c_{33} - c_{44}) = (c_{13} + c_{44})^2$.

Solving for the unknown coefficient c_{13},

$$c_{13} = \pm\sqrt{(2\rho v^2 - c_{11} - c_{44})(2\rho v^2 - c_{33} - c_{44})} - c_{44}.$$

To determine the polarization directions for these two waves we solve the two equations for U_3/U_1:

$$\frac{U_3}{U_1} = -\frac{1}{2}\left[\frac{c_{11} - c_{33}}{c_{13} + c_{44}}\right] \pm \sqrt{\frac{(c_{11} - c_{33})^2}{4(c_{13} + c_{44})^2} + 1}.$$

Consider the quantity $c_{11} - c_{33}$ which appears in both terms. If the hexagonal crystal was elastically isotropic, then $c_{11} = c_{33}$, and $U_3/U_1 = \pm 1$. When $U_3 = U_1$ then the vibration direction is parallel to the wave normal ($N_1 = N_3$). When $U_3 = -U_1$ the vibration direction is perpendicular to the wave normal, making it a transverse vibration. Thus these two situations correspond to a pure longitudinal wave and a pure shear wave.

In most hexagonal crystals, however, $c_{11} - c_{33} \neq 0$, and the vibration directions of the two waves will not be purely longitudinal or purely transverse. One will be quasilongitudinal and the other will be quasitransverse. The quasilongitudinal wave will generally be faster than the quasitransverse wave. The third wave traveling in this direction is a pure shear wave polarized parallel to $Z_2 = [120]$.

For all four wave normals considered in this example, there was a longitudinal (or quasilongitudinal) wave and two transverse (or quasitransverse waves) waves which usually travel at a slower velocity. This is true in general for crystals.

If the wave (or phase) velocity is plotted as a function of the wave normal, a triple surface is obtained corresponding to the three waves in any direction.

The phase velocity surfaces in α-quartz (point group 32) are shown in Fig. 23.4. Note that the surfaces viewed along $Z_1 = [100]$ and $Z_3 = [001]$ have two and threefold symmetry in accordance with Neumann's Law which says that the symmetry of a physical property must include the symmetry of the point group.

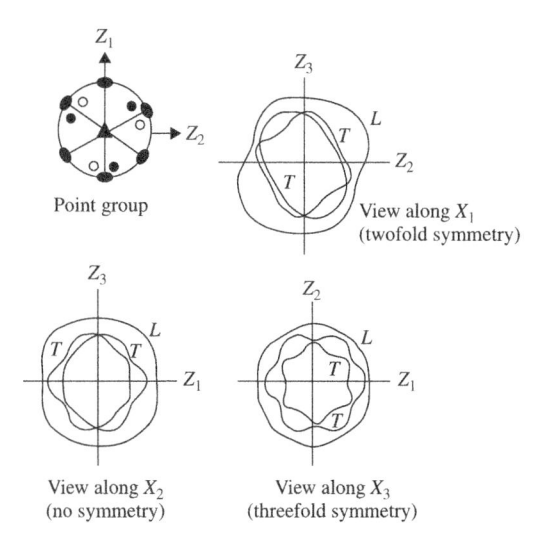

Fig. 23.4 Wave velocity surfaces for α-quartz (SiO$_2$) at room temperature. One surface (usually the fastest wave) corresponds to longitudinal polarization (or nearly so) and two to transverse waves (labeled T). Note that the two transverse waves travel with the same velocity along Z_3, the so-called "acoustic" axis.

23.3 Matrix representation

The previous section on hexagonal crystals outlines some of the basic ideas, but now we need to generalize the discussion of acoustic waves to treat all directions in all crystals, even those of lowest symmetry. When written in matrix form, triclinic crystals have 36 elastic constants, 21 of which are independent. The Christoffel Equation can be rewritten as

$$(c_{ijkl}N_jN_l - \rho v^2 \delta_{ik})U_i = 0,$$

where δ_{ik} is the Kronecker delta. To express the equation in matrix form we introduce a set of Christoffel matrix coefficients C_{ik}.

$$C_{ik} \equiv \frac{1}{\rho} c_{ijkl} N_j N_l.$$

In the general case of a triclinic crystal,

$$C_{ik} = \frac{1}{\rho}\big[c_{i1k1}N_1^2 + c_{i2k2}N_2^2 + c_{i3k3}N_3^2 + (c_{i2k3} + c_{i3k2})N_2N_3$$
$$+ (c_{i1k3} + c_{i3k1})N_1N_3 + (c_{i1k2} + c_{i2k1})N_1N_2\big].$$

To illustrate the matrix method, we write out C_{22} and C_{13}. The tensor stiffness c_{ijkl} are converted to matrix stiffnesses c_{mn} where $m, n = 1\text{--}6$.

$$C_{22} = \frac{1}{\rho}\big[c_{66}N_1^2 + c_{22}N_2^2 + c_{44}N_3^2 + (c_{24} + c_{24})N_2N_3$$
$$+ (c_{46} + c_{46})N_1N_3 + (c_{26} + c_{26})N_1N_2\big]$$

$$C_{13} = \frac{1}{\rho}\big[c_{15}N_1^2 + c_{46}N_2^2 + c_{35}N_3^2 + (c_{36} + c_{45})N_2N_3$$
$$+ (c_{13} + c_{55})N_1N_3 + (c_{14} + c_{56})N_1N_2\big].$$

The other four Christoffel coefficients (C_{11}, C_{33}, C_{12}, and C_{23}) are evaluated in a similar way. All six can be combined in the Christoffel matrix given in Table 23.1.

Table 23.1 Matrix components of the Christoffel Tensor $C_{ik} = (1/\rho)c_{ijkl}N_jN_l$. These coefficients are used to calculate the wave velocities of sound waves in triclinic crystals. N_1, N_2, and N_3 are the direction cosines of the wave normal

	N_1^2	N_2^2	N_3^2	N_2N_3	N_3N_1	N_1N_2
C_{11}	11	66	55	2 × 56	2 × 15	2 × 16
C_{22}	66	22	44	2 × 24	2 × 46	2 × 26
C_{33}	55	44	33	2 × 34	2 × 35	2 × 45
C_{23}	56	24	34	23 + 44	36 + 45	25 + 46
C_{13}	15	46	35	36 + 45	13 + 55	14 + 56
C_{12}	16	26	45	25 + 46	14 + 56	12 + 66

Knowing the elastic constants of a crystal, it is a straightforward matter to evaluate Christoffel matrix coefficients. For the solution of the Christoffel Equation to be nonzero, the determinant of the coefficients of the equation must vanish. This means that $|C_{ik} - v^2\delta_{ik}| = 0$.

Written as a matrix,

$$\begin{vmatrix} C_{11} - v^2 & C_{12} & C_{13} \\ C_{12} & C_{22} - v^2 & C_{23} \\ C_{13} & C_{23} & C_{33} - v^2 \end{vmatrix} = 0.$$

When multiplied out,

$$(C_{11} - v^2)(C_{22} - v^2)(C_{33} - v^2) - C_{23}^2(C_{11} - v^2) - C_{13}^2(C_{22} - v^2)$$
$$- C_{12}^2(C_{33} - v^2) + 2C_{12}C_{13}C_{23} = 0.$$

This is a cubic equation in v^2 so there are three waves traveling in the direction $\vec{N}$ with wave normal direction cosines N_1, N_2, and N_3. For the general triclinic case the three waves will travel with different speeds v_I, v_II, and v_III. Each of the so-called isonormal waves will have its own vibration direction U_I, U_II, and U_III. In the general case, two of the waves will be quasitransverse and one quasilongitudinal. The vibration directions are determined by substituting the solutions for v in the Christoffel Equations.

23.4 Isotropic solids and pure mode directions

Polycrystalline solids, glasses, and normal liquids have isotropic (spherical) symmetry, point group $\infty\infty m$. The nonzero elastic constants are $c_{11} = c_{22} = c_{33}$, $c_{12} = c_{13} = c_{23}$, and $c_{44} = c_{55} = c_{66} = \frac{1}{2}(c_{11} - c_{12})$.

As an illustration of the Christoffel matrix method, we evaluate acoustic waves in an isotropic solid. Since all directions are the same in an isotropic material, we arbitrarily select the wave normal along Z_1 for which $N_1^2 = 1$, $N_2^2 = N_3^2 = 0$, $N_1N_2 = N_2N_3 = N_3N_1 = 0$. The Christoffel coefficients are

$$C_{11} = \frac{c_{11}}{\rho}, \quad C_{22} = \frac{c_{44}}{\rho}, \quad C_{33} = \frac{c_{44}}{\rho}, \quad C_{23} = C_{13} = C_{12} = 0.$$

The determinant used to obtain the phase velocities is $(C_{11} - v^2)(C_{22} - v^2) \times (C_{33} - v^2) = 0$, leading to the result that there is a pure longitudinal wave with velocity $v_L = \sqrt{c_{11}/\rho}$ and two pure degenerate shear waves with velocity $v_T = \sqrt{c_{44}/\rho}$.

Acoustic waves in isotropic materials are always pure modes in the sense that the particle velocity is always either parallel or perpendicular to the wave normal $\vec{N}$. Pure modes are so much easier to handle than quasilongitudinal or quasitransverse waves, that it is interesting to inquire where they are located in solids of lower symmetry.

As pointed out earlier, for any given wave normal, there are always three solutions to the Christoffel Equation. If there are no symmetry restrictions, the three waves travel with different velocities with different vibration directions. The three vibration directions are perpendicular to one another. The vibration direction closest to the wave normal is called the quasilongitudinal wave. The other two are called quasitransverse. When no mirror planes or rotation axes are present, there are no pure modes. This is the case for triclinic crystals in point groups 1 or $\bar{1}$. Since elastic vibrations are centrosymmetric, the inversion center makes no difference.

Next consider a monoclinic crystal belonging to point group m. If the wave normal is perpendicular to the mirror plane, two of the vibration directions must be parallel to the mirror, and the other is parallel to the wave normal and perpendicular to the mirror plane. In other words, all three waves are pure vibration modes.

If the wave normal is parallel to the mirror then two of the vibration directions may also lie in the plane without violating Neumann's Principle. One wave will be the quasilongitudinal wave, the other is quasitransverse. The third wave is a pure transverse mode with the vibration direction perpendicular to the mirror plane. It is required to be perpendicular to the other two vibration directions.

When the wave normal is in any other orientation it is not a symmetry direction and its vibration directions are not restricted by Neumann's Law. In this case the three waves are quasitransverse and quasilongitudinal as in the triclinic system.

Similar symmetry arguments apply to acoustic waves in crystals belonging to other symmetry groups. Restrictions are placed on the vibration directions whenever the wave normal is either parallel or perpendicular to a mirror plane or rotational symmetry axis. The key idea is the orientation of the wave normal relative to the symmetry element. If the wave normal possesses a certain symmetry then the vibration directions of the three associated acoustic waves must conform to this symmetry.

Table 23.2 identifies the pure modes and degenerate modes for various wave normal orientations. When the wave normal is parallel to a rotation axis of threefold symmetry or higher, the two transverse shear modes are degenerate.

Pure modes may propagate along nonsymmetry directions when certain relationships between elastic constants are satisfied. In hexagonal crystals, for example, when the wave normal makes an angle θ with the sixfold symmetry axis, where θ is given by

$$\cot^2 \theta = \frac{c_{11} - 2c_{44} - c_{13}}{c_{33} - 2c_{44} - c_{13}},$$

the shear wave is polarized perpendicular to the wave normal.

258 Acoustic waves I

Table 23.2 The symmetry of the wave normal determines which acoustic waves are pure longitudinal (P) and which are quasilongitudinal (Q), and which transverse waves are pure (P) and which are not (Q). The transverse waves traveling along high symmetry directions are pure waves with the same velocity (P = P)

Special orientation of wave normal	Longitudinal wave	Two transverse waves
None	Q	Q Q
Parallel to a mirror plane or perpendicular to a 2-, 3-, 4-, 6-, or ∞-fold rotation axis	Q	P Q
Parallel to twofold axis or perpendicular to a mirror plane	P	P P
Parallel to 3-, 4-, 6-, or ∞-fold rotation axis	P	P = P

Problem 23.1

The mineral aragonite ($CaCO_3$) is orthorhombic, point group *mmm*, with twofold symmetry axes along [100], [010], and [001]. The density of aragonite is 2700 kg/m^3. The following acoustic wave velocities were measured. Determine as many stiffness coefficients as possible from these data. Which elastic constants remain to be determined? Suggest acoustic experiments that could be used to complete the stiffness measurements.

Wave normal	Polarization orientation	Speed [km/s]
[100]	[100]	7.68
[100]	[010]	3.96
[100]	[001]	3.08
[010]	[010]	5.68
[010]	[100]	3.96
[010]	[001]	3.90
[001]	[001]	5.59
[001]	[100]	3.08
[001]	[010]	3.90

23.5 Phase velocity and group velocity

The group velocity of an elastic wave is the velocity of energy flux. In isotropic materials the energy flow is parallel to the wave normal, but in anisotropic media the two vectors are often oriented differently. This can create complications in experiments where the acoustic beam reflects from side faces of the crystal (Fig. 23.5) causing multiple scattering and confusion in measuring elastic constants.

The phase velocities of zinc are plotted as a function of wave normal in Fig. 23.6. As pointed out in Section 13.6, zinc has a hexagonal close-packed crystal structure, but the unit cell has a rather large c/a ratio. Zn–Zn bonds in

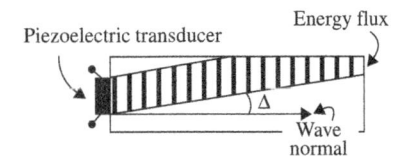

Fig. 23.5 Beam divergence takes place when the energy flow is not parallel to the wave normal. Difficulties arise when the divergence angle Δ becomes large.

the close-packed [001] plane are therefore the shortest and strongest. As a result the structure is stiffer in the Z_1 and Z_2 directions, and more compliant along Z_3, the sixfold symmetry axis. Stiffness coefficient c_{33} is therefore smaller than c_{11} and c_{22}, and longitudinal acoustic waves along Z_3 are slower than in the perpendicular directions. This is quite apparent in the phase velocity surface.

Another result of the elastic anisotropy is beam divergence. Divergence angles Δ of almost 40° are observed for zinc single crystals (Fig. 23.6(b)). Symmetry requires that $\Delta = 0$ along $Z_3 = [001]$ and $Z_1 = [100]$. Comparing Fig. 23.6(a) and (b), it is apparent that the divergence increases rapidly for those angles where the wave velocity is changing rapidly with direction.

The divergence angles and group velocities can be obtained in a simple geometric procedure by replotting the wave (phase) velocity surface as a slowness surface in which $1/v_p$ is plotted as a function of the wave normal. The energy flow is always normal to the slowness surface, as indicated in Fig. 23.7(a). Divergence angles for the three waves associated with a given wave normal are evaluated geometrically by drawing the normals to the wave surfaces. After measuring the three Δ values, the group velocities are obtained from the relation

$$v_g = \frac{v_p}{\cos \Delta}$$

as shown in Fig. 23.7(b).

The proof of this procedure involves calculating the sum of the kinetic and potential energies of an elastic wave, and leads to the so-called *Acoustic Poynting Vector*, which defines the energy flow direction.

In the case of cubic crystals, the divergence angle Δ depends strongly on the elastic anisotropy factor $A = 2c_{44}/(c_{11} - c_{12})$. If $A = 1$, then $\Delta = 0$ and the energy flow is parallel to the wave normal.

Problem 23.2
As an exercise, carry out this procedure for the wave velocity surface of zinc (Fig. 23.6(a)). Using this drawing, plot out one quadrant of the corresponding slowness surface. To avoid confusion, it is best to make separate plots for the quasilongitudinal, pure transverse, and quasitransverse waves. Draw tangents to the surface for orientation angles $\theta = 10°$, 45°, and 70°. The energy flow directions for these wave normals will be perpendicular to the tangent lines. Measure the corresponding Δ angles and compare with the values in Fig. 23.6(b). Compute the corresponding group velocities from $v_g = v_p/\cos \Delta$.

Acoustic anisotropy occurs in crystals with anisotropic bonding. Metallic bismuth has a layer-like structure with easy cleavage between the layers. Bismuth atoms form three strong bonds to nearby neighbors in a puckered layer oriented perpendicular to Z_3, the threefold symmetry axis. As a result, the structure is stiffest in the Z_1–Z_2 plane, and the longitudinal acoustic waves are fastest in these directions (Fig. 23.8).

Paratellurite (TeO$_2$) is another interesting example with exceptionally fast shear waves. Normally, longitudinally polarized waves travel at about twice the speed of shear waves, but in paratellurite one of the shear wave velocities along $Z_1 = [100]$ exceeds that of the longitudinal wave (Fig. 23.9).

This effect has been attributed to the Te^{4+} ion in TeO$_2$ which has a distorted coordination because of its lone-pair electronic configuration. The crystals also

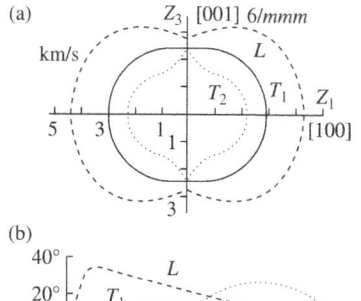

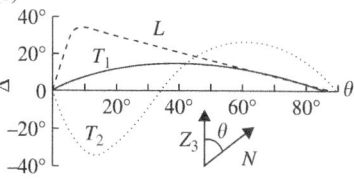

Fig. 23.6 (a) The acoustic phase velocity surface for zinc single crystals. Waves are fastest in the hexagonal (001) plane where the bond lengths are shorter. L refers to the longitudinal wave, T_1 the pure transverse wave, and T_2 the quasitransverse wave. (b) Divergence angles Δ for acoustic waves in single crystal zinc, plotted as a function of θ, the angle between the wave normal and Z_3, the sixfold symmetry axis. Note that Δ is large for the quasilongitudinal wave (L) and the quasitransverse wave (T_2) near $\theta = 10°$ where the phase velocity changes rapidly with angle.

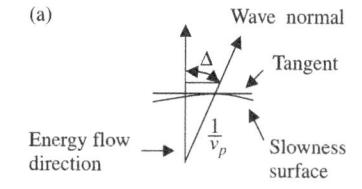

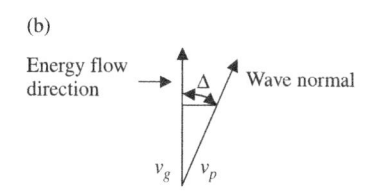

Fig. 23.7 (a) The slowness surface is used to determine the directions of energy flow. For a given wave normal, the flow is perpendicular to the tangent to the surface. The divergence angle Δ is the angle between the wave normal and the energy flow direction. (b) The speed of the energy flow (the group velocity v_g) is obtained from the phase velocity v_p and the divergence angle Δ.

Fig. 23.8 Metallic bismuth has a layer structure belonging to trigonal point group $\bar{3}m$. The longitudinal (L) and transverse waves (T_1 and T_2) polarized parallel to the strongly bonded (001) planes are faster than those polarized parallel to the [001] = Z_3 axis.

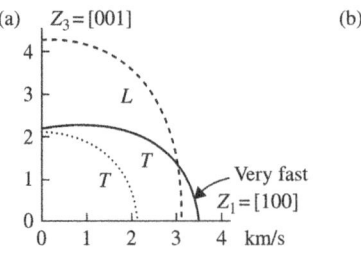

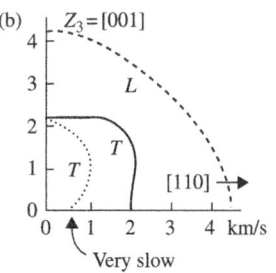

Fig. 23.9 Paratellurite (tetragonal, point group 422) has an unusual wave velocity surface with very fast and very slow shear waves.

exhibit extremely slow shear waves. Transversely polarized waves along [110] are about five times slower than those along [100]. In this case the phase velocities are only about 600 m/s. By way of comparison the speed of sound in air under normal conditions is about 340 m/s. Even slower speeds of 150 m/s have been observed in low density aerogels and polymer foams.

Acoustic waves II

24

Acoustic impedance, acoustic losses, acoustic waves in piezoelectric solids, and surface waves are discussed in this chapter, along with a number of nonlinear acoustic phenomena.

24.1	Acoustic impedance	261
24.2	Ultrasonic attenuation	262
24.3	Physical origins of attenuation	264
24.4	Surface acoustic waves	265
24.5	Elastic waves in piezoelectric media	266
24.6	Nonlinear acoustics	270

24.1 Acoustic impedance

The reflection and transmission of acoustic waves across a boundary is governed by acoustic impedance. One of the most important boundary value problems in acoustics concerns a plane wave incident upon a planar surface, dividing one medium from another. In the general case of an anisotropic medium, the incident beam consists of three waves (one quasilongitudinal, two quasitransverse), each traveling at a different velocity. Each of the three incident waves will be refracted and reflected at the boundary. If the second medium is also anisotropic, each incident wave will generate three reflected waves and three refracted waves, a total of 27 waves in all. Wave propagation in a polycrystalline solid where there are many grain boundaries becomes very complicated.

The simpler case of a pure longitudinally-polarized wave at normal incidence to the boundary provides insight into the more general problem. In this case the reflection and transmission coefficients are governed by the relatively simple acoustic impedance parameter $(\rho c)^{1/2} = \rho v$, where ρ is the density, c the stiffness coefficient, and v the phase velocity. The reflection coefficient R at the interface between medium I and medium II is

$$R = \frac{(\rho v)_{II} - (\rho v)_{I}}{(\rho v)_{II} + (\rho v)_{I}}.$$

The MKS unit for acoustic impedance is the Rayl (=kg/m^2 s). A typical value for a solid is about 10^7 rayls (Table 24.1).

In many acoustic applications it is desirable to reduce reflection by matching the acoustic impedance of the two media. Lithium tantalate transducers are well-matched to iron, for example. Sound transmission from the transducer to the medium can be enhanced with composite materials or with graded coupling layers. Backing materials are often selected to promote reflection. In this case acoustic impedances are mismatched. Tungsten and air are two commonly used backing materials.

In an isotropic material the acoustic impedance is $(\rho c_{11})^{1/2}$ for longitudinal waves and $(\rho c_{44})^{1/2}$ for shear waves. For anisotropic materials the wave velocities and acoustic impedance change with direction as indicated earlier.

Table 24.1 Longitudinal acoustic impedance (ρv) values in megarayls = 10^6 kg/m^2

Metals		Nonmetals	
Tungsten	105	LiTaO$_3$	47
Platinum	89	Al$_2$O$_3$	44
Gold	62	ZnO	36
Nickel	53	Pb(Zr,Ti)O$_3$	35
Iron	47	LiNbO$_3$	34
Brass	36	PbNb$_2$O$_6$	21
Silver	36	SiO$_2$	15
Lead	22	Polymers	1–4
Aluminum	17	Water	1.5
		Air	4×10^{-4}

Table 24.2 Damping decrement of common polycrystalline metals. The fraction of vibration energy lost per cycle is given in percent

Lead	1.44%	Copper	0.70	Molybdenum	0.10
Tin	1.08	Iron	0.40	Magnesium	0.04
Nickel	0.93	Zinc	0.15	Aluminum	0.01

24.2 Ultrasonic attenuation

Damping takes place in all real materials. For metals, the fraction of vibration energy lost per cycle is typically around 1%. Light metals like Al and Mg have much lower losses than Sn and Pb (Table 24.2). Elastic damping is significantly smaller in insulators and semiconductors.

In most materials acoustic losses can be described by a viscous damping term in the constitutive equation. Hooke's Law is modified to include a viscosity coefficient η:

$$X_{ij} = c_{ijkl}x_{kl} + \eta_{ijkl}\frac{\partial x_{kl}}{\partial t}.$$

In tensor form the viscosity coefficients are components of a fourth rank tensor like the elastic constants and transform in the usual way. Coefficients in the new coordinate system are related to those in the old through the product of four direction cosines.

$$\eta'_{ijkl} = a_{im}a_{jn}a_{ko}a_{lp}\eta_{mnop}.$$

The tensor coefficients can be rewritten as a 6×6 matrix in which $\eta_{1111} = \eta_{11}$, $\eta_{1122} = \eta_{12}$, $\eta_{1212} = \eta_{66}$, etc. In the MKS system, viscosity coefficients are expressed in N s/m^2 but much of the viscosity literature is in centipoise = 10^{-3} N s/m^2. Typical numerical values cover about three orders of magnitude from 100 cP for metals and amorphous materials to about 0.1 cP for the best single crystal oxides.

In wave phenomena, damping is normally described by an attenuation factor α. The amplitude of the wave decreases steadily in a lossy media. For a one-dimensional system the particle displacement takes the form

$$u = A \exp(-\alpha Z) \exp i(\omega t - kZ),$$

where A is the amplitude, ω the angular frequency, and k is the wave vector. The resulting strain field is

$$x = \frac{\partial u}{\partial Z} = -i(k - i\alpha)A \exp(i\omega t) \exp(-i(k - i\alpha)Z).$$

For a longitudinal wave traveling along the [100] direction of a cubic crystal, the stress field is

$$X_1 = c_{11}x_1 + \eta_{11}\frac{\partial x_1}{\partial t}$$

$$= i(k_1 - i\alpha_1)(c_{11} + i\omega\eta_{11})A \exp(i\omega t) \exp(-i(k_1 - i\alpha_1)Z_1).$$

Substituting into the equation of motion

$$\frac{\partial X_1}{\partial Z_1} = \rho \frac{\partial^2 u_1}{\partial t^2}$$

gives the dispersion relation

$$(k - i\alpha)^2(c_{11} + i\omega\eta_{11}) = \rho\omega^2.$$

Separating this equation into real and imaginary parts gives

$$c_{11}(k_1^2 - \alpha_1^2) + 2\alpha_1 k_1 \omega \eta_{11} = \rho\omega^2$$

$$i((k_1^2 - \alpha_1^2)\omega\eta_{11} - 2\alpha_1 k_1 c_{11}) = 0.$$

Rearranging these equations and making use of the fact that $\omega\eta_{11} \ll c_{11}$ even for very high frequencies, leads to the wave number

$$k = \omega \left(\frac{\rho}{c_{11}}\right)^{1/2} \left(1 + \frac{3}{8}\left(\frac{\omega\eta_{11}}{c_{11}}\right)^2\right)^{-1/2}$$

and the attenuation coefficient

$$\alpha = \left(\frac{\omega^2}{2}\right)\left(\frac{\rho}{c_{11}}\right)^{1/2}\left(\frac{\eta_{11}}{c_{11}}\right).$$

The result is that introducing loss leads to a small decrease in k and a small increase in the phase velocity $v = \omega/k$. A more important result is that attenuation is proportional to the square of the frequency. As a consequence, high quality low-loss single crystals must be used in applications above 100 MHz. The attenuation per wavelength is $\alpha\lambda = 2\pi\alpha/k$. For the longitudinal wave along [100] in a cubic crystal this is proportional to $c_{11}/\omega\eta_{11}$, the so-called acoustic Q. For a shear wave in the same direction $Q = c_{44}/\omega\eta_{44}$.

As just shown, the amplitude of a wave traveling along the Z-axis decreases exponentially. For two points separated by a distance ΔZ the attenuation in amplitude is $e^{-\alpha\Delta Z}$. In experimental work it is more common to measure attenuation on a logarithmic scale by taking the natural logarithm: attenuation $= \alpha \Delta Z$ [nepers]. A more commonly used scale is based on the decibel [dB]. In the decibel scale,

$$\text{attenuation} = 20(\log e)\alpha\Delta Z \text{ [dB]}.$$

$$\alpha \text{ [dB/m]} = 8.686\alpha \text{ [nepers/m]}.$$

Experimental values for several single crystals are listed in Table 24.3. The attenuation factors in oxide crystals are generally lower than those in semiconductors and metals. YAG (Yttrium Aluminum Garnet = $Y_3Al_5O_{12}$) is used in acoustic delay line devices because of its acoustic transparency.

Problem 24.1
To illustrate the relationship between the viscosity coefficient η and the attenuation α, calculate the viscosity coefficients for MgO from the data in Table 24.3. The density of magnesium oxide is 3650 kg/m^3, and its elastic constants are $c_{11} = 28.6 \times 10^{10}$, $c_{12} = 8.7 \times 10^{10}$, and $c_{44} = 14.8 \times 10^{10}$ N/m^2.

24.3 Physical origins of attenuation

A wide variety of acoustic loss mechanisms have been proposed, but at room temperature thermal effects dominate in single crystals. Longitudinal waves consist of alternating regions of compression and rarefaction (Fig. 24.1(a)). Since solids warm under pressure, this leads to localized temperature gradients and an internal heat dissipation mechanism known as the *thermoelastic effect*. The effect is absent in shear waves because no changes in volume are involved in shear motions. A simple correlation exists between thermoelastic attenuation and thermal conductivity. Semiconductors and metals have higher losses than oxide insulators because of their high thermal conductivity (Table 24.3).

Table 24.3 Attenuation coefficients for cubic single crystals at 1 GHz. Values at other frequencies may be estimated according to the $\alpha \sim f^2$ dependence

Crystal	Propagation direction	Polarization direction	Attenuation α [dB/m]
MgO	[100]	[100]	330
	[100]	[010]	40
SrTiO$_3$	[100]	[100]	600
Y$_3$Fe$_5$O$_{12}$	[100]	[100]	200
	[100]	[010]	34
Y$_3$Al$_5$O$_{12}$	[100]	[100]	20–32
	[100]	[010]	110
Ge	[100]	[100]	2300
	[100]	[010]	1000
Si	[100]	[100]	1000
	[111]	[111]	650
Al	[110]	[110]	7500
Cu	[100]	[100]	27,000
Au	[110]	[110]	20,000

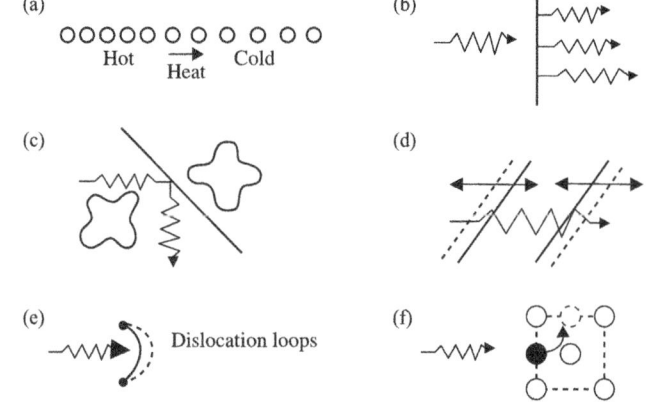

Fig. 24.1 Six of the dissipation mechanisms leading to acoustic attenuation are pictured here. (a) Heat conduction from regions of compression to regions of rarefaction. (b) Interaction between acoustic wave and thermal phonons in the Akhieser Effect. (c) Scattering at grain boundaries in polycrystalline materials. (d) Quivering domain walls in ferroic solids. (e) Movements in dislocation loops. (f) Point defect losses.

A second type of phonon damping involves interactions between the acoustic beam and the thermal phonons associated with lattice vibrations. Attenuation arises because the frequency spectrum of thermal phonons is modified by the strain field of the ultrasonic wave through lattice anharmonicity. This damping mechanism is known as the *Akhieser Effect* (Fig. 24.1(b)) The materials with the lowest losses have strong bonds with low atomic numbers, high Debye temperatures and complex structures. In addition to the garnets, minerals like topaz ($Al_2SiO_4F_2$), beryl ($Be_3Al_2Si_6O_{18}$), and tourmaline ($NaMg_3Al_6B_3Si_6O_{27}(OH)_4$) satisfy these criteria, and all are found to exhibit very low losses.

Four other causes of loss are also illustrated in Fig. 24.1. Damping is high in polycrystalline solids because of multiple scattering at the grain boundaries. Changes in grain orientation at the boundary lead to elastic and acoustic impedance mismatch, as pointed out in Section 24.1. Other types of losses are associated with porosity, domain walls, dislocations, and point defects. All are capable of interacting with acoustic waves, but some of these mechanisms are not viscous in nature and will not lead to an ω^2 dependence.

24.4 Surface acoustic waves

Technology based on elastic surface waves has led to the development of compact and inexpensive signal-processing components. Surface waves are used in delay lines, filters, and more sophisticated devices. The size and weight savings can be as much as 10^5, corresponding to the ratio of the velocity of light to the speed of sound. For instance, an acoustic path of 1 cm delays the signal several microseconds—equivalent to a kilometer of coaxial cable or waveguide. Surface waves are always accessible for signal processing, a clear advantage over bulk waves. In the 1 GHz range, surface wave wavelengths are a few microns, on the same scale as silicon microcircuits. The waves can be focused, channeled, sensed, mixed, and are compatible with integrated circuit technology.

Rayleigh waves traveling on the surface of a crystalline solid cause a point on the surface to undergo an elliptical motion with both vertical and horizontal amplitude, like a ripple on a pond. Most of the energy is confined to a surface layer one wavelength thick. Increasing the frequency increases the power density giving larger signals (Fig. 24.2).

In a Rayleigh surface-wave device, an interdigital transducer converts an electromagnetic signal to an elastic surface wave. Displacement amplitudes for Rayleigh waves decrease exponentially with depth and are confined to within one acoustic wavelength of the surface. At 10^3 MHz the acoustic wavelength is about 4μ. An interdigital transducer consists of a thin-film metal grating with half-wavelength spacing deposited on a piezoelectric substrate. Electrode patterns are produced with the same photolithographic processes used for planar integrated circuits. A voltage pulse applied to the grating produces a localized mechanical strain that propagates along the surface.

For a Rayleigh wave device with interdigital electrode spacing d, the operating frequency is v_{saw}/d. As with bulk wave devices, the velocity of the surface acoustic wave (v_{saw}) is determined by the elastic constants and the density. In general, however, the velocities are slightly lower because of the proximity of the air boundary.

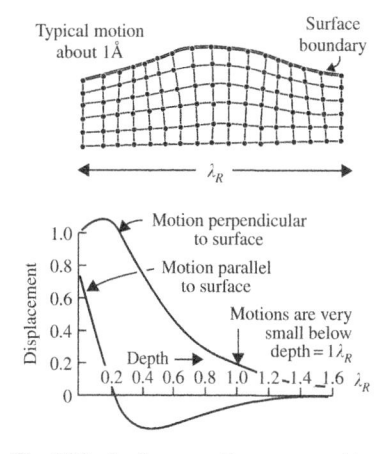

Fig. 24.2 Surface acoustic waves combine shear and longitudinal motions with typical amplitudes of about 1 Å (=10^{-10} m).

Fig. 24.3 Rayleigh wave speeds are comparable to those of bulk shear waves. Speeds are fastest in materials like beryllium with low density and high stiffness.

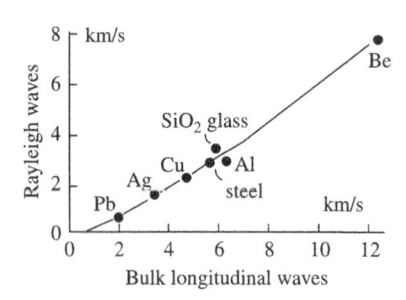

Table 24.4 Rayleigh wave velocities and electromechanical coupling factors for surface acoustic wave substrates

	v_{saw}	k^2
LiNbO$_3$	3.4 km/s	0.05
SiO$_2$ (quartz)	3.1	0.001
Bi$_{12}$GeO$_{20}$	1.7	0.01

For cubic crystals, the speeds are controlled by stiffness coefficients c_{11}, c_{12}, and c_{44}. Consider a plate cut parallel to the (100) face with surface acoustic waves traveling parallel to [010]. A bulk longitudinal wave traveling in this direction would have a speed $v = (c_{11}/\rho)^{1/2}$. For the surface wave the speed is reduced by a factor R,

$$v_{saw}^2 = \frac{Rc_{11}}{\rho},$$

where R is determined by the transcendental relation

$$\left(\frac{1-c_{11}R}{c_{44}}\right)\left(\frac{1-R-c_{12}^2}{c_{11}}\right)^2 = R^2(1-R).$$

The net result is that Rayleigh waves travel at a speed slightly lower than bulk shear waves (compare the Rayleigh wave speeds in Fig. 24.3 with the shear wave speeds in Fig. 23.1). Both are about half the speed of bulk longitudinal waves.

Surface acoustic wave devices require piezoelectric substrates to couple the electric signal into Rayleigh wave motion. Because of the availability of good quality crystals much of the work has been done on quartz, but materials with higher piezoelectric coupling have lower insertion loss and wider bandwidth. Lithium niobate, lithium tantalate, and bismuth germanium oxide have larger piezoelectric constants than quartz, but zero-temperature coefficient cuts are required for signal processing. A positive temperature coefficient for at least one elastic stiffness constant is needed for temperature compensation. Only quartz and tellurium dioxide have thus far yielded compensated cuts for surface wave applications and neither is strongly piezoelectric. Rayleigh wave speeds and electromechanical coupling factors are listed in Table 24.4.

24.5 Elastic waves in piezoelectric media

Electromechanical coupling occurs through the third rank piezoelectric coefficients d_{ijk}. The stresses and strains generated by acoustic waves produce electric polarization through the direct piezoelectric effect. In the reverse direction, electric fields produce stresses and strains through the converse piezoelectric effect. As a result the acoustic waves traveling in piezoelectric crystals require a modified Christoffel Equation.

Wave propagation in piezoelectric media involves a coupled solution of the mechanical equations of motion and Maxwell's equations for electromagnetic waves. In general, the coupling leads to five waves: three acoustic waves and two electromagnetic waves. In this discussion we are mainly concerned with the acoustic waves consisting of one longitudinal and two transverse waves.

The equation of state in a piezoelectric medium can be divided into terms involving the applied electric field (E) and the mechanical strain (x) with stress (X) and electric displacement (D) as dependent variables. We begin by writing the terms in matrix form.

The electric displacement (D) originates from two terms. First an experiment in which the crystal is mechanically clamped ($x = 0$) and an electric field is applied. Normally this measurement is carried out at high frequencies. In this case $(D) = (\varepsilon^x)(E)$. The second term comes from an experiment in which there is no applied electric field ($E = 0$). A mechanical stress (X) then produces an electric displacement $(D) = (P) = (d)(X)$.

In matrix form the sum of these two terms is

$$\begin{matrix} 3 \times 1 & & 3 \times 3 & 3 \times 1 & & 3 \times 6 & 6 \times 1 \\ (D) & = & (\varepsilon^x) & (E) & + & (d) & (X). \end{matrix}$$

The stress (X) can be converted to strain (x) through the stiffness coefficients (c^E) measured under constant field conditions. This is done by measuring the elastic constants on a fully electroded sample which short-circuits any piezoelectric effect. The result is

$$(D) = (\varepsilon^x)(E) + (d)(c^E)(x) = \begin{matrix} 3 \times 3 & 3 \times 1 & & 3 \times 6 & 6 \times 1 \\ (\varepsilon^x) & (E) & + & (e) & (x). \end{matrix}$$

The (e) coefficients are the so-called piezoelectric stress coefficients. The matrices for (e) coefficients are almost identical in form to those of the (d) coefficients tabulated in Section 12.3. There are small changes for point groups 3, 32, 3m, $\bar{6}$, and $\bar{6}m2$ where factors of 2 appear. The factors of 2 are dropped for the (e) matrices. In point group 32, for instance, $e_{26} = -e_{11}$ whereas $d_{26} = -2d_{11}$. All factors of 2 are removed from the other (e) matrices as well.

Next visualize two experiments in which strain is measured first in the absence of electric field, and then in the absence of applied mechanical stress. The first experiment is carried out on a fully electroded crystal and gives $(x) = (s^E)(X)$. The second experiment is done on a partially electroded, unclamped crystal resulting in a piezoelectric strain $(x) = (d^t)(E)$. The total strain is

$$\begin{matrix} 6 \times 1 & & 6 \times 6 & 6 \times 1 & & 6 \times 3 & 3 \times 1 \\ (x) & = & (s^E) & (X) & + & (d^t) & (E). \end{matrix}$$

Strain is converted to stress by multiplying through by (c^E):

$$(c^E)(x) = (c^E)(s^E)(X) + (c^E)(d^t)(E).$$

Since $(c^E)(s^E) = 1$, and $(e^t) = (c^E)(d^t)$, we obtain the total stress as

$$\begin{matrix} 6 \times 1 & & 6 \times 6 & 6 \times 1 & & 6 \times 3 & 3 \times 1 \\ (X) & = & (c^E) & (x) & - & (e^t) & (E). \end{matrix}$$

Poled ferroelectric ceramics of PZT or barium titanate have piezoelectric stress coefficients of about 10 C/m². For normal piezoelectrics like quartz the e coefficients are two orders of magnitude smaller near 0.1 C/m².

In full tensor form the equations of the state are

$$X_{ij} = c_{ijkl}x_{kl} - e_{mij}E_m$$

and

$$D_m = e_{mkl}x_{kl} + \varepsilon_{mj}E_j.$$

Using Newton's second law, the acceleration components are given by

$$\rho\ddot{u}_i = \frac{\partial X_{ij}}{\partial Z_j}$$

$$= c_{ijkl}\frac{\partial x_{kl}}{\partial Z_j} - e_{mij}\frac{\partial E_m}{\partial Z_j}.$$

With no free charges present, the corresponding Maxwell's Equation for the electromagnetic waves is

$$\frac{\partial D_m}{\partial Z_j} = 0 = e_{jkl}\frac{\partial x_{kl}}{\partial Z_j} + \varepsilon_{mj}\frac{\partial E_m}{\partial Z_j}.$$

Remembering that $x_{kl} = \partial u_k/\partial Z_l$ and $E_i = -\partial \phi/\partial Z_i$, we can rewrite the equations in terms of the displacement u_k and the electric potential ϕ.

$$\rho\ddot{u}_i = c_{ijkl}\frac{\partial^2 u_k}{\partial Z_j \partial Z_\ell} + e_{mij}\frac{\partial^2 \phi}{\partial Z_j \partial Z_m}$$

$$0 = e_{jkl}\frac{\partial^2 u_k}{\partial Z_j \partial Z_\ell} + \varepsilon_{mj}\frac{\partial^2 \phi}{\partial Z_j \partial Z_m}.$$

For plane waves the solutions take the form

$$u_k = A_k \exp i(\omega t - k_i Z_i) \quad \text{and} \quad \phi = \Phi \exp i(\omega t - k_i Z_i).$$

The solutions for u_k lead to the three acoustic waves while those for ϕ correspond to the electromagnetic waves. Substituting these expressions into the previous equations,

$$\rho\omega^2 u_i = c_{ijkl}k_j k_l u_k + e_{mij}k_j k_m \phi$$

$$0 = e_{jkl}k_l k_j u_k - \varepsilon_{mj}k_j k_m \phi.$$

Since we are mainly interested in the acoustic waves we eliminate ϕ and solve for u. The wave vector components $k_i = (\omega/v)N_i$ where v is the velocity, ω the angular frequency and N_i the direction cosines of the wave normal. The velocities are then obtained from

$$\left[\frac{1}{\rho}c_{ijkl}N_j N_\ell + \frac{1}{\rho}\frac{(N_m e_{mij}N_j)(N_j e_{jk\ell}N_\ell)}{N_m \varepsilon_{mj}N_j} - v^2\delta_{ik}\right]u_k = 0.$$

The first term corresponds to the Christoffel Equation in Section 23.1, and the second term to the piezoelectric interaction.

To solve this equation, we follow a similar procedure to the matrix representation used to solve the generalized Christoffel Equation (Section 23.2).

24.5 Elastic waves in piezoelectric media

Table 24.5 Components of the matrix $C_i = e_{mij}N_mN_j$ used in determining the phase velocities in piezoelectric crystals. C is the abbreviated dielectric term $C = \rho\varepsilon_{mj}N_mN_j$

	N_1^2	N_2^2	N_3^2	N_2N_3	N_1N_3	N_1N_2
C_1	11	26	35	25 + 36	15 + 31	16 + 21
C_2	16	22	34	24 + 32	14 + 36	12 + 26
C_3	15	24	33	23 + 24	13 + 35	14 + 25
C	11	22	33	2 × 23	2 × 13	2 × 12

In matrix notation

$$C_{ik} \equiv \frac{1}{\rho}c_{ijkl}N_jN_\ell$$

and the components are written out in full in Table 23.1. For the piezoelectric interaction term we introduce a second matrix operator.

$$C_i \equiv N_m e_{mij} N_j.$$

From which the corrections can be obtained. The components of the C_i matrix are listed in Table 24.5. As an illustration,

$$C_1 = e_{11}N_1^2 + e_{26}N_2^2 + e_{35}N_3^2 + (e_{25} + e_{36})N_2N_3$$
$$+ (e_{15} + e_{31})N_1N_3 + (e_{16} + e_{21})N_1N_2.$$

Note that the piezoelectric stress coefficients e_{ijk} (where $i, j, k = 1$–3) have been converted to matrix coefficients e_{im} where $i = 1$–3 and $m = 1$–6.

Also included in Table 24.5 is the shortened form of the electric permittivity term

$$C \equiv \rho N_m \varepsilon_{mj} N_j.$$

The full expression for the modified Christoffel Equation is

$$\left(C_{ik} + \left(\frac{C_iC_k}{C}\right) - v^2\delta_{ik}\right)u_k = 0.$$

Since the piezoelectric correction is generally small, it constitutes a correction $\Delta C_{ik} = C_iC_k/C$, and is often written as

$$(C_{ik} + \Delta C_{ik})u_k = v^2 u_i.$$

The phase velocities v are obtained from this equation.

Problem 24.2
Write out the modified Christoffel Equation for point group $\bar{4}3m$. For this point group the piezoelectric (e) matrix is similar to the corresponding (d) matrix. Examine the solutions for waves traveling along the [100] and [110] directions.

The piezoelectric stiffening term is quite substantial for some waves but is totally absent for others. In LiNbO$_3$, for instance, the electromechanical factor for longitudinal waves traveling along $Z_3 = [001]$ is

$$\left(\frac{e_{33}^2}{c_{33}^E \varepsilon_{33}^x}\right)^{1/2} = 0.163$$

but for shear waves in same direction there is no piezoelectric coupling.

Portions of the lithium niobate slowness surfaces are shown in Fig. 24.3 where $1/v_p$ is plotted for wave normals in the Z_2–Z_3 plane. In point group $3m$ this is a mirror plane and Z_3 is the threefold symmetry axis. The slowness surfaces with and without the piezoelectric contribution are compared in Fig. 24.4(b) and (a).

The pointed out in Section 23.5, the slowness surfaces can be used to determine the energy flow directions. The ray directions are perpendicular to tangents drawn to the surface.

A comparison of three piezoelectric crystals is given in Table 24.6. Quartz is a weak piezoelectric, lithium niobate an average piezoelectric, and PZT-5 a strong piezoelectric. As expected, the changes in elastic stiffness ($\Delta c/c = e^2/c\varepsilon$) and acoustic phase velocities ($\Delta v/v = e/\sqrt{c\varepsilon}$) are much larger in PZT. Of course even in PZT there are some types of waves for which the piezoelectric stress coefficient is zero. Under these conditions the acoustic waves are unaffected by the piezoelectric effect.

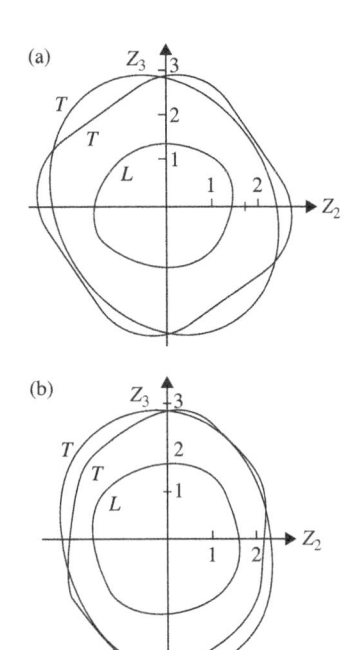

Fig. 24.4 Slowness surfaces for LiNbO$_3$ plotted (a) without the piezoelectric terms, and (b) with the piezoelectric stiffening. The $(1/v_p)$ values are in units of 10^{-4} s/m.

24.6 Nonlinear acoustics

A quiet revolution is taking place in ultrasound technology that involves departures from the normal equations of state governing the Christoffel Equation. For linear, low-loss centrosymmetric solids, Hooke's Law is normally sufficient:

$$X_{ij} = c_{ijkl} x_{kl}.$$

But as discussed earlier in this chapter, attenuation in acoustic beams takes place when losses are present. For *viscous media* the equation of state becomes

$$X_{ij} = c_{ijkl} x_{kl} + \eta_{ijkl} \frac{\partial x_{kl}}{\partial t}.$$

These losses increase rapidly with frequency, generally as ω^2, and make many acoustic materials unusable in the GHz region.

Elastic waves in *piezoelectric* media were discussed in the previous section using a modified constitutive equation:

$$X_{ij} = c_{ijkl} x_{kl} - e_{mij} E_m.$$

The electric field components E_m make sizable changes in the wave velocities whenever the piezoelectric coefficients e_{mij} are large.

Similar effects occur in crystals with large *electrostrictive effects*. For relaxor ferroelectrics like lead magnesium niobate (PMN) the electrostriction

Table 24.6 Fractional changes in the elastic stiffness and acoustic phase velocity brought about by piezoelectric coupling

	Quartz 32	LiNbO$_3$ $3m$	PZT-5 ∞m
$\hat{N}$	[100]	[001]	[001]
$\hat{U}$	L	L	L
e [C/m^2]	0.171	1.3	23.3
c [N/m^2]	8.67×10^{10}	2.45×10^{10}	11.7×10^{10}
ε [F/m]	$4.5\varepsilon_0$	$29\varepsilon_0$	$1470\varepsilon_0$
$\Delta c/c = e^2/c\varepsilon$	0.0085	0.027	0.357
$\Delta v_p/v_p = e/\sqrt{c\varepsilon}$	0.09	0.16	0.60

coefficients are large even above the Curie point where the crystal structure is centrosymmetric and the piezoelectric effect is absent. In this case the constitutive equation involves two fourth rank tensors:

$$X_{ij} = c_{ijkl}x_{kl} - e_{mnij}E_m E_n.$$

The strong coupling between E^2 and mechanical stress leads to second harmonic signals analogous to those observed in optics.

Magnetostrictive materials have a similar effect with stress proportional to H^2, or if used under DC bias, will exhibit a *piezomagnetic* term with $X \sim H(0)H(\omega)$.

Geophysicists and oceanographers are concerned with the behavior of sound waves deep within the earth's mantle, and deep in the ocean. Temperature and pressure changes control the elastic constants and speed of sound. Most materials soften with increasing temperature and stiffen with increasing pressure. Temperature and pressure both increase with depth in the earth's crust, but the pressure effect dominates, leading to an increase in sound velocity. Changes such as this come from *elastic nonlinearity* and higher order elastic constants.

$$X_{ij} = c_{ijkl}x_{kl} + c_{ijklmn}x_{kl}x_{mn}.$$

Polymers, seawater, and the human body are noticeably nonlinear elastically. During the past few years, a number of complex nonlinear phenomena have been investigated for use in harmonic imaging, parametric arrays, elastography, sonoelasticity, and time-reversed acoustics. In these applications the transducer may be called upon to transmit at one frequency and receive at another, or to simultaneously transmit at two frequencies and receive at a sum or difference frequency, or to interrogate at one frequency and deliver focused high power at another, or to interrogate at high frequency while continuously repositioning the transducer at a much lower frequency, or to transmit in narrow beam patterns while receiving in an omnidirectional mode. As an example, a *parametric acoustic array* emits two high intensity waves at frequencies ω_1 and ω_2. The modulated beam is projected into an acoustically nonlinear medium that then produces sum $(\omega_1 + \omega_2)$ and difference $(\omega_1 - \omega_2)$ frequencies. In viscous, lossy media such as seawater or the human body, the higher frequencies ω_1, ω_2, and $\omega_1 + \omega_2$ are absorbed while the low frequency acoustic wave $(\omega_1 - \omega_2)$ is transmitted. The net result is a narrow beam at low frequency with a wide bandwidth suitable for signal processing.

Acoustoelectricity is another interesting nonlinear effect which results from interactions between acoustic waves and alternating electric currents $J_i(\omega)$. The interaction term takes the form of a third rank relationship between stress and current density:

$$X_{ij} = c_{ijkl}x_{kl} + \rho_{mij}J_m.$$

Since third rank tensors disappear in centrosymmetric media, and metals are acoustically lossy, the prime candidates for acoustoelectric compounds are acentric semiconductors. ZnO, CdS, GaAs, and other II–VI and III–V compounds that crystallize in the wurtzite or sphalerite structures, which are acentric. Piezoelectric semiconductor devices made from these materials amplify or absorb ultrasonic waves, employing the principle shown in Fig. 24.5. The acoustic wave generated by the transducer produces local mechanical strain with regions of compression and extension, which in turn produce electric fields through the direct piezoelectric effect. The local electric field has the

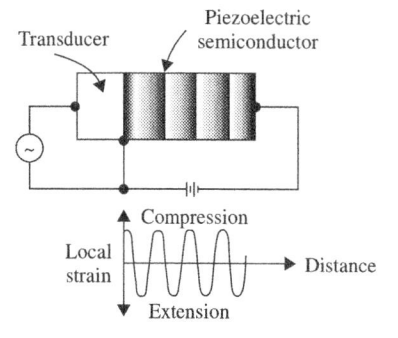

Fig. 24.5 The acoustoelectric experiment in which longitudinal acoustic waves interact with conduction electrons through the converse piezoelectric effect.

periodicity of the acoustic wave and moves through the semiconductor where it interacts with conduction electrons, causing bunching of the charge carriers. As the acoustic wave moves, the electrons are dragged along and meet resistance, dissipating energy as heat. To maintain their velocity, the electrons extract energy from the local electric field which attenuates the acoustic wave.

Amplification rather than attenuation may result if an electric field is simultaneously applied to the semiconductor (Fig. 24.5). The electric field supplies energy to the conduction electrons, and if the field is adjusted to give electron velocities somewhat greater than the acoustic velocity, energy is transferred to the acoustic wave.

Acoustic activity is another odd-rank nonlinear acoustic effect. This is the acoustic analog to optical activity in which the plane of polarization is rotated as the acoustic beam traverses through the crystal. The constitutive equation takes the form

$$X_{ij} = c_{ijkl} x_{kl} + b_{ijklm} \frac{\partial x_{kl}}{\partial Z_m}.$$

Being a fifth rank tensor, acoustic activity disappears in centric media.

The effect has been observed in α- and β-quartz at frequencies near 30 GHz. Quartz belongs to acentric point group 32 which is both optically active and acoustically active. Experiments were carried out along Z_3, the acoustic axis. Transverse waves traveling in this direction are normally degenerate (see Fig. 23.4) but at high frequencies the degeneracy is lifted causing right- and left-handed circularly-polarized waves to travel at different velocities (Fig. 24.6). The fast and slow waves are reversed when the experiments are done on right- and left-handed crystals.

The underlying structure–property relation can be understood by examining the crystal structures of right- and left-handed quartz (Fig. 16.17). The structures contain Si–O–Si–O helices spiraling up the trigonal Z_3-axis. Circularly-polarized shear waves that stay in phase with the spiral will travel at a slower velocity than those which do not. The interaction becomes more important at high frequencies where the acoustic wave length is closer to the unit cell dimension ($c = 5.48$ Å).

Acoustic activity is analogous to optical activity (Chapter 30). Both effects are caused by spatial dispersion of the constitutive parameters due to local fields or forces on the atomic scale. This leads to frequency-dependent behavior of the dielectric permittivity and the elastic compliance, which in turn lead to optical activity and acoustic activity. In quartz, acoustic activity is larger than optical activity. For transverse acoustic waves in the GHz region, the plane of polarization rotates about 10^7 degrees/m. This is about 100 times larger than

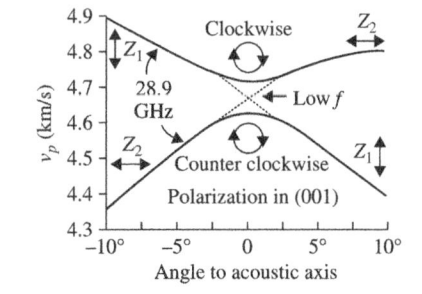

Fig. 24.6 A small portion of the phase velocity surface of quartz showing that right- and left-circularly polarized transverse waves along Z_3 travel at different velocity. The effect is largest at high frequencies and provides a clear demonstration of acoustic activity.

24.6 Nonlinear acoustics

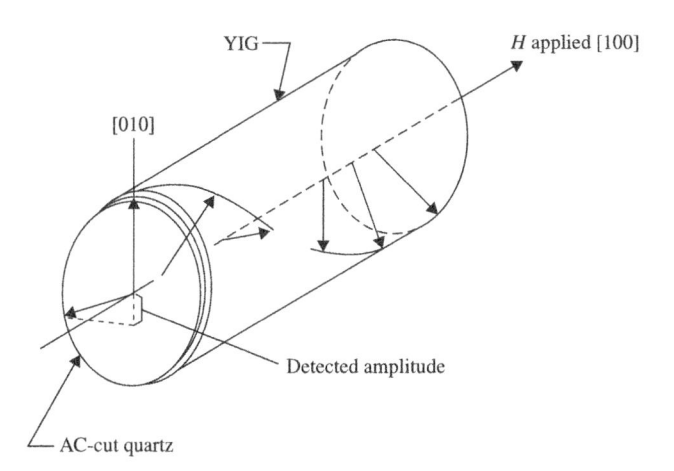

Fig. 24.7 The Acoustic Faraday Effect involves a rotation of a plane-polarized acoustic wave in a magnetoelastic crystal. Under a pulsed field, an AC-cut transducer generates a transversely polarized wave, and later detects the reflected return wave. No rotation occurs unless a magnetic field is applied to align the domains within the ferromagnetic yttrium iron garnet crystal. Losses are low for acoustic waves traveling along [100] and polarized parallel [010]. (See Table 24.3.)

optical activity in the visible region. In both cases, the activity grows larger as c/λ gets bigger in keeping with its underlying origin of spatial dispersion.

There is yet another nonlinear effect in acoustics that also comes from dispersion. In this case, it is *temporal* dispersion rather than spatial dispersion. This is caused by interactions between the wave (optical or acoustic) with magnetic spin waves in the presence of a DC magnetic field. As the acoustic or optic wave propagates parallel to the field, the plane of polarization rotates about the field direction. The optical effect was discovered by Michael Faraday in 1845, and is known as the *Faraday Effect*. The corresponding *Acoustic Faraday Effect*, sometimes called the *magnetoelastic effect*, has been investigated more recently. The rotation angle is proportional to the thickness of the sample and to the magnitude of the magnetic field parallel to the wave normal. At first glance, the effect appears similar to optical or acoustic activity, but the effects are different both in origin and in experiment. In the optical Faraday Effect and the acoustic Faraday Effect, a wave traveling in the $+Z_3$ direction, parallel to the applied magnetic field, experiences a clockwise rotation of the plane of polarization. For the $-Z_3$ direction, antiparallel to the field, the rotation is counterclockwise. Thus the rotation is doubled when the wave is reflected. For acoustic activity or optical activity this is not true. In this case the rotation always proceeds in the same sense relative to the wave normal. Therefore the reflected wave cancels the original rotation when the wave normal is reversed.

Yttrium iron garnet (YIG = $Y_3Fe_5O_{12}$) is an ideal candidate for magnetoelastic experiments because of its low ultrasonic losses. At 10 MHz, the mechanical Q of YIG is about 10^7 which is comparable to the very best quartz crystals. A typical experiment is illustrated in Fig. 24.7. An AC-cut quartz transducer is bonded to one face of a cylindrical YIG crystal. A DC magnetic field is directed along the length of the crystal parallel to the acoustic beam. As described previously in Section 13.10, AC-cut quartz generates a pure transverse vibration that is often used in the ultrasonic measurement of elastic constants. As the transversely-polarized wave proceeds through the YIG crystal parallel to the magnetic field, the plane of polarization rotates through an angle θ of the Acoustic Faraday Effect. The wave travels down the rod and is reflected at the opposite end. The reflection travels back and excites the quartz transducer which now acts as a receiver. The polarization of the returning shear wave is rotated through an angle 2θ from its initial orientation.

25 Crystal optics

25.1 Electromagnetic waves 274
25.2 Optical indicatrix and refractive index measurements 276
25.3 Wave normals and ray directions 278
25.4 Structure–property relationships 280
25.5 Birefringence and crystal structure 282

Calcite ($CaCO_3$) is a beautiful transparent mineral that readily cleaves into rhombohedra. Images viewed through a thin slice of a cleaved calcite crystal are split into two images, an effect known as double refraction, or birefringence. Birefringence is the most obvious manifestation of optical anisotropy in crystals. For any given wave normal, there are two light waves, transversely polarized in mutually perpendicular directions, traveling with different velocities, and consequently different refractive indices.

Double refraction is caused by dielectric anisotropy. For transparent crystals like calcite, the magnetic susceptibility is small and the permeability $\mu \cong \mu_0$, the permeability of free space. In this class of materials the refractive index $n = c/v = \sqrt{K}$ where c is the speed of light in vacuum, v the velocity of light in the crystal, and K is the dielectric constant measured at optical frequencies. Refractive indices of transparent materials lie between 1 and 3.

Electromagnetic waves differ from acoustic waves in that there are, for a given wave normal, two waves rather than three. In the acoustic case there are, in general, two quasitransverse waves and a quasilongitudinal wave.

25.1 Electromagnetic waves

Starting with Maxwell's Equations and the material constitutive relations, the propagation of electromagnetic waves through transparent crystals are described in terms of the refractive indices, wave normals, and polarization directions.

Written in tensor notation Maxwell's Equations for a nonmagnetic, transparent insulator take the form

$$\varepsilon_{ijk} \frac{\partial H_k}{\partial Z_j} = \frac{\partial D_i}{\partial t}$$

$$\varepsilon_{ijk} \frac{\partial E_k}{\partial Z_j} = -\mu_0 \frac{\partial H_i}{\partial t}$$

$$\frac{\partial D_i}{\partial Z_i} = 0$$

$$\mu_0 \frac{\partial H_i}{\partial Z_i} = 0.$$

The accompanying constituent equations for such a medium are $D_i = \varepsilon_{ij} E_j$ and $B_i = \mu_0 H_i$. In these expressions, E_k and H_k are the components of the electric field and magnetic field associated with the electromagnetic wave; D_i and B_i are the resulting electric displacement and magnetic induction vectors in the transparent insulator, t is time, Z_i coordinate, and ε_{ijk} the rotation tensor. $\varepsilon_{ijk} = 0$ unless i, j, k are all different, $\varepsilon_{ijk} = +1$ if i, j, k are in cyclic order, and -1 if they are in anticyclic order.

To describe the behavior of light waves in anisotropic media, we examine monochromatic plane waves with angular frequency ω and wave vector k. The electric and magnetic fields take the form

$$E_k = E_{0k} \exp[i(\omega t - k_i Z_i)]$$

and

$$H_k = H_{0k} \exp[i(\omega t - k_i Z_i)],$$

where E_{0k} and H_{0k} are the field amplitudes.

Substituting the expressions for E_k and H_k into the first two Maxwell Equations and replacing the wave vector $k_i = N_i \omega/v$, we find that

$$\varepsilon_{ijk} N_j E_k = \mu_0 v H_i$$

and

$$\varepsilon_{ijk} N_j H_k = -v D_i.$$

From the first of these two equations it can be seen that the magnetic field vector H is perpendicular to the wave normal N and also to the electric field E. The second equation shows that the electric displacement vector D is perpendicular to the wave normal N and the magnetic field H. Note also that the electric field vector E is, in general, *not* perpendicular to N but is coplanar with D and N (Fig. 25.1). The magnetic induction vector B is parallel to H for the transparent insulators under consideration.

To introduce refractive indices into the wave equation, we begin by solving for the electric displacement. The result is

$$D_i = \frac{1}{\mu_0 v^2}(E_i - N_i(E_j N_j)) = \varepsilon_{ij} E_j$$

$$= \frac{N_i(E_j N_j)}{(1/\varepsilon_{ii}) - \mu_0 v^2}.$$

Remembering that D and N are perpendicular to one another,

$$D_i N_i = 0 = \frac{N_i^2 (E_j N_j)}{(1/\varepsilon_{ii}) - \mu_0 v^2}.$$

Dividing through by $(E_j N_j)/\mu_0$ gives the wave velocity surfaces:

$$\frac{N_i^2}{(1/\mu_0 \varepsilon_{ii}) - v^2} = \frac{N_1^2}{v_1^2 - v^2} + \frac{N_2^2}{v_2^2 - v^2} + \frac{N_3^2}{v_3^2 - v^2} = 0,$$

where $v_1^2 = 1/\mu_0 \varepsilon_{11}$, $v_2^2 = 1/\mu_0 \varepsilon_{22}$, and $v_3^2 = 1/\mu_0 \varepsilon_{33}$. In terms of the principal refractive indices $v_1 = c/n_1$, $v_2 = c/n_2$, and $v_3 = c/n_3$, where c is the speed of light in vacuum equal to $(\mu_0 \varepsilon_0)^{-1/2}$.

To plot the phase velocity surfaces as a function of the wave normal, we rewrite the equation as

$$N_1^2(v_2^2 - v^2)(v_3^2 - v^2) + N_2^2(v_1^2 - v^2)(v_3^2 - v^2) + N_3^2(v_1^2 - v^2)(v_2^2 - v^2) = 0.$$

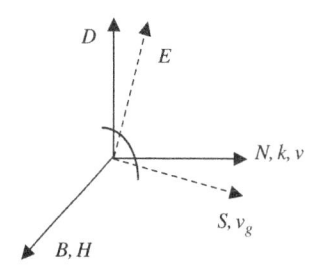

Fig. 25.1 The relative orientation of the electric, magnetic, and wave normal directions for electromagnetic waves in transparent media. S is the Poynting vector perpendicular to E and H which defines the energy flow direction. v_g is the group velocity.

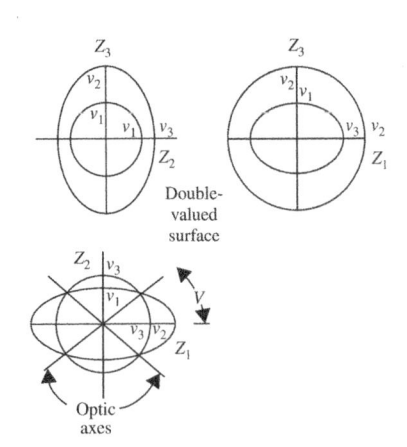

Fig. 25.2 Phase velocity (v) plotted as a function of wave normal (N), assuming $n_1 > n_3 > n_2$. It is a double-valued surface in every direction except for the two optic axes where the birefringence is zero. The angle V is determined from the expression $\tan^2 V = (v_2^2 - v_3^2)/(v_3^2 - v_1^2)$.

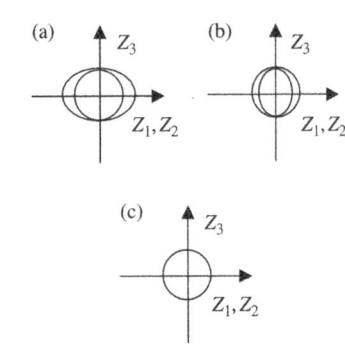

Fig. 25.3 Phase velocity surfaces for uniaxial and isotropic media. (a) For a uniaxially positive crystal with $n_3 > n_1 = n_2$, the surfaces consist of a sphere inside an oblate ellipsoid of revolution. The two surfaces intersect tangentially at the optic axis Z_3. (b) For a uniaxial negative crystal with $n_1 = n_2 > n_3$, the reverse is true with a prolate ellipsoid of revolution inside a sphere. (c) The wave surface of cubic crystals and other isotropic media is a single sphere.

For waves traveling along Z_1, the wave normal is specified by the direction cosines $N_1 = 1$, $N_2 = N_3 = 0$, and wave velocity equation becomes $(v_2^2 - v_3^2)(v_3^2 - v^2) = 0$, for which the roots are $v = \pm v_2$ and $v = \pm v_3$.

The resulting wave velocity surface is plotted in Fig. 25.2. For all directions except the two optic axes there are two values of v corresponding to two waves with different polarization directions. In general there will be a fast wave and a slow wave giving rise to double refraction. The difference in refractive indices Δn is called the *birefringence*.

The wave velocity surface shown in Fig. 25.2 is quite general since no assumptions were made regarding the symmetry of the crystal. The only assumption was that Z_1, Z_2, and Z_3 are the principal axes of the optical permittivity tensor. The intrinsic symmetry of the optical wave surface is *mmm* with mirror planes perpendicular to the three principal axes. Neumann's Law states that the symmetry of a physical property (as represented by the wave velocity surface) must include the symmetry of the point group of the material. In this case, it is true for triclinic, monoclinic, and orthorhombic crystals since the wave surface symmetry *mmm* includes the symmetry of point groups 1, $\bar{1}$, 2, m, $2/m$, 222, $mm2$, and mmm. Crystals belonging to these point groups will be *optically biaxial*.

For higher symmetry point groups, it is necessary to modify the shape of the wave surface. Trigonal, tetragonal, and hexagonal crystals have a rotational symmetry axis (3, 4, or 6) along Z_3 which is not included in point group *mmm*. This places a restriction on the wave velocities, and makes $v_1 = v_2$ to give the surfaces circular symmetry when viewed along Z_3. Neumann's law is satisfied by making Z_3 the optic axis. Trigonal, tetragonal, and hexagonal crystals are *optically uniaxial* rather than biaxial. This is also true for nematic liquid crystals and other textured materials containing a single ∞-fold axis. The wave surfaces for uniaxial crystals are illustrated in Fig. 25.3.

Further constraints are required for cubic crystals and for the two Curie Groups with more than one ∞-fold axis. All cubic crystals possess threefold rotation axes along the $\langle 111 \rangle$ body diagonal directions. For the wave velocity surface in Fig. 25.2 this is only possible if $v_1 = v_2 = v_3$, converting the doubly-valued surface into a single sphere (Fig. 25.3). Two ∞-fold axes in different directions lead to the same result. When the wave surface is a sphere, all directions are optic axes, and the material is *optically isotropic*. Birefringence disappears. The symmetry discussion is summarized in Table 25.1.

25.2 Optical indicatrix and refractive index measurements

In addition to the wave velocity surface, there is another way of plotting the optical properties that is simpler and more informative. The *optical indicatrix* is a single-valued surface which specifies the relationships between refractive indices, wave normals, polarization directions, and energy flow directions (= ray directions).

The optical indicatrix or index ellipsoid is generated by the equation

$$\frac{Z_1^2}{n_1^2} + \frac{Z_2^2}{n_2^2} + \frac{Z_3^2}{n_3^2} = 1,$$

25.2 Optical indicatrix and refractive index measurements

Table 25.1 Effect of symmetry on the optical properties of crystals and textured media

Properties	Symmetry groups
Biaxial $n_1 \neq n_2 \neq n_3$	1, $\bar{1}$, 2, m, $2/m$, 222, $mm2$, mmm
Uniaxial $n_1 = n_2 \neq n_3$	3, $\bar{3}$, 32, $3m$, $\bar{3}m$, 4, $\bar{4}$, $4/m$, 422, $\bar{4}2m$, $4mm$, $4/mmm$, 6, $\bar{6}$, $6/m$, 622, $6mm$, $\bar{6}m2$, $6/mmm$, ∞, $\infty 2$, ∞/m, ∞m, ∞/mm
Isotropic $n_1 = n_2 = n_3$	23, $m3$, $\bar{4}3m$, 432, $m3m$, $\infty\infty$, $\infty\infty m$

where Z_1, Z_2, and Z_3 are the principal axes, and n_1, n_2, and n_3 are the principal refractive indices. Fig. 25.4 shows the optical indicatrix of a biaxial crystal. It is a general ellipsoid with $n_1 \neq n_2 \neq n_3$ representative of the optical properties of triclinic, monoclinic, and orthorhombic crystals. For biaxial crystals, three or more measurements are required to specify the optical indicatrix. In orthorhombic crystals the principal axes coincide with the crystallographic axes [100], [010], and [001] so only three measurements are required.

In monoclinic crystals, one principal axis is parallel to $Z_2 = [010]$, the twofold symmetry axis. Z_1 and Z_3, the other two principal axes, lie in the (010) plane so that four measurements are required to specify the three principal refractive indices and the angle between [100] and Z_1.

Six measurements are needed for triclinic crystals: three for the principal refractive indices and three to specify the orientation of the principal axes. The general procedure for determining these values follows the same procedure described earlier (Section 9.4) for dielectric constants.

As indicated in Table 25.1, only two measurements are needed for trigonal, tetragonal, and hexagonal crystals. The two principal indices of refraction are for light waves polarized parallel and perpendicular to the optic axis Z_3. These are the so-called extraordinary (n_e) and ordinary (n_o) refractive indices shown in the uniaxial indicatrix in Fig. 25.5. The birefringence is $\Delta n = n_e - n_o$. The equation describing the uniaxial indicatrix is $(Z_1^2 + Z_2^2)/n_o^2 + Z_3^2/n_e^2 = 1$.

Cubic crystals and glasses are optically isotropic. The indicatrix is therefore a sphere of radius n and only one refractive index measurement is required. It should be remembered, however, that the refractive index depends on wavelength because of dispersion. This will be discussed in the next chapter.

It is relatively easy to measure refractive indices. Measurements can be carried out in several ways, but for simplicity it is usually done with index oils and a polarizing microscope (Fig. 25.6). The essential components include a light source, a polarizer and analyzer, and several lenses to provide magnification and to view optical interference figures in convergent light.

To determine the refractive index of a cubic crystal, a small crystal is placed on a glass slide and immersed in an index oil of known refractive index. The slide is positioned on the sample stage and examined in monochromatic light. For cubic crystals it is not necessary to polarize the light. If the refractive indices of the crystal and oil are identical, the crystal boundary disappears. When the two indices are not matched, there will be a distinct boundary line. Using commercially-available index liquids, the test is repeated with other oils until a match is obtained. For uniaxial crystals (Fig. 25.5) the values of n_e and n_o are measured separately. A small crystal is positioned on the stage with the light beam oriented perpendicular to the optic axis. By using the polarizer and rotating the optical stage, the crystal is examined in light polarized perpendicular to the

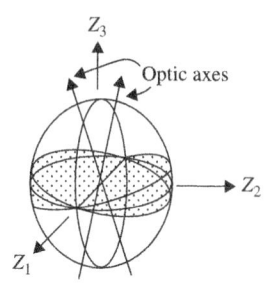

Fig. 25.4 Optical indicatrix for biaxial crystals in which the refractive index is plotted as a function of polarization (electric displacement) direction. The indicatrix is a general ellipsoid with intercepts of n_1, n_2, and n_3 along the principal axes Z_1, Z_2, and Z_3, respectively. The optic axes are oriented perpendicular to the two circular cross-sections of the ellipsoid.

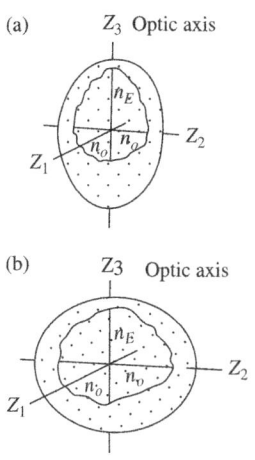

Fig. 25.5 Optical indicatrix for uniaxial crystals. (a) The uniaxial positive indicatrix is a prolate ellipsoid of revolution with $n_e > n_o$. (b) For uniaxial negative crystals the indicatrix is an oblate ellipsoid of revolution with $n_e < n_o$.

278   Crystal optics

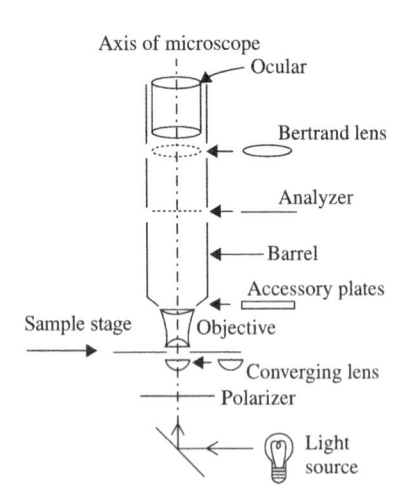

Fig. 25.6 The polarizing microscope is used to determine the refractive indices of crystals.

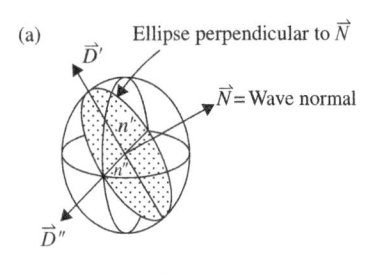

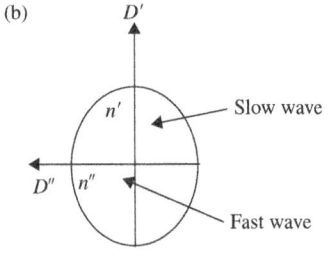

Fig. 25.7 (a) Biaxial indicatrix showing are arbitrary wave normal $\vec{N}$, and (b) the elliptical section perpendicular to $\vec{N}$. The vibration directions ($\vec{D}'$ and $\vec{D}''$) of the two waves traveling along $\vec{N}$, are oriented along the major and minor axis of the ellipse. Their phase velocities are c/n' and c/n'', respectively.

optic axis. The refractive index of the ordinary wave (n_o) is determined in this position by comparing it with various index liquids. To obtain n_e, the crystal is rotated to the position in which the optic axis is parallel to the polarization direction. A match with the index liquids is carried out again.

The procedure for biaxial crystals (Fig. 25.4) is more complicated and usually involves the use of interference figures obtained in convergent light. First the principal axes (Z_1, Z_2, and Z_3) are located from the interference figures and then the refractive indices (n_1, n_2, and n_3) are measured in sections cut perpendicular to the principal axes.

25.3 Wave normals and ray directions

Next we consider optical waves traveling in an arbitrary direction specified by a wave normal $\vec{N}$. Its relation to the generalized indicatrix of a biaxial crystal is illustrated in Fig. 25.7. In this case the indicatrix is a triaxial ellipsoid with intercepts n_1, n_2, and n_3 along the principal axes (Fig. 25.4). In all directions except the two optic axes there will be a fast wave and a slow wave (Fig. 25.2). As proven earlier in Section 25.1, the polarization vectors (D' and D'') are perpendicular to the wave normal N. The elliptical section of the indicatrix perpendicular to the wave normal has a major and a minor axis (Fig. 25.7(b)) corresponding to D' and D''. The refractive indices obtained from elliptical section determine the phase velocities. The speeds of the fast and slow waves are c/n'' and c/n', respectively. For the wave normal N, the birefringence is $\Delta n = n' - n''$.

In addition to birefringence, there is another important feature of optical anisotropy having to do with the flow of energy. As with acoustic waves (see Figs. 23.5 and 23.6) the direction of energy flow does not always coincide with the wave normal. For acoustic waves the two vectors may diverge by angles of 30° or more, presenting considerable difficulties in the study of elastic waves in anisotropic crystals. Fortunately the divergence angles (Δ) are usually smaller in optic experiments, even in highly anisotropic crystals like calcite.

The direction of energy flow, the ray direction, is determined by the *Poynting Vector* $\vec{S} = \vec{E} \times \vec{H}$ as illustrated in Fig. 25.1. Since the vectors $\vec{D}$ and $\vec{H}$ are perpendicular to the wave normal $\vec{N}$, then the divergence angle Δ between $\vec{S}$ and $\vec{N}$ will be equal to the angle between $\vec{E}$ and $\vec{D}$. In other words the divergence is directly controlled by the dielectric anisotropy. For a transparent insulator the optical dielectric constant is n^2. The anisotropy in n^2 is seldom larger than 25%, and is usually more like 1%, whereas the anisotropy in elastic stiffnesses exceeds 200% in layer structures such as mica and graphite, and certain hexagonal close-packed metals like zinc and cadmium. The divergence angle in optical waves is usually less than 5°.

The Nicol Prism polarizer used in polarizing microscopes provides a good example of optical divergence. Nicol Prisms are made from calcite crystals which have an exceptionally large birefringence: $n_e = 1.486$, $n_o = 1.658$. The prism is made from two calcite crystals glued together with a transparent cement. As shown in Fig. 25.8(a), the ordinary and extraordinary beams diverge on entering the crystal and the ordinary wave is totally reflected at the interface. The transmitted extraordinary wave provides the desired polarized light beam.

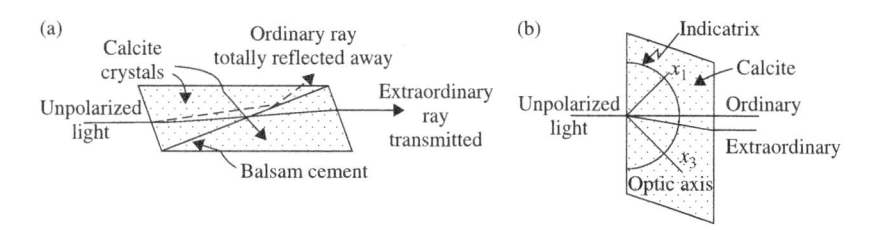

Fig. 25.8 (a) The Nicol Prism polarizer and (b) the divergence of ordinary and extraordinary beams in calcite.

Because of its large birefringence, calcite crystals can provide divergence angles as large as $6°16'$ (Fig. 25.8(b)). The ray direction for the ordinary wave is always parallel to the wave normal but the extraordinary ray is not. Quartz has a much smaller birefringence than calcite. The maximum divergence in quartz is only $20'$. For uniaxial crystals like calcite and quartz, the maximum angular separation is given by $\tan^2 \Delta = (n_e^2 - n_o^2)/2n_o n_e$.

The divergence of ordinary and extraordinary light waves in uniaxial crystals is easily visualized with the optical indicatrix. Since $n \sim 1/v$, the indicatrix is a slowness surface similar to those discussed for acoustic waves in Section 23.5. Energy flow directions are determined by the tangents to the slowness surface. The ray directions for an optically positive uniaxial crystal are shown in Fig. 25.9. The wave normal defines the wave front and the two polarization directions. $\vec{D}_e$, the polarization vector for the extraordinary wave lies in the plane of the drawing and is perpendicular to $\vec{N}$. For a uniaxial positive crystal this is the slow wave. $\vec{D}_0$, the polarization vector for the ordinary wave, is perpendicular to the plane of the drawing, and also to the wave normal $\vec{N}$. In this case, the ordinary wave is the fast wave.

The ray direction for the extraordinary ray is obtained by drawing a tangent to the indicatrix, as shown in Fig. 25.9. The ray forms an angle Δ with the wave normal, and the group velocity $v_g = v_p/\cos \Delta$. The Poynting vector $\vec{S}_e$ of the extraordinary wave is parallel to the ray direction and its electric field vector $\vec{E}_e$ is perpendicular to $\vec{S}_e$.

For the ordinary wave, the polarization vector $\vec{D}_0$ is already tangential to the indicatrix because the uniaxial indicatrix is an ellipsoid of revolution. Therefore the ray direction is parallel to the wave normal and perpendicular to the wave front. Its group velocity v_g and phase velocity (c/n_o) are equal. The divergence angle $\Delta = 0$ for the ordinary wave, which is why it is called an ordinary wave.

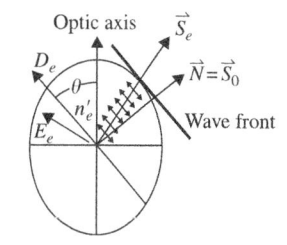

Fig. 25.9 Ray direction and wave normals in a uniaxial crystal with $n_e > n_o$. For a general wave normal there is a slow extraordinary wave with Poynting vector $\vec{S}_e$ and a fast ordinary wave with a Poynting vector $\vec{S}_0$ parallel to $\vec{N}$. The phase velocity of the extraordinary wave is c/n'_e where $n'_e = (n_o^2 \sin^2 \theta + n_e^2 \cos^2 \theta)^{1/2}$ and for the ordinary wave it is c/n_o.

Problem 25.1

Sodium nitrite (NaNO$_2$) is orthorhombic with refractive indices $n_1 = 1.340$, $n_2 = 1.425$, and $n_3 = 1.655$ along the principal axes.

Write out the equations for the wave velocity surfaces and the indicatrix. Simplify these expressions for Z_1–Z_2, Z_1–Z_3, and Z_2–Z_3 planes. Draw the three projections for the indicatrix and wave velocity surfaces showing the location of the optic axes. What are the speeds of waves along the optic axes and the principal axes?

Problem 25.2

A hexagonal crystal with refractive indices n_o and n_e is cut so that the optic axis is perpendicular to the surface. Show that for a beam of light with an angle of incidence θ, the angle of refraction ϕ_e of the extraordinary ray is given by

$$\tan \phi_e = \frac{n_o \sin \theta}{n_e \sqrt{n_e^2 - \sin^2 \theta}}.$$

25.4 Structure–property relationships

The refractive index for transparent materials is equal to the ratio of the speed of light in vacuum to that in the material. Because of their low densities, gases have refractive indices near 1, while for liquids and solids n ranges between 1.3 and 3 (Table 25.2). The magnitude of n is determined chiefly by the density of packing and the polarizability of the ions. Densely packed arrays of highly polarizable groups result in large refractive indices.

The polymorphs of SiO_2 demonstrate the relationship between packing density and refractive index (Fig. 25.10). Silica glass, cristobalite, and tridymite are high temperature forms with low densities and low refractive indices. Quartz is the stable form at room temperature and one atmosphere pressure.

Table 25.2 Typical refractive indices for inorganic minerals

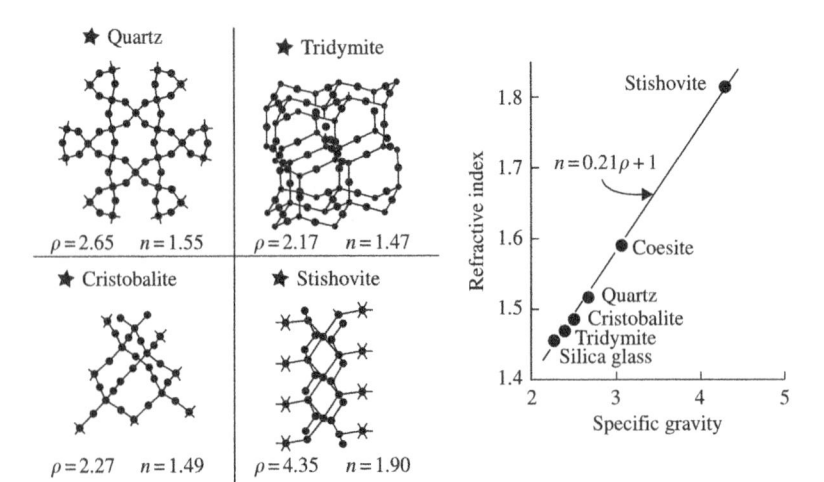

	n_1	n_2	n_3
Sulfur (S) mmm	1.96	2.01	2.25
Diamond (C) $m3m$	2.42	2.42	2.42
Sphalerite (ZnS) $\bar{4}3m$	2.37	2.37	2.37
Cinnabar (HgS) 32	2.91	2.91	3.27
Halite (NaCl) $m3m$	1.54	1.54	1.54
Fluorite (CaF_2) $m3m$	1.43	1.43	1.43
Ice (H_2O) $\bar{3}m$	1.309	1.309	1.311
Periclase (MgO) $m3m$	1.74	1.74	1.74
Corundum (Al_2O_3) $\bar{3}m$	1.77	1.77	1.76
Quartz (SiO_2) 32	1.54	1.54	1.55
Rutile (TiO_2) $4/mmm$	2.62	2.62	2.90
Calcite ($CaCO_3$) $\bar{3}m$	1.66	1.66	1.49

Fig. 25.10 The crystal structures, densities and refractive indices of several polymorphs of silica illustrating the dependence of n on ρ.

★ Quartz: $\rho = 2.65$, $n = 1.55$
★ Tridymite: $\rho = 2.17$, $n = 1.47$
★ Cristobalite: $\rho = 2.27$, $n = 1.49$
★ Stishovite: $\rho = 4.35$, $n = 1.90$

$n = 0.21\rho + 1$

Table 25.3 Refractive coefficients for some common oxide constituents for use in the Gladstone–Dale relation. (After Larsen and Berman)

	Molecular weight	k		Molecular weight	k
H_2O	18	0.34	Y_2O_3	226	0.14
Li_2O	30	0.31	La_2O_3	326	0.15
Na_2O	62	0.18	Bi_2O_3	464	0.16
K_2O	94	0.19	CO_2	44	0.22
BeO	25	0.24	SiO_2	60	0.21
MgO	40	0.20	TiO_2	80	0.40
CaO	56	0.23	ZrO_2	123	0.20
SrO	104	0.14	SnO_2	151	0.15
BaO	153	0.13	N_2O_5	108	0.24
PbO	223	0.15	P_2O_5	142	0.19
B_2O_3	70	0.22	Nb_2O_5	268	0.30
Al_2O_3	102	0.20	SO_3	80	0.18

Coesite and stishovite are dense polymorphs formed under high pressure. The average refractive index is plotted against specific gravity in Fig. 25.10. The slope of the line (0.21) gives the refractive coefficient for SiO_2 used in the Gladstone–Dale relation.

The empirical Gladstone–Dale relation is useful in predicting the average refractive indices of oxides:

$$n = 1 + \rho \sum_i p_i k_i,$$

where ρ is density, and p_i and k_i are the weight fraction and refractive coefficient of the ith component. An abbreviated table of the refractive coefficients is given in Table 25.3. To illustrate its use we calculate the predicted refractive index for pyrope garnet, $Mg_3Al_2Si_3O_{12}$. The molecular weight is 402 and the specific gravity 3.56. The constituent oxides are $3MgO + Al_2O_3 + 3SiO_2$, for which the weight fractions $3 \times 40/402$, $102/402$, and $3 \times 60/402$. Substituting in the Gladstone–Dale formula using the k values in Table 25.3 gives

$$n = 1 + 3.56(0.3 \times 0.20 + 0.253 \times 0.20 + 0.447 \times 0.21) = 1.71,$$

in good agreement with the observed value 1.72. The equation gives agreement to within 5% for a wide range of oxides. It works well because of the additivity of atomic polarizabilities which change little from compound to compound.

Note that the oxides of the isoelectronic ions Na^+, Mg^{2+}, Al^{3+}, and Si^{4+} all have about the same refractive power. Therefore the refractive indices of minerals containing these ions are principally dependent on the density of packing. This trend is illustrated with a number of complex oxides in Fig. 25.11.

A few of the constituent oxides in Table 25.4 possess unusually large refractive coefficients. The k value for TiO_2 is twice that of most others, for instance. This is one of the oxides with electronic transitions in the very near ultraviolet, a transition from 2p level of O^{2-} to the 3d orbital of Ti^{4+}. The presence this low-lying excited state augments the electronic polarizability which in turn contributes to the refractive index, the dielectric constant, and leads to ferroelectricity in a number of titanates.

So far we have considered only oxides. The influence of other anions on the refractive index can be seen from alkali halides with the rocksalt structure. Table 25.4 compares the indices of fluorides, chlorides, bromides, and iodides. In every case the iodides are the largest and the fluorides the smallest. Larger

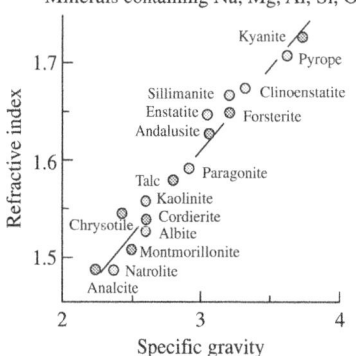

Fig. 25.11 Refractive index-density relation for a number of oxide minerals containing Na, Mg, Al, and Si. For anisotropic minerals the refractive indices are averaged.

282 Crystal optics

Table 25.4 Refractive indices of cubic alkali halide crystals. The larger, heavier anions like I⁻ generally have larger n values because of their higher polarizabilities

	n	Anions
Fluorides		
LiF	1.38	⑩ F⁻
NaF	1.33	
KF	1.36	
RbF	1.39	
Chlorides		
LiCl	1.66	⑱ Cl⁻
NaCl	1.54	
KCl	1.49	
RbCl	1.49	
Bromides		
LiBr	1.78	㊱ Br⁻
NaBr	1.62	
KBr	1.56	
RbBr	1.55	
Iodides		
LiI	1.95	�554 I⁻
NaI	1.71	
KI	1.68	
RbI	1.65	

anions with more loosely bound electrons have higher polarizability and larger refractive indices.

25.5 Birefringence and crystal structure

The birefringence of crystals can be visualized using the indicatrix, an ellipsoid showing the variation of the refractive indices with vibration direction. Each radius vector from the center to the surface represents a vibration direction whose length measures the index of refraction of a wave polarized parallel to the radius vector.

In transparent materials, refractive index is proportional to the square root of the electronic polarization. The latter is in turn proportional to the polarizabilities of the ions in the crystal and also the local electric field. To illustrate, consider the triatomic molecule in Fig. 25.12. In (a) the electric vector of the light waves is parallel to the length of the molecule and in (b) it is perpendicular. The electric dipole moment of the atom is equal to the product of the local field and polarizability. The local field is the vector sum of the applied field E and the dipole fields associated with the neighboring atoms. In the parallel orientation (a) the dipole field of the neighboring atom enhances E so that both atoms are polarized more, giving rise to larger dipole moments and a larger refractive index for this polarization direction. The opposite effect occurs in (b). Here the dipole field is in opposition to E, reducing the dipole moments and refractive index. Thus waves polarized parallel to the molecule travel slower than waves polarized perpendicular to the molecule, creating birefringence in crystals with aligned molecules.

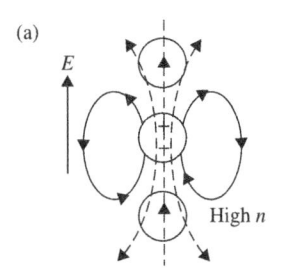

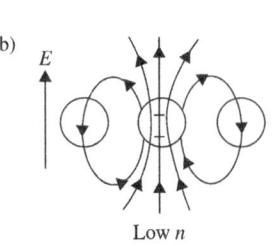

Fig. 25.12 The electric field of the light wave is modified by the local dipole fields from nearby atoms. (a) When the dipole fields enhance the driving field E, the light wave is slowed down, and the refractive index is raised. (b) When the dipole fields partially cancel the driving field, the velocity is increased, and refractive index is reduced. Molecular arrangments such as these are the cause of birefringence in crystals.

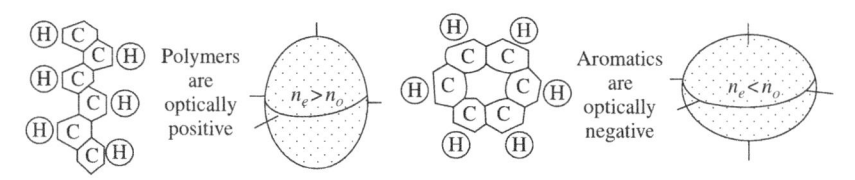

Fig. 25.13 When the molecular groups in a crystal or textured material are parallel to one another, the shape of the optical indicatrix resembles the shape of the molecule. It is for this reason (a) that aligned polymers and nematic liquid crystals are often optically positive while (b) aromatic crystals with planar ring structures are optically negative.

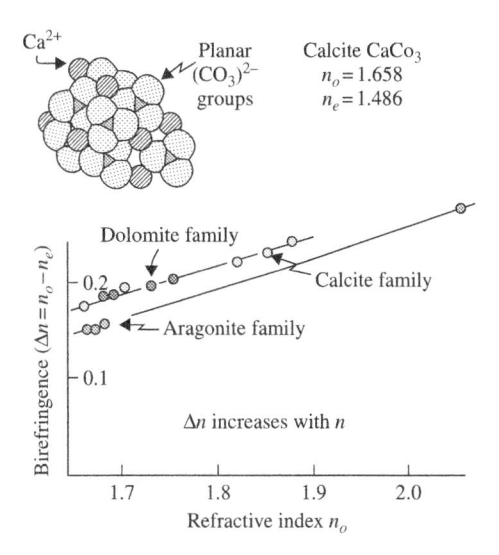

Fig. 25.14 Calcite crystals with aligned carbonate groups are optically negative. Other carbonate minerals such as dolomite and aragonite have similar structures and similar birefringence.

Birefringence in many compounds containing molecular groups can be explained in this way. The basic relationship between crystal structure and birefringence is illustrated in Fig. 25.13. For linear or near-linear groups, positive birefringence occurs when they are parallel to another, and negative when they lie in a plane perpendicular to a common direction. Under these circumstances the shape of the indicatrix mimics the shape of the molecular groups.

The classical examples of birefringence are the minerals calcite ($CaCO_3$) and rutile (TiO_2). Calcite and other carbonate crystals such as aragonite ($CaCO_3$) and dolomite ($CaMg(CO_3)_2$) all contain triangular $(CO_3)^{2-}$ anions with carbon bonded to three oxygens. In these minerals the flat carbonate groups are aligned in layers giving rise to an oblate optical indicatrix with strong negative birefringence (Fig. 25.14).

When the planar molecular groups are not parallel to one another but are parallel to a common direction, positive birefringence results with the vibration direction of large refractive index along the common direction. The rare carbonate mineral vaterite is an example.

Rutile and its polymorphs have tetragonal crystal structures with chains of edge-sharing TiO_6 octahedra parallel to Z_3, the fourfold symmetry axis. The resulting anisotropy makes rutile, tellurite (TeO_2), and cassiterite (SnO_2) very birefringent. Because of the chain-like structure they have a uniaxial positive indicatrix (Fig. 25.15).

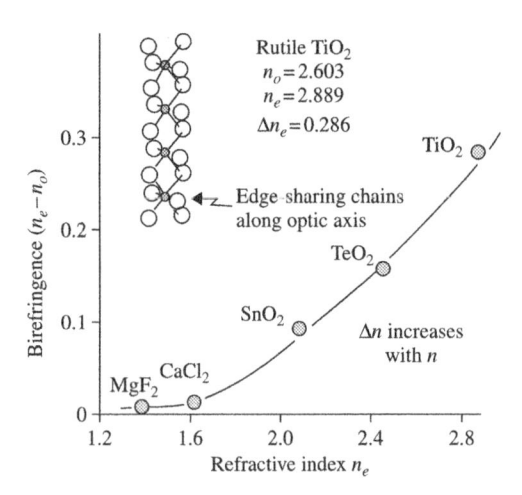

Fig. 25.15 Rutile and isomorphic compounds have strong birefringence with $n_e > n_o$. The prolate indicatrix has the same orientation as the chains in the crystal structure.

Table 25.5 Correlation between molecular shape and optic sign

	Molecular shape	n_1	n_2	n_3	Optic sign
$Ca(OCl)_2 \cdot 3H_2O$	Linear molecule	1.535	1.535	1.630	+
$NaNO_2$	Obtuse V-shaped	1.340	1.425	1.655	+
$KClO_3$	Low pyramid	1.410	1.517	1.524	−
$C_{10}H_8$	Flat with parallel planes	1.422	1.775	1.932	−

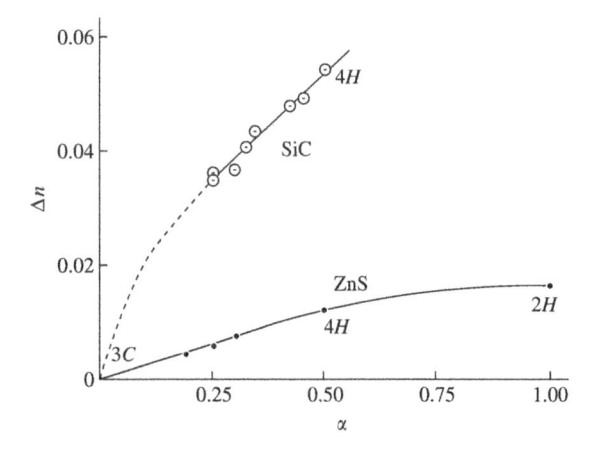

Fig. 25.16 Birefringence Δn plotted against hexagonality α for several ZnS and SiC polytypes. Birefringence decreases as the fraction of cubic close packed layers increases because cubic crystals are optically isotropic.

Calcite and rutile are both uniaxial, but biaxial crystals show similar correlations between optic sign and crystal structure. Examples are given in Table 25.5. When flat molecules are parallel to one another, a large negative birefringence results with the vibration direction of lowest index perpendicular to the plane of the molecules, as in naphthalene. Its other two refractive indices are considerably larger.

The final examples concern the polytypes of silicon carbide (SiC) and zinc sulfide (ZnS). These structures are intermediate between the cubic sphalerite structure and the hexagonal wurtzite structure, so that the various polytypes can be characterized by parameter α which refers to fraction of layers stacked in hexagonal close-packed order, Thus $\alpha = 0$ for the 3C polytype sphalerite and $\alpha = 1$ for wurtzite (2H) with all other polytypes in between 0 and 1.

Several of the physical properties of these compounds can be correlated with α, including absorption edges, color, and birefringence (Fig. 25.16). As might be expected the birefringence disappears as α approaches zero since the structures are becoming more cubic.

26 Dispersion and absorption

26.1 Dispersion 286
26.2 Absorption, color, and dichroism 288
26.3 Reflectivity and luster 291
26.4 Thermo-optic effect 292

In this chapter we discuss *dispersion*, the dependence of the refractive index on wavelength, and *absorption* that is related to the imaginary part of the refractive index. The variation of the refractive index with temperature (thermo-optic effect) is also described.

In terms of their directional behavior, all three of these effects can be represented by second rank tensors because of their relationship to the optical dielectric constant K. The thermo-optic coefficients relate ΔK_{ij} to a temperature change ΔT. Dispersion refers to the frequency dependence of the optic dielectric constant $K_{ij}(\omega)$, and absorption to the imaginary part of the complex dielectric constant $K_{ij}^* = K_{ij}' - iK_{ij}''$.

26.1 Dispersion

Refractive index depends on wavelength, giving rise to dispersion. For most transparent substances, n increases as λ decreases; refractive indices for violet light are generally a few percent larger than those for red. Dispersion is caused by electronic transitions in the ultraviolet region. When the photon energy approaches the value required for transition, the electrons in the solid undergo wide excursions, producing large polarizability and large refractive indices.

In many crystals dispersion is small throughout the visible range from 0.4 to 0.7 μm. Diamond is the most striking example with a change in refractive index from 2.465 in the violet to 2.407 in the red. The refracted rays from brilliant-cut diamonds show a remarkable rainbow of colors in bright sunshine.

Dispersion in diamond is compared with silicon and germanium in Fig. 26.1. All three have large refractive indices and very large dispersion because their band gaps are near by. Many semiconductors and ferroelectric oxides have band gaps in the near ultraviolet or in the visible range. These materials are of special interest in optical applications because of their strong interactions with light. Examples will be discussed later in the chapters on acousto-optics, electro-optics, and nonlinear optics.

For oxides the refractive index is inversely related to the band gap E_g:

$$n^2 \cong 1 + \frac{15}{E_g},$$

where E_g is expressed in electron-volts. Other classes of materials have somewhat different constants. If an oxide is to be transparent throughout the visible range, the equation states that the band gap must be at least $hc/4000 \text{ Å} \cong 3$ eV,

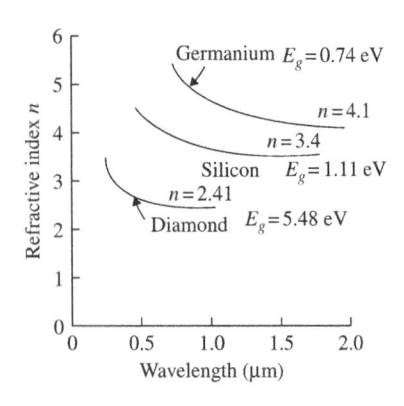

Fig. 26.1 Refractive indices for cubic C, Si, and Ge in the visible and near infrared range. Ge and Si are nearly opaque in the visible region because of their smaller band gaps.

giving $n \cong 2.5$. To obtain higher refractive indices, the minimum wavelength must be raised, closing the window. The shortest wavelength for which transmission is desired determines the maximum refractive index.

Among oxides, rutile (TiO_2) and other titanates have band gaps near 3 eV and refractive indices larger than two. They also have large dispersion, like diamond, which can give rise to interesting optical effects. Brookite, a polymorph of TiO_2, is orthorhombic and therefore optically biaxial. In blue light the two optic axes lie in the (100) plane with $n_2 > n_1 \geq n_3$. As the wavelength is increased into the yellow range, n_1 decreases faster than n_3 and there is a critical wavelength where $n_2 > n_1 = n_3$. At this wavelength, brookite is optically uniaxial with [010] as the optic axis. On further increasing the wavelength into the red, the crystal again becomes biaxial with $n_2 > n_3 \geq n_1$, but now the two optic axes lie in the (001) plane. The dispersion curves are shown in Fig. 26.2(a).

A second interesting example is the tetragonal sulfide $AgGaS_2$. Like diamond and rutile, it has large refractive indices and large dispersion effects. On increasing the wavelength from blue to red it changes from uniaxial positive ($n_e > n_o$) to isotropic ($n_e = n_o$) to uniaxial negative ($n_o > n_e$). In both these cases the apparent optical symmetry changes with wavelength.

Problem 26.1
Make sketches of the indicatrix and wave velocity surfaces of brookite and silver gallium sulfide. Show how the surfaces change going from green to orange wavelengths.

More normal dispersion behavior is shown in Fig. 26.3. Cadmium sulfide (CdS) is hexagonal, point group 6mm, with the wurtzite structure. It has strong dispersion near the absorption edge at 0.5 μm but then flattens out in the near infrared. The birefringence changes very little with wavelength, decreasing only slightly from $n_e - n_o = 0.018$ at 0.6 μm to 0.013 at 10.6 μm.

Dispersion curves like this are important to the designers of lenses and other optical components. The variation of refractive indices are carefully measured and mathematically modeled to enhance or eliminate chromatic effects in optical systems.

One of the expressions most often used to fit the dispersion curves is the Sellmeier equation

$$n^2 = A + \frac{B}{C - \lambda^2} + D\lambda^2.$$

The coefficients for uniaxial $LiNbO_3$ (trigonal, 3m) are listed in Table 26.1.

Dispersion is largest near absorption bands. On the short wavelength side, dispersion is caused by electronic transitions in the near ultraviolet, while at

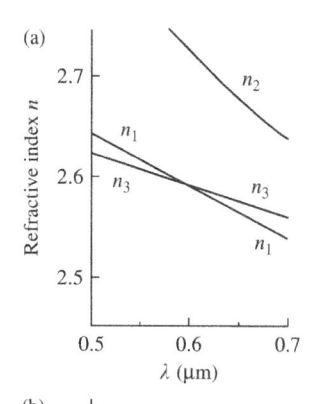

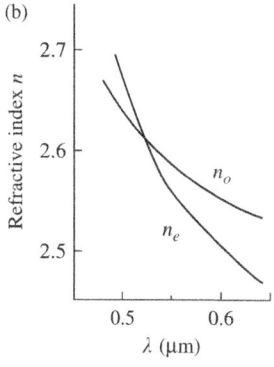

Fig. 26.2 Dispersion in (a) Brookite, TiO_2, and (b) silver gallium sulfide, $AgGaS_2$, showing how optical symmetry can change with frequency. In both crystals the absorption edge overlaps the visible range creating color, large refractive indices and large dispersion effects.

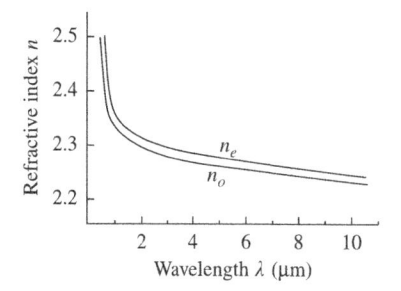

Fig. 26.3 Refractive indices of CdS plotted as a function of wavelength into the infrared range. In this case dispersion has little effect on the birefringence.

Table 26.1 Sellmeier coefficients for $LiNbO_3$ at 20°C. Wavelength λ is expressed in microns

	n_o	n_e
A	4.9048	4.582
B	0.11768	0.099169
C	0.0475	0.044432
D	0.027169	0.02195

288 *Dispersion and absorption*

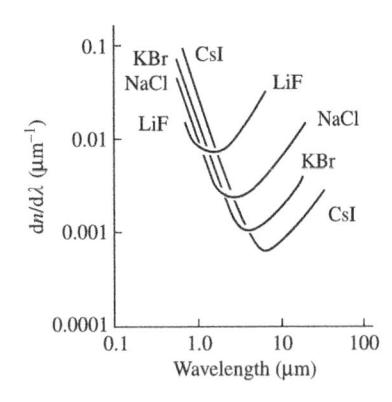

Fig. 26.4 Dispersion in four alkali halide crystals in the infrared range. CsI and other high molecular weight halides have the widest transmission range and the lowest dispersion.

long wavelengths dispersion increases again because of the infrared vibration bands. Dispersion passes through a minimum in the optical window between these two absorption bands. Refractive index measurements on alkali halide crystals demonstrate this point (Fig. 26.4).

As discussed in Chapter 30, dispersion can be very helpful in the phase-matching methods used in nonlinear optics. To generate strong second harmonic signals it is necessary keep the fundamental (frequency ω) and harmonic (2ω) waves in phase. This is generally accomplished by matching the refractive index of the harmonic $n(2\omega)$ with that of the fundamental $n(\omega)$. Because of dispersion, this is difficult in an optically isotropic medium, but it is sometimes possible in uniaxial crystals in which the birefringence exceeds the dispersion.

For a $Y_{3-x}Nd_xAl_5O_{12}$ (neodymium YAG) laser, the fundamental wavelength is at 1.06 μm and the second harmonic at 0.53 μm. Therefore the difference in refractive indices between these two wavelengths is critical. In low refractive index crystals like quartz $n(\omega) - n(2\omega)$ is about 0.01, and about ten times larger in ferroelectric crystals such as $LiNbO_3$ and $PbTiO_3$.

26.2 Absorption, color, and dichroism

The attractive colors of gemstones and other colored crystals are usually due to absorption bands in the visible spectrum, though diffraction effects are occasionally important, as in opal. The visible spectrum extends from 4000 to 7000 Å; in this region electronic transitions are the most important. There are four common types: (1) internal transitions within transition-metal, rare-earth, or other ions with incomplete electron shells; (2) charge transfer processes in which an electron is transferred from one ion to another; (3) electronic transitions associated with crystal imperfections; and (4) band gap transitions—intrinsic coloration found in many semiconductor compounds. Types (1)–(3) are usually associated with small amounts of dopants, impurities or defects, whereas (4) is a bulk property.

The following examples will serve to illustrate the various types and the cause of color in crystals. The color spectrum and their wavelengths are given in Table 26.2. When an absorption band removes a certain color from the transmitted beam, the eye collects the remaining colors and produces the complementary color to the one removed. Thus when red light is absorbed—as by Fe^{2+} in olivine—the crystal appears green, the complementary color. In oxides and fluorides containing transition-metal ions, the absorption bands and color depend principally on the nature of the ion and secondarily on the strength and symmetry of the crystal field. The electron configuration of the ion is important. Thus Fe^{3+} with five 3d electrons and a relatively stable half-filled shell, has a far different spectrum than the $3d^6$ configuration of Fe^{2+}. Chrysoberyl $(Al, Fe^{3+})_2BeO_4$ has almost the same structure as olivine $(Mg, Fe^{2+})_2SiO_4$ yet the spectra are vastly different. Trivalent iron absorbs in the violet region, giving chrysoberyl a pale yellow complementary color.

Absorption experiments are generally carried out on an optical or infrared spectrometer. A thin sample of thickness t is inserted into the optical beam and the intensity of the transmitted light I is compared with the incident intensity I_0. Corrections are made for the light lost by reflection. The absorption coefficient α

Table 26.2 The spectral colors, their wavelengths, and the complementary colors

λ (Å)	Spectral color	Complementary color
4100	Violet	Lemon-yellow
4300	Indigo	Yellow
4800	Blue	Orange
5000	Blue-green	Red
5300	Green	Purple
5600	Lemon-yellow	Violet
5800	Yellow	Indigo
6100	Orange	Blue
6800	Red	Blue-green

is determined from the relationship.

$$I = I_0 \exp(-\alpha t).$$

Measurements on anisotropic crystals are carried out using polarized light waves transmitted along principal axes. Experiments are repeated for different wavelengths to obtain the absorption spectrum.

Some crystals absorb light differently for different polarization directions. The effect is called *pleochroism* in biaxial crystals and *dichroism* in uniaxial crystals. Cubic crystals do not show differential light absorption under normal circumstances. In a few minerals such as cordierite, pleochroism is apparent even in unpolarized light (Fig. 26.5). Cordierite is orthorhombic (point group *mmm*). Its chemical composition is $(Mg_{2-x}Fe_x)Al_4Si_5O_{18} \cdot nH_2O$. The broad peaks in the absorption spectra are attributed to Fe^{2+} and the narrow lines to overtone vibrations from oriented water molecules trapped in cages within the crystal structure. The absorption coefficients and colors are remarkably different for light polarized in different directions. There has been speculation that the Vikings used cordierite as a navigation aid in far northern waters where the sun is often below the horizon. By using the stone to detect polarized scattered light, they may have been able to locate the sun's position.

Absorption anisotropy also appears in uniaxial crystals. The mineral tourmaline (class $3m$) absorbs light strongly for vibration directions perpendicular to the trigonal axis but only weakly when the electric vector is parallel to c. For many years mineralogists used tourmaline "tongs" made of crossed tourmaline crystals to examine minerals in polarized light. The dichroism of tourmaline arises from Fe^{2+} (t^4e^2) to Fe^{3+} (t^3e^2) charge transfer transitions. Di- and trivalent iron occupy octahedral sites in tourmaline with the octahedra clustered in flat triangular arrangements separated by borosilicate groups. The octahedra share edges in planes parallel to (0001) so that when the electric vector of the light wave is perpendicular to c, the electrons move between octahedra, hopping from one cation to the next, and absorbing light (Fig. 26.6).

Crystal fields are important in determining the colors of transition-metal compounds because of the electronic transition within the incomplete 3d shell. Transition metal ions are usually found in tetrahedral or octahedral coordination in oxides and fluorides. The two coordinations show different spectra and different colors because the electronic ground state is often different for tetrahedral and octahedral fields. Co^{2+}, for example, has a $e^4t_2^3$ ground state in tetrahedral environment and $t_{2g}^5 e_g^2$ symmetry in an octahedral field. Not only the symmetry

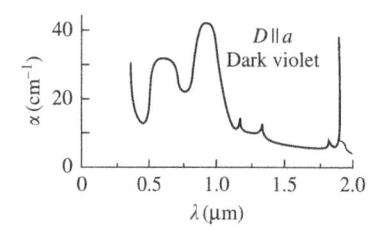

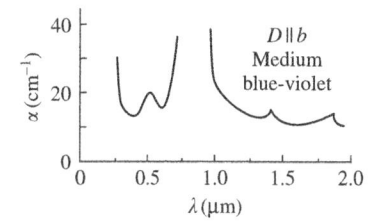

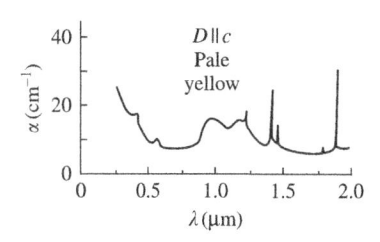

Fig. 26.5 Pleochroism in the mineral cordierite. Light polarized parallel to the orthorhombic [100] and [010] directions are more strongly absorbed than along [001].

290 Dispersion and absorption

(a) [Figure: Schorl tourmaline absorbance spectrum showing E⊥C Ordinary wave and E∥C Extraordinary wave, Absorbance vs Wavelength (μm) from 0.4 to 2.0]

(b) [Figure: Trigonal 3m structure showing Fe octahedra, Charge transfer spectra, Hopping electrons $Fe^{2+} \rightleftarrows Fe^{3+} + e^-$, Iron octahedra share edges in (00.1) plane]

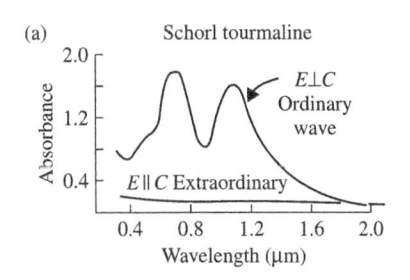

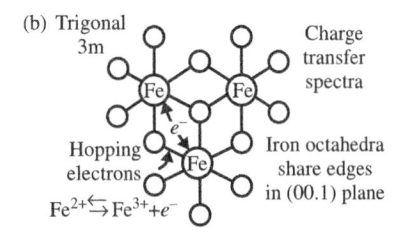

Fig. 26.6 Dichroism in tourmaline. (a) Light polarized perpendicular to the optic axis is much more strongly absorbed that the extraordinary ray polarized parallel to c. (b) The effect is attributed to charge transfer between iron ions lying in the (001) plane.

of the field is important, but its strength as well. Chromium bearing oxides illustrate this point: chromia, emerald, and most chromium containing oxides are green, but ruby, spinel and a few others are red. Alexandrite is red or green, depending on the source of illumination. The intense absorption bands of Cr^{3+} shift toward higher energy for the larger crystal fields (Fig. 26.7). Large crystal fields are caused by close neighbors. In ruby the Cr–O distances are about 0.1 Å shorter than in Cr_2O_3, and as a result, the important $^4A_2 \rightarrow {}^4T_1$ and $^4A_2 \rightarrow {}^4T_2$ absorption bands shift about 400 Å, changing the color from red to green.

In complex crystals such as these it is important for the experimentalist to identify the origin of absorption and color. This is often done by preparing crystals with different dopant levels. In the case of ruby, growing crystals of $Al_{2-x}Cr_xO_3$ with different chromium content. By plotting the intensity of various absorption peaks as a function of the percentage chromium, one can determine which spectra are associated with single chromium ions, which are associated with pairs of chromiums, and which have different origins. The absorption data in Fig. 26.8 illustrate this point.

In organic dichroic materials, internal absorption takes place when the electric polarization vectors are parallel to the direction of chromophoric groups, along chemical linkages such as

$$-N=N- \qquad =C\begin{matrix} C=C \\ \\ C=C \end{matrix} C=$$

If such groups are arranged parallel to one another in the crystal, or a textured material, strong dichroism results. The oriented dye molecules used in polaroid films and liquid crystal displays are good examples (Section 30.5).

Colors are also associated with imperfections in solids. When colorless alkali halide crystals are irradiated or heated in the alkali vapor they change color: NaCl becomes yellow, KBr blue, and KCl magenta. The most common type of coloring imperfection is the F-center (Farbzentrum) in which single electrons are trapped in anion vacancies. Visible radiation is absorbed by the trapped electron whose energy states can be approximated by hydrogen-like wave functions.

Undoped semiconductor crystals are colored according to the band gap. Si and Ge are metallic-looking because all visible radiation promotes electrons

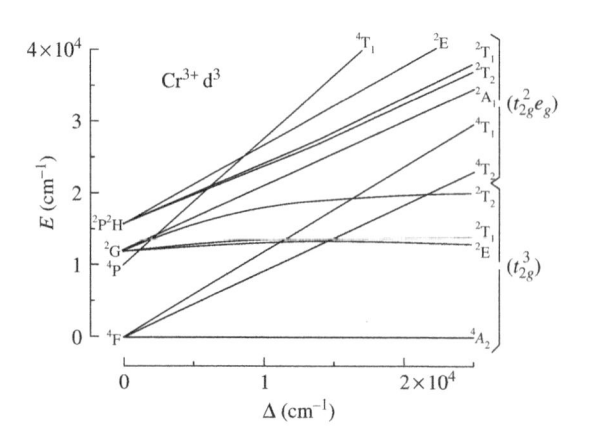

[Figure: Energy level diagram for $Cr^{3+} d^3$, $E (cm^{-1})$ vs $\Delta (cm^{-1})$, showing levels $^4T_1$, ^2E, $^2T_1$, $^2T_2$, $^2A_1$, $(t_{2g}^2 e_g)$, $^4T_1$, $^4T_2$, $^2T_2$, $^2T_1$, ^2E, (t_{2g}^3), $^4A_2$, and $^2P^2H$, ^2G, ^4P, ^4F]

Fig. 26.7 Energy-level diagrams for octahedrally coordinated Cr^{3+} plotted as a function of the crystal-field parameter Δ.

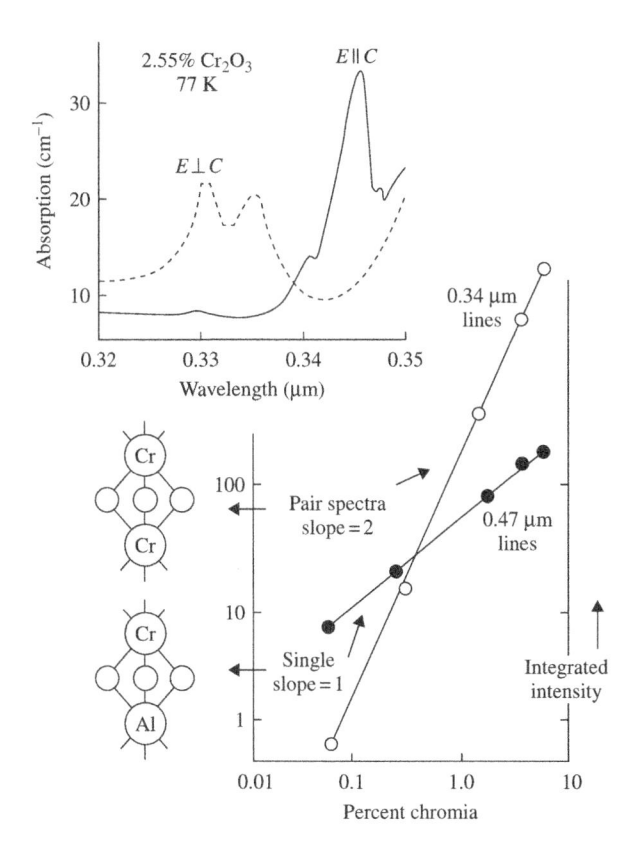

Fig. 26.8 The dichroic absorption peaks of ruby are caused by electronic transitions in chromium. Normal crystal field spectra are linearly proportional to the Cr content while those associated with pairs of Cr ions have absorption coefficients proportional to the square of the Cr content.

to the conduction band. CdS ($E_g = 2.45$ eV) is yellow because blue and violet can promote electrons but the longer wavelengths cannot. Hence blues are absorbed and the crystal appears yellow, the complementary color.

26.3 Reflectivity and luster

Refractive index determines the intensity of reflected light from a surface. The luster of a solid refers to its appearance in reflected light. For normal incidence on a smooth surface, the ratio of reflected intensity to incident intensity is

$$R = \frac{(n-1)^2 + n^2 k^2}{(n+1)^2 + n^2 k^2}.$$

The reflective index is n and k the absorption index. For transparent materials, $k = 0$ and R depends on n only. Only about 2% of the light is reflected for low-index solids, giving them a glassy-like appearance. The high reflectivity of diamond ($n = 2.41$, $R = 17\%$) imparts a high luster to the stone.

The luster of various solids are compared in Table 26.3. Many fluorides and polymers have low refractive indices ($n \sim 1.3$) and low luster making them useful as nonreflective coatings. Silica and common silicate glasses ($n \sim 1.5$) have a vitreous luster, while zircon ($n \sim 2.0$), zirconia ($n \sim 2.2$), calcium titanate ($n \sim 2.4$) and rutile ($n \sim 2.7$) have much larger reflection coefficients. The zirconates and titanates are widely used in glazes, enamels, and paints for

Dispersion and absorption

Table 26.3 Reflection coefficients for several types of luster, together with mineral examples

	R (%)	Example
Transparent crystals		
Subvitreous	<4	Fluorite (CaF$_2$)
Vitreous	4–8	Topaz (Al$_2$SiO$_4$F$_2$)
Subadamantine	8–14	Zircon (ZrSiO$_4$)
Adamantine	14–21	Diamond (C)
Adamantine splendent	>21	Cinnabar (HgS)
Opaque crystals		
Submetallic	<20	Ilmenite (FeTiO$_3$)
Metallic	20–50	Molybdenite (MoS$_2$)
Metallic splendent	50	Gold (Au)

this reason. Because of their high refractive indices and high reflectivities they have excellent covering power.

The high reflectivity of gold and other metals comes from their large absorption coefficients that in turn are caused by high electrical conductivity. Light waves interact strongly with conduction electrons. The effect of electrical conductivity (σ) on the reflection coefficient (R) is quantified through the Hagen–Rubens Law:

$$R = 1 - \left(\frac{16\varepsilon_0 f}{\sigma_0}\right)^{1/2},$$

where f is the frequency and σ_0 is the DC conductivity. R values at 3 μm for several metals, semimetals, and semiconductors are shown in Fig. 26.9. At high frequency metals become transparent.

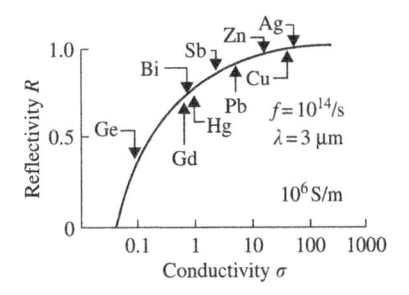

Fig. 26.9 Reflection coefficients for various conductors at $f = 10^{14}$ Hz. The curve follows the Hagen–Rubens relation.

26.4 Thermo-optic effect

Thermo-optic coefficients relate changes in the optical indicatrix ΔB_{ij} to changes in temperature ΔT. Since temperature is a scalar, the thermo-optic effect is a symmetric second rank tensor like the dielectric constant. The same symmetry restrictions apply (Table 9.1).

The temperature dependence of the refractive index is generally small, except near phase transformations. The situation is analogous to low frequency dielectrics. For silica and alumina, the permittivity is nearly independent of temperature (Fig. 9.13), but ferroelectrics exhibit enormous changes near T_c (Fig. 9.14).

Similar effects can occur in other regions of the electromagnetic spectrum with other variables. For example, the low frequency dielectric constants of tetragonal KH$_2$PO$_4$ change anisotropy with temperature (Fig. 9.10). At room temperature $K_{11} > K_{33}$ and at low temperatures $K_{33} > K_{11}$. At the crossover, near 200 K, KH$_2$PO$_4$ crystals are dielectrically isotropic with $K_{11} = K_{33}$.

In the optical range the dielectric constant $K = n^2$, and its temperature dependence is $(dK/dT) = 2n(dn/dT)$. This quantity is plotted as a function of wavelength in Figs. 26.10 and 26.11 for three transparent materials. Silica glass has a low refractive index ($n \sim 1.45$) and a small temperature coefficient ($dn/dT \sim 10^{-5}$). As pointed out in Section 25.4, the refractive index of common oxides increases with density (see Figs. 25.10 and 25.11). Because of thermal expansion, density decreases with increasing temperature, and therefore refractive index decreases too. Thermal expansion makes a small negative contribution to the temperature coefficient of refractive index dn/dT. This effect is often overwhelmed by changes in the electronic band gap or by phase changes. These effects can be either positive or negative, depending on the nature of the energy levels or on the location of the phase transformation.

Thermo-optic coefficients for LiNbO$_3$ and LiIO$_3$ are shown in Fig. 26.11. These coefficients are about an order of magnitude larger than in silica glass. Both crystals are uniaxial positive, but the changes in birefringence with temperature and with wavelength are small. The thermo-optic coefficients of LiNbO$_3$ and LiIO$_3$ have opposite sign, possibly because of different physical origins. On heating, the refractive indices of lithium niobate increases rather rapidly in the visible range. This causes a troublesome thermal drift in high power laser experiments. Both SiO$_2$ and LiNbO$_3$ have positive values of dn/dT. If the band gap of these oxides decreases with increasing temperature, the ultraviolet absorption edge will move toward the visible range, raising the refractive index.

The negative thermo-optic effect of LiIO$_3$ could be explained by an absorption edge shift *away* from the visible range. A second possible explanation is its low melting point that could lead to larger thermal expansion coefficients and negative values of dn/dT.

Most alkali halides have negative dn/dT values ranging from about 10^{-5}/K for LiF to about 10^{-4}/K for CsI. In these crystals, the large thermal expansion coefficients control the thermo-optic effect. Stiffer materials with lower thermal expansion coefficients usually have positive dn/dT values controlled by the electronic band gap. The refractive indices of MgO, ZnS, SiC and diamond all increase with temperature.

One of the most interesting thermo-optic phenomena is the *Mitscherlisch Effect*. At room temperature, the mineral gypsum (CaSO$_4$·2H$_2$O, monoclinic 2/m) is optically biaxial positive. When heated, the optic angle gets smaller and at 90°C, 2V goes to zero causing the crystal to become optically uniaxial for wavelengths near 0.589 μm. Above this temperature the gypsum crystal is again biaxial but with the optic plane rotated by 90°. The condition for the Mitscherlisch Effect is that $n_3 > n_2 \approx n_1$. At the crossover temperature, $n_2 = n_1$. Because of dispersion, the crossover temperature will depend on wavelength.

Even more spectacular thermo-optic phenomena can occur when all three refractive indices are approximately equal. Orthorhombic Cs$_2$SeO$_4$ is especially interesting. For temperatures between 0°C and 250°C, each of the three orthorhombic axes becomes the acute bisectrix, and the optic plane changes from (100) to (010) to (001).

Problem 26.2
Make a drawing of how the refractive indices of Cs$_2$SeO$_4$ must change with temperature to explain the optical observations. Draw the indicatrix for each of the three temperature ranges.

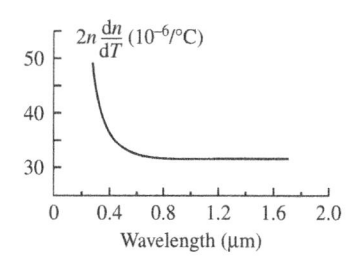

Fig. 26.10 Temperature coefficient of the optical dielectric constant plotted as a function of wavelength for silica glass (vycor). Note the large changes in the thermo-optic coefficient near the ultraviolet absorption edge where dispersion is also large. Small shifts in the band gap E_g can lead to large changes in dK/dT.

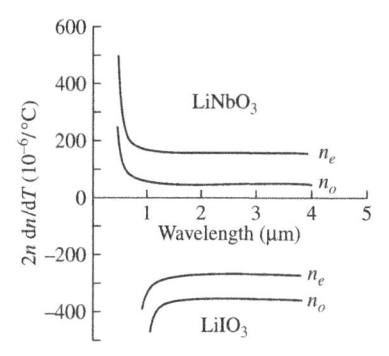

Fig. 26.11 Thermo-optic effects in lithium niobate and lithium iodate.

27 Photoelasticity and acousto-optics

27.1 Basic concepts 294
27.2 Photoelasticity 295
27.3 Static photoelastic measurements 296
27.4 Acousto-optics 298
27.5 Anisotropic media 300
27.6 Material issues 300

The change in refractive indices with mechanical stress, the photoelastic effect, is used in analyzing stress patterns in engineering components of complicated shape. It is also important in acousto-optic devices, optical switches, modulators, and scanners in which ultrasonic waves modulate the refractive index, producing an optical grating.

27.1 Basic concepts

Materials with sizable photoelastic coefficients (p) are required to enhance the interaction between mechanical strain x and refractive index n. Changes in the indicatrix are given by

$$\Delta\left(\frac{1}{n^2}\right) = px.$$

These quantities are actually tensors but are treated as scalars in the following discussion which is concerned with the magnitudes of the photoelastic coefficients p, and not their variation with direction.

Unlike the linear electro-optic effect, photoelasticity occurs in all symmetry classes and is *not* a null property. Photoelastic coefficients are dimensionless because strain and refractive index are dimensionless. For most oxides and halides, $p_{max} \cong 0.2$. The maximum values of p measured for other materials range from 0.1 to 0.6.

To gain a clearer understanding of the effects of stress on refractive index, consider the effect of hydrostatic pressure on a cubic crystal. The Lorenz–Lorentz equation is valid for many cubic materials:

$$\frac{n^2-1}{n^2+1} = KN\alpha.$$

K is a proportionality constant, N the number of molecules per unit volume, and α the polarizability per molecule. Differentiating with respect to the density ρ, gives

$$\frac{dn}{d\rho} = \frac{(n^2-1)(n^2+2)}{6n\rho}\left(1 + \frac{\rho}{\alpha}\frac{d\alpha}{d\rho}\right).$$

This leads to an estimated photoelastic coefficient

$$p = \frac{(n^2-1)(n^2+2)}{3n^4}\left(1 + \frac{\rho}{\alpha}\frac{d\alpha}{d\rho}\right).$$

It is apparent that the refractive indices depend on pressure through both the density ρ and the polarizability α. As the pressure increases, the atoms are packed closer together causing an increase in the refractive index. Pressure changes may also change the atomic polarizabilities so that $d\alpha/d\rho \neq 0$. When a solid is compressed, the electrons are bonded more tightly, reducing the polarizability. Thus the term $d\alpha/d\rho$ is negative, and the resulting change in refractive index opposes the effect of a volume decrease. The two effects are of the same order of magnitude, so that the refractive index increases with pressure in some crystals (Al_2O_3), decreases in some (MgO), and is almost constant in others ($Y_3Al_5O_{12}$).

27.2 Photoelasticity

The photoelastic effect, sometimes called the piezo-optic effect, concerns the changes in optical indicatrix with mechanical stress or strain. For an arbitrary coordinate system (Z_1, Z_2, Z_3) the indicatrix takes the form $B_{ij}Z_iZ_j = 1$ or

$$B_{11}Z_1^2 + B_{22}Z_2^2 + B_{33}Z_3^2 + 2B_{23}Z_2Z_3 + 2B_{13}Z_1Z_3 + 2B_{12}Z_1Z_2 = 1.$$

Changes in the indicatrix under stress X_{kl} or strain x_{kl} are represented through the linear relations

$$\Delta B_{ij} = \pi_{ijkl} X_{kl}$$

or

$$\Delta B_{ij} = p_{ijkl} x_{kl}.$$

The coefficients π_{ijkl} and p_{ijkl} are fourth polar rank tensors, sometimes referred to as the piezo-optic and elasto-optic effects, respectively. Since stress and strain are related to one another through the elastic constants, the π and p coefficients are also. In what follows we shall work mainly with the elasto-optic tensor p_{ijkl}.

Both the indicatrix coefficients and the strain tensors are symmetric, so that $B_{ij} = B_{ji}$, $x_{kl} = x_{lk}$, and $p_{ijkl} = p_{jikl} = p_{ijlk} = p_{jilk}$. This means that 36 coefficients in a 6 × 6 matrix are sufficient to describe the linear photoelastic effect:

$$\begin{pmatrix} \Delta B_1 \\ \Delta B_2 \\ \Delta B_3 \\ \Delta B_4 \\ \Delta B_5 \\ \Delta B_6 \end{pmatrix} = \begin{pmatrix} p_{11} & p_{12} & p_{13} & p_{14} & p_{15} & p_{16} \\ p_{21} & p_{22} & p_{23} & p_{24} & p_{25} & p_{26} \\ p_{31} & p_{32} & p_{33} & p_{34} & p_{35} & p_{36} \\ p_{41} & p_{42} & p_{43} & p_{44} & p_{45} & p_{46} \\ p_{51} & p_{52} & p_{53} & p_{54} & p_{55} & p_{56} \\ p_{61} & p_{62} & p_{63} & p_{64} & p_{65} & p_{66} \end{pmatrix} \begin{pmatrix} x_1 \\ x_2 \\ x_3 \\ x_4 \\ x_5 \\ x_6 \end{pmatrix},$$

where $\Delta B_1 = \Delta B_{11}$, $\Delta B_4 = \Delta B_{23}$, etc., and $x_1 = x_{11}$, $x_4 = 2x_{23}$, etc.

For the general case of an unstrained crystal, referred to its principal axes, the indicatrix is governed by the equation

$$B_1^0 Z_1^2 + B_2^0 Z_2^2 + B_3^0 Z_3^2 = 1.$$

Referring to Section 25.2, $B_1^0 = 1/n_1^2$, $B_2^0 = 1/n_2^2$, $B_3^0 = 1/n_3^2$ for biaxial crystals. For uniaxial crystals $B_1^0 = B_2^0 = 1/n_o^2$, and $B_3^0 = 1/n_e^2$, and for isotropic media $B_1^0 = B_2^0 = B_3^0 = 1/n^2$.

Triclinic crystals require the measurement of 36 elasto-optic coefficients but the number of coefficients is far smaller in higher symmetry crystals.

The matrices for the elasto-optic effect are identical to those of the electrostriction and magnetostriction effects (Table 15.5).

Photoelasticity is a fourth rank polar tensor that, like elasticity, is present in all point groups. The elasto-optic matrix coefficients p_{mn} are equal to the corresponding tensor coefficients p_{ijkl}, such that $p_{11} = p_{1111}$, $p_{12} = p_{1122}$, $p_{66} = p_{1212}$, etc. This is not true for the piezo-optic coefficients π_{mn}. In this case $\pi_{mn} = \pi_{ijkl}$ if $n = 1, 2,$ or 3, but $\pi_{mn} = 2\pi_{ijkl}$ if $n = 4, 5,$ or 6.

As mentioned earlier, the piezo-optic and elasto-optic coefficients are interrelated through the elastic constants:

$$p_{mn} = \pi_{mp} c_{pn} \quad \text{and} \quad \pi_{mn} = p_{mp} s_{pn},$$

where c_{pn} and s_{pn} are the elastic stiffnesses and elastic compliances written in matrix notation. Note that the elastic constant matrices are required to be symmetric, but the photoelastic matrices are not. The energy argument does not apply, so unless required by symmetry, $\pi_{mn} \neq \pi_{nm}$ and $p_{mn} \neq p_{nm}$.

Elasto-optic coefficients for several isotropic and uniaxial materials are compiled in Table 27.1.

27.3 Static photoelastic measurements

For many years, mechanical engineers have use the photoelastic effect to measure stress and strain patterns in complex designs. Fig. 27.1 shows a coupling hook for railway cars which are subject to enormous tensile stresses. To test the design, a model is made of a transparent material such as plexiglass, and then placed under tensile load. For a flat design, the stresses are all in the Z_1–Z_2 plane. In the photoelastic experiments (Fig. 27.2) the light beam is in the perpendicular direction along Z_3.

Plexiglass is optically isotropic so there are only two independent elasto-optic coefficients p_{11} and p_{12}. The shear coefficient $p_{44} = \frac{1}{2}(p_{11} - p_{12})$. The strains at any given point in the test specimen are x_1, x_2, and x_6. The resulting changes in the indicatrix are given by

$$\begin{pmatrix} \Delta B_1 \\ \Delta B_2 \\ \Delta B_3 \\ \Delta B_4 \\ \Delta B_5 \\ \Delta B_6 \end{pmatrix} = \begin{pmatrix} p_{11} & p_{12} & p_{12} & 0 & 0 & 0 \\ p_{12} & p_{11} & p_{12} & 0 & 0 & 0 \\ p_{12} & p_{12} & p_{11} & 0 & 0 & 0 \\ 0 & 0 & 0 & p_{44} & 0 & 0 \\ 0 & 0 & 0 & 0 & p_{44} & 0 \\ 0 & 0 & 0 & 0 & 0 & p_{44} \end{pmatrix} \begin{pmatrix} x_1 \\ x_2 \\ 0 \\ 0 \\ 0 \\ x_6 \end{pmatrix}.$$

Based on this expression, the equation of the indicatrix under strain is

$$\left(\frac{1}{n^2} + p_{11}x_1 + p_{12}x_2\right) Z_1^2 + \left(\frac{1}{n^2} + p_{12}x_1 + p_{11}x_2\right) Z_2^2$$
$$+ \left(\frac{1}{n^2} + p_{12}x_1 + p_{12}x_2\right) Z_3^2 + 2\left(\frac{1}{2}\right)(p_{11} - p_{12})Z_1 Z_2 = 1$$

or

$$B_{11}Z_1^2 + B_{22}Z_2^2 + B_{33}Z_3^2 + B_{12}Z_1 Z_2 = 1.$$

A Mohr circle construction can be used to determine the principal axes in the Z_1–Z_2 plane (Fig. 27.3). The angle between the measurement axes and the

Table 27.1 Room-temperature elasto-optic coefficients (dimensionless) (full matrices for each point group are listed in Table 15.4)

Isotropic ($\infty\infty m$) glasses and polymers

	SiO$_2$	As$_2$S$_3$	Polystyrene	Lucite
λ (μm)	0.630	1.150	0.633	0.633
p_{11}	0.121	0.308	±0.30	±0.30
p_{12}	0.270	0.299	±0.31	±0.28

Cubic crystals (point groups $\bar{4}3m, m3m$)

	GaP	GaAs	β-ZnS	Y$_3$Al$_5$O$_{12}$
λ (μm)	0.630	1.150	0.633	0.633
p_{11}	−0.151	−0.165	0.091	−0.029
p_{12}	−0.082	−0.140	−0.01	−0.0091
p_{44}	−0.074	−0.072	0.075	−0.0615

Tetragonal crystals (point groups $\bar{4}2m, 422, 4/mm$)

	NH$_4$H$_2$PO$_4$	KH$_2$PO$_4$	TeO$_2$	TiO$_2$
λ (μm)	0.589	0.589	0.633	0.514
p_{11}	0.319	0.287	0.0074	−0.001
p_{12}	0.277	0.282	0.187	0.113
p_{13}	0.169	0.174	0.340	−0.167
p_{31}	0.197	0.241	0.0905	−0.106
p_{33}	0.167	0.122	0.240	−0.064
p_{44}	−0.058	−0.019	−0.17	0.0095
p_{66}	−0.091	−0.064	−0.0463	−0.066

Trigonal crystals (32, 3m, $\bar{3}m$)

	Al$_2$O$_3$	LiNbO$_3$	LiTaO$_3$	α-SiO$_2$
λ (μm)	0.644	0.633	0.633	0.589
p_{11}	−0.23	−0.026	−0.081	0.16
p_{12}	−0.03	0.09	0.081	0.27
p_{13}	0.02	0.133	0.093	0.27
p_{14}	0.00	−0.075	−0.026	−0.03
p_{31}	−0.04	0.179	0.089	0.29
p_{33}	−0.20	0.071	−0.044	0.10
p_{41}	0.01	−0.151	−0.085	−0.047
p_{44}	−0.10	0.146	0.028	−0.079

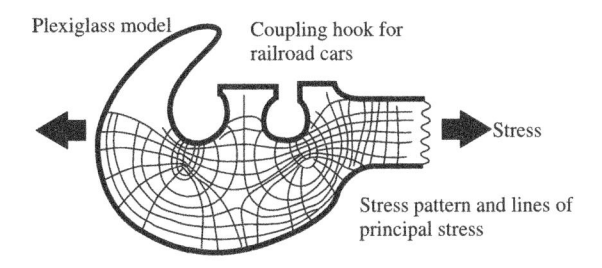

Fig. 27.1 Transparent model of a coupling hook under tensile stress showing the regions of maximum strain where fracture is likely to occur. Photoelastic experiments are used to determine the directions and magnitudes of the strain.

principal axes (θ) and the difference between the principal strains ($x_1 - x_2$) are determined from the construction.

The principal advantage of these photoelastic experiments is that they provide a rapid way of locating regions of stress concentration. They provide a useful way of verifying calculations carried out by finite element analysis on complex engineering designs.

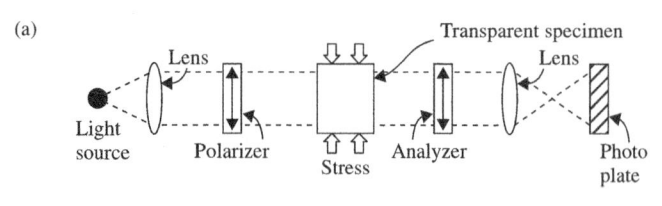

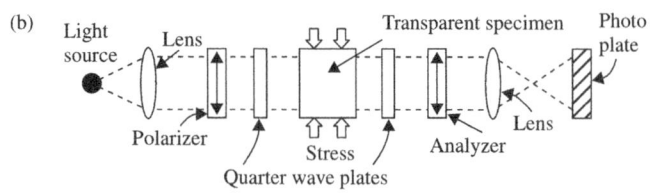

Fig. 27.2 A photoelastic polariscope is used to map the stress and strain configuration in engineering structures. (a) Directions of principal strain (isotropic lines) are determined by examining the specimen between crossed polarizers. (b) Isochromatic lines give the difference in strain level between the two principal strains (x_1 and x_2) at any point. These patterns are obtained by inserting quarter wave plates into the beam.

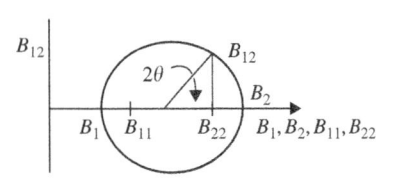

Fig. 27.3 Principal axes for the strain at any point in the plexiglass model are obtained by a Mohr circle construction by plotting the strained indicatrix components.

27.4 Acousto-optics

There are a number of acousto-optic devices involving the interaction of light waves with sound waves in a photoelastic medium. Acousto-optic phenomena have been used to build spectral analyzers, beam deflectors, optical modulators, and tunable filters. In many of these devices, laser beams in the optical range interact with elastic waves in the microwave region through Bragg diffraction.

As an example, consider a longitudinal sound wave in an isotropic medium (point group $\infty\infty m$) such as silica glass. The elasto-optic coefficients p_{11} and p_{12} are given in Table 27.1, and the elasto-optic matrix in Table 15.4. Let Z_3 be the direction of the wave normal, and the vibration direction since the wave is longitudinal. The strain field associated with the acoustic plane wave will be periodic along Z_3:

$$x_3 = A \sin(\Omega t - K Z_3),$$

where A is the amplitude of the wave, Ω the angular frequency, and K the wave vector. (The symbols ω and k are reserved for the optical beam). The resulting changes in the optical indicatrix are given by

$$\begin{pmatrix} \Delta B_1 \\ \Delta B_2 \\ \Delta B_3 \\ \Delta B_4 \\ \Delta B_5 \\ \Delta B_6 \end{pmatrix} = \begin{pmatrix} p_{11} & p_{12} & p_{12} & 0 & 0 & 0 \\ p_{12} & p_{11} & p_{12} & 0 & 0 & 0 \\ p_{12} & p_{12} & p_{11} & 0 & 0 & 0 \\ 0 & 0 & 0 & p_{44} & 0 & 0 \\ 0 & 0 & 0 & 0 & p_{44} & 0 \\ 0 & 0 & 0 & 0 & 0 & p_{44} \end{pmatrix} \begin{pmatrix} 0 \\ 0 \\ x_3 \\ 0 \\ 0 \\ 0 \end{pmatrix},$$

where $p_{44} = \frac{1}{2}(p_{11} - p_{12})$. This leads to a new uniaxial ellipsoid governed by the equation

$$\left(Z_1^2 + Z_2^2\right)\left(\frac{1}{n^2} + p_{12} A \sin(\Omega t - K Z_3)\right)$$
$$+ Z_3^2 \left(\frac{1}{n^2} + p_{11} A \sin(\Omega t - K Z_3)\right) = 1.$$

The important point is that, in the presence of an acoustic wave, the isotropic medium becomes a dynamic anisotropic medium. In effect it is converted to a volume grating with a periodicity $\Lambda = 2\pi/K$ along the wave normal Z_3. Optical waves can be diffracted by this grating.

The acoustic wave creates a perturbation that is periodic both in space and time. However, the speed of the acoustic wave (typically 10^3 m/s) is many orders of magnitude smaller than that of an optic wave (typically 10^8 m/s). Thus the acoustic grating is essentially stationary with respect to the optical beam. Accompanying the sound wave is a periodic change in refractive index which scatters the optical waves. The refractive index change is caused by the local changes in density and polarizability, as explained in Section 27.1.

The acousto-optic scattering process is pictured in Fig. 27.4. Consider a $\lambda = 0.5$ μm light wave in silica glass ($n = 1.5$) with a phase velocity $v = 2 \times 10^8$ m/s and an angular frequency $\omega = 2.5 \times 10^{15}$ Hz. A 2 GHz acoustic wave in silica has a speed $V = 6$ km/s and a wavelength $\Lambda = 3$ μm. The Bragg diffraction condition for these two waves is

$$2\Lambda \sin \theta = \lambda,$$

which is satisfied for a scattering angle of about $5°$. The diffracted optical beam is shifted in frequency by an amount equal to the sound frequency, which is a relatively small change. During the diffraction process the optical photon increases its energy by absorbing an acoustic phonon.

Energy and momentum are conserved during the scattering of light waves by the acoustic grating. For the isotropic medium just considered, the energies of the incident and scattered photons are $\hbar\omega$ and $\hbar\omega'$, where $\hbar$ is Planck's constant divided by 2π. The phonons associated with the acoustic wave have energy $\hbar\Omega$. Conservation of energy requires that

$$\omega' = \omega + \Omega.$$

If the direction of the acoustic beam is reversed $\omega' = \omega - \Omega$ and a new phonon is created as another is destroyed.

The momentum of the incident photon is $\hbar\bar{k}$ where $|k| = 2\pi/\lambda$ is the wave vector parallel to the wave normal. For the scattered photon and the acoustic phonon, the momenta vectors are $\hbar\bar{k}'$ and $\hbar\bar{K}$, respectively. Momentum conservation requires that $\bar{k}' = \bar{k} + \bar{K}$.

For an isotropic solid, optical beams travel with the same velocity in all directions. This condition is altered only slightly in the presence of the low energy phonons. This means that

$$\omega' \cong \omega \quad \text{and} \quad |\bar{k}'| \cong |\bar{k}|.$$

Therefore the momentum conservation condition leads to the Bragg diffraction condition shown in Fig. 27.5(a). In this case the scattered angle θ' and the angle of incidence θ are nearly equal. The magnitude of K is then $2k \sin \theta$ which is equivalent to the Bragg equation $2\Lambda \sin \theta = \lambda$.

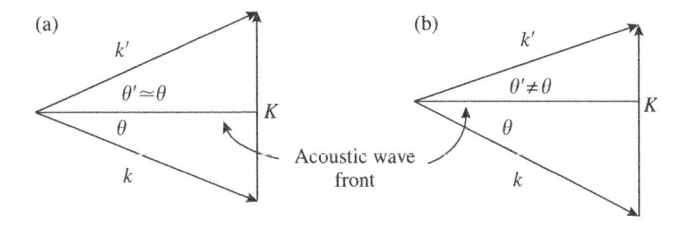

Fig. 27.4 Bragg diffraction of optical waves from an ultrasonic acoustic wave in the microwave range.

Fig. 27.5 Momentum conservation conditions in (a) an isotropic solid and (b) an anisotropic solid.

27.5 Anisotropic media

For an anisotropic solid, the optical beams traveling in different directions generally have different velocities and different refractive indices. Let n and n' represent the refractive indices of the incident and diffracted beams. In anisotropic crystals such as calcite and rutile, n and n' can differ by 10%. Under these conditions the angles θ and θ' between the photon beams and the ultrasonic wave front are no longer equal. The diffraction conditions are obtained from the triangle specifying momentum conservation (Fig. 27.5(b)). For the incident photon

$$2k \sin \theta = K - \frac{(k'^2 - k^2)}{K},$$

and for the diffracted photon

$$2k' \sin \theta' = K + \frac{(k'^2 - k^2)}{K}.$$

Rewriting these equations in terms of wavelengths and refractive indices and remembering that $\omega' \cong \omega$, the diffraction conditions for anisotropic media become

$$\sin \theta = \left(\frac{1}{2n}\right)\left[\left(\frac{\lambda}{\Lambda}\right) - \left(\frac{\Lambda}{\lambda}\right)(n'^2 - n^2)\right]$$

and

$$\sin \theta' = \left(\frac{1}{2n'}\right)\left[\left(\frac{\lambda}{\Lambda}\right) + \left(\frac{\Lambda}{\lambda}\right)(n'^2 - n^2)\right].$$

When $n' = n$ these expressions revert to the Bragg equation with $\theta' = \theta$. For a uniaxial crystal, the maximum and minimum values of n' and n would correspond to n_e and n_o, and the acousto-optic experiment would often involve both ordinary and extraordinary rays with different polarization directions.

27.6 Material issues

The intensity of the diffracted light depends on the path length within the sample, the optical wavelength, and a material figure of merit M.

$$M = \frac{n^6 \bar{p}^2}{\rho V^3}.$$

n is the refractive index, V the acoustic wave velocity, ρ the density, and $\bar{p}$ the effective elasto-optic coefficient. $V, \bar{p}, n$, and M all depend on the configuration of the experiment: sample orientation, direction of propagation, and the polarization states of the optical and ultrasonic waves. Therefore the numerical values in Table 27.2 are only approximate. Nevertheless some general conclusions can be drawn regarding the types of material that give the largest acousto-optic interactions.

The dominant factors are the refractive index n and the acoustic velocity V. Slow waves—both optic and acoustic—enhance the interaction time and the diffracted intensities. Common oxides are generally mechanically stiff and have fast acoustic waves. Therefore the figures of merit for silica, alumina, lithium niobate, and most other oxides are small. An exception is paratellurite,

Table 27.2 Representative figures of merit for acousto-optic materials (all are expressed in units of 10^{-15} MKS)

Amorphous		Tetragonal	
SiO_2	1.5	$NH_4H_2PO_4$	6.4
As_2S_3	433	KH_2PO_4	3.8
Polystyrene	120	TeO_2	800
		TiO_2	3.9

Cubic		Trigonal	
GaP	45	Al_2O_3	0.34
GaAs	104	$LiNbO_3$	7.0
β-ZnS	3.4	$LiTaO_3$	1.4
$Y_3Al_5O_{12}$	0.07		

TeO_2, which has an exceedingly slow shear wave along [110]. The acoustic wave velocity surfaces for TeO_2 were shown in Fig. 23.9. Transverse acoustic waves generally travel slower than longitudinal waves in most materials, giving them an advantage in acousto-optic experiments. Polymers, organic crystals, and other low-melting compounds generally have weaker chemical bonds and slower acoustic waves. Polystyrene, water, and iodic acid (HIO_3) all have slow acoustic waves and good M values.

Refractive indices are also very important. Common oxides and common salts like SiO_2 and LiF suffer here as well. Heavy metal sulfides such as CdS and As_2S_3 have very high refractive indices because of their proximity to an absorption band.

Elasto-optic coefficients (Table 27.1) show a great deal of scatter with few real trends. As pointed out in Section 27.1 there are conflicting causes for the photoelastic coefficients which means they can be positive, negative, or near zero. Because of the scatter, orientation of the wave normals and polarization directions are very important. In yttrium aluminum garnet, for instance, the figure of merit for longitudinal acoustic waves along [110] have a much higher figure of merit than those traveling along [100]. A listing of acousto-optic figures of merit is presented in Table 27.2.

28 Electro-optic phenomena

28.1 Linear electro-optic effect 303
28.2 Pockels Effect in KDP and ADP 304
28.3 Linear electro-optic coefficients 308
28.4 Quadratic electro-optic effect 309

Optical beams can be controlled by manipulating the refractive indices and absorption coefficients with applied electric fields. In communication systems electro-optic effects are used in phase and amplitude modulation, in beam deflectors, and in tunable filters.

Three such effects are illustrated in Fig. 28.1. Lead lanthanum zirconate titanate (PLZT) is a transparent electroceramic that can be prepared in several different ferroelectric forms with large electro-optic coefficients. When prepared in a normal ferroelectric form it can be used in two different ways. A light-tunable shutter is constructed by coating a multidomain ceramic of PLZT with a photoconducting layer and transparent electrodes (Fig. 28.1(a)). A bias voltage on the electrodes is transferred to the ceramic when the photoconductor is illuminated. The electric field alters the domain structure and the degree of light scattering, controlling the intensity of light.

Fully poled ferroelectric ceramics exhibit the linear electro-optic effect (Fig. 28.1(b)) Using planar electrodes the PLZT is poled perpendicular to the optical beam. Polarizer and analyzer are positioned in the ±45° positions, and light intensity is controlled by altering the birefringence with an electric field.

The third experiment utilizes a pseudo-cubic PLZT composition with a large quadratic electro-optic effect (Fig. 28.1(c)). No poling is required in this case.

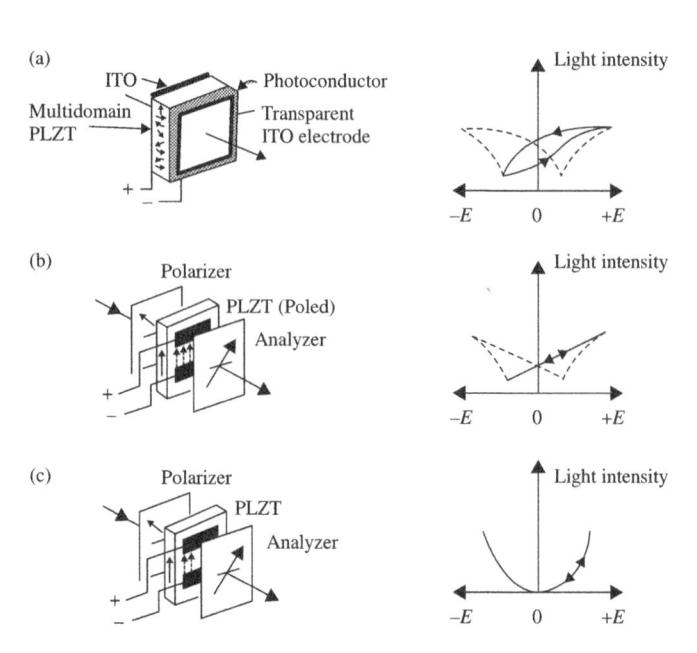

Fig. 28.1 Three ways of controlling light intensity are: (a) by light scattering from domain walls, (b) by the linear electro-optic effect, and (c) by the quadratic electro-optic effect. Various compositions of transparent PLZT ceramics made of ferroelectric (Pb, La)(Zr, Ti)O_3 have been used in these experiments (Haertling).

With polarizer and analyzer again in the $\pm 45°$ positions, the transmitted light intensity is proportional to E^2 rather than E.

28.1 Linear electro-optic effect

Linear and quadratic electro-optic coefficients are defined in terms of the field-induced changes in the optical indicatrix:

$$B_{ij}(E) - B_{ij}(0) = \Delta B_{ij} = r_{ijk}E_k + R_{ijkl}E_k E_l.$$

Since the indicatrix components B_{ij} are dimensionless, and the applied electric field components E_k and E_l are measured in volts/meter, the units of the linear (r_{ijk}) and quadratic (R_{ijkl}) electro-optic coefficients are m/V and m²/V², respectively.

The linear electro-optic effect is a third rank polar tensor known as the *Pockels Effect*. Since the indicatrix components are symmetric ($\Delta B_{ij} = \Delta B_{ji}$) the tensor can be written as a 6 × 3 matrix with 18 Pockels coefficients:

$$\Delta B_i = r_{ij}E_j \quad (i = 1\text{--}6, \ j = 1\text{--}3).$$

No factors of 2 appear in the relations between the tensor coefficients and the matrix components. Thus, for example, $r_{222} = r_{22}$ and $r_{123} = r_{213} = r_{63}$.

For triclinic crystals (point group 1) there are 18 linear electro-optic coefficients but symmetry greatly reduces this number. In the next section we consider the electro-optic effect in KH$_2$PO$_4$ (KDP), so we begin by deriving the Pockels matrix for point group $\bar{4}2m$. Two symmetry elements are required to generate this group, a fourfold inversion axis along Z_3 and a twofold axis parallel to Z_1. The direct inspection method can be used for these symmetry elements. For $\bar{4} \parallel Z_3$, $1 \to -2 \to -1$, $2 \to 1 \to -2$, $3 \to -3 \to 3$. The tensor coefficients and their corresponding matrix coefficients transform as follows:

$$111 \to -222 \to -111, \quad r_{11} = r_{22} = 0$$
$$112 \to 221 \to -112, \quad r_{12} = r_{21} = 0$$
$$113 \to -223 \to 113, \quad r_{13} = -r_{23}$$
$$331 \to -332 \to -331, \quad r_{31} = r_{32} = 0$$
$$333 \to -333, \quad r_{33} = 0$$
$$231 \to 132 \to 231, \quad r_{41} = r_{52}$$
$$233 \to 133 \to -233, \quad r_{43} = r_{53} = 0$$
$$121 \to 212 \to -121, \quad r_{61} = r_{62} = 0$$
$$123 \to 213 \to 123, \quad r_{63}.$$

For $2 \parallel Z_1$, $1 \to 1$, $2 \to -2$, $3 \to -3$ and the remaining nonzero coefficients transform as follows.

$$113 \to -113, \quad r_{13} = r_{23} = 0$$
$$231 \to 231, \quad r_{41} = r_{52}$$
$$232 \to -232, \quad r_{42} = r_{51} = 0$$
$$123 \to 123, \quad r_{63}.$$

Therefore only three nonzero electro-optic coefficients remain for point group $\bar{4}2m$: $r_{41} = r_{52}$ and r_{63}. The experiment described in the next section utilizes r_{63}.

Electro-optic coefficients for the 32 crystallographic point groups and seven Curie groups are given in Table 28.1. Except for a few factors of 2, the linear electro-optic matrices are identical in form to those used for the converse piezoelectric effect. Like the pyroelectric and piezoelectric effects, the Pockels Effect disappears in all centrosymmetric point groups. Again from the direct inspection method, an inversion center takes $1 \to -1$, $2 \to -2$, and $3 \to -3$. Therefore for any polar third-rank tensor subscript, $ijk \to -ijk = 0$. The same argument holds for other odd-rank polar tensors. They too disappear in centric groups. Pockels coefficients are also zero in two noncentrosymmetric point groups 432 and Curie group $\infty\infty$.

In the absence of an electric field, the optical indicatrix is an ellipsoid

$$\frac{Z_1^2}{n_1^2} + \frac{Z_2^2}{n_2^2} + \frac{Z_3^2}{n_3^2} = 1,$$

where n_1, n_2, and n_3 are the refractive indices associated with the principal axes Z_1, Z_2, Z_3. In terms of the dielectric impermittivity coefficients, $B_{11}(0) = 1/n_1^2$, $B_{22}(0) = 1/n_2^2$, and $B_{33}(0) = 1/n_3^2$. All other $B_{ij}(0)$ terms are zero.

When an electric field E_k is applied to the crystal, the ellipsoid is modified to read,

$$\left(\frac{1}{n_1^2} + r_{1k}E_k\right)Z_1^2 + \left(\frac{1}{n_2^2} + r_{2k}E_k\right)Z_2^2 + \left(\frac{1}{n_3^2} + r_{3k}E_k\right)Z_3^2$$
$$+ 2r_{4k}E_k Z_2 Z_3 + 2r_{5k}E_k Z_1 Z_3 + 2r_{6k}E_k Z_1 Z_2 = 1.$$

We have assumed the quadratic electro-optic effect to be negligibly small. New principal axes are required whenever the cross-terms involving r_{4k}, r_{5k}, and r_{6k} are present. A new set of principal axes are obtained by coordinate rotation.

As explained in the next section, the preferred light path in most electro-optic experiments is along an optic axis. The presence of birefringence makes it difficult to observe the electro-optic effect.

28.2 Pockels Effect in KDP and ADP

Large transparent crystals of potassium dihydrogen phosphate (KDP = KH_2PO_4) and ammonium dihydrogen phosphate (ADP = $NH_4H_2PO_4$) can be grown from water solution. KDP and ADP are good examples of a hydrogen-bonded ferroelectric and antiferroelectric, respectively, but the electro-optic experiments to be described here are carried out near room temperature where the crystals are above T_c in the paraelectric state. The point group is $\bar{4}2m$.

The crystal structures of KDP and ADP consist of PO_4 phosphate groups bonded together by K^+ or NH_4^+ ions and hydrogen bonds (Fig. 28.2). At room temperature the hydrogen atoms are disordered, occupying two sites with equal probability. Both structures undergo phase transitions at low temperatures, with the protons ordering in double potential wells.

Table 28.1 Linear electro-optic matrices for the 32 crystal classes and seven Curie groups

1
$$\begin{pmatrix} r_{11} & r_{12} & r_{13} \\ r_{21} & r_{22} & r_{23} \\ r_{31} & r_{32} & r_{33} \\ r_{41} & r_{42} & r_{43} \\ r_{51} & r_{52} & r_{53} \\ r_{61} & r_{62} & r_{63} \end{pmatrix}$$

2
$$\begin{pmatrix} 0 & r_{12} & 0 \\ 0 & r_{22} & 0 \\ 0 & r_{32} & 0 \\ r_{41} & 0 & r_{43} \\ 0 & r_{52} & 0 \\ r_{61} & 0 & r_{63} \end{pmatrix}$$

m
$$\begin{pmatrix} r_{11} & 0 & r_{13} \\ r_{21} & 0 & r_{23} \\ r_{31} & 0 & r_{33} \\ 0 & r_{42} & 0 \\ r_{51} & 0 & r_{53} \\ 0 & r_{62} & 0 \end{pmatrix}$$

222
$$\begin{pmatrix} 0 & 0 & 0 \\ 0 & 0 & 0 \\ 0 & 0 & 0 \\ r_{41} & 0 & 0 \\ 0 & r_{52} & 0 \\ 0 & 0 & r_{63} \end{pmatrix}$$

mm2
$$\begin{pmatrix} 0 & 0 & r_{13} \\ 0 & 0 & r_{23} \\ 0 & 0 & r_{33} \\ 0 & r_{42} & 0 \\ r_{51} & 0 & 0 \\ 0 & 0 & 0 \end{pmatrix}$$

3
$$\begin{pmatrix} r_{11} & -r_{22} & r_{13} \\ -r_{11} & r_{22} & r_{13} \\ 0 & 0 & r_{33} \\ r_{41} & r_{51} & 0 \\ r_{51} & -r_{41} & 0 \\ -r_{22} & -r_{11} & 0 \end{pmatrix}$$

32
$$\begin{pmatrix} r_{11} & 0 & 0 \\ -r_{11} & 0 & 0 \\ 0 & 0 & 0 \\ r_{41} & 0 & 0 \\ 0 & -r_{41} & 0 \\ 0 & -r_{11} & 0 \end{pmatrix}$$

3m
$$\begin{pmatrix} 0 & -r_{22} & r_{13} \\ 0 & r_{22} & r_{13} \\ 0 & 0 & r_{33} \\ 0 & r_{51} & 0 \\ r_{51} & 0 & 0 \\ -r_{22} & 0 & 0 \end{pmatrix}$$

4, 6, ∞
$$\begin{pmatrix} 0 & 0 & r_{13} \\ 0 & 0 & r_{13} \\ 0 & 0 & r_{33} \\ r_{41} & r_{51} & 0 \\ r_{51} & -r_{41} & 0 \\ 0 & 0 & 0 \end{pmatrix}$$

$\bar{4}$
$$\begin{pmatrix} 0 & 0 & r_{13} \\ 0 & 0 & -r_{13} \\ 0 & 0 & 0 \\ r_{41} & -r_{51} & 0 \\ r_{51} & r_{41} & 0 \\ 0 & 0 & r_{63} \end{pmatrix}$$

422, 622, ∞2
$$\begin{pmatrix} 0 & 0 & 0 \\ 0 & 0 & 0 \\ 0 & 0 & 0 \\ r_{41} & 0 & 0 \\ 0 & -r_{41} & 0 \\ 0 & 0 & 0 \end{pmatrix}$$

4mm, 6mm, ∞m
$$\begin{pmatrix} 0 & 0 & r_{13} \\ 0 & 0 & r_{13} \\ 0 & 0 & r_{33} \\ 0 & r_{51} & 0 \\ r_{51} & 0 & 0 \\ 0 & 0 & 0 \end{pmatrix}$$

$\bar{4}2m$
$$\begin{pmatrix} 0 & 0 & 0 \\ 0 & 0 & 0 \\ 0 & 0 & 0 \\ r_{41} & 0 & 0 \\ 0 & r_{41} & 0 \\ 0 & 0 & r_{63} \end{pmatrix}$$

$\bar{6}$
$$\begin{pmatrix} r_{11} & -r_{22} & 0 \\ -r_{11} & r_{22} & 0 \\ 0 & 0 & 0 \\ 0 & 0 & 0 \\ 0 & 0 & 0 \\ -r_{22} & -r_{11} & 0 \end{pmatrix}$$

$\bar{6}m2$
$$\begin{pmatrix} 0 & -r_{22} & 0 \\ 0 & r_{22} & 0 \\ 0 & 0 & 0 \\ 0 & 0 & 0 \\ 0 & 0 & 0 \\ -r_{22} & 0 & 0 \end{pmatrix}$$

$\bar{4}3m$, 23
$$\begin{pmatrix} 0 & 0 & 0 \\ 0 & 0 & 0 \\ 0 & 0 & 0 \\ r_{41} & 0 & 0 \\ 0 & r_{41} & 0 \\ 0 & 0 & r_{41} \end{pmatrix}$$

Most other groups *centric*:
$\bar{1}, 2/m,$
$mmm, \bar{3}, \bar{3}m$
$4/m, 4/mmm$
$6/m, 6/mmm$
$m3, m3m, \infty/m$
$\infty/mm, \infty\infty m$
acentric: 432, ∞∞

$$\begin{pmatrix} 0 & 0 & 0 \\ 0 & 0 & 0 \\ 0 & 0 & 0 \\ 0 & 0 & 0 \\ 0 & 0 & 0 \\ 0 & 0 & 0 \end{pmatrix}$$

Below the Curie point, the hydrogens in KH_2PO_4 are in an ordered arrangement with two hydrogens near every PO_4 group. KH_2PO_4 polarizes along the c crystallographic axis with P_s either parallel or antiparallel to c, forming 180° domains. In domains with the spontaneous polarization parallel to c, the protons at the base of the tetrahedra move close and the upper ones move away (Fig. 28.2(a)). Applying an electric field parallel to $-c$ switches P_s and the lower protons move away while the upper protons move close.

$NH_4H_2PO_4$ is nearly isomorphous with KH_2PO_4 but the proton ordering is different. The symmetry of KH_2PO_4 changes from $\bar{4}2m$ to $mm2$ at the transition, whereas the ammonium salt transforms from $\bar{4}2m$ to 222. At room temperature the acid hydrogens in $NH_4H_2PO_4$ are disordered, as in KH_2PO_4, and at low temperatures they adopt the arrangement shown in Fig. 28.2(b). In this case, one lower H^+ and one upper H^+ move close to each PO_4 group, canceling any shifts along c. Ammonium dihydrogen phosphate does not polarize spontaneously at the transition, and is, therefore, not a ferroelectric. It is called an antiferroelectric because of the antiparallel shifts, and because it is closely related to ferroelectric KH_2PO_4. ADP is ferroelastic and potentially ferrobielectric.

In any electro-optic experiment, the first questions concern the orientation directions. How should the light beam be aligned and what are the preferred directions for the electric field and the optical polarizers?

Electro-optic experiments are generally done along an optic axis because the induced birefringence is much smaller than the standing birefringence. By way of illustration, the intrinsic birefringence in KDP is $\Delta n = 0.04$. The linear electro-optic coefficients in crystals are in the range 10^{-12} to 10^{-10} m/V. To determine how the indicatrix coefficients depend on the electric field, we recognize that the unmodified B coefficient is $1/n^2$, so $dB/dn = -2n^{-3}$. This means that the field-induced birefringence is

$$\Delta n(E) = \left(-\frac{n^3}{2}\right)(\Delta B) = -\left(-\frac{n^3}{2}\right)rE.$$

For KDP, $n \sim 1.5$ and $r \sim 10^{-11}$ m/V, and even under a sizeable field of 10^6 V/m, $\Delta n(E) \sim 10^{-5}$ which is three orders of magnitude smaller than the natural birefringence in most anisotropic crystals.

To avoid the standing birefringence, the light beam must be aligned along an optic axis. Any direction in a cubic crystal will satisfy this requirement. Optic axis directions in uniaxial or biaxial media will also eliminate birefringence, but in biaxial crystals the optic axis directions change with wavelength and temperature, and are awkward to work with.

The orientation of the field depends on which electro-optic coefficients are available. For KDP and ADP (point group $\bar{4}2m$), the nonzero coefficients are $r_{41} = r_{52}$ and r_{63}. The crystals are optically uniaxial with $n_1 = n_2 = n_o$ and $n_3 = n_e$ and the optic axis along Z_3, and we are most concerned with the section of the indicatrix for which $Z_3 = 0$.

With no field, the indicatrix is

$$\frac{1}{n_o^2}(Z_1^2 + Z_2^2) + \frac{1}{n_e^2}Z_3^2 = 1.$$

The section perpendicular to Z_3 is a circle of radius n_o, the ordinary refractive index (Fig. 28.3(b)). In the presence of an electric field the indicatrix is altered to

$$\frac{1}{n_o^2}(Z_1^2 + Z_2^2) + \frac{1}{n_e^2}Z_3^2 + 2r_{41}(E_1 Z_2 Z_3 + E_2 Z_1 Z_3) + 2r_{63}E_3 Z_1 Z_2 = 1.$$

The $Z_3 = 0$ section perpendicular to the light path is

$$\frac{1}{n_o^2}(Z_1^2 + Z_2^2) + 2r_{63}E_3 Z_1 Z_2 = 1.$$

Based on this argument, the only effective direction for applying the electric field is along Z_3. This results in an elliptical section (Fig. 28.3(b)) with the principal axes aligned parallel to the [110] and [$\bar{1}$10] directions of the tetragonal KDP crystal.

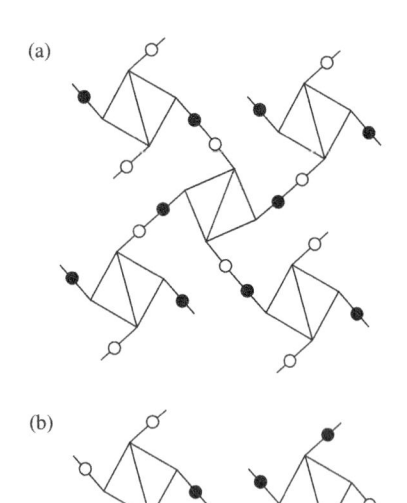

Fig. 28.2 Ordering of hydrogen ions on the O–H–O bonds in (a) ferroelectric KH_2PO_4 and (b) antiferroelectric $NH_4H_2PO_4$. Full and empty proton sites are represented by solid and open circles, respectively. Most of the electro-optic experiments are carried out at room-temperature where the protons are disordered.

The final point concerns the orientation of the polarizer and analyzer. Since the goal is to modulate the intensity of the light beam with an electric field, the polarizer and analyzer are crossed to assure extinction. With the polarizer along $Z_1 = [100]$ and the analyzer along $Z_2 = [010]$, the intensity is zero when $E = 0$. When the field E_3 is switched on the crystal brightens as the principal axes an aligned along [110] and [$\bar{1}$10]. The brightness will be proportional to E_3.

For maximum brightness, calculate the voltage that produces a half-wavelength path difference. This is the so-called half-wavelength voltage $V_{\lambda/2}$ determined from the condition

$$\Delta n t = \frac{\lambda}{2},$$

where t is the crystal thickness and Δn is the birefringence induced by the applied electric field. Referring to the new principal axes for KDP (Fig. 28.3(b)), the maximum (n'_1) and minimum (n'_2) refractive indices along [110] and {$\bar{1}$10} are given by

$$\left(\frac{1}{n'_1}\right)^2 = \left(\frac{1}{n_o^2}\right)^2 + r_{63} E_3$$

and

$$\left(\frac{1}{n'_2}\right)^2 = \left(\frac{1}{n_o^2}\right)^2 - r_{63} E_3.$$

Combining these two terms and remembering that $r_{63} E_3$ is only a small correction gives the field-induced birefringence

$$\Delta n = n'_1 - n'_2 \cong r_{63} E_3 n_o^3.$$

Substituting Δn into the half-wavelength path difference condition gives the half-wavelength voltage.

$$V_{\lambda/2} = E_3 t = \lambda / 2 r_{63} n_o^3.$$

For KDP, $r_{63} = 10.3$ pm/V and $n_o = 1.5115$ at a wavelength of 0.546 μm. The corresponding value of $V_{\lambda/2}$ is 7.8 kV.

Notice that the half-wave voltage is directly proportional to wavelength and inversely proportional to the product rn^3. To reduce the applied voltage it is advantageous to use crystals with large refractive indices and large electro-optic coefficients. Table 28.2 compares $V_{\lambda/2}$ values for several crystals and transparent ceramics. Many of the better electro-optic materials operate near an absorption edge (to increase n) or near a ferroelectric transition (to increase r), or both.

The temperature dependence of the KDP coefficient r_{63} illustrates this point. As KDP is cooled toward the paraelectric–ferroelectric phase transition r_{63} increases by two orders of magnitude (Fig. 28.4).

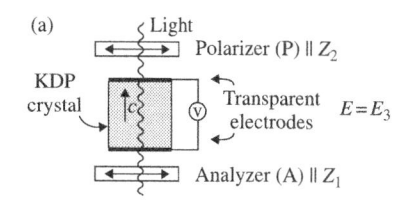

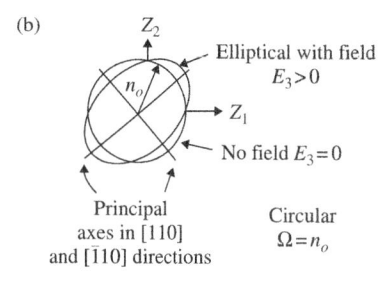

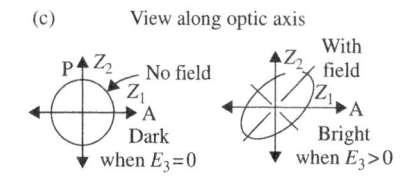

Fig. 28.3 Electro-optic experiment in KDP, ADP, and other crystals belonging to point group $\bar{4}2m$. (a) The light beam is directed along $Z_3 = [001]$ to avoid standing birefringence. The electric field must also be parallel to Z_3 to produce a linear electro-optic effect. Polarizer and analyzer are aligned along Z_1 and Z_2, respectively, to modulate the intensity of the light beam. (b) Sections of the optical indicatrix perpendicular to the optic axis, with and without an applied electric field. (c) Positions of polarizer and analyzer viewed along Z_3. When $E = 0$ the light is blocked, but brightens in the presence of a field E_3.

Table 28.2 Half-wave voltages for single crystals and poled ceramics exhibiting the Pockels Effect

SiO$_2$ (α-quartz)	290 kV	ZnSe	7.1 kV
NaClO$_3$	200	LiNbO$_3$	2.8
Bi$_4$Ge$_3$O$_{12}$	73	NaBa$_2$Nb$_5$O$_{15}$	2.1
Bi$_{12}$GeO$_{20}$	6	Pb(Mg,Nb,Ti)O$_3$	1.0
KH$_2$PO$_4$	7.8	(Pb,La)(Zr,Ti)O$_3$	0.1

28.3 Linear electro-optic coefficients

Poled PLZT has a sizeable Pockels Effect when prepared as a transparent ceramic by hot-pressing. The point group of a poled ceramic is ∞m with five electro-optic coefficients $r_{13} = r_{23}$, r_{33}, and $r_{51} = r_{42}$. If as in Fig. 28.1(b), the optical beam is directed along Z_1 and the electric field along Z_3, the principal axes will remain unchanged but the birefringence will change from $\Delta n(0) = (n_e - n_o)$ to $\Delta n(E) = (n_e - n_o) - \frac{1}{2}(n_e^3 r_{33} - n_o^2 r_{13})E_3$. The change in birefringence is quite large compared to most other electro-optic media and leads to a very low half-wave voltage (Table 28.2).

Electro-optic coefficients for a number of single crystals are collected in Table 28.3. For non-ferroelectric crystals the r values are generally in the range 1–10 pm/V. Ferroelectrics can have much larger values.

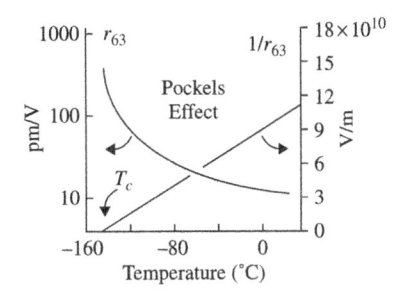

Fig. 28.4 The temperature dependence of r_{63} and $1/r_{63}$ for KH_2PO_4. The ferroelectric phase transformation leads to large dipolar response to electric fields, with $r \sim (T - T_c)^{-1}$.

Problem 28.1
Ferroelectric crystals are often used in electro-optic devices because only small control voltages are required. The refractive indices and electro-optic coefficients of $KNbO_3$ for $\lambda = 0.63$ μm are listed in Table 28.3. Determine the orientations of the two optic axes. An electro-optic experiment is performed with both the light beam and the applied electric field along an optic axis. Write out equations governing the indicatrix with and without an electric field. How should the polarizer and analyzer be positioned to maximize the electro-optic effect?

The Pockels coefficients in Table 28.3 depend on the temperature of the crystal, the wavelength of light, and the frequency of the applied electric field. The temperature dependence is small in most crystals, often within experimental error, but can be very large near phase transformations. As pointed out

Table 28.3 Linear electro-optical coefficients measured at room temperature in the visible and near infrared range

Crystal	λ (μm)	n	r (pm/V)
CdTe ($\bar{4}3m$)	1.0	2.84	$r_{41} = 4.5$
GaAs ($\bar{4}3m$)	1.15	3.43	$r_{41} = 1.43$
GaP ($\bar{4}3m$)	0.63	3.32	$r_{41} = -0.97$
ZnS ($\bar{4}3m$)	0.5	2.42	$r_{41} = 1.81$
ZnSe ($\bar{4}3m$)	0.55	2.66	$r_{41} = 2.0$
ZnTe ($\bar{4}3m$)	0.59	3.06	$r_{41} = 4.51$
CdS ($6mm$)	1.15	$n_o = 2.32$	$r_{13} = 3.1, r_{33} = 3.2$
		$n_e = 2.34$	$r_{51} = 2.0$
KH_2PO_4 ($\bar{4}2m$)	0.55	$n_o = 1.507$	$r_{41} = 8.77$
		$n_e = 1.467$	$r_{63} = 10.3$
$NH_4H_2PO_4$ ($\bar{4}2m$)	0.55	$n_o = 1.522$	$r_{41} = 23.76$
		$n_e = 1.478$	$r_{63} = 8.56$
Ba Sr$_3$Nb$_8$O$_{24}$ ($4mm$)	0.63	$n_o = 2.3117$	$r_{13} = 67, r_{33} = 1340$
		$n_e = 2.2987$	$r_{51} = 42$
LiNbO$_3$ ($3m$)	0.63	$n_o = 2.286$	$r_{13} = 9.6, r_{22} = 6.8$
		$n_e = 2.200$	$r_{33} = 30.9, r_{51} = 32.6$
α-HIO$_3$ (222)	0.63	$n_1 = 1.8365$	$r_{41} = 6.6$
		$n_2 = 1.984$	$r_{52} = 7.0$
		$n_3 = 1.96$	$r_{63} = 6.0$
KNbO$_3$ ($mm2$)	0.63	$n_1 = 1.8365$	$r_{13} = 28, r_{23} = 1.3$
		$n_2 = 1.984$	$r_{33} = 64, r_{42} = 380$
		$n_3 = 1.96$	$r_{51} = 105$

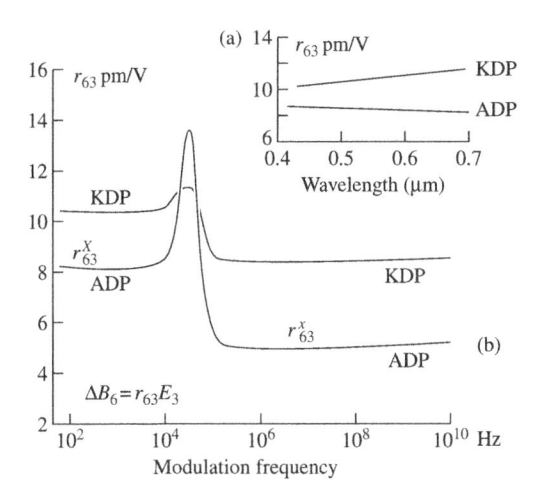

Fig. 28.5 (a) Pockels coefficient r_{63} of KDP and ADP plotted as a function of wavelength. (b) Effect of modulation frequency on r_{63} for KDP and ADP showing the difference in the coefficient measured at constant stress (r_{63}^X) and constant strain (r_{63}^x). At still higher frequencies above the infrared range, where optic and acoustic infrared modes are clamped, only electronic contributions to the Pockels coefficient remain.

previously for KDP, the r coefficient follows a Curie–Weiss law near the Curie temperature (Fig. 28.4).

The dependence of r on optical wavelength are similar in size to the dispersion of refractive indices. Data for KDP and ADP are shown in Fig. 28.5(a).

The effect of the modulating frequency follows the earlier discussion of the dielectric constant (Section 9.1). Polarization mechanisms drop out, one by one, as frequency increases. When the modulating frequency is low, the crystal deforms through piezoelectric coefficients d_{ijk}. The electrically induced strains affect the refractive indices through the elasto-optic coefficients p_{ijkl}. At modulating frequencies above the piezoelectric resonances, the crystal does not have time to deform but there is, nevertheless, an electro-optic effect. This is the constant strain coefficient r_{ijk}^x which is sometimes referred to as the "true" or "primary" electro-optic coefficient. The low-frequency (stress-free) coefficient r_{ijk}^X is equal to the sum of the strain free coefficient and the piezoelectric contribution:

$$r_{ijk}^X = r_{ijk}^x + p_{ijlm}d_{lmk}.$$

The second term is sometimes called the "false" or "secondary" electro-optic coefficient, but it is not really false. It is an important part of the Pockels Effect at low frequency. When written in matrix form, the equation becomes

$$r_{mn}^X = r_{mn}^x + p_{mp}d_{pn}^t,$$

where $m, p = 1$–6, and $n = 1$–3.

The piezoelectric contribution drops out in the microwave region, leaving only r_{mn}^x. Since the piezoelectric and elasto-optic coefficients can be either positive or negative, the change in Pockels coefficient can also be positive or negative. Measured values for the r_{63} coefficient of KDP and ADP are shown in Fig. 28.5(b).

28.4 Quadratic electro-optic effect

The quadratic electro-optic effect (the *Kerr Effect*) relates the changes in the optical indicatrix ΔB_{ij} to the square of the applied electric field. In tensor form

$$B_{ij}(E) - B_{ij}(0) = \Delta B_{ij} = R_{ijkl}E_k E_l.$$

310 Electro-optic phenomena

The Kerr Effect tensor R_{ijkl} is analogous to the electrostriction and magnetostriction tensors in that it relates a symmetric second rank tensor to the product of two first-rank tensors (see Sections 15.3 and 15.4). It is a fourth rank tensor, but not a symmetric fourth rank tensor like the elastic constants, since the energy argument does not apply. It is convenient to rewrite the quadratic electro-optic coefficients as a 6×6 matrix. Remembering that the equation of the indicatrix is that of an ellipsoid,

$$B_{ij}Z_iZ_j = B_{11}Z_1^2 + B_{22}Z_2^2 + B_{33}Z_3^2 + 2B_{12}Z_1Z_2 + 2B_{13}Z_1Z_3 + 2B_{23}Z_2Z_3$$
$$= B_1Z_1^2 + B_2Z_2^2 + B_3Z_3^2 + 2B_6Z_1Z_2 + 2B_5Z_1Z_3 + 2B_4Z_2Z_3.$$

There are no factors of 2 involved in going from the tensor coefficients B_{ij} ($i, j = 1$–3) to the matrix coefficients B_i ($i, j = 1$–6). This holds true for the changes as well, $\Delta B_{ij} = \Delta B_i$. Using the notation employed earlier (Section 15.3) for electrostriction, $E_1E_1 = E_1^2$, $E_1E_2 = E_6^2$, $E_1E_3 = E_5^2$, etc., gives the matrix form

$$\begin{array}{ccc} 6 \times 1 & & 6 \times 6 \quad 6 \times 1 \\ (\Delta B) & = & (R) \quad (E^2) \end{array}$$

Thus there are 36 terms in the Kerr Effect matrix for a triclinic crystal. Matrices for the other point groups are identical to those for the magnetostriction effect in Table 15.4. Symmetry greatly reduces the number of coefficients.

The quadratic Kerr Effect is generally smaller than the linear Pockels Effect so experiments are usually performed on centrosymmetric solids and liquids where the linear effect is absent.

Kerr cells employing liquid media have been used as high speed electro-optic shutters to provide chopped light sources at frequencies of 10^9–10^{10} Hz. The shutters consist of polarizer, analyzer, and a transparent liquid under a transverse field (Fig. 28.6(a)). The quadratic electro-optic coefficient is largest for molecules with large electric dipole moments such as nitrobenzene (Fig. 28.6(b)).

In the presence of an electric field the symmetry of a liquid changes from $\infty\infty m$ to ∞m. Point group ∞m corresponds to an uniaxial indicatrix with the optic axis Z_3 directed along the applied electric field. The light beam is perpendicular to the electric field and the optic axis. Let Z_1 be the direction of the optical beam.

The indicatrix of an isotropic solid or liquid is

$$B_i(0) = \frac{1}{n^2}(Z_1^2 + Z_2^2 + Z_3^2) = 1.$$

where n is the refractive index. When an electric is applied along Z_3 the indicatrix components change by

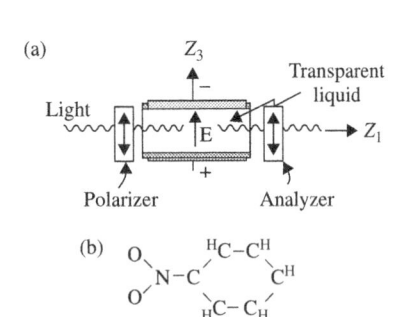

Fig. 28.6 The quadratic electro-optic effect is used in the Kerr cell. (a) a transverse electric field partially aligns the molecules of the liquid. This introduces an optical birefringence for light waves polarized parallel and perpendicular to the electric field. (b) Polar liquids such as nitrobenzene have sizeable Kerr Effects.

$$\begin{pmatrix} \Delta B_1 \\ \Delta B_2 \\ \Delta B_3 \\ \Delta B_4 \\ \Delta B_5 \\ \Delta B_6 \end{pmatrix} = \begin{pmatrix} R_{11} & R_{12} & R_{12} & 0 & 0 & 0 \\ R_{12} & R_{11} & R_{12} & 0 & 0 & 0 \\ R_{12} & R_{12} & R_{11} & 0 & 0 & 0 \\ 0 & 0 & 0 & R_{44} & 0 & 0 \\ 0 & 0 & 0 & 0 & R_{44} & 0 \\ 0 & 0 & 0 & 0 & 0 & R_{44} \end{pmatrix} \begin{pmatrix} 0 \\ 0 \\ E_3^2 \\ 0 \\ 0 \\ 0 \end{pmatrix} = \begin{pmatrix} R_{12}E_3^2 \\ R_{12}E_3^2 \\ R_{11}E_3^2 \\ 0 \\ 0 \\ 0 \end{pmatrix},$$

Table 28.4 Quadratic electro-optic coefficients $R_{11}-R_{12}$ (in m^2/V^2) and Kerr constants K (in m/V^2) for several liquids and solids. The ferroelectrics are measured in the cubic state

	λ (μm)	n	$R_{11}-R_{12}$	K
Benzene	0.55	1.503	1.6×10^{-21}	4.9×10^{-15}
CS_2	0.55	1.633	1.0×10^{-20}	3.9×10^{-14}
CCl_4	0.63	1.456	1.5×10^{-22}	7.4×10^{-16}
Water	0.59	1.333	2.5×10^{-20}	5.1×10^{-14}
Nitrotoluene	0.59	1.548	4.5×10^{-19}	1.4×10^{-12}
Nitrobenzene	0.59	1.552	7.6×10^{-19}	2.4×10^{-12}
$BaTiO_3$	0.63	2.42	2.3×10^{-15}	2.6×10^{-9}
$KNb_{0.37}Ta_{0.63}O_3$	0.63	2.29	2.9×10^{-15}	2.8×10^{-9}
$SrTiO_3$	0.63	2.38	3.1×10^{-17}	3.3×10^{-11}
$(Pb_{0.88}La_{0.08})(Ti_{0.35}Zr_{0.65})O_3$	0.55	2.45	1.8×10^{-15}	2.4×10^{-9}

where $2R_{44} = R_{11} - R_{12}$. Adding $\Delta B_i(E)$ to $B_i(0)$ gives the indicatrix equation of the liquid in the presence of an electric field.

$$\left(\frac{1}{n^2} + R_{12}E_3^2\right)(Z_1^2 + Z_2^2) + \left(\frac{1}{n^2} + R_{11}E_3^2\right)Z_3^2 = 1.$$

Since the light beam is parallel to Z_1, the elliptical section perpendicular to Z_1 determines the birefringence. The refractive indices along the major and minor axes are

$$n_o = n - \frac{1}{2}n^3 R_{12}E_3^2 \quad \text{and} \quad n_e = n - \frac{1}{2}n^3 R_{11}E_3^2$$

and the resulting birefringence is

$$\Delta n = n_e - n_o = \frac{1}{2}n^3(R_{12} - R_{11})E_3^2 = -n^3 R_{44}E_3^2.$$

Quadratic electro-optic coefficients $R_{11}-R_{12}$ for several solids and liquids are given in Table 28.4. The coefficients for barium titanate and other ferroelectric solids are several orders of magnitude larger than the liquids. Polar liquids like water have larger electro-coefficients than carbon tetrachloride and other nonpolar liquids. Also listed in Table 28.4 are the so-called Kerr constants K. They are related to the quadratic electro-optic coefficients by the equation

$$K = \frac{(R_{12} - R_{11})n^3}{2\lambda}.$$

Like the linear effect, the quadratic electro-optic effect can be divided into low and high frequency contributions. In tensor form,

$$R_{ijkl}^X = R_{ijkl}^x + p_{ijmn}M_{mnkl},$$

where p_{ijmn} and M_{mnkl} are the elasto-optic and electrostriction tensors, respectively. R_{ijkl}^X, the quadratic electro-optic coefficient is measured with a low-frequency electric field under stress-free conditions. The strain-free quadratic coefficient R_{ijkl}^X is measured at high frequencies above mechanical resonance. The difference between the two comes from the field-induced electrostrictive strain coupled to the elasto-optic effect. The latter effect is sometimes referred to as the "secondary" or "false" quadratic electro-optic effect.

Problem 28.2
Which of the Kerr Effect coefficients of rutile (point group $4/mmm$) can be determined with the light beam along the optic axis? How should the electric field be oriented for each measurement?

Nonlinear optics

29

In most dielectrics, the linear relation between electric polarization and applied electric field is accurately obeyed even for fairly large fields of 10^7 V/m. The reason is that the atomic displacements are extremely small, in the range of nuclear sizes—millions of times smaller than the size of atoms. Though non-linear effects such as electrostriction have been known for some time, it was not until the invention of the laser that sufficiently large optical fields became available to produce sizeable nonlinear optical effects. The induced polarization P can be written as a power series in an electric field,

$$P = \chi E + dE^2 + \cdots,$$

where χ is the linear electric susceptibility, and the higher-order terms lead to nonlinear effects such as second harmonic generation (Fig. 29.1).

The electric field associated with the incident light is sinusoidal, $E = E_0 \sin \omega t$, and when E is substituted in the expression for P, a power series in $\sin \omega t$ results. The second term is $dE_0^2 \sin^2 \omega t = \frac{1}{2} dE_0^2 (1 - \cos 2\omega t)$, which includes a component of polarization with twice the frequency of the impressed field E. This rapidly oscillating induced dipole moment is the source of second harmonic light. The intensity of the light depends on the size of d, the second order coefficient.

29.1	Structure–property relations	313
29.2	Tensor formulation and frequency conversion	315
29.3	Second harmonic generation	316
29.4	Phase matching	318
29.5	Third harmonic generation	322

29.1 Structure–property relations

Crystal symmetry is a major factor in the second-order effect. The one-dimensional polar chain in Fig. 29.2 illustrates the origin of the quadratic term. When the applied field is directed to the left, the ions and bonding electrons are in very close contact and the displacements will be small because of short range repulsive forces. These forces do not oppose motion in the opposite direction, so that fields directed to the right give larger motions and larger polarizations. A centric chain does not show this effect. Such a chain can give rise to odd-order terms producing saturation but not to even power terms in the $P(E)$ relation.

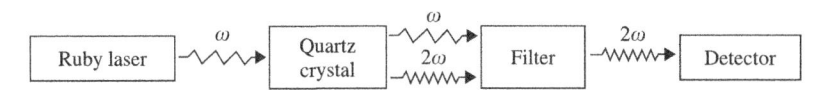

Fig. 29.1 Schematic representation of the second harmonic generation experiment. Intense red light from ruby laser is incident upon piezoelectric quartz, generating second harmonic blue light.

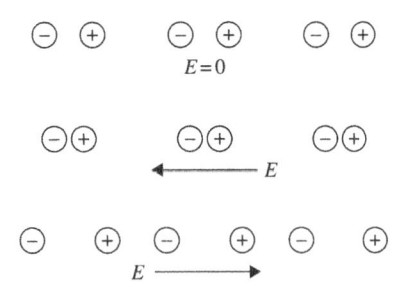

Fig. 29.2 A one-dimensional polar chain in fields directed to the left and right, showing the origin of nonlinear optical effects.

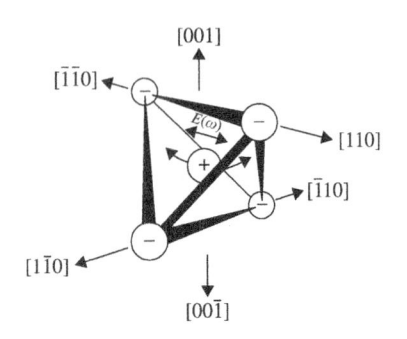

Fig. 29.3 Displacement of a tetrahedrally-coordinated cation in response to an alternating electric field. The path is curved upward because of the attraction to anions, giving rise to optical nonlinearity.

This means that centric crystals are useless as second harmonic generators. In fact, the second harmonic experiment is a good test for the existence of inversion symmetry. A strong signal is proof of the absence of a center of symmetry because intense SHG is possible only in acentric crystals. One approach to obtaining acentric crystals is the use of the acentric molecules with permanent distortions caused by nonbonded electrons. HIO_3 and $LiIO_3$ crystals contain IO_6 octahedra with large trigonal distortions and are promising nonlinear optical materials.

Quartz is acentric but is not an outstanding second harmonic generator. The best SHG materials have large refractive indices. According to *Miller's Rule* the SHG coefficients are proportional to $(n^2 - 1)^3$. Increasing n from 1.5 to 2.5 increases d by two orders of magnitude. Thus the titanates and niobates are excellent nonlinear optical materials, and narrow band gap semiconductors are also outstanding because of the inverse relation between n and E_g. InSb has a refractive index of 3.5 at 1.06μ, and an SHG signal more than a thousand times greater than quartz.

The physical origin of optical nonlinearity in cubic crystals can be demonstrated with the zincblende structure. In cubic ZnS, each atom is tetrahedrally coordinated to four nearest neighbors of opposite charge. The tetrahedral edges are parallel to ⟨110⟩ directions as shown in Fig. 29.3.

Consider an electromagnetic wave of frequency ω propagating through the crystal in a [$\bar{1}$10] direction and polarized alternately along [110] and [$\bar{1}\bar{1}$0]. The ion at the center of the tetrahedron responds to the electric force by initially moving in the direction of the field. But as the ion leaves the equilibrium position in either the [110] or [$\bar{1}\bar{1}$0] direction, it is attracted to the negatively charged ions above as the distance to one of these ions decreases. The result is a displacement which tends to follow the electric vector of the light wave but is curved upwards at the extremes.

Two components can be identified when this curved motion is observed along the direction of the light beam. One is a motion along [110] parallel to the driving field E, and vibrating with the same frequency ω. But, there is also a small component of displacement parallel to [001] and vibrating at frequency 2ω. For every cycle of $E(\omega)$, the ion will twice reach maximum excursion along [001]. Thus a light wave traveling along [$\bar{1}$10] and polarized parallel to [110] generates a second harmonic polarized along [001]. This is the nonlinear optic coefficient d_{14} for zincblende crystals in point group $\bar{4}3m$.

As explained later, birefringence can be useful in SHG materials. Greatly amplified harmonics are possible if the velocities of the fundamental and harmonic waves are made equal. This kind of phase matching is possible if the difference in refractive index due to dispersion can be matched by birefringence. Noncritical phase-matching in uniaxial crystals is the best because energy flow directions are also coincident, thereby eliminating walk-off problems.

Optical images can be stored in crystals as phase gratings which can be written and read by laser beams. The holograms are written by liberating free charge carriers from traps with incident light. Electrons diffuse from the illuminated regions to the darker regions producing space-charge electric fields which in turn modulate the refractive index through the electro-optic effect. The refractive index of the crystal is, therefore, modulated according to the optical image, forming a phase grating. In crystals such as $LiNbO_3$ the image can be fixed by gentle heating, allowing positive ions to diffuse to the regions of negative space charge, neutralizing the local electric fields. After cooling the crystal, the field is restored and the phase grating fixed by uniform illumination. This makes the electron configuration uniform but leaves behind a nonuniform charge distribution which forms the phase grating.

29.2 Tensor formulation and frequency conversion

Crystal optics involves solutions to Maxwell's Equation for wave propagation under various boundary conditions. Normally it is assumed that the electric displacement D and electric polarization P are linearly proportional to the electric field E through the electric permittivity ε and electric susceptibility χ. This assumption breaks down when describing the properties of dielectrics under large electric fields. Nonlinear effects can be described by expanding the polarization in a power series.

$$P_i = \chi_{ij} E_j + \chi_{ijk} E_j E_k + \chi_{ijkl} E_j E_k E_l + \cdots.$$

The lead term is the electric susceptibility which is related to the optical permittivity, $\chi_{ij} = \varepsilon_{ij} - \varepsilon_0$. The second and third terms are the focus of this chapter. The second order nonlinear coefficient χ_{ijk} is closely related to the linear electro-optic effect. It is responsible for second harmonic generation (SHG), for the production of sum and difference frequencies, and for parametric amplification and generation. The third order term χ_{ijkl} is a fourth rank tensor that gives rise to third harmonic generation (THG), Raman and Brillouin scattering, and optical phase conjugation.

The various phenomena related to the quadratic term become clearer when the tensor relations are written as a function of frequency, ω.

$$P_i(\omega) = \chi_{ijk} E_j(\omega_1) E_k(\omega_2).$$

If the fields are sinusoidal with time,

$$E_j(\omega_1) = E_{0j}(\omega_1) \cos \omega_1 t$$
$$E_k(\omega_2) = E_{0k}(\omega_2) \cos(\omega_2 t + \phi),$$

where E_0 represents the wave amplitudes and ϕ is the phase difference between the two driving fields. The product of the two fields can be written in terms of sum and difference frequencies

$$E_j(\omega_1) E_k(\omega_2) = \frac{1}{2} E_{0j}(\omega_1) E_{0k}(\omega_2) \{ \cos[(\omega_1 - \omega_2)t - \phi]$$
$$+ \cos[(\omega_1 + \omega_2)t + \phi] \}.$$

This means the resulting polarization $P_i(\omega)$ that comes from the quadratic term has either a difference $(\omega_1 - \omega_2)$ or a sum $(\omega_1 + \omega_2)$ frequency. The third rank tensor χ_{ijk} connecting the two driving fields and the polarization is nonzero in only two cases:

$$\chi_{ijk}(\omega_1 - \omega_2, \omega_1, \omega_2) \quad \text{and} \quad \chi_{ijk}(\omega_1 + \omega_2, \omega_1, \omega_2).$$

Sum and difference waves are used in frequency-conversion processes which include second harmonic generation (Fig. 29.1) and a number of parametric

Fig. 29.4 Three examples of second-order optical nonlinearity in crystals. (a) A parametric amplifier augments the intensity of an input signal by drawing energy from a laser. (b) Parametric oscillators make use of a resonant cavity tuned to frequencies ω_1 and ω_2. (c) Frequency upconversion is accomplished by combining two low frequency photons to generate a higher energy photon. All three experiments utilize transparent noncentrosymmetric crystals as the nonlinear (NL) mixer.

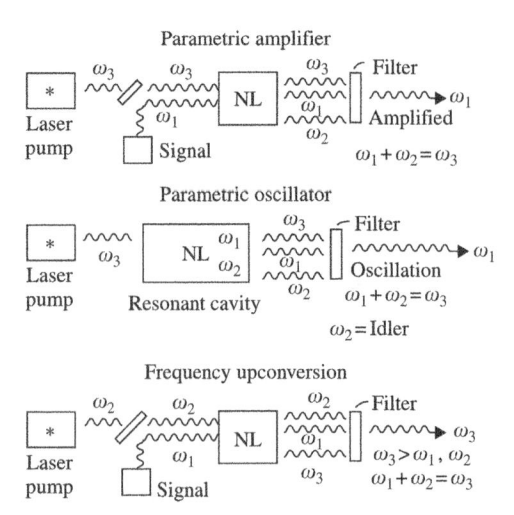

phenomena (Fig. 29.4). The parametric frequency-conversion processes are used primarily for generation of radiation at new wavelengths, although some of the interactions have also been used to amplify input signals.

Fig. 29.4(a) shows the basic configuration of a *parametric amplifier*. An input signal at frequency ω_1 is incident on a nonlinear optic crystal together with an intense laser beam at frequency ω_3. Wave mixing occurs inside the crystal with the laser beam transferring energy to the input signal and to a second wave satisfying the relation $\omega_2 = \omega_3 - \omega_1$. The amplification of the input signal ω_1 is accompanied by the generation of an idler wave at frequency ω_2. Parametric gains are often relatively modest so this process is not widely used.

Parametric oscillators (Fig. 29.4(b)) use a similar configuration to generate new frequencies. A pump wave at frequency ω_3 supplies energy to optical waves at frequencies ω_1 and ω_2 for which $\omega_3 = \omega_1 + \omega_2$. In this experiment, the nonlinear crystal is positioned within an optical resonator that is resonant at the signal frequency (ω_1), or the idler frequency (ω_2), or both. An important advantage of the parametric oscillator is that the resonant cavity can be tuned continuously over a wide range of frequencies.

Frequency upconversion (Fig. 29.4(c)) adds a signal frequency (ω_1) and a strong laser beam (ω_2) to produce a higher frequency signal $\omega_3 = \omega_1 + \omega_2$. Upconversion offers a way to detect infrared signals in wavelength ranges where detectors are slow or inefficient. *Parametric downconversion* works on a principle similar to upconversion.

29.3 Second harmonic generation

If a noncentrosymmetric crystal is illuminated by an intense laser beam, the quadratic polarization term leads to a constant component and a second harmonic. The sum and difference terms become $\chi_{ijk}(2\omega, \omega, \omega)$ and $\chi_{ijk}(0, \omega, \omega)$.

Since χ_{ijk} is a polar third rank tensor like piezoelectricity and the linear electro-optic effect, second harmonic generation (SHG) does not occur in centrosymmetric crystals. When measured by second harmonic experiments, the χ_{ijk} coefficients are replaced by d_{ijk} coefficients, where $2d_{ijk} = \chi_{ijk}$.

The factor of 2 appears because $\omega_1 = \omega_2 = \omega$ and the field components $E_j(\omega)$ and $E_k(\omega)$ can be interchanged.

$$P_i(2\omega) = d_{ijk} E_j(\omega) E_k(\omega).$$

Written in matrix form, the nonlinear polarization becomes

$$\begin{pmatrix} P_1 \\ P_2 \\ P_3 \end{pmatrix} = \begin{pmatrix} d_{11} & d_{12} & d_{13} & d_{14} & d_{15} & d_{16} \\ d_{21} & d_{22} & d_{23} & d_{24} & d_{25} & d_{26} \\ d_{23} & d_{32} & d_{33} & d_{34} & d_{35} & d_{36} \end{pmatrix} \begin{pmatrix} E_1^2 \\ E_2^2 \\ E_3^2 \\ 2E_2 E_3 \\ 2E_1 E_3 \\ 2E_1 E_2 \end{pmatrix}.$$

The contracted d_{ij} matrix obeys the same symmetry restrictions as the direct piezoelectric coefficients (Table 12.1). The only differences are the factors of 2 appearing in point groups 3, 32, 3m, $\bar{6}$, and $\bar{6}m2$. The factors of 2 are absent in the SHG matrices.

There is also a close relation between the SHG coefficients and those of the linear electro-optic coefficients r_{ijk}. Both are third rank tensors which interconnect electric fields and optical behavior. The SHG matrices appear like the transpose of the electro-optic matrices, with the same set of coefficients for each point group symmetry. The magnitudes of the coefficients are also related, but not precisely because the measurement conditions are different. Referred to principal axes, the relationship is

$$d_{ijk} = \frac{-\varepsilon_{ii} \varepsilon_{jj} r_{ijk}}{4 \varepsilon_0}.$$

However, the applied electric fields in the Pockels experiments are generally at much lower frequencies than the optical frequencies involved in second harmonic generation. Therefore the polarization contributions are different. Ionic motions influence the electro-optic coefficients but not the SHG constants.

Table 29.1 contains the second harmonic coefficients for a number of acentric crystals used in nonlinear optics. Crystals with large refractive indices generally have large d coefficients but some are not transparent in the visible region. Others cannot be used because phase-matching is impossible. Common oxides like quartz and tourmaline have very modest values but $BaTiO_3$ and other ferroelectric oxides with large refractive indices have more useful SHG coefficients in accordance with Miller's Rule.

Nonlinear optic coefficients show dispersion similar to refractive indices. Fig. 29.5 shows the variation of the d_{36} coefficients of ADP and KDP over optical and near infrared wavelengths.

According to Miller's Rule, The SHG coefficients are proportional to $(n^2 - 1)^3$. For KDP and ADP the change in $(n^2 - 1)^3$ from 0.6 to 1.3 μm for n_e and n_o is about 15%. The nonlinear coefficient d_{36} decreases by about the same percentage over this wavelength range.

Problem 29.1
Plot out the nonlinear optic coefficients of $LiNbO_3$ in the Z_1–Z_3 and Z_2–Z_3 planes using the coefficients in Table 29.1. Make the corresponding drawings for the Pockels Effect discussed in the previous chapter. Discuss the similarities and differences between these two third-rank properties of $LiNbO_3$.

Table 29.1 Magnitudes of nonlinear optical constants measured in units of 10^{-23} F/V (wavelengths λ refer to the fundamental frequency and are in μm)

Cubic $\bar{4}3m$ (Zincblende structure)			
III–V	GaP	GaAs	GaSb
λ	10.6	10.6	10.6
d_{14}	51	119	345
II–VI	ZnS	ZnSe	ZnTe
λ	10.6	10.6	10.6
d_{14}	27	71	80
I–VII	CuCl	CuBr	CuI
λ	1.06	1.06	1.06
d_{14}	8.6	9.2	5.6

Hexagonal $6mm$ (Wurtzite structure)			
	CdS	CdSe	ZnO
λ	10.6	10.6	1.06
d_{31}	23	26	2.4
d_{33}	39	49	8.0
d_{15}	26	28	2.7

Tetragonal $4mm$ (Ferroelectric oxides)			
	$BaTiO_3$	$PbTiO_3$	$SrBaNb_4O_{12}$
λ	1.06	1.06	1.06
d_{31}	20	52	5.9
d_{33}	7.8	10.3	15.7
d_{15}	20	47	8.3

Tetragonal $\bar{4}2m$			
	KH_2PO_4	$NH_4H_2PO_4$	$AgGaSe_2$
λ	1.06	1.06	10.6
d_{36}	0.56	0.68	60

Trigonal $3m$			
	$LiNbO_3$	$LiTaO_3$	Tourmaline
λ	1.06	1.06	1.06
d_{31}	5.2	1.6	0.20
d_{33}	30	23	0.69
d_{22}	3.6	2.4	0.10
d_{15}	6.8		0.32

Trigonal 32			
	α-quartz	Se	Te
λ	1.06	1.06	1.06
d_{11}	0.44	86	814
d_{14}	0.004		

Orthorhombic $mm2$ (Ferroelectric oxides)		
	$Ba_2NaNb_5O_{15}$	$KNbO_3$
λ	1.06	1.06
d_{31}	18	13
d_{32}	18	16
d_{33}	24	24
d_{15}	18	14
d_{14}	17	16

29.4 Phase matching

Constructive interference is required to maximize the intensity of the second harmonic wave. The technique most widely used in keeping the fundamental (ω) and harmonic (2ω) in phase takes advantage of the natural birefringence of anisotropic crystals. The two waves travel with the same velocity if the refractive indices are equal, $n(2\omega) = n(\omega)$. Dispersion makes this difficult since normally

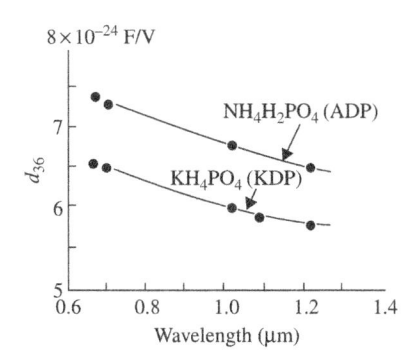

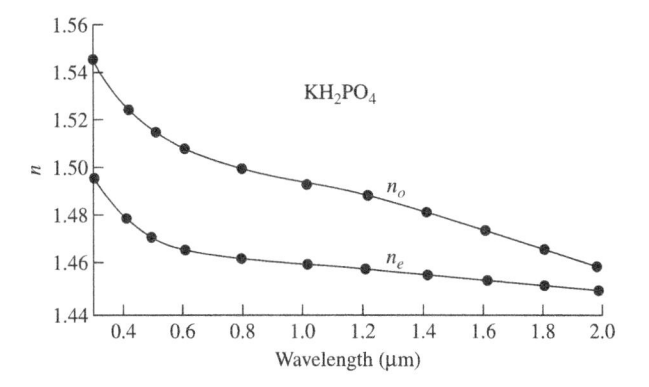

Fig. 29.5 Dispersion of the d_{36} coefficient of ammonium dihydrogen phosphate and potassium dihydrogen phosphate.

Fig. 29.6 Optical dispersion in tetragonal potassium dihydrogen phosphate (KDP). The ordinary wave refractive index n_o is larger than n_e and shows greater dispersion. The resulting birefringence can be used to produce second harmonic phase matching.

the refractive index of the harmonic exceeds that of the fundamental by several percent. Fig. 29.6 shows the changes in n_e and n_o for KDP over the visible and near IR range. It is interesting to note that KDP is uniaxial negative with $n_o > n_e$ and that the largest dispersion occurs in n_o. The ordinary refractive index n_o is controlled by polarization in the (001) plane which is where the hydrogen bonds undergo an order–disorder transformation (Fig. 28.2). The primary changes in bonding are in (001) despite the fact that the ferroelectric polarization is in the perpendicular direction along [001].

To equalize the refractive indices of the fundamental and the harmonic, we make use of the fact that the speed of the extraordinary wave depends on the direction of the wave normal. If the wave normal forms an angle θ with respect to the optic axis (Z_3), the refractive index of the extraordinary wave is given by

$$\frac{1}{n_e^2(\theta)} = \frac{\cos^2\theta}{n_o^2} + \frac{\sin^2\theta}{n_e^2}.$$

The variation of the n_e with angle makes it possible to match the refractive indices of the fundamental and the harmonic at an angle θ_m for which

$$n_e(2\omega, \theta_m) = n_o(\omega).$$

The phase-matching angle θ_m is given by

$$\frac{1}{n_e^2(2\omega, \theta_m)} = \frac{1}{n_o^2(\omega)} = \frac{\cos^2\theta_m}{n_o^2(2\omega)} + \frac{\sin^2\theta_m}{n_e^2(2\omega)}.$$

Solving for θ_m gives the relation

$$\sin^2\theta_m = \frac{(1/n_o^2(\omega)) - (1/n_o^2(2\omega))}{(1/n_e^2(\omega)) - (1/n_o^2(2\omega))}.$$

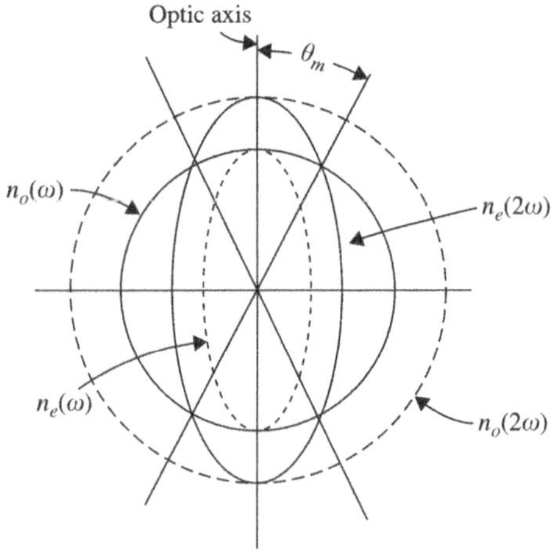

320 *Nonlinear optics*

Fig. 29.7 Refractive index ellipsoids for the fundamental (ω) and harmonic (2ω) waves in a uniaxial negative crystal such as KDP. Phase matching between the ordinary wave at frequency ω and the extraordinary wave at 2ω occurs at the angle θ_m.

The critical angle for a uniaxial negative crystal is illustrated in Fig. 29.7. Phase matching is only possible in crystals with low dispersion and relatively large birefringence.

The refractive indices of KDP (Fig. 29.6) are suitable for phase matching. For a Nd-glass laser ($\lambda = 1.06$ μm) the second harmonic is in the visible range at 0.53 μm. The refractive indices for these wavelengths are $n_o(\omega) = 1.4942$, $n_e(2\omega) = 1.4712$. Substituting these values into the condition for phase matching gives a critical angle of $\theta_m = 42°$.

In this case the extraordinary wave of the second harmonic was matched with the ordinary wave of the fundamental, or $(2\omega, \omega, \omega) = (e, o, o)$ for short. Another way of phase matching is $(2\omega, \omega, \omega) = (e, o, e)$. In this method, the two components of the input field $E_j(\omega)$ and $E_k(\omega)$ correspond to different polarization directions, one is an ordinary wave polarized perpendicular to the optic axis and the other is an extraordinary wave polarized in the plane containing the wave normal and the optic axis. For (e, o, e) phase matching the critical angle is obtained from the equation

$$2\left[\frac{\cos^2\theta_m}{n_o^2(2\omega)} + \frac{\sin^2\theta_m}{n_e^2(2\omega)}\right]^{1/2} = n_o(\omega) + \left[\frac{\cos^2\theta_m}{n_o^2(\omega)} + \frac{\sin^2\theta_m}{n_e^2(\omega)}\right]^{1/2}.$$

Thus, in general, there is more than one way to satisfy the phase-matching condition.

The (e, o, o) and (e, o, e) phase-matching conditions just described apply to uniaxial negative crystals such as KDP with $n_o > n_e$. For uniaxial positive crystals with $n_e > n_o$, the second harmonic must be an ordinary wave, and the phase-matching schemes are (o, e, e) and (o, e, o). It should be remembered that phase-matching is not always possible. If the dispersion over the wavelength interval between ω and 2ω exceeds the birefringence it is impossible to make $n(\omega) = n(2\omega)$. Birefringence is zero in cubic crystals and isotropic glasses. Therefore phase matching is nearly impossible unless there is an absorption band between ω and 2ω. In that case there might be an accidental match of the fundamental and harmonic refractive indices (see Chapter 26).

For a uniaxial crystal, the phase-matching condition is satisfied for any wave normal in the cone of directions forming an angle $\theta = \theta_m$ with the optic axis Z_3, but there are two other conditions that must also be met. There must be a nonlinear optic coefficient connecting $P_i(2\omega)$ to the product $E_j(\omega)E_k(\omega)$. The third condition is that the polarization direction of the wave with doubled frequency must be perpendicular to the polarization direction of the fundamental, and both polarization directions are perpendicular to the wave normal.

For KDP in point group $\bar{4}2m$, the only nonlinear optic coefficients are $d_{14} = d_{25}$ and d_{36}. Therefore the nonlinear polarization components referred to the principal axes ($Z_1 = [100]$, $Z_2 = [010]$, and $Z_3 = [001]$) are

$$P_1(2\omega) = 2d_{14}E_2(\omega)E_3(\omega),$$
$$P_2(2\omega) = 2d_{14}E_3(\omega)E_1(\omega),$$

and

$$P_3(2\omega) = 2d_{36}E_1(\omega)E_2(\omega).$$

The (e, o, o) phase-matching scheme utilizes an ordinary wave at the fundamental frequency. The polarization direction of the fundamental is therefore in the (001) plane perpendicular to the optic axis making $D_3(\omega) = 0$. Since the birefringence of KDP is small, $E_3(\omega)$ is nearly parallel to $D_3(\omega)$, and $E_3(\omega) \cong 0$ as well. Therefore the harmonic components $P_1(2\omega) = P_2(2\omega) \cong 0$, leaving only $P_3(2\omega)$. To maximize P_3, the $E_1(\omega)E_2(\omega)$ product must be maximized. This is accomplished by orienting the field along a [110] direction where $E_1 = E_2 = E/\sqrt{2}$. Therefore $P_3(2\omega) = d_{36}E^2(\omega)$. The component of P_3 in the polarization direction of the harmonic is $d_{36}E^2 \sin\theta_m$ where E is directed along [110] and θ_m is the critical angle for phase-matching. θ_m is about 42° for a Nd-glass laser ($\lambda = 1.06$ μm) and about 50° for a ruby laser ($\lambda = 0.69$ μm).

Fig. 29.8 shows the polarization direction of the harmonic wave and the wave normal plotted in the (110) plane of KDP. The polarization direction of the fundamental wave is in the [110] direction perpendicular to the (110) plane. Since the harmonic is an extraordinary wave, its E vector is not parallel to the polarization vector. This means that the energy flow is not parallel to the wave normal causing a walk-off problem. However the birefringence of KDP is not large and therefore the problem is not a serious one.

Several of the acentric point groups are unsuitable for SHG because it is difficult to obtain both phase-matching and a large nonlinear coefficient. Uniaxial negative crystals in point groups 422, 622, and $\infty2$, and uniaxial positive crystals in 4mm, 6mm, and ∞m are found to be totally unsuitable, while others are generally very small. The most useful crystal groups are those like $\bar{4}2m$ where the phase-matching condition can be satisfied in orientations where the nonlinear coupling coefficients are large. In addition to ADP and KDP, several sulfide and selenide crystals also belong to $\bar{4}2m$ and have proven useful in nonlinear optic experiments. Point group 3m with crystals such as LiNbO$_3$ and Ag$_3$AsS$_3$ (proustite) is also widely used. Heavy metal sulfides, selenides and tellurides have large SHG coefficients and are also transparent in much of the infrared range where CO$_2$ lasers operate.

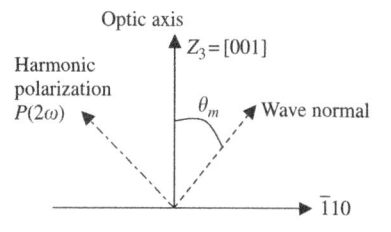

Fig. 29.8 Phase-matched directions for the wave normal and harmonic polarization directions in KDP. The fundamental polarization is in the [110] direction perpendicular to all four of the vectors shown in the drawing.

Problem 29.2
Phase matching conditions for uniaxial negative crystals were discussed in the text. What are the corresponding relations between $n_o(\omega)$, $n_e(\omega)$, $n_o(2\omega)$ and $n_e(2\omega)$ for uniaxial positive crystals? Illustrate the phase matching condition with wave velocity surfaces. Show what the surfaces look like when dispersion is very large and phase matching is impossible.

29.5 Third harmonic generation

Third harmonic generation (THG) is one of the higher order effects arising from the third order susceptibility relation

$$P_i = \chi_{ijkl} E_j E_k E_l.$$

Since χ_{ijkl} is a fourth rank polar tensor it is found in all 32 crystal classes and all seven Curie groups. All permutations of E_j, E_k, and E_l are indistinguishable so

$$\chi_{ijkl} = \chi_{ijlk} = \chi_{iljk} = \chi_{ilkj} = \chi_{ikjl} = \chi_{iklj}.$$

This reduces the number of coefficients from 81 to 30, making it possible to represent third order effects by a 3×10 matrix:

$$\begin{pmatrix} P_1 \\ P_2 \\ P_3 \end{pmatrix} = \begin{pmatrix} \chi_{11} & \chi_{12} & \chi_{13} & \chi_{14} & \chi_{15} & \chi_{16} & \chi_{17} & \chi_{18} & \chi_{19} & \chi_{10} \\ \chi_{21} & \chi_{22} & \chi_{23} & \chi_{24} & \chi_{25} & \chi_{26} & \chi_{27} & \chi_{28} & \chi_{29} & \chi_{20} \\ \chi_{31} & \chi_{32} & \chi_{33} & \chi_{34} & \chi_{35} & \chi_{36} & \chi_{37} & \chi_{38} & \chi_{39} & \chi_{30} \end{pmatrix} \times \begin{pmatrix} E_1^3 \\ E_2^3 \\ E_3^3 \\ 3E_2 E_3^2 \\ 3E_2^2 E_3 \\ 3E_1 E_3^2 \\ 3E_1^2 E_3 \\ 3E_1 E_2^2 \\ 3E_1^2 E_2 \\ 6E_1 E_2 E_3 \end{pmatrix}.$$

In matrix form $\chi_{im} = \chi_{ijlk}$ where $i, j, k, l = 1$–3, and m is used to represent the following tensor subscripts.

m =	1	2	3	4	5	6	7	8	9	0
jkl =	111	222	333	233	223	133	113	122	112	123

Symmetry further reduces the number of coefficients. Applying Neumann's Law lowers the number of coefficients from 30 down to 9 in cubic crystals, with only two independent matrix coefficients in the highest symmetry cubic classes (Table 29.2).

Further simplification makes use of the *Kleinmann approximation* in which the tensor coefficients are symmetric for all permutations. This means that the first subscript for the polarization can be exchanged with any of the three field subscripts. For example, $\chi_{1233} = \chi_{2133} = \chi_{3123}$. In matrix form this means that $\chi_{14} \cong \chi_{26} \cong \chi_{30}$. The Kleinmann approximation has been verified by experiment. In the most general case (triclinic crystals) this reduces the number of independent coefficients from 30 to 15.

As indicated in Table 29.2, cubic crystals belonging to point group 432, $\bar{4}3m$, and $m3m$ have just nine nonzero coefficients. Only two of these are independent χ_{11} and χ_{16}. The other matrix components are

$$\chi_{11} = \chi_{22} = \chi_{33}$$

$$\chi_{16} = \chi_{18} = \chi_{24} = \chi_{29} = \chi_{35} = \chi_{37}.$$

Table 29.2 Effect of symmetry on third harmonic generation coefficients. The total number of nonzero matrix coefficients is listed for each crystallographic point group with the number of independent coefficients in parentheses. Point groups that also exhibit second harmonic generation are underlined

Triclinic <u>1</u>, $\bar{1}$	30(30)
Monoclinic <u>2</u>, <u>m</u>, 2/m	16(16)
Orthorhombic <u>222, mm2</u>, mmm	9(9)
Trigonal <u>3</u>, $\bar{3}$	25(10)
<u>32, 3m</u>, $\bar{3}m$	14(5)
Tetragonal <u>4</u>, $\bar{4}$, 4/m	15(8)
<u>422, $\bar{4}2m$, 4mm</u>, 4/mmm	9(5)
Hexagonal <u>6</u>, $\bar{6}$, 6/m	15(6)
<u>622, $\bar{6}m2$, 6mm</u>, 6/mmm	9(4)
Cubic m3, <u>23</u>	9(3)
432, <u>$\bar{4}3m$</u>, m3m	9(2)

Table 29.3 Third order matrix coefficients for liquids, glasses, and cubic crystals. χ_{11} and χ_{16} values are in units of 10^{-33} Fm/V^2

	χ_{11}		χ_{11}	χ_{16}
Cubic crystals				
LiF	3	Y$_3$Al$_5$O$_{12}$	22	
NaCl	21	SrTiO$_3$	4400	
KCl	24	CaF$_2$	5	2
KBr	37	SrF$_2$	5	4
KI	3	BaF$_2$	11	4
MgO	12	CdF$_2$	18	6
Liquids		*Glasses*		
CS$_2$	1300	SiO$_2$	5	
CCl$_4$	19	Others	6–37	
C$_6$H$_6$	69			

All others are zero. A few numerical values for χ_{11} and χ_{16} are collected in Table 29.3.

Liquids and glasses with isotropic ($\infty\infty m$) properties also possess third order nonlinearities. For these materials there is only one independent coefficient:

$$\chi_{11} = \chi_{22} = \chi_{33} = 3\chi_{16} = 3\chi_{18} = 3\chi_{24} = 3\chi_{29} = 3\chi_{35} = 3\chi_{37}.$$

All other matrix components are zero. Representative χ_{11} values are included in Table 29.3.

Note that polar high-index liquids like carbon disulfide and high refractive index oxides like strontium titanate ($n \sim 2.4$) have much higher THG coefficients in keeping with Miller's Rule. This is borne out further in Fig. 29.9 where nonlinearity is plotted as a function of refractive index. A number of different glass compositions have been tested for use in optical beam control. Under intense illumination the refractive index changes according to the relation

$$n = n_o + n_2 I,$$

Fig. 29.9 Silica and other optical glasses are widely used in optical communication systems. The nonlinear coefficient n_2 relates the change in refractive index $\Delta n = n - n_o = n_2 I$ where I is in W/m^2 and n_2 in m^2/W. High refractive index glasses generally have higher nonlinear effects.

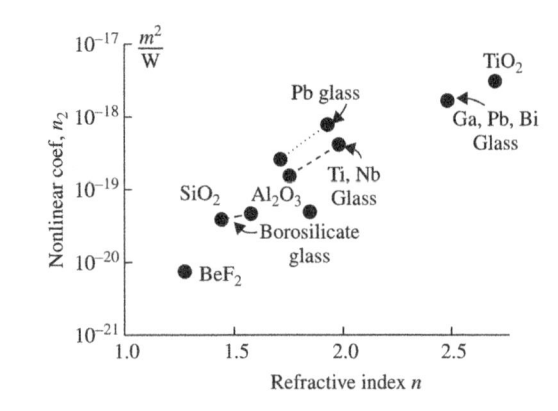

where n_2 is a nonlinear optic coefficient and I is the optical beam intensity measured in W/m^2. The nonlinear optic coefficient n_2 is proportional to the third order coefficient χ_{11} ($=\chi_{1111}$).

Phase matching for third harmonics is carried out in a similar manner to second harmonic generation. Birefringence can sometimes compensate for dispersion. Three types of phase-matching can be used for uniaxial negative crystals. For THG the three possibilities are $(3\omega, \omega, \omega, \omega) = (e,o,o,o)$, (e,o,o,e), and (e,o,e,e) where e and o stand for extraordinary and ordinary waves, respectively. The phase-matching angle θ_m for the (e,o,o,o) scheme is given by

$$\sin^2 \theta_m = \frac{\left(1/n_o^2(\omega)\right) - \left(1/n_o^2(3\omega)\right)}{\left(1/n_e^2(3\omega)\right) - \left(1/n_o^2(3\omega)\right)}.$$

Many of the most important applications of third-order nonlinearity involve phase conjugation by four-wave mixing. Phase conjugate optics makes use of nonlinear optic techniques for the real-time processing of electromagnetic waves. Among the many practical applications are image transmission, pulse compression, and image processing in optical communication systems.

In four-wave mixing, the amplitude of the induced polarization at frequency $\omega_1 = \omega_2 + \omega_3 - \omega_4$ is related to the electric fields of three input waves:

$$P_i(\omega_1) = \chi_{ijkl}(-\omega_1, \omega_2, \omega_3, -\omega_4) E_j(\omega_2) E_k(\omega_3) E_l(\omega_4).$$

Fig. 29.10 illustrates the basic experiment. A nonlinear medium is pumped simultaneously by two intense plane waves traveling in opposite directions. Pump waves 1 and 2 operate at frequency ω. Two other waves, 3 and 4, are also present in the nonlinear medium. They are traveling in a different direction than waves 1 and 2. Wave 3 is an input wave at frequency ω. When coupled through the third-order nonlinear coefficient χ_{ijkl}, waves 1, 2, and 3 generate a fourth wave of frequency ω. The amplitude of the fourth wave is the complex conjugate of the input signal wave 3.

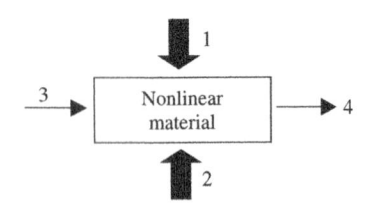

Fig. 29.10 Four-wave mixing experiment used in phase conjugate optics. Nonlinear media such as CS$_2$ with large third-order coefficients couple pumping waves 1 and 2 to the input signal 3 to generate the output wave 4.

Optical activity and enantiomorphism

30

When plane-polarized light enters a crystal it divides into right- and left-handed circularly polarized waves. If the crystal possesses handedness, the two waves travel with different speeds, and are soon out of phase. On leaving the crystal, the circularly polarized waves recombine to form a plane polarized wave, but with the plane of polarization rotated through an angle αt. The crystal thickness t is in mm, and α is the optical activity coefficient expressed in degrees/mm. The polarization vector of the combined wave can be visualized as a helix, turning $\alpha°$/mm path length in the optically-active medium. Because of the low symmetry of a helix, optical activity is not observed in many high symmetry crystals. Point groups possessing a center of symmetry are inactive.

30.1	Molecular origins	325
30.2	Tensor description	327
30.3	Effect of symmetry	329
30.4	Relationship to enantiomorphism	331
30.5	Liquids and liquid crystals	333
30.6	Dispersion and circular dichroism	337
30.7	Electrogyration, piezogyration, and thermogyration	340

30.1 Molecular origins

In relating α to crystal chemistry it is convenient to divide optically-active materials into two categories: Those which retain optical activity in liquid form, and those which do not. It has long been known that optically-active solutions crystallize to give optically-active solids. This follows from the fact that molecules lacking mirror or inversion symmetry can never crystallize in a pattern containing such symmetry elements. Thus one way of obtaining optically-active materials is to begin with optically-active molecules, as in Rochelle salt, tartaric acid and cane sugar. Few of these crystals are very stable, however, and the optical activity coefficients are usually small, typically 2°/mm.

The same is true of many inorganic solids, though they are seldom optically active in the liquid state. For $NaClO_3$ and $MgSO_4 \cdot 7H_2O$, α is about 3°/mm. Quartz and selenium, however, have coefficients an order of magnitude larger (Fig. 30.1(a)), showing the importance of helical structures to optical activity. Both compounds crystallize as right- and left-handed forms in space groups $P3_12$ and $P3_22$, with helices spiraling around the trigonal screw axes.

Quartz contains nearly regular SiO_4 tetrahedra with Si–O distances of 1.61 Å. Levorotatory quartz belongs to space group $P3_12$ and contains right-handed helices; enantiomorphic dextrorotatory quartz crystallizes in $P3_22$. Trigonal selenium also contains helical chains (Fig. 30.1(b)). Se–Se distances along the chain are 2.32 Å, much less than the shortest distance to atoms in neighboring chains, 3.46 Å.

When a solid contains helices it is obvious why right- and left-circularly polarized light travel at different speeds causing optical activity. Optical activity

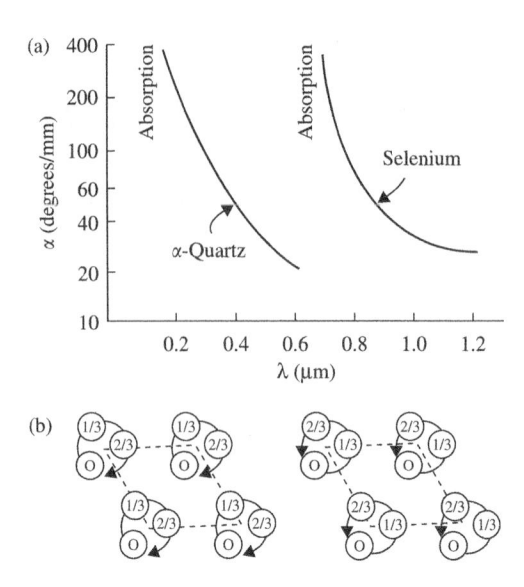

Fig. 30.1 (a) Dispersion of the principal optical-activity coefficient of quartz and selenium. The plane of polarization is rotated in opposite directions for right- and left-handed crystals. (b) Both structures contain helical chains along the trigonal c-axis.

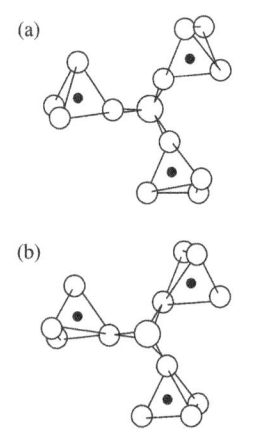

Fig. 30.2 Domain states in the optically-active ferroelectric $Pb_5Ge_3O_{11}$. In the 0° domain (a) Pb bonds up, producing positive polarization and right-handed rotatory power. It bonds down in the 180° domain (b), reversing the polarization and the handedness. An alternating electric field reverses both the spontaneous polarization and the optical activity coefficient.

is a spatial dispersion effect in which the dipolar fields of nearby atoms influence the local fields caused by the incident optical wave. The wave with the same handedness as the structure does more work because the electric vector continuously polarizes matter, and the wave therefore travels slower.

There are several ways to enhance this effect to obtain larger activity coefficients. Atoms with large polarizabilities can be expected to give larger interactions with the light waves. Acentric crystals with large refractive indices interact strongly with light waves. $AgGaS_2$ ($n \sim 2.5$, $\alpha \sim 500°$/mm) and TeO_2 ($n \sim 2.2$, $\alpha \sim 200°$/mm) have much larger rotatory power than low refractive index compounds. Since polarizabilities increase near an absorption edge, the optical activity coefficients also increase (Fig. 30.1(a)).

Another important parameter is the pitch of the helix. In most inorganic crystals the helix pitch is 10 Å or less, far smaller than the wavelength of visible light. Helices with larger pitches give greater rotatory power, as has been demonstrated with liquid crystals. Cholesteric liquid crystals possess optical activity coefficients as large as 10^5°/mm. A critical parameter determining the magnitude of α is the ratio of the pitch of the spiral to the wavelength of the electromagnetic wave.

Ambidextrous behavior is observed in lead germanate, $Pb_5Ge_3O_{11}$, an unusual crystal which exhibits reversible optical rotatory power. Below 177°C, it is a ferroelectric in which a dextro–levo conversion accompanies reversal of the spontaneous polarization. The rotatory power is 5°35'/mm at 6328 Å, large enough for opto-electronic devices. The molecular origin of the effect is illustrated in Fig. 30.2. In the high-temperature prototype structure, some of the lead ions in lead germanate are coordinated to six oxygens arranged in a trigonal prism. On transforming to the ferroelectric state, the lead ions are displaced forming short bonds to three oxygens and creating spontaneous polarization. Four oxygens are bonded to germanium, and the GeO_4 tetrahedra are twisted when Pb displaces. The twists impart a handedness to the molecular configuration, causing optical activity. The molecular groups resemble a three-bladed airplane propeller of variable pitch.

30.2 Tensor description

In an optically-active medium, right- and left-handed circularly-polarized waves have slightly different velocities and refractive indices. Let $n(\text{R})$ and $n(\text{L})$ represent the refractive indices of the two waves. When a plane-polarized wave enters such a medium it divides into the right- and left-handed waves that then recombine into a plane-polarized wave as they emerge from the crystal. The polarization direction of the emerging wave is rotated through an angle ϕ given by

$$\phi = \frac{\pi t}{\lambda}(n(\text{L}) - n(\text{R})) = \alpha t,$$

where t is the thickness of the crystal and λ is the wavelength in free space.

It is important to estimate the size of the birefringence $n(\text{L}) - n(\text{R})$ associated with optical activity. The optical activity coefficient α is linearly proportional to this birefringence.

$$n(\text{L}) - n(\text{R}) = \frac{\alpha \lambda}{\pi}.$$

For quartz, $\alpha = 18.8°/\text{mm} = 0.328$ rad/mm at $\lambda = 0.63$ μm, giving $n(\text{L}) - n(\text{R}) = 6.6 \times 10^{-5}$. This is small compared to the intrinsic birefringence $n_e - n_o = 1.553 - 1.544 = 0.009$, or about 1%, a typical number for many anisotropic materials. For this reason it is difficult to observe optical activity in the presence of birefringence. Most experiments are carried out along optic axes, or in optically isotropic materials, where the birefringence is zero.

An optically-active fluid or solid is called *right-handed* (dextrorotatory) if the sense of rotation of the plane of polarization is *counterclockwise* as viewed looking into the optical beam toward the light source. If the rotation is clockwise, the crystal is called *left-handed* (levorotatory). Many chemical compounds exist in both right- and left-handed forms.

The sign of the optical activity coefficient α is *positive* if the rotation is right-handed, and *negative* if it is left-handed. This means that α is an axial zero-rank tensor, sometimes called a pseudoscalar. Under a transformation (a) that involves a change in handedness, α reverses sign. Therefore $\alpha' = |a|\alpha$ where $|a| = \pm 1$.

To determine how optical activity varies with direction, we first consider a crystal without optical activity. The optical properties are represented by the indicatrix, an ellipsoid in which refractive index is plotted as a function of polarization direction (Section 25.2). For a given wave normal $\vec{N}$, there are two optical waves with refractive indices n' and n'' corresponding to a fast wave and a slow wave. The values of n' and n'' are the major and minor axes of the ellipse perpendicular to the wave normal $\vec{N}$ (Section 25.3). The wave velocity surface for these two waves is

$$((v')^2 - v^2)((v'')^2 - v^2) = 0,$$

which leads to

$$(n^2 - (n')^2)(n^2 - (n'')^2) = 0,$$

since $v = c/n$. These equations apply to an arbitrary wave normal in an optically inactive medium.

For an optically active medium, a small correction is required. A gyration vector G is introduced for which

$$(n^2 - (n')^2)(n^2 - (n'')^2) = G^2.$$

Optical activity is easiest to observe near an optic axis where the birefringence is small and

$$n' \cong n'' \cong \bar{n} = \sqrt{n'n''}.$$

Under these conditions, $(n^2 - \bar{n}^2)^2 = G^2$ and $n^2 - \bar{n}^2 = \pm G$.

Since G is very small, $n = \bar{n} \pm G/2\bar{n}$. For the right- and left-circularly polarized waves, $n(L) = \bar{n} + G/2\bar{n}$ and $n(R) = \bar{n} - G/2\bar{n}$.

The optical activity coefficient α is related to the gyration vector G through the equation

$$\alpha = \frac{\phi}{t} = \left(\frac{\pi}{\lambda}\right)(n(L) - n(R)) = \left(\frac{\pi}{\lambda}\right)\left(\frac{G}{\bar{n}}\right).$$

To determine how α and the gyration coefficient G depend on direction, we make use of two experimental observations: (1) the optical activity coefficient changes sign when the handedness of the crystal changes, and (2) if the wave normal is reversed, the optical activity remains the same, both in magnitude and sign. Based on these two facts, G is an axial tensor like α, and transforms between new and old coordinate systems as $G' = |a|G$. Its directional dependence can be described by a power series.

$$G = g_i N_i + g_{ij} N_i N_j + g_{ijk} N_i N_j N_k + \cdots,$$

where N_i, N_j, and N_k are direction cosines of the wave normal $\vec{N}$. Since the optical activity is unchanged when the wave normal is reversed,

$$G(N_i) = G(-N_i) = -g_i N_i + g_{ij} N_i N_j - g_{ijk} N_i N_j N_k + \cdots.$$

Therefore all odd power terms are zero. Keeping only the lowest power term $g_{ij} N_i N_j$, the gyration tensor transforms as

$$G' = |a|G = |a|g_{ij} N_i N_j.$$

Since the wave normal is a polar first rank tensor $N_i = a_{ki} N'_k$ and $N_j = a_{lj} N'_l$.

$$G' = |a|G = |a|g_{ij} a_{ki} N'_k a_{\ell j} N'_\ell = g'_{k\ell} N'_k N'_\ell.$$

Therefore $g'_{kl} = |a|a_{ki} a_{\ell j} g_{ij}$. The directional part of the gyration tensor, g_{ij}, transforms as an axial second rank tensor.

The optical activity coefficient is

$$\alpha = \frac{\pi}{\lambda \bar{n}} G = \frac{\pi}{\lambda \bar{n}} g_{ij} N_i N_j.$$

30.3 Effect of symmetry

Optical activity is a null property that disappears in centrosymmetric point groups. In matrix form, g_{ij} transforms as follows.

$$\underset{(g')}{3 \times 3} = |a| \underset{(a)}{3 \times 3} \underset{(g)}{3 \times 3} \underset{(a)_t}{3 \times 3}.$$

For a center of symmetry,

$$(a) = \begin{pmatrix} -1 & 0 & 0 \\ 0 & -1 & 0 \\ 0 & 0 & -1 \end{pmatrix} = (a)_t \quad \text{and} \quad |a| = -1.$$

Therefore, by Neumann's Law,

$$(g') = (-1)(-1)(g)(-1) = -(g) = 0.$$

Eleven of the 32 crystal classes, and three of the seven limiting groups contain inversion symmetry, and are optically inactive.

Cubic sodium chloride (NaCl, point group $m3m$) is centrosymmetric and optically inactive, but cubic sodium chlorate (NaClO$_3$, point group 23) is optically active in all directions. The two independent symmetry elements for point group 23 are a twofold axis along $Z_1 = [100]$ and a threefold rotation axis along the body-diagonal direction [111]. For the twofold axis,

$$(a) = \begin{pmatrix} 1 & 0 & 0 \\ 0 & -1 & 0 \\ 0 & 0 & -1 \end{pmatrix} = (a)_t \quad \text{and} \quad |a| = +1.$$

$$(g') = (+1) \begin{pmatrix} 1 & 0 & 0 \\ 0 & -1 & 0 \\ 0 & 0 & -1 \end{pmatrix} \begin{pmatrix} g_{11} & g_{12} & g_{13} \\ g_{21} & g_{22} & g_{23} \\ g_{31} & g_{32} & g_{33} \end{pmatrix} \begin{pmatrix} 1 & 0 & 0 \\ 0 & -1 & 0 \\ 0 & 0 & -1 \end{pmatrix}$$

$$= \begin{pmatrix} g_{11} & -g_{12} & -g_{13} \\ -g_{21} & g_{22} & g_{23} \\ -g_{31} & g_{32} & g_{33} \end{pmatrix} = \begin{pmatrix} g_{11} & g_{12} & g_{13} \\ g_{21} & g_{22} & g_{23} \\ g_{31} & g_{32} & g_{33} \end{pmatrix}$$

by Neumann's Law. Therefore $g_{12} = g_{13} = g_{21} = g_{22} = 0$.

For the threefold axis along [111],

$$(g') = (+1) \begin{pmatrix} 0 & 1 & 0 \\ 0 & 0 & 1 \\ 1 & 0 & 0 \end{pmatrix} \begin{pmatrix} g_{11} & 0 & 0 \\ 0 & g_{22} & g_{23} \\ 0 & g_{32} & g_{33} \end{pmatrix} \begin{pmatrix} 0 & 0 & 1 \\ 1 & 0 & 0 \\ 0 & 1 & 0 \end{pmatrix}$$

$$= \begin{pmatrix} g_{22} & g_{23} & 0 \\ g_{32} & g_{33} & 0 \\ 0 & 0 & g_{11} \end{pmatrix} = \begin{pmatrix} g_{11} & 0 & 0 \\ 0 & g_{22} & g_{23} \\ 0 & g_{32} & g_{33} \end{pmatrix}.$$

by Neumann's Law. Therefore $g_{23} = g_{32} = 0$ and $g_{11} = g_{22} = g_{33}$.
The resulting matrix for point group 23 is

$$(g') = \begin{pmatrix} g_{11} & 0 & 0 \\ 0 & g_{11} & 0 \\ 0 & 0 & g_{11} \end{pmatrix}.$$

330 Optical activity and enantiomorphism

Table 30.1 Gyration tensor for the 32 crystal classes and 7 texture groups

1	$\begin{pmatrix} g_{11} & g_{12} & g_{13} \\ g_{12} & g_{22} & g_{23} \\ g_{13} & g_{23} & g_{33} \end{pmatrix}$	2	$\begin{pmatrix} g_{11} & 0 & g_{13} \\ 0 & g_{22} & 0 \\ g_{13} & 0 & g_{33} \end{pmatrix}$
m	$\begin{pmatrix} 0 & g_{12} & 0 \\ g_{12} & 0 & g_{23} \\ 0 & g_{23} & 0 \end{pmatrix}$	222	$\begin{pmatrix} g_{11} & 0 & 0 \\ 0 & g_{22} & 0 \\ 0 & 0 & g_{33} \end{pmatrix}$
$mm2$	$\begin{pmatrix} 0 & g_{12} & 0 \\ g_{12} & 0 & 0 \\ 0 & 0 & 0 \end{pmatrix}$	3, 32 4, 422 6, 622 $\infty, \infty 2$	$\begin{pmatrix} g_{11} & 0 & 0 \\ 0 & g_{11} & 0 \\ 0 & 0 & g_{33} \end{pmatrix}$
$\bar{4}$	$\begin{pmatrix} g_{11} & g_{12} & 0 \\ g_{12} & -g_{11} & 0 \\ 0 & 0 & 0 \end{pmatrix}$	$\bar{4}2m$	$\begin{pmatrix} g_{11} & 0 & 0 \\ 0 & -g_{11} & 0 \\ 0 & 0 & 0 \end{pmatrix}$
23, 432, $\infty\infty$	$\begin{pmatrix} g_{11} & 0 & 0 \\ 0 & g_{11} & 0 \\ 0 & 0 & g_{11} \end{pmatrix}$	All others	$\begin{pmatrix} 0 & 0 & 0 \\ 0 & 0 & 0 \\ 0 & 0 & 0 \end{pmatrix}$

Optical activity matrices for other point groups are collected in Table 30.1. The number of measurements required to specify the full gyration tensor range from six in triclinic point group 1 to one in optically-active liquids (Curie group $\infty\infty$) and cubic crystals (crystal classes 23 and 432). The g_{ij} matrix is symmetric because the direction cosines of the wave normal, N_i and N_j can be interchanged.

Problem 30.1
Optical activity coefficients disappear in all centrosymmetric point groups. They also disappear in the following acentric classes: $3m, \bar{6}, 6mm, 4mm, \bar{6}m2, \bar{4}3m$, and ∞m. Prove why this is so.

Symmetric tensors can be transformed to principal axes, thereby reducing the number of coefficients. When rotated to principal axes, the six coefficients of a triclinic crystal are reduced to three. The procedure is the same as that outlined earlier for dielectrics. The gyration tensor is

$$G = g_{ij}N_iN_j = g_{11}N_1^2 + g_{22}N_2^2 + g_{33}N_3^2.$$

For uniaxial crystals like quartz belonging to point group 32, $g_{11} = g_{22}$. In spherical coordinates, $N_1 = \sin\theta\cos\Phi$, $N_2 = \sin\theta\sin\Phi$, $N_3 = \cos\theta$, and $G = g_{11}\sin^2\theta + g_{33}\cos^2\theta$. The values measured for right-handed quartz at 0.51 μm are $g_{11} = -5.82 \times 10^{-5}$ and $g_{33} = +12.96 \times 10^{-5}$. Signs are reversed for left-handed quartz.

The gyration surface of quartz is plotted in Fig. 30.3. Since the signs of g_{11} and g_{33} are opposite, there will be a critical angle θ for which $G = 0$. The angle is given by $\tan^2\theta_0 = -g_{33}/g_{11}$. For quartz, the directions of zero gyration form a cone about the optic axis with $\theta_0 = 56°10'$.

Optical activity in biaxial crystals is more complex because there are two optic axes and they do not coincide with symmetry directions. As an example, consider monoclinic point group 2. Two different situations arise. In one case, the twofold symmetry axis lies in the same plane as the two optic axes (Fig. 30.4(a)) and in the other case, it is perpendicular to the two optic axes (Fig. 30.4(b)).

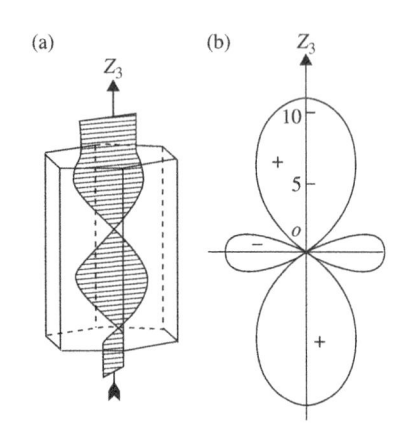

Fig. 30.3 Optical activity in α-quartz. (a) When looking toward the light source a right-handed crystal rotates the plane of polarization in a counter-clockwise direction. (b) The gyration surface is plotted as a function of wave normal for right-handed quartz. Signs are reversed for a left-handed crystal. Units of g are 10^{-5}. At 20°C and 0.589 μm quartz rotates the plane of polarized light by 21°40'/mm about the optic axis Z_3.

Fig. 30.4 Optical activity in monoclinic crystals belonging to point group 2. (a) When the twofold axis lies in the plane of the two optic axes, they both exhibit the same optical activity coefficient. (b) When the twofold axis is perpendicular to the two optic axes, the optical activity is different. In (a) the two optic axes are related by 180° rotation around the twofold axis Z_2, but in (b) they are not.

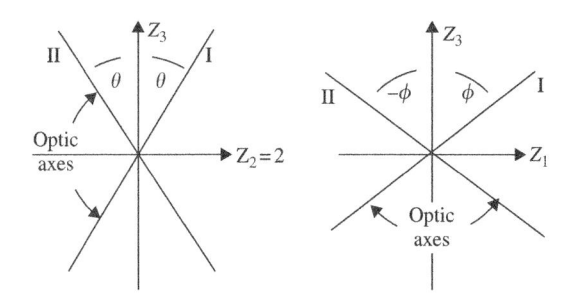

For monoclinic point group 2 the gyration coefficient is

$$G = g_{11}N_1^2 + g_{22}N_2^2 + g_{33}N_3^2 + 2g_{13}N_1N_3.$$

Referring to Fig. 30.4(a), the gyration for optic axis I ($N_1 = 0$, $N_2 = \sin\theta$, $N_3 = \cos\theta$) is $G = g_{22}\sin^2\theta + g_{33}\cos^2\theta$. For optic axis II ($N_1 = 0$, $N_2 = -\sin\theta$, $N_3 = \cos\theta$), $G = g_{22}\sin^2\theta + g_{33}\cos^2\theta$. Both axes exhibit the same rotation.

A different result is obtained for the configuration in Fig. 30.4(b) where the twofold axis Z_2 is perpendicular to the optic axes. Now optic axes I and II are not related to one another by symmetry and they have different optical activity coefficients. For optic axis I ($N_1 = -\sin\phi$, $N_2 = 0$, $N_3 = \cos\phi$), $G = g_{11}\sin^2\phi + g_{33}\cos^2\phi - 2g_{13}\sin\phi\cos\phi$. For optic axis II ($N_1 = \sin\phi$, $N_2 = 0$, $N_3 = \cos\phi$), $G = g_{11}\sin^2\phi + g_{33}\cos^2\phi + 2g_{13}\sin\phi\cos\phi$. The two optic axes may be quite different in this case.

Examples of both types are known. The two optic axes in tartaric acid both have a rotatory power of $+10.8°$/mm. In cane sugar, one axis is $-1.6°$/mm and the other is $+5.4°$/mm. Statistically the first case is more common since the twofold axis has equal probability of aligning along any of the three principal axes. Only one-third of the many crystals in point group 2 have Z_2 perpendicular to the optic axes.

30.4 Relationship to enantiomorphism

Enantiomorphic crystals and molecules do not contain mirror or inversion symmetry. Chemists refer to these compounds as *chiral*, coming from the Greek word for "hand".

Eleven of the 32 crystal classes and three of the seven Curie groups are enantiomorphic (Table 30.2). Four null properties associated with noncentrosymmetric point groups are compared in this table. Piezoelectricity, the linear electro-optic (Pockels) effect and second harmonic coefficients occur in 20 of 21 acentric crystal classes. Pyroelectricity is observed in 10 of these 20 classes. All pyroelectric crystals are also piezoelectric. Fifteen of the acentric classes are optically active and 11 of the 15 are also enantiomorphic. All enantiomorphic classes are optically active.

Enantiomorphic crystals can sometimes be distinguished by morphology. Right- and left-handed crystals appear as mirror images when the appropriate faces appear. The *d*- and *l*-sodium ammonium tartrate crystals studied by Louis Pasteur are shown in Fig. 30.5(a). He went on to demonstrate the relationship between enantiomorphism and optical activity. In many organic compounds the chirality extends down to the molecular scale, where the molecules are

Table 30.2 Properties associated with acentric crystals and textures: piezoelectricity (P), pyroelectricity (P*), optical activity (O), and enantiomorphism (E)

Crystal classes					Curie groups				
1	O	E	P	P*	∞	O	E	P	P*
2	O	E	P	P*	$\infty 2$	O	E	P	
m	O		P	P*	∞m			P	P*
222	O	E	P		$\infty\infty$	O	E		
mm2	O		P	P*					
3	O	E	P	P*					
32	O	E	P						
3m			P	P*					
4	O	E	P	P*					
$\bar{4}$	O		P						
422	O	E	P						
4mm			P	P*					
$\bar{4}2m$	O		P						
6	O	E	P	P*					
$\bar{6}$			P						
622	O	E	P						
6mm			P	P*					
$\bar{6}m2$			P						
23	O	E	P						
432	O	E							
$\bar{4}3m$			P						

Crystal classes

Acentric symmetry — 21
Optical activity — 15
Piezoelectricity — 20
Enantiomorphism — 11
Pyroelectricity — 10

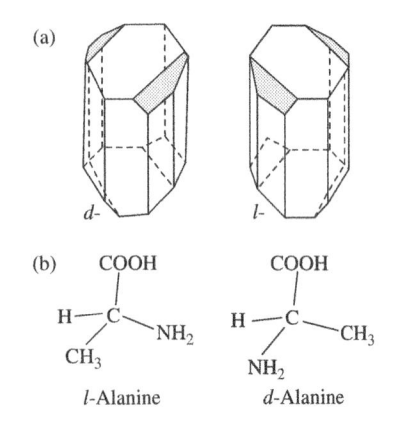

Fig. 30.5 Enantiomorphism often appears in crystal morphology and molecular structure. (a) d- and l-sodium ammonium tartrate crystals are mirror images of one another, as are (b) the molecular structures of d- and l-alanine, $NH_2CH(CH_3)COOH$.

right- and left-handed. The structures of d- and l-alanine (Fig. 30.5(b)) are a good example. Pasteur was impressed by the fact that most natural products, cellulose, sugar, quinine, and turpentine exist in only one handedness. In many ways this dissymmetry was the beginning of biochemistry.

The relationship between optical activity and enantiomorphism is very close because optical activity is difficult to observe in all the symmetry groups that are not enantiomorphic. Of the 18 point groups (15 crystal classes and 3 Curie groups) that exhibit optical activity, all but four are enantiomorphic. The exceptions are m, $mm2$, $\bar{4}$, and $\bar{4}2m$. These four groups possess either mirror planes or inversion operations that are not allowed in chiral groups.

At first glance, it is rather puzzling that optical activity can occur in these point groups with mirror and inversion symmetry since optical rotation is normally pictured as a helix (Fig. 30.3), and a helix does not contain these types of symmetry. To understand why mirrors are sometimes allowed in optically active crystals, consider the orientation of the mirror plane relative to the helix. Mirror planes cannot exist parallel or perpendicular to the helix, but they can intercept the helix at a different angle. This does not violate the symmetry of the helix, but it generates another helix of the opposite handedness (Fig. 30.6).

This leads to a biaxial crystal in which one optic axis is right-handed and the other left-handed. In monoclinic point group m, principal axis Z_2 is perpendicular to the mirror plane as shown in Fig. 30.6(b). If we assume that the principal refractive indices are $n_3 > n_2 > n_1$, then Z_1 is perpendicular to the two optic axes and Z_3 lies midway between them. The nonzero optical activity coefficients for point group m are $g_{12} = g_{21}$ and $g_{23} = g_{32}$. The resulting gyration equation is $G = 2g_{12}N_1N_2 + 2g_{23}N_2N_3$. The direction cosines for optic axis I are $N_1 = 0$, $N_2 = \sin\theta$, $N_3 = \cos\theta$, giving $G = 2g_{23}\sin\theta\cos\theta$. For optic axis II, $N_1 = 0$, $N_2 = -\sin\theta$, $N_3 = \cos\theta$, and $G = -2g_{23}\sin\theta\cos\theta$. Thus the optical activity coefficients of the two optic axes are equal and opposite

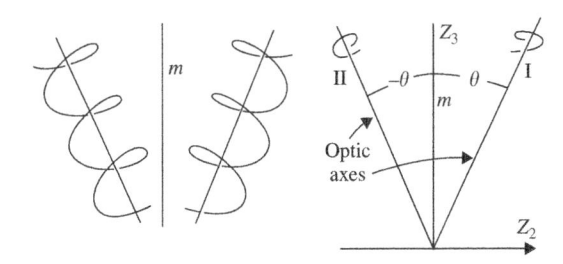

Fig. 30.6 Point group m is optically active but is not enantiomorphic. (a) A mirror plane turns a right-handed helix into a left-handed helix. With axial orientation in (b), one optic axis has right-handed optical activity, and the other left.

in sign as predicted by the symmetry argument. And since the optical activity is equally balanced between right- and left-handed behavior, the crystal is not chiral. Similar arguments can be developed for point group $mm2$, $\bar{4}$, and $\bar{4}2m$ that are also optically active but not enantiomorphic. In $\bar{4}2m$, the only two nonzero gyration coefficients are $g_{11} = -g_{22}$. Therefore $G = g_{11}N_1^2 - g_{11}N_2^2$. The standard setting for $\bar{4}2m$ is $Z_1 = [100]$ and $Z_2 = [010]$, with the optic axis along $Z_3 = [001]$. Hence the crystal shows right-handed optical activity along $[100]$ and left-handed optical activity along $[010]$, but the optic axis $[001]$ is not optically active.

Birefringence makes it very difficult to observe optical activity in point group $\bar{4}2m$. A solution to this problem was found by taking advantage of the unusual dispersion curves in silver gallium sulfide crystals. $AgGaS_2$ changes optic sign in the visible range and is therefore optically isotropic at the crossover wavelength (Fig. 26.2(b)). This makes it possible to measure the optical activity coefficients along $[100]$ and $[010]$.

30.5 Liquids and liquid crystals

Water is a transparent liquid with a modest refractive index (1.33) and zero optical activity. Molecules of H_2O (molecular symmetry $mm2$) possess no handedness and are randomly oriented in the liquid state resulting in optical isotropy (spherical symmetry $\infty\infty m$). Dissolving 10 weight% d-glucose ($C_6H_{12}O_6$) in water gives a weak optical activity of $0.055°/mm$. In order for a fluid to be optically active its constituent molecules must individually rotate the plane of polarized light since, in a liquid, there is no long-range order. The random orientation does not lead to cancellation of the effect because the handedness of a chiral molecule like d-glucose is not dependent on direction. The symmetry group of a chiral liquid is $\infty\infty$. Saccharimeters have long been used to determine the sugar content of various fluids. The magnitude of the optical rotation is proportional to the amount of dissolved sugar. A rather long optical path, typically 10 cm or more, is required for an accurate reading because the rotatory powers are small. Rotations for optically active organic liquids are generally in the 0.01 to $1°/mm$ range.

Much larger optical effects are observed in liquid crystals. Anisotropic molecular interactions lead to several types of *mesophases* with properties (and symmetries) intermediate between solids and liquids (Fig. 30.7). *Crystals* have translational periodicity and orientational order. *Plastic crystals* have translational periodicity but no orientation order. Molecular compounds such as adamantane and WCl_6 with near spherical shape and weak intermolecular

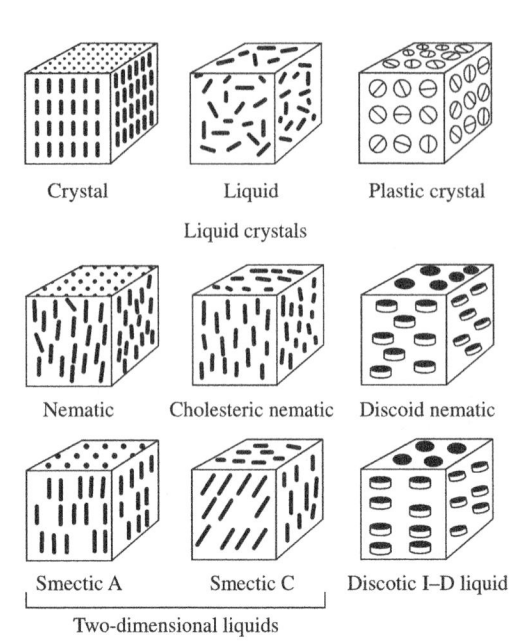

Fig. 30.7 Four states of matter. When crystals melt they normally lose both translational and orientational order. But sometimes the melting process proceeds in stages through various mesophases showing partial order. Liquid crystals retain some of the orientational order and often show pronounced anisotropy. Plastic crystals lose orientational order while retaining periodicity. The name comes from their butter-like mechanical properties.

bonds fall into this category. *Liquid crystals* have partial orientational order but lack full translational periodicity. *Liquids* possess neither translational periodicity nor orientational order.

Several different types of liquid crystals have been identified (Fig. 30.7). The molecules tend to be long and slender or disk-like in shape. Shape anisotropy promotes the stability of liquid crystal phases. A *smectic liquid crystal* is closest in structure to a crystal. The molecules organize themselves in layers with one-dimensional translational periodicity as well as orientational order. In smectic A liquid crystals the molecules are oriented perpendicular to the layers, but in smectic C they are tilted. The molecules tend to be long and thin with floppy ends (Fig. 30.8(a)).

Nematic liquid crystals possess orientational order but lack translational periodicity. The elongated molecules are parallel to a common direction but are not organized into layers. As a result they are more disordered than smectic liquid crystals and usually occur at high temperature. p-Azoxyanisole, a long molecule with a stiff central portion, is a typical nematic liquid crystal (Fig. 30.8(b)).

In *cholesteric liquid crystals* the molecular orientation is twisted from one layer to the next. They form helical structures with a pitch that depends on temperature. The nematic and cholesteric phases have the greatest number of optical applications. Their strongly anisotropic electro-optic properties form the basis for many types of optical displays used in television, computers, and wristwatches. Cholesteric liquid crystals are sometimes referred to as chiral nematic liquid crystals.

Discotic liquid crystals have quite different molecular geometries. As the name implies, the molecules have a flat, disk-like shape (Fig. 30.8(d)). The disks are sometimes stacked like pancakes to form a one-dimensional liquid, while at higher temperature they remain parallel to one another but lose the columnar order.

Fig. 30.8 Four liquid crystal compounds exhibiting (a) smectic, (b) nematic, (c) cholesteric, and (d) discoidal behavior. Note the rather limited temperature ranges.

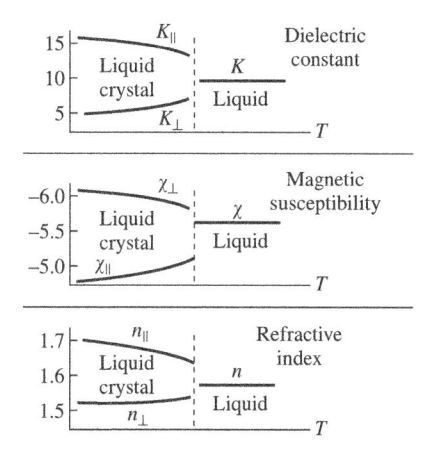

Fig. 30.9 Typical anisotropies developed at the liquid–liquid crystal phase transition.

Because of their orientational order, liquid crystals conform to various Curie group symmetries. Many possess cylindrical symmetry (∞/mm) but some have ferroelectric properties with conical symmetry (∞m) in the poled state. Cholesteric liquid crystals are optically active with chiral symmetry ($\infty 2$) and a few are ferroelectric as well, reducing the symmetry to ∞. The only Curie group that is not well represented in liquids and liquid crystals is ∞/m. Ferroliquid crystals containing magnetic particles come close (Fig. 4.2).

The aligned molecules in liquid crystals give rise to highly anisotropic physical properties (Fig. 30.9). The anisotropy in dielectric constant $\Delta K = K_\parallel - K_\perp$ can be very large since the chemical bonding is quite different parallel and perpendicular to the alignment direction. The dielectric constants and refractive indices are generally higher parallel to the alignment direction because the local dipole fields reinforce the applied field in this direction (see Section 25.5).

This is not true for magnetic susceptibility. Liquid crystals tend to be diamagnetic with $\chi_\perp > \chi_\parallel$. Here the aromatic rings play a dominant role just as they do in most strongly diamagnetic organic crystals (Section 14.7). In liquid crystals the aromatic rings are usually parallel to the alignment direction giving the maximum susceptibility in the perpendicular directions.

336 *Optical activity and enantiomorphism*

Nematic and smectic liquid crystals are optically uniaxial positive with the optic axis parallel to the elongated molecules. Liquid crystals have a number of useful electro-optic effects. Under an applied field the molecules tend to rotate into the field direction because of the dielectric anisotropy. The time constant involved in the reorientation is often in the millisecond range, depending on the viscosity.

A twisted nematic mode is used in many liquid crystal displays. A thin layer of liquid crystal is aligned in a 90° helical pattern by surface treatment of the electrodes (Fig. 30.10). When viewed between crossed polarizers in zero field, the display is bright. An applied field aligns many of the molecules into the field direction, removing the original 90° twist. The crossed polarizers then darken the display.

Cholesteric liquid crystals with long helical pitch have very large optical activity coefficients. Fig. 30.11 shows the optical rotatory effect in a mixture of two cholesteric liquids. The α values are strongly temperature because of thermal instability of liquid crystals. This makes cholesteric compounds useful as temperature sensors in medical diagnosis and "mood" rings. Constructive interference takes place when the pitch of the helix equals the wavelength of light, resulting in bright reflected colors.

The strong temperature dependence is a disadvantage in other display applications. To overcome this difficulty, mixtures of liquid crystals are used to broaden the mesophase stability range.

Dichroic dyes such as anthroquinone and azoquinone are often added to nematic mixtures to produce colored liquid crystal displays. When the electric

Fig. 30.10 The twisted nematic liquid crystal display works well in ambient illumination. Surface treatment leads to an artificial optical activity by aligning molecules in a 90° rotation. When switched from the "off" to "on" position with an applied voltage, the reflected light intensity is altered as the molecules align with the field.

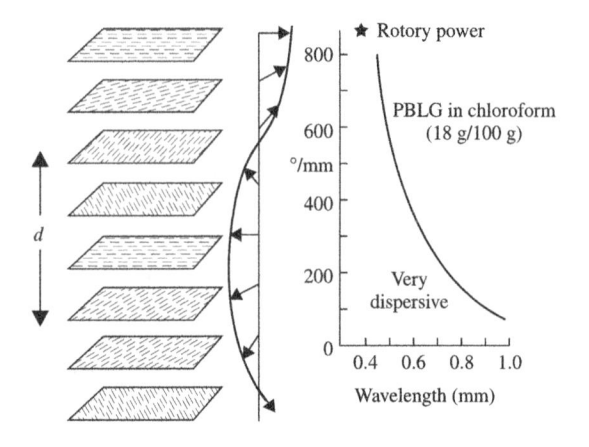

Fig. 30.11 Cholesteric liquid crystals sometimes have long helical pitch comparable to the wavelength of light resulting in very large optical rotation coefficients. The pitch is very sensitive to temperature causing a change in sign and selective reflection at the crossover point where $t = \lambda$. The data are from a mixture of cholesteryl chloride and cholesteryl myristate measured near room temperature.

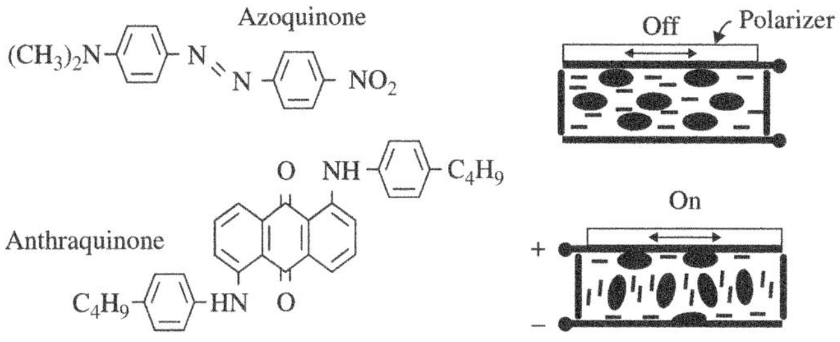

Fig. 30.12 Dichroic dyes are used in guest-host LCDs to produce color changes. Elongated dye molecules are re-oriented under electric field along with the host liquid crystal. The change in orientation causes a change in optical absorption.

field is switched on the elongated dye molecules change orientation with the host nematic liquid crystal molecules. A pronounced color change accompanies the reorientation of the dichroic dye (Fig. 30.12).

Ferroliquid crystals are an interesting family of complex magnetic fluids in which elongated single-domain magnetic particles are suspended in a liquid crystalline carrier. The family includes ferrocholesteric, ferronematic, and ferrosmectic ferroliquids. Being a composite medium, ferroliquid crystals adopt some features of both parent phases. From the liquid crystal side, they inherit orientational molecular order, optical birefringence, and optical activity. From the magnetic fluid side, they exhibit high magnetic susceptibility and the possibility of spontaneous magnetization. But the real novelty of ferroliquid crystals is the presence of a strong interaction between the suspended magnetic particles and the liquid crystal matrix. A number of different orientational transitions can take place in an applied magnetic field, including *Frederik's Transitions*, in which a realignment of the magnetic vector causes a reorientation of the nematic host molecule.

Problem 30.2
As discussed in the text, liquids and liquid crystals belong to several different Curie groups including $\infty\infty m$, $\infty\infty$, ∞/mm, ∞m, $\infty 2$, and ∞. Normally these liquids are nonmagnetic, so they also possess $1'$, the time reversal operator (Section 14.2). Ferroliquids are composite liquids consisting of liquid crystals with dispersed magnetic particles. The magnetization vectors (magnetic group ∞/mm') of the particles align in a strong magnetic field. Using the Curie Principle of symmetry superposition, what magnetic groups represent these composites? Assume that the magnetic field is either parallel or perpendicular to the alignment axis of the liquid crystal.

30.6 Dispersion and circular dichroism

In 1812 Biot noted an increase in the angle of rotation as the color of light changed from red to violet. Quantitative measurements led to the Law of Inverse Squares relating the optical activity coefficient, α, to the wavelength, λ; α is proportional to $1/\lambda^2$. For normal dispersion, as in quartz, $\alpha\lambda^2$ is approximately constant over the visible range. More precise measurements over wider wavelength ranges led to Drude's formula in which α is proportional to $1/(\lambda^2 - \lambda_0^2)$. The coefficient λ_0 corresponds to the absorption band, as shown in Fig. 30.1. Further modifications are required when more than one absorption edge is involved.

The optical activity coefficient of quartz is accurately represented by the equation

$$\alpha = \frac{9.5639}{\lambda^2 - 0.0127493} - \frac{2.3113}{\lambda^2 - 0.000974} - 0.1905,$$

where λ is expressed in microns and α in degrees per millimeter. This formula fits the data from $\lambda = 3.2$ μm in the infrared to 0.2 μm in the near ultraviolet. α increases from 0.5 to over 250°/mm in this wavelength range.

Most organic compounds show similar behavior approximating the inverse square relationship. In this class of normal dispersive media, the rotatory power increases progressively with diminishing wavelength. The rotatory power α, and its first and second derivatives, $d\alpha/d\lambda$ and $d^2\alpha/d\lambda^2$, remain constant in sign throughout the region of transparency.

But not all optically-active substances behave normally. *Anomalous rotatory dispersion* occurs in liquids and solids with optically active absorption bands and is often accompanied by circular dichroism. Anomalous dispersion involves a reversal of sign with $\alpha = 0$ at the point of reversal, peaks in the rotatory dispersion curve when $d\alpha/d\lambda = 0$, and inflection points where $d^2\alpha/d\lambda^2 = 0$. The dispersion curve of potassium chromium tartrate (Fig. 30.14) illustrates the anomalous behavior. In the near infrared, α is about $+1°$/mm, rising to a peak in the visible, then passes through zero to a negative peak before reversing again. There are different causes for these anomalies. In transparent media the effect may be due to presence of both right- and left-handed molecules with different dispersion curves. This can occur in crystals or in solutions. The classic example is tartaric acid. The right- and left-handed portions of complex molecules can be responsible for such behavior. In other optically active materials the anomalies are related to circular dichroism.

The discovery of circular dichroism may be regarded as a sequel to the discovery of dichroism. Biot discovered dichroism in crystals of tourmaline in 1815. As pointed out in Section 26.2, the ordinary ray in tourmaline is absorbed much more strongly than the extraordinary ray. In an isotropic optically-active medium, there are also two kinds of rays, a right circularly-polarized ray and a left circularly-polarized ray. The two rays travel with different velocities leading to circular double refraction. It is therefore natural to ask if the two circularly-polarized rays are absorbed differently. Circular dichroism was observed in amethyst (purple quartz) in 1860, and later in a number of optically-active colored solutions. Unequal absorption is observed in solutions of potassium copper d-tartrate, with an equal but opposite effect in potassium copper l-tartrate. No dichroism is observed in a racemic mixture of the d- and l-tartrates. These phenomena and the accompanying anomalous dispersion near the absorption peak are known as the *Cotton Effect*.

The Cotton coefficient governing circular dichroism is proportional to the difference in intensity for the right- and left-circularly polarized rays, $I(R) - I(L)$. It is related to the imaginary part of the optical activity coefficient g_{ij} and can therefore be represented by an axial second rank tensor.

When a plane polarized ray passes through an optically-active medium that absorbs the two circular components differently, not only is the plane of polarization rotated, but the ray becomes elliptically polarized (Fig. 30.13).

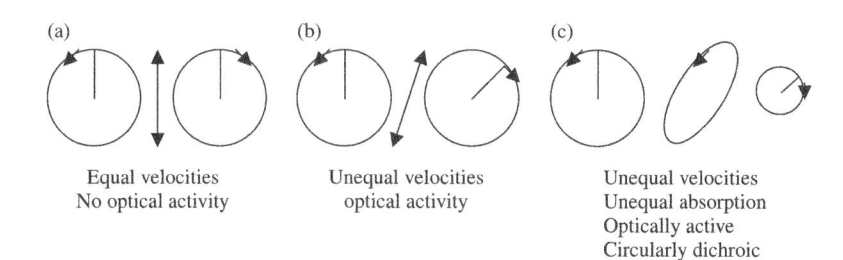

Fig. 30.13 Behavior of polarized light in (a) inactive medium, (b) optically-active medium, and (c) optically-active and circularly dichroic medium.

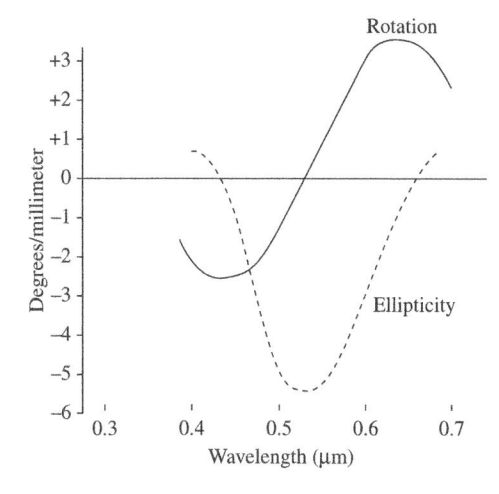

Fig. 30.14 Anomalous rotatory dispersion and elliptical polarization in potassium chromium tartrate (Cotton).

The major and minor axes of the resultant elliptical polarization are equal to the sum and difference, respectively, of the amplitudes of the right- and left-circularly polarized waves. The angle through which the major axis has been rotated is equal to half the difference in phase between the two rays, and the sign of the ellipse is the same as that of the circular component of greatest intensity.

To fully characterize a circularly dichroic material, it is necessary to make two measurements: (1) the rotation angle between the incident vibration direction and the major axis of the ellipse, and (2) the circular dichroism, that can be described either as the ellipticity of the emergent light, or as the difference in intensity of the right- and left-circularly polarized beams. Measurements on potassium chromium tartrate are shown in Fig. 30.14. Electronic transitions in the 3d-shell of chromium causes a strong absorption band in the optical range resulting in anomalous dispersion and circular dichroism. Measurements on a number of different colored solutions containing chiral molecules led to the following conclusions: (1) ellipticity reaches a maximum near where the absorption is most intense, and (2) the rotatory power changes sign at the same wavelength where the ellipticity is largest. Optically-active absorption bands such as this are responsible for anomalous dispersion and for circular dichroism. *Bruhat* devised a rule that summarizes the behavior of polarized light in an optically active medium: "On the red side of the absorption band, the ray that is less absorbed is propagated with the greater velocity; on the violet side the reverse is the case."

It should be noted, however, that not all absorption bands are optically active. Cotton showed that circular dichroism and anomalous dispersion do not occur in colored solutions in which the optical activity and absorption band are caused by different molecules. For example, no change in rotatory power was observed in sugar solutions colored with magenta dye, nor was the transmitted light elliptically polarized.

30.7 Electrogyration, piezogyration, and thermogyration

External fields or forces will alter the optical activity effect. Under an applied electric field, the change in the gyration coefficient can be written as a power series:

$$\Delta g_{ij} = g_{ij}(E) - g_{ij}(0) = A_{ijk}E_k + A_{ijkl}E_k E_l + \cdots .$$

To describe the effect of symmetry on A_{ijk} and A_{ijkl} we must first determine how they transform. When written in the new coordinate system,

$$\Delta g'_{ij} = |a|a_{ik}a_{jl}\Delta g_{kl} = |a|a_{ik}a_{jl}A_{klm}E_m$$
$$= |a|a_{ik}a_{jl}A_{klm}a_{nm}E'_n = A'_{ijn}E'_n.$$

Therefore the *linear electrogyration* coefficients A_{ijn} transform as an axial third rank tensor:

$$A'_{ijn} = |a|a_{ik}a_{jl}a_{nm}A_{klm}.$$

In a similar way, it can be shown that the *quadratic electrogyration* coefficients A_{ijkl} constitute a fourth rank axial tensor which transforms as

$$A'_{ijkl} = |a|a_{im}a_{jn}a_{kp}a_{lo}A_{mnpo}.$$

In matrix form the A_{ijk} coefficients can be written as a 6×3 matrix resembling the piezoelectric effect.

$$\begin{array}{ccc} 6 \times 1 & 6 \times 3 & 3 \times 1 \\ (\Delta g) = & (A) & (E). \end{array}$$

However, the effect of symmetry is quite different because electrogyration is an axial tensor whereas piezoelectricity is a polar tensor. Piezoelectricity disappears in centrosymmetric media but electrogyration does not. As pointed out earlier, the birefringence associated with optical activity is small compared with standing birefringence, therefore the electrogyration experiment should be carried out along an optic axis or an optically isotropic material. Moreover it would be best to avoid competing electro-optic and piezoelectric effects by using a centrosymmetric crystal. The experiment shown in Fig. 30.15 is a simple one in which the electric field and optical beam are parallel to one another along an optic axis. This set of conditions requires a nonzero coefficient A_{33} in a uniaxial

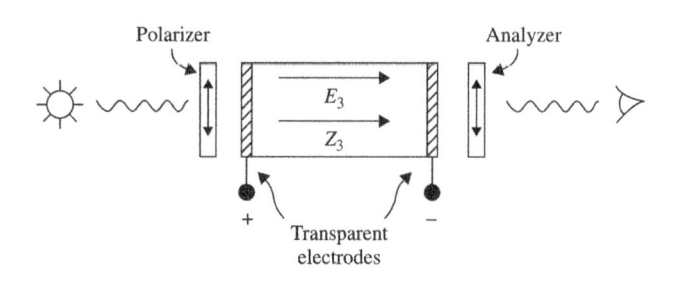

Fig. 30.15 Electrogyration experiment on a uniaxial crystal with the optical light path and electric field directed along the optic axis $Z_3 = [001]$.

centrosymmetric point group. Only point groups $\bar{3}$, $4/m$, and $6/m$ satisfy these criteria.

Measurements have been carried out on crystals of lead molybdate (point group $4/m$). In the experiment the coefficients of interest are A_{333} and A_{3333}, the first- and second-order electrogyration constants.

$$\Delta g_{33} = g_{33}(E) - g_{33}(0) = A_{333}E_3 + A_{3333}E_3^2.$$

Optical activity is absent in point group $4/m$ so $g_{33}(0) = 0$. The effect of $4/m$ symmetry on A_{333} and A_{3333} are easily determined by the direct inspection method. The fourfold symmetry axis along Z_3 takes $1 \rightarrow 2 \rightarrow -1$, and $3 \rightarrow 3$. There is no handedness change so $|a| = +1$. Therefore $333 \rightarrow 333$ and $3333 \rightarrow 3333$. There is no effect on either A_{333} or A_{3333}. For the mirror plane perpendicular to Z_3, $1 \rightarrow 1$, $2 \rightarrow 2$, and $3 \rightarrow -3$. In this case there is a handedness change, $|a| = -1$. Under the mirror operation, $333 \rightarrow (-1)(-3)(-3)(-3) \rightarrow 333$, and $3333 \rightarrow (-1)(-3)(-3)(-3)(-3) \rightarrow -3333$. By Neumann's Law, A_{333} is allowed but $A_{3333} = 0$. For PbMoO$_4$ a rotation of $1°$ was obtained for an applied field of 5 kV/mm and a wavelength of 0.633 μm, confirming the existence of A_{333}.

The *piezogyration effect* describes the influence of mechanical stress on optical activity.

$$\Delta g_{ij}(X) = g_{ij}(X) - g_{ij}(0) = C_{ijkl}X_{kl} + C_{ijklmn}X_{kl}X_{mn} + \cdots.$$

The linear piezogyration coefficients C_{ijkl} is a fourth rank axial tensor like the quadratic electrogyration effect. The quadratic piezogyration effect C_{ijklmn} is a sixth rank axial tensor. Both of these effects disappear in centrosymmetric materials. Even rank axial tensors and odd-rank polar tensors disappear when inversion symmetry is present. Bear in mind, however, that the situation changes when dealing with magnetic symmetry and time reversal.

Thermogyration, the influence of temperature on optical activity, can be described as a power series in temperature. All the coefficients in $\Delta g_{ij}(T)$ are second rank axial tensors like g_{ij} since $T, T^2, T^3, \ldots$ are all scalars.

The rotatory power α often changes rapidly near phase transitions. As pointed out in Section 30.5, α is very sensitive to temperature in liquid crystals. Rotations of the order of 10,000°/mm in cholesteric liquid crystals decrease to very small values when heated into the normal liquid state. The large rotatory power of liquid crystals is caused by the helical packing of molecules rather than the chiral nature of individual molecules.

The α–β transition in quartz provides another example. The optical activity coefficient increases by 16% when heated from room temperature through the phase transformation at 573°C. In this case the change is due to thermal expansion and the internal straightening of the Si–O–Si bonds to form more perfect helices. The fractional change in rotatory power, $(1/\alpha)(d\alpha/dT)$, is about 0.00015 K^{-1} near room temperature.

31 Magneto-optics

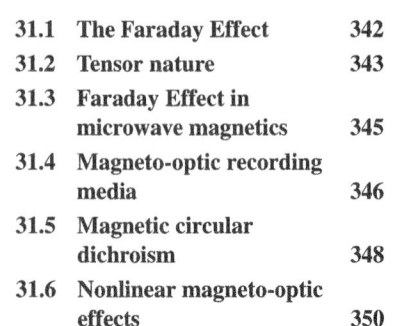

31.1	The Faraday Effect	342
31.2	Tensor nature	343
31.3	Faraday Effect in microwave magnetics	345
31.4	Magneto-optic recording media	346
31.5	Magnetic circular dichroism	348
31.6	Nonlinear magneto-optic effects	350
31.7	Magnetoelectric optical phenomena	351

The magneto-optic properties of interest are the Faraday Effect, Kerr Rotation, and the Cotton–Mouton Effect. In 1846, Michael Faraday discovered that when linearly polarized light passes through glass in the presence of a magnetic field, the plane of polarization is rotated. The *Faraday Effect* is now used in a variety of microwave and optical devices. Normally the Faraday experiment is carried out in transmission, but rotation also occurs in reflection, the so-called Kerr Rotation that is used in magneto-optic disks with Mbit storage capability. Other magneto-optic phenomena of less practical interest include the *Cotton–Mouton Effect*, a quadratic relationship between birefringence and magnetic field, and *magnetic circular dichroism* that is closely related to the Faraday Effect. A number of nonlinear optical effects of magnetic or magnetoelectric origin are also under study. Almost all these magnetooptical effects are caused by the splitting of electronic energy levels by a magnetic field. This splitting was first discovered by the Dutch physicist Zeeman in 1896, and is referred to as the *Zeeman Effect*.

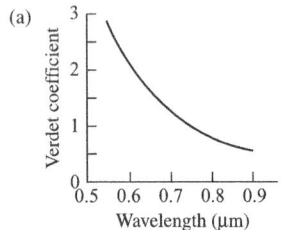

31.1 The Faraday Effect

When linearly polarized light travels parallel to a magnetic field, the plane of polarization is rotated through an angle ψ. It is found that the angle of rotation is given by

$$\psi(\omega) = V(\omega)Ht,$$

where H is the applied magnetic field, t is the sample thickness, ω is the angular frequency of the electromagnetic wave, and $V(\omega)$ is the *Verdet coefficient*. Faraday rotation is observed in nonmagnetic materials as well as in ferromagnets. The Verdet coefficient of a commercial one-way glass is plotted as a function of wavelength in Fig. 31.1(a). Corning 8363 is a rare earth borate glass developed to remove reflections from optical systems. A polarized laser beam is transmitted through the glass parallel to the applied magnetic field. The plane of polarization is rotated 45° by the Faraday Effect. The transmitted beam passes through the analyzer that is set at 45° to the polarizer. But the reflected waves coming from the surface of the glass and from the analyzer are rotated another 45° as they return toward the laser. This puts them 90° out of phase with the polarizer and blocks their return. The experimental configuration is illustrated in Fig. 31.1(b). The Verdet coefficient in the rare-earth glass

Fig. 31.1 Faraday Effect in a one-way glass. (a) The Verdet coefficient for a commercial rare-earth borate glass (Corning 8363). (b) One-way optical system utilizing the Faraday Effect in 8363 glass. Reflected waves are rotated 90° and blocked by the polarizer.

is large compared to other glasses and liquids (Table 31.1) but small compared to magnetic materials.

At first glance the Faraday Effect appears similar to optical activity since both involve the rotation of the plane of polarized light, but they differ in both theory and experiment. The two experiments are compared in Fig. 31.2. Optical activity originates from a helical internal structure and does not involve any external field or forces. The sense of rotation bears a fixed relation to the wave normal, such that when a beam of light is reflected back on itself, the net rotation is zero. The Faraday Effect behaves quite differently. In this case the rotation depends on the direction of the wave normal and the direction of the magnetic field. Both are reversed on reflection, so that rotation continues in the same direction, doubling the rotation.

The underlying origins of the two effects are also quite different. Faraday rotation and optical activity arise from two different types of dispersion. *Temporal* or *frequency dispersion* is responsible for the Faraday Effect. It is produced by interactions between the electromagnetic wave and moving electrical charge, generally in the form of circulating currents or magnetic spins. The Faraday Effect requires the presence of a DC magnetic field. Resonances associated with the charge carrier motion cause a frequency-dependent response to the field, and this appears as temporal dispersion in the electric permittivity or magnetic permeability.

Optical activity arises from a different kind of dispersion, *spatial dispersion*. The local field at any position in a solid depends not only on the E and H-fields of the electromagnetic wave but also on the induced dipole fields coming from neighboring atoms. In other words the dielectric response is partly nonlocal. When the atomic arrangement has a handedness this can lead to different behavior of left- and right-handed circularly polarized waves. When these two waves recombine on exiting the crystal, the result is optical activity and circular dichroism.

Table 31.1 Verdet coefficients measured at 0.589 μm and 20° in degrees/G mm

Water (H_2O)	2.2×10^{-5}
Fluorite (CaF_2)	1.5×10^{-6}
Diamond (C)	2.0×10^{-5}
Silica glass (SiO_2)	2.8×10^{-5}
Carbon disulfide (CS_2)	7.0×10^{-5}

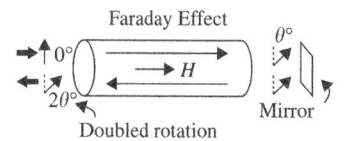

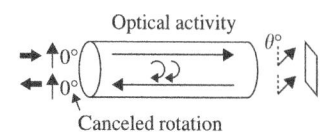

Fig. 31.2 Rotatory effects associated with optical activity and the Faraday Effect. When polarized light transverses the specimen in forward and reverse paths, the net rotation cancels for optical activity and doubles for the Faraday Effect.

31.2 Tensor nature

In the presence of a magnetic field, all materials exhibit Faraday rotation, regardless of symmetry. Unlike optical activity, it is not restricted to acentric point groups. As pointed out earlier, the angle of rotation is proportional to the path length t and to the size of the applied field H.

Relatively few measurements have been made on anisotropic crystals. Natural birefringence makes the experiments difficult in any direction other than optic axes. For angles close to the optic axes, the usual procedure is to measure the Faraday rotation parallel and antiparallel to the applied magnetic field. Averaging the two measurements eliminates some of the errors.

In nonmagnetic materials the Faraday Effect is very weak, usually much weaker than optical activity and natural birefringence. It can only be seen in non-enantiomorphic media which are either isotropic or oriented along an optic axis.

In tensor form the Faraday rotation is given by

$$\psi = V_{ij} t N_i H_j,$$

where ψ is the angle of rotation, t is the specimen thickness, N_i are the direction cosines of the wave normal, H_j are the components of DC magnetic field, and V_{ij} is Faraday tensor. In magnetic materials the Faraday Effect is much larger and is proportional to the magnetization of the material.

To determine how the Faraday tensor transforms we consider how ψ, t, N_i, and H_j transform. The rotation ψ follows a helical path through the medium. It has a handedness and is therefore a pseudoscalar (axial zero rank tensor) which transforms as $\psi' = |a|\psi$. Magnetic field H is a first rank axial tensor, as are magnetic induction B and magnetization I. All three are also subject to the time reversal operator, as discussed in Chapter 14. They transform as $H'_i = \pm |a|a_{ij}H_j$ where the $\pm$ signs refer to time reversal and $|a|$ is the handedness change. The wave normal N_i is a polar first rank tensor (=polar vector) but is subject to time reversal. When time is reversed the direction of the wave is reversed. Therefore N transforms as $N'_i = \pm a_{ij}N_j$. The thickness t is a true scalar so $t' = t$. Remembering that the direction cosine subscripts are reversed in going from the new to the old systems, $\psi' = |a|\psi$, $t' = t$, $N_i = \pm a_{ki}N'_k$ and $H_j = \pm |a|a_{lj}H'_l$.

The Faraday coefficients V_{ij} transform according to the following derivation.

$$\psi' = |a|\psi = |a|V_{ij}tN_iH_j = |a|V_{ij}t'(\pm a_{ki}N'_k)(\pm |a|a_{lj}H'_l)$$
$$= a_{ki}a_{lj}V_{ij}t'N'_kH'_l = V'_{kl}t'N'_kH'_l.$$

Therefore the Faraday coefficients constitute a polar second rank tensor relating the rotation angle ψ to the wave normal and magnetic field:

$$V'_{kl} = a_{ki}a_{lj}V_{ij}.$$

Like other polar second rank tensors the Faraday Effect is present in all 32 crystal classes and all seven Curie groups. The number of independent coefficients ranges from nine in the two triclinic classes to one in the five cubic classes and Curie groups $\infty\infty m$ and $\infty\infty$. The matrix is not required to be symmetric by energy arguments, so for low symmetry classes $V_{ij} \neq V_{ji}$.

The matrices for Faraday coefficients are identical to those of the thermoelectric coefficients listed in Table 21.4. For cubic and isotropic materials the only nonzero coefficients are $V_{11} = V_{22} = V_{33}$. This is the same as the Verdet coefficient described in Section 31.1.

There is an *inverse Faraday Effect* in transparent materials. Electrodynamic theory tells us that a rotating electric field acts as a magnetic field. From this it follows that a solid is magnetized when exposed to an intense circularly-polarized electromagnetic wave. Strong inverse Faraday Effects have been observed in crystals doped with paramagnetic ions such as $Cd_{1-x}Mn_xTe$. The dominant mechanism appears to be the alignment of paramagnetic ion spins by polarized charge carriers.

Problem 31.1
The Faraday rotation angle ψ depends on the angle θ between the wave normal and the direction of the magnetic field. Numerical values for the Verdet coefficient of several materials are given in Table 31.1. Estimate ψ for silica glass in a magnetic field of 10^4 G and plot ψ as a function of θ for a specimen 1 mm

thick. Compare the angle of rotation with that measured for the optical activity effect in a quartz plate 1 mm thick.

31.3 Faraday Effect in microwave magnetics

Faraday rotation is used in the processing of electromagnetic waves in the microwave region between 1 and 100 GHz. Strong interactions take place between the microwaves and precessing electronic spins in magnetized ferrites.

Consider a magnetic material with a saturation magnetization I_s aligned with a strong DC magnetic field H along the Z_3 direction. If the alignment of I_s is disturbed, it precesses about H with all the electron spins precessing together. The precession comes about because of the angular momentum of each electron. If the spins and the magnetization are deflected by an angle θ, a restoring force $\mu H \sin \theta$ acts to oppose the change, but the magnetization will not return immediately to the field direction. Instead it will precess about Z_3 with an angular frequency $\omega_0 = \gamma H$, where γ is the gyromagnetic ratio. For electron spins, $\gamma = 35$ kHz m/A. When viewed along Z_3, the magnetization I_s rotates in a clockwise direction about H (Fig. 31.3(a)). These rotational motions are the cause of the temporal dispersion mentioned previously.

When a plane-polarized microwave enters a solid it becomes two circularly polarized waves, one right-handed and one left-handed. As the two waves travel along Z_3, they interact differently with the precessing spins. The left-handed wave rotates in the clockwise direction like the precessional motion of the electron spins. It rotates in phase with $I \sin \theta$ for $\omega < \omega_0$, and as ω increases, θ becomes larger. In contrast, the right-handed counterclockwise microwave does not couple to the clockwise precession. In practice, it is customary to plot the microwave permeability as a function of the DC magnetic field, keeping the microwave frequency fixed. The real (μ') and imaginary (μ'') parts of the permeability are plotted in Fig. 31.3(b). Absorption occurs at the resonance condition $H_0 = \omega/\gamma$. The left-handed wave is absorbed but the right-handed circularly-polarized microwave is not.

Microwave devices are generally operated in the low-loss region below resonance where the permeabilities $\mu'(L)$ and $\mu'(R)$ are quite different. This means that the two waves will travel at different velocities since $v = c/\sqrt{\bar{\mu}'\bar{\varepsilon}'}$ where c is the speed of light in vacuum, and $\bar{\mu}'$ and $\bar{\varepsilon}'$ are the relative permeabilities and permittivities at microwave frequencies. Below resonance, $\bar{\mu}'(L) > \bar{\mu}'(R)$ so that $v(R) > v(L)$. As a result the plane of polarization of the incident microwave traveling along Z_3 will be rotated clockwise. A typical value of the Faraday coefficient is 10^4 degrees/m, a much larger rotation than those observed on nonmagnetic materials.

In microwave systems the Faraday Effect is used to accept or reject plane polarized waves. Isolators, gyrators, phase shifters, and circulators all make use of the effect. Ferrites with the garnet structure have good electrical resistivity that lowers the Eddy current losses so common in other magnetic materials. Though cubic, the garnet structure is rather complex with three types of cation sites (Fig. 31.4). Gadolinium iron garnet ($Gd_3Fe_5O_{12}$) is a typical magnetic garnet. Three out of five Fe^{3+} ions are in tetrahedral positions, with the remainder in octahedral coordination. Gadolinium is in dodecahedral coordination with four oxygen neighbors at 2.38 Å and four more at 2.44 Å.

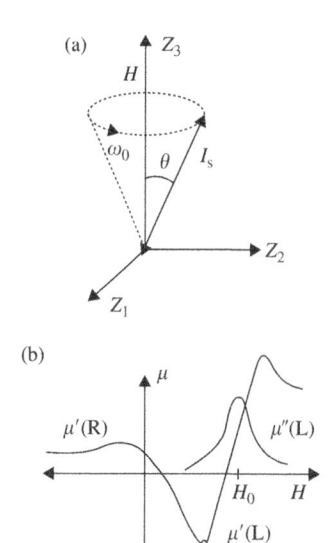

Fig. 31.3 (a) Precession of the magnetization I about the applied DC field H. (b) Relative permeability of a ferrite at a fixed microwave frequency. Real ($\bar{\mu}'$) and imaginery ($\bar{\mu}''$) permeabilities are plotted as a function of the DC field H for right (R)- and left (L)-handed circularly-polarized waves.

Ferric iron is octahedrally coordinated to six oxygens at 2.00 Å, slightly larger than the tetrahedral Fe–O distance of 1.88 Å. Every oxygen is coordinated to two yttrium and two iron, one octahedral and one tetrahedral. Iron atoms in the octahedral and tetrahedral sites are coupled antiferromagnetically by the superexchange mechanism. Gadolinium is weakly coupled to the net moment of the iron atoms, again antiferromagnetically. Because of its large moment and weak coupling, gadolinium contributes little to the saturation magnetization at high temperatures, but tends to dominate at low temperatures. Between these two extremes the magnetization changes sign, the so-called compensation temperature (Fig. 14.9).

As explained earlier, the operation of these microwave devices depends on ferrimagnetic resonance, in which energy from a circularly polarized radio-frequency wave sustains precession of unpaired electrons about an applied static magnetic field. In these devices it is important to control the magnetization and its variation with temperature, as well as the resonance and remanence properties, and the high-power characteristics.

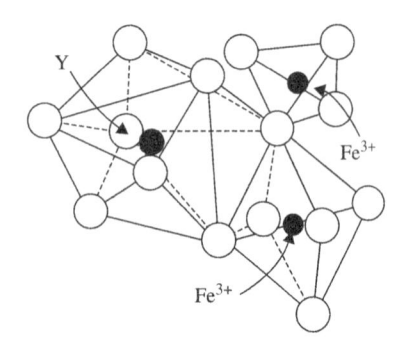

Fig. 31.4 A portion of the Yttrium Iron Garnet (YIG) structure illustrating the coordination of the various cations with oxygen.

31.4 Magneto-optic recording media

Ferrimagnetic garnets also have technological importance in the visible and near infrared range. Compared to other magnetic materials, the garnets have exceptionally low optical absorption coefficients. The ratio of the Faraday rotation ψ_F to the absorption coefficient α is often used as an engineering figure of merit for magneto-optic devices operated in a transmission mode. Magnetic metals like Fe and Co have low ψ_F/α values because of their huge absorption coefficients. Other more transparent ferromagnetic crystals like $CrBr_3$ have very low Curie temperatures.

Bismuth-substituted garnets are among the most promising magneto-optic materials in the visible wavelength region. The Faraday rotation spectrum of polycrystalline $Gd_{3-x}Bi_xFe_5O_{12}$ in the visible region is shown in Fig. 31.5. A magnetic field is not required in ferrimagnetic materials such as this. Once the magnetization has been saturated, the internal magnetic field creates the Faraday rotation. The units are therefore degrees/m.

Problem 31.2
Since the best magneto-optic materials lack optical transparency, the figure of merit for most technical applications is not the magnitude of the Faraday rotation, but the ratio of Faraday rotation to the absorption coefficient α, ψ_F/α expressed in degrees/dB. The largest value measured at room temperature is about 9.6 degrees/dB for $Gd_2BiFe_5O_{12}$ at 0.7 μm. Using the data in Fig. 31.5, estimate the Faraday rotation at 0.7 μm for a crystal 1 mm thick. How does the rotation change when the magnetization is reversed, as in 180° domains? Compare these rotation angles with the optical activity coefficients for several of the chiral crystals and liquid crystals in Chapter 30.

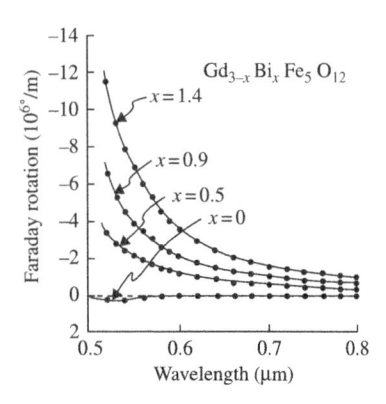

Fig. 31.5 Faraday rotation in $Gd_{3-x}Bi_x$-Fe_5O_{12} plotted as a function of wavelength for several bismuth compositions. The high dispersion has been attributed to charge transfer effects.

The development of magneto-optic memory systems and of magneto-optic light modulators has revived interest in the Faraday Effect and related phenomena. Magneto-optic recording was first introduced in 1957 using MnBi films

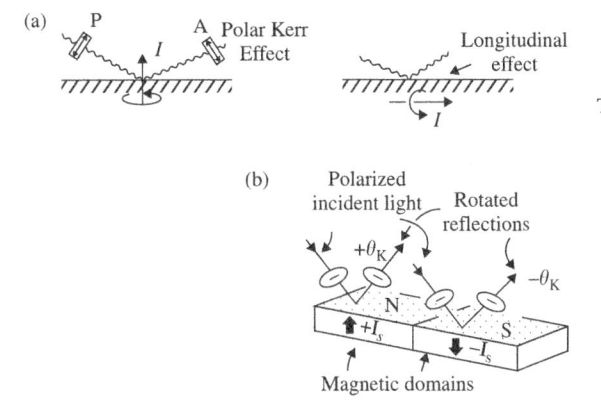

Fig. 31.6 (a) In the magneto-optic Kerr Effect polarized light is reflected from the surface of a magnetic material. Different orientations of the magnetization are used in the polar-, longitudinal-, and transverse-Kerr effects. (b) Most optical recording disks utilize the polar Kerr Effect in which a change in rotation of the polarized light is measured for 180° domains.

magnetized perpendicular to the surface. The films are interrogated optically by reflecting polarized light from the surface. Rotation can be observed in reflection as well as transmission, the so-called *Kerr Rotation*. The Kerr Effect can be seen in three different orientations (Fig. 31.6(a)) but the polar configuration is generally preferred for high storage density. Optical rotation is reversed for reflections from 180° magnetic domains (Fig. 31.6(b)).

Magneto-optic disks are widely used in the information and entertainment industries. They combine the high-bit density of optical recording with the best features of magnetic storage, permanence, and erasability. Information is stored through the mechanism of Curie point writing in which light is absorbed in a magnetic film, thereby heating the film. Regions of the film under intense illumination revert to the paramagnetic state when heated above the Curie point. When cooled in a suitable magnetic field, the paramagnetic region returns to a magnetized state, but the direction of magnetization is changed. The resulting domain pattern mimics the incident light beam. Information is read out by the Kerr Effect.

Requirements for a useful storage medium are high bit-density, high write-sensitivity and high read-out efficiency. Bit-density is determined by minimum domain size, so that a large magnetocrystalline anisotropy is needed for thin-film configurations. Important material parameters for Curie-point writing include low thermal conductivity, a convenient Curie point, and high optical absorption

Large heavy atoms favor low thermal conductivity as well as large Faraday rotation. Large magneto-optic effects are necessary for read-out efficiency. To avoid deleterious birefringence effects, it is advantageous to direct the light beam along an optic axis. This is easiest in optically isotropic media.

The spin-obit and charge transfer phenomena found in heavy elements with mixed orbitals give rise to large Faraday rotations. The magneto-optic Kerr effects in magnetic metals and intermetallic compounds are significantly larger than in most ferrites.

Most of these requirements were met by MnBi and EuO but grain boundaries turned out to be a fatal flaw. Large background noise limited the use of polycrystalline thin films. The introduction of amorphous intermetallic films with no grain boundaries solved this problem. Amorphous thin films of rare earth-transition metal alloys are made by a sputtering process. Among these alloys, $Tb_{20}Fe_{74}Co_6$ is widely used as a magneto-optic disk material. As pointed out in Section 15.6, $TbFe_2$ crystals have large magnetostriction coefficients and large magnetic anisotropy because of spin-orbit coupling in terbium. Both contribute to the uniaxial anisotropy K_u that makes it possible to magnetize the films in upward and downward directions normal to the surface. The criterion for perpendicular magnetization is $K_u > 2\pi I_s^2$. To reduce I_s the alloy composition is adjusted to be near a compensation point where the Tb and Fe magnetic moments nearly cancel. This is possible because terbium and iron magnetize in opposite directions as in magnetostrictive (Tb, Dy) Fe_2 (Fig. 15.10). Similar effects are observed in rare-earth iron garnets (Fig. 14.9). Information storage in magneto-optic devices of this type are soon projected to reach the 1 Gbit/cm^2 range.

A figure of merit often adopted for magneto-optic memories is $R\sin^2\theta_K$ where R is the reflectivity and θ_K is the Kerr rotation. Cerium-based compounds such as CeBi and CeSb have figures of merit three orders of magnitude larger than TbFeCo and other alloys, but only at very low temperatures under very high fields. The enhanced magneto-optic effects have been attributed to 4f to 5d electronic transitions.

31.5 Magnetic circular dichroism

Magnetic circular dichroism is similar in origin to the Faraday Effect. In a ferromagnet, two circularly polarized waves propagate with different refractive indices and *different absorption coefficients*. After traveling a distance t in the sample, there appears a phase difference between the right- and left-handed waves given by

$$\psi = \frac{\pi t}{\lambda}(n(L) - n(R)) = \theta_F t,$$

where $n(L)$ and $n(R)$ are the refractive indices of the two circularly polarized electromagnetic waves, and θ_F is the Faraday coefficient of a fully magnetized material in degrees/m.

If the two circularly polarized waves are absorbed at different rates their relative amplitudes will change as well. The magnetic circular dichroism is defined as the difference of the absorption coefficient α for the right- and left-circularly polarized light:

$$\Delta\alpha = \alpha(L) - \alpha(R).$$

$\Delta\alpha$ and θ_F are both strong functions of frequency ω. Both are governed by the same type of tensor relations.

When no magnetic field is applied, the intensity of light transmitted through a sample of thickness t is

$$I = I_0 \exp(-\alpha t),$$

where α is the absorption coefficient and I_0 is input intensity. In the presence of a magnetic field or a magnetization, the absorption coefficients for right- and

31.5 Magnetic circular dichroism

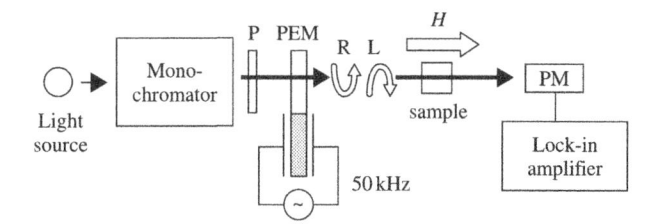

Fig. 31.7 Measurement system for magnetic circular dichroism. Alternating right- and left-circularly polarized light beams are generated using a polarizer (P) and a photoelastic modulator (PEM). A photomultiplier (PM) and a lock-in amplifier are used in the detection system.

left-circularly polarized waves, $\alpha(L) - \alpha(R)$, are different, leading to a difference in intensity $I(R) - I(L)$. The magnetic circular dichroism, expressed in degrees per gauss, is given by

$$\psi = \left(\frac{90}{\pi}\right) \frac{(I(R) - I(L))}{(I(R) + I(L))}.$$

An experimental arrangement for studying magnetic circular dichroism is shown in Fig. 31.7. Experiments are generally carried out over the visible and near infrared range using transmission geometry. The polarization is driven between right- and left-handed circularly polarized states at a frequency of 50 kHz using a photoelastic modulator. Transmitted light is detected with a photomultiplier and a lock-in amplifier. A 7% change in transmitted light intensity corresponds to one degree of magnetic circular dichroism.

Magnetic circular dichroism is used in the study of dilute magnetic semiconductors such as $Cd_{1-x}Mn_xTe$. This class of materials has attracted considerable interest because of their magnetic, electronic, and optical properties. The valence band structure of a magnetic semiconductor is strongly influenced by the d-orbital electronic states of the magnetic ions. Magneto-optic experiments have made important contributions in clarifying the interactions between the sp^3 bonding states of the tetrahedrally-coordinated semiconductor atoms and the d-levels of the transition-metal elements.

The magnetic circular dichroism spectrum of manganese telluride is shown in Fig. 31.8. The large negative peak near 3.4 eV and the weaker positive peak at 3.6 eV are attributed to optical band-gap transitions in the semiconductor. The structure around 3.1 eV comes from the d–d transitions in the manganese ions. Changes in the dichroic peak intensities take place with temperature and composition as the samples undergo magnetic phase transformations. Two of the exciting areas of research concern the superlattice structures and quantum wells which appear in dilute magnetic semiconductors.

Apart from the interesting scientific questions, there are practical reasons for investigating dilute magnetic semiconductors. New application areas combining "band-gap engineering" and "spin engineering" are expected to emerge. One example is the optical isolator utilizing $Cd_{1-x-y}Hg_xMn_yTe$ crystals as amplifiers in Er-doped optical fiber communication systems.

Magnetooptical luminescence is defined in a manner analogous to magnetic circular dichroism. Luminescence originates from electronic transitions between excited states and lower energy ground states. The difference in intensity between right- and left-circularly polarized luminescence spectra provides information about the magnetization state of the excited ions.

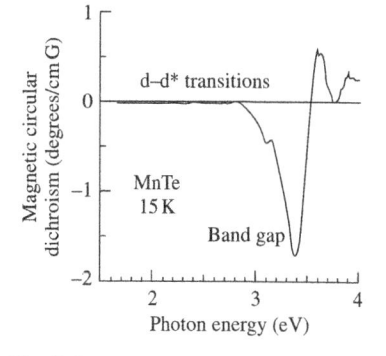

Fig. 31.8 Magnetic circular dichroism spectrum of cubic MnTe with the zincblende structure. The films are antiferromagnetic below 67 K.

31.6 Nonlinear magneto-optic effects

The effect of magnetic fields on refractive indices can be determined by expanding the indicatrix components in a power series.

$$\Delta B_{ij} = B_{ij}(H) - B_{ij}(0) = q_{ijk}H_k + q_{ijkl}H_kH_l + \cdots .$$

Since ΔB_{ij} is a polar second rank tensor and H_k is an axial first rank tensor, q_{ijk} is a third rank axial tensor, and q_{ijkl} is a fourth rank polar tensor. Third rank axial tensors have been discussed several times previously. The Hall Effect, the piezomagnetic effect, and the electrogyration effect were all axial tensors of rank three. For glasses, liquids, and high symmetry cubic crystals, the only nonzero coefficients are $q_{123} = q_{231} = q_{312} = -q_{213} = -q_{321} = -q_{132}$. But the ΔB_{ij} tensor must be symmetric, therefore $q_{ijk} = q_{jik} = 0$. The linear dependence of refractive indices on magnetic field is zero, except for the Faraday Effect.

When the light beam is perpendicular to the magnetic field, the Faraday Effect also disappears. Under these circumstances the nonlinear quadratic term dominates. This is the *Cotton–Mouton Effect* which is similar to the quadratic electro-optic effect (the Kerr Effect) but much weaker. In both cases the birefringence is proportional to the square of the field. The Kerr Effect equations in Section 28.4 apply equally well to the Cotton–Mouton Effect, replacing E^2 with H^2. The experimental arrangement shown in Fig. 31.9 is used to measure the Cotton–Mouton Effect in liquids. The biggest effects are observed in aromatic liquids like benzene and chloroform which have low viscosity and sizable diamagnetic susceptibilities. As the ring-like structures align in the magnetic field the symmetry of the liquid changes from $\infty\infty m1'$ to ∞/mm', introducing birefringence. Even larger Cotton–Mouton effects are observed in liquid crystals and magnetic liquids (Section 30.5).

In magnetically-ordered crystals the symmetry changes at the Curie temperature T_c. For YIG ($Y_3Fe_5O_{12}$) the change is from cubic point group $m3m1'$ above 560 K to trigonal point group $\bar{3}m'$ below. Unpaired spins are directed along one of the body-diagonal directions in the low-temperature ferrimagnetic state. Very small changes in the unit cell dimensions and atomic positions take place at the phase transition. As a result, a small optical birefringence appears in the YIG crystals. To measure the birefringence, a small magnetic field is applied along [111], converting the crystal into a single domain state through domain wall motion. The birefringence $\Delta n = n_\parallel - n_\perp$ is then obtained with the wave normal perpendicular to the field direction. The refractive indices $n_\parallel$ and $n_\perp$ are measured with light polarized parallel and perpendicular to the applied field. For YIG, the *magnetic linear birefringence* (Δn) at 295 K is 5.16×10^{-5}, measured in the near infrared at a wavelength of 1.15 μm. Other rare earth garnets give similar birefringences of $2-10 \times 10^{-5}$.

By way of comparison, the corresponding birefringence in ferroelectric crystals is much larger. The cubic to tetragonal in barium titanate generates a birefringence of about 5×10^{-2}, about three orders of magnitude greater than the ferrimagnetic garnets.

Optical birefringence occurs in antiferromagnetic crystals as well. The domain changes in ferrobimagnetic NiO (Fig. 15.16) were visualized in this way. The absorption coefficients of crystals with long range magnetic order show similar behavior, an effect sometimes referred to as *magnetic linear*

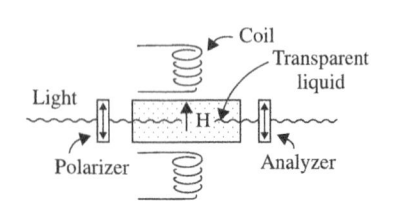

Fig. 31.9 The Cotton–Mouton Effect, a weak nonlinear magneto-optic phenomenon, can be observed in polarized light with the magnetic field orthogonal to the light beam. The birefringence Δn is proportional to H^2.

dichroism. In these experiments, measurements are carried out with the light beam perpendicular to the applied magnetic field. The difference in absorption coefficients $\Delta\alpha = \alpha_\| - \alpha_\perp$ is measured with light polarized parallel and perpendicular to the field.

Other interesting nonlinear phenomena also occur in metals and inorganic crystals with long range magnetic order. Nonlinear optical properties of magnetic origin are sometimes quite different from those with electric origin. Those driven by the magnetic vector of the light wave can be distinguished experimentally from those driven by the electric field vector. Consider the measurement of refractive indices or absorption coefficients of a uniaxial crystal. Experiments can be carried out in polarized light with three different orientations of the electric vector (E), the magnetic vector (H), and the wave normal (N):

	E	H	N
π-spectra	$\| c$	$\perp c$	$\perp c$
σ-spectra	$\perp c$	$\| c$	$\perp c$
α-spectra	$\perp c$	$\perp c$	$\| c$

In most transparent nonmagnetic materials, the α-spectra is identical with the σ-spectra, and we conclude that the properties are due to electric dipole interactions with the electric vector of the light wave. Magnetic interactions dominate when the α- and π-spectra are identical.

The best examples of magnetic dipole transitions are found in the microwave and far infrared spectrum of transparent iron garnets. As discussed earlier in Section 31.3, it is associated with the gyromagnetic contribution to the Faraday Effect. A characteristic feature of the gyromagnetic Faraday Effect in the infrared is that the rotation of the plane of polarization is independent of frequency. At higher frequencies near the visible range, the Faraday Effect increases rapidly because of electric dipole contributions associated with charge transfer transitions (Fig. 31.5). The Faraday coefficients in the infrared are smaller, generally in the 40–70 degrees/cm range.

31.7 Magnetoelectric optical phenomena

Some crystals exhibit both magnetic- and electric dipole behavior. Antiferromagnetic chromium oxide provides an interesting example of mixed dipole excitation. The magnetoelectric effect discussed in Section 14.9 provides the coupling mechanism between electric fields and magnetization.

The crystal structure of Cr_2O_3 is isomorphous with α-Al_2O_3 with chromium ions octahedrally coordinated to hexagonally close-packed oxygen ions. Near room temperature at 307 K, Cr_2O_3 undergoes a phase transformation from the high temperature paramagnetic state to a low temperature antiferromagnetic state in which the unpaired 3d electrons of Cr^{3+} align along the threefold symmetry axis in an antiparallel fashion (Fig. 14.6). The symmetry group changes from $\bar{3}m1'$ above T_N to $\bar{3}'m'$ below T_N.

Nonlinear optical properties of nonmagnetic crystals were described in Chapter 29. For second harmonic generation the electric polarization vector of the second harmonic ($P_i(2\omega)$) is related to the electric field components of the fundamental $(E_j(\omega))$ and $(E_k(\omega))$ through the SHG coefficients d_{ijk}:

$$P_i(2\omega) = d_{ijk}E_j(\omega)E_k(\omega).$$

Polarization and electric field are both polar first rank tensors, and therefore d_{ijk} is a polar third rank tensor which, like the piezoelectric effect, disappears in centrosymmetric media. Cr_2O_3 possesses a center of symmetry above T_N, so the second harmonic signals disappear in the paramagnetic state.

Electric dipoles couple the electric vector E to the electric vector P. This is the normal way in which second harmonic waves are generated, but it is not the only way. There can also be coupling between magnetization (I_i) and electric field components (E_j, E_k):

$$I_i(2\omega) = d_{ijk}^m E_j(\omega)E_k(\omega).$$

The superscript m indicates that the interaction between E and I proceeds through magnetic dipoles rather than through electric dipoles. To distinguish the two types of SHG effects we use d_{ijk}^e for the electric dipole coefficients and d_{ijk}^m for the magnetic dipole coefficients. As pointed out previously, d_{ijk}^e is a polar third rank tensor, but d_{ijk}^m is not. Since magnetization is an axial first rank tensor, and E_j and E_k are polar first rank tensors, d_{ijk}^m is an axial third rank tensor. Axial third rank tensors do not disappear in centrosymmetric classes, including point group $\bar{3}m1'$, and therefore the second harmonic signals from Cr_2O_3 above T_N come from magnetic dipole interactions.

The temperature dependence of the second harmonic signals for Cr_2O_3 single crystals is shown in Fig. 31.10. The electric dipole signal drops to zero at T_N, but the magnetic dipole signal does not.

The experiment was carried out with the fundamental and harmonic wave normals parallel to the threefold symmetry axis ($c = [001] = Z_3$). The incoming fundamental is polarized parallel to the twofold symmetry axis ($a = [100] = Z_1$). With $E_j = E_k = E_1$, the SHG relations become

$$P_i(2\omega) = d_{i11}^e E_1^2$$

for the electric dipole harmonic, and

$$I_i(2\omega) = d_{i11}^m E_1^2$$

for the magnetic dipole harmonic.

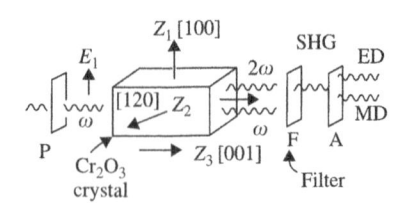

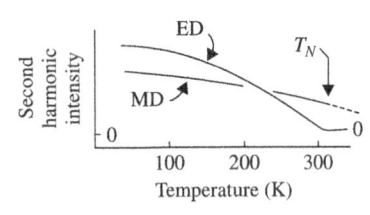

Fig. 31.10 Second harmonic experiment on single-crystal Cr_2O_3 using linearly polarized light at 2.6 eV. Incoming light is polarized (P) parallel to Z_1 and propagates along Z_3. Two types of harmonic waves are generated in the antiferromagnetic state below T_N. Electric dipole response comes from polarization component $P_1(2\omega) = d_{111}^e E_1^2(\omega)$, and magnetic dipole response $I_2(\omega) = d_{211}^m E_1^2(\omega)$. The electric dipole (ED) harmonic is observed by setting the analyzer (A) parallel to Z_1. For the magnetic dipole harmonic, the analyzer is oriented parallel to Z_2. The ED harmonic disappears at T_N but MD does not.

As can be seen in Fig. 31.10, both harmonics are observed below T_N in the antiferromagnetic state. The magnetic and electric dipole contributions are distinguished by polarizing the harmonic wave. Below T_N the symmetry group is $\bar{3}'m'$. Using Neumann's Law it is easily demonstrated that $d_{211}^e = 0$ and $d_{111}^m = 0$. Therefore second harmonic light polarized parallel to Z_1 comes from the electric dipole term, while second harmonic wavelengths polarized parallel to Z_2 have a magnetic dipole origin.

Antiferromagnetic domains in Cr_2O_3 (Fig. 14.15) can be observed using the interference patterns between the two types of second harmonic waves. The technique works best when the fundamental wavelength corresponds to one of the absorption bands in chromium oxide crystal field spectrum in this case the intense $^4A_{2g} \rightarrow {}^4T_{2g}$ transition (Fig. 26.7). Second harmonic generation is a nonlinear optical process involving the simultaneous absorption of two photons from the fundamental wave, followed by the emission of the photon with frequency 2ω. The presence of an intense absorption band aids in the conversion process.

Nonlinear magnetoelectric and magneto-optics have yet to find an important practical application, but a number of interesting experiments are in progress. Thus both magnetically-induced second harmonic generation (MSHG) and electrically-induced SHG (ESHG) take place in chromium oxide. MSHG has also been observed in $BiFeO_3$.

32 Chemical anisotropy

32.1 Crystal morphology 354
32.2 Growth velocity 356
32.3 Crystal growth and crystal structure 358
32.4 Surface structures and surface transformations 360
32.5 Etch figures and symmetry relations 361
32.6 Micromachining of quartz and silicon 363
32.7 Tensor description 366

Chemical anisotropy concerns the ways in which crystals grow or dissolve in different directions. It is an appropriate subject to end this book because it brings together the oldest and the newest parts of crystal physics. Long, long ago mineralogists described the shapes of natural crystals and noted correlations with cleavage, hardness, and other physical properties. Chemical etching was another favorite topic in classical crystal physics that has undergone a recent revival because of the interest in the micromachining of semiconductor devices.

Chemical anisotropy involves the interaction of a crystal with a chemically active environment that promotes dissolution or growth. For this reason it is primarily a surface property, rather than a bulk property of the crystal. This is one of the reasons why chemical anisotropy is not normally included in crystal physics books. The other reason is that rates of growth and dissolution depend on the chemical nature of the environment much more than the bulk properties of crystals do. Nevertheless, this is an important subject in contemporary crystal physics. Surfaces become more and more important as the scale of engineered devices grows smaller. The crystal physics of surface properties is a natural extension of classical crystal physics. It is a topic still in its infancy.

32.1 Crystal morphology

Under favorable conditions, crystal growth takes place in such a way that the external surface is bounded by a set of plane faces. The preferred shape of rocksalt family crystals is a cube bounded by six symmetry-related {100} faces. For diamond, an octahedral shape with eight {111} faces often appears. Quartz, calcite, and rutile belong to lower symmetry crystal systems with more anisotropic morphologies. Quartz crystals are often elongated along the c-axis with a hexagonal cross-section bounded by six {100} faces while the ends are terminated by six {101} and six {011} faces. Calcite tends to form rhombohedra with six faces shaped like parallelograms. Rutile (TiO$_2$) crystals are often elongated along the c-axis forming slender needles. The needles have a square cross-section consisting of four {110} faces terminated by eight small {111} faces. The symmetry group is $4/mmm$.

The morphology of the mineral dioptase is illustrated in Fig. 32.1. Dioptase (CuSiO$_3 \cdot$H$_2$O) is a relatively rare mineral formed during the weathering of copper sulfide ore. The emerald colored crystals are interesting because they represent a beautiful example of trigonal point group $\bar{3}$ that has a single threefold

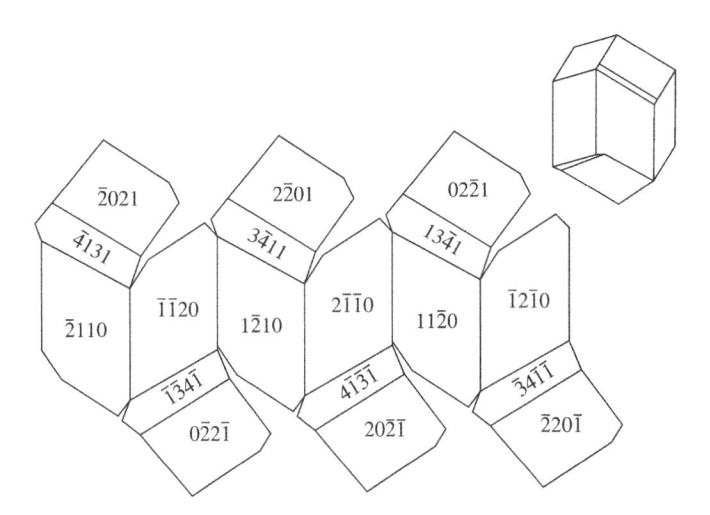

Fig. 32.1 The mineral dioptase is formed as trigonal crystals with six prism faces $\{2\bar{1}\bar{1}0\}$ and six rhombodral faces $\{2\bar{2}01\}$. The presence of the smaller $\{13\bar{4}1\}$ faces signifies that the true symmetry is trigonal point group $\bar{3}$. A perspective drawing is shown in the upper right corner. The plane view can be used to assemble a three-dimensional model.

inversion axis along the c-axis [001]. Specimens showing the full morphological development are widely sought by mineral collectors.

Three types of faces are commonly observed on dioptase crystals. All faces are labeled on the plan view in Fig. 32.1. The major faces are the six hexagonal prism faces so common in many trigonal and hexagonal crystals. These are the $\{2\bar{1}\bar{1}0\}$ faces on the plan view that are parallel to the [001] symmetry axis. The other two types of faces are rhombohedral forms that are very common in other trigonal crystals like calcite. These are the six $\{2\bar{2}01\}$ faces and the six $\{13\bar{4}1\}$ faces on the dioptase plan view. The general form $\{13\bar{4}1\}$ is especially interesting because it illustrates the true point group symmetry $\bar{3}$. The relative position of these faces shows that other types of symmetry elements such as mirror planes and twofold rotation axes are absent in dioptase.

Unfortunately most crystals do not show the full point group symmetry illustrated in the drawings. If a general form is not present, the morphological symmetry of the crystal may be higher than the true structural symmetry as determined by X-ray diffraction. In dioptase, if only the larger $\{2\bar{1}\bar{1}0\}$ and $\{2\bar{2}01\}$ faces are present, the apparent symmetry of dioptase would be $\bar{3}m$ rather than $\bar{3}$, with twofold axes and mirror planes present. The general form $\{13\bar{4}1\}$ is required to identify the true symmetry.

Classical crystallography is based on the constancy of interfacial angles. In crystallography, the interfacial angle is defined as the angle between the normals to the two faces. Before the advent of X-ray diffraction, these angles were used to identify symmetry groups and gather crystallographic data. Groth's "Chemische Kristallographie" and Dana's "System of Mineralogy" are the classic compendia.

For hexagonal or trigonal crystals like dioptase, the angle Φ between the normals to $(h_1 k_1 l_1)$ and $(h_2 k_2 l_2)$ is given by

$$\cos \Phi = \frac{h_1 h_2 + k_1 k_2 + \frac{1}{2}(h_1 k_2 + h_2 k_1) + (3a^2/4c^2)l_1 l_2}{[(h_1^2 + k_1^2 + h_1 k_1 + (3a^2/4c^2)l_1^2)(h_2^2 + k_2^2 + h_2 k_2 + (3a^2/4c^2)l_2^2)]^{1/2}},$$

where a and c are the lattice parameters of the unit cell. Formulae for other crystal systems are given in the International Tables for X-ray Crystallography.

Problem 32.1
Calculate the angle between the $(13\bar{4}1)$ and $(11\bar{2}0)$ faces of dioptase. The unit cell dimensions are $a = 14.565$ and $c = 7.775$ A. Remember that the third Miller index for hexagonal crystals is redundant ($i = -h-k$).

Problem 32.2
Using an optical goniometer, the angles between crystal faces can be measured quite accurately. The following data were collected for a trigonal crystal of corundum (Al_2O_3). Calculate the c/a axial ratios and compare them with the value measured by X-ray diffraction ($a = 4.758$, $c = 12.991$ A).

$$(0001)-(11\bar{2}3) = 61°11'$$
$$(0001)-(11\bar{2}1) = 79°37'$$
$$(0001)-(01\bar{1}2) = 57°35'$$
$$(0001)-(10\bar{1}1) = 72°23'$$

32.2 Growth velocity

The general rule relating morphology to growth velocity is that "fast-growing faces disappear, slow-growing faces remain." For rocksalt, the slow-growing {100} faces form cubes. The fast-growing {111} and {110} faces grow out and disappear.

Quantitative measurements of growth velocities are monitored by grinding single crystals into spheres and immersing them in saturated solutions. As the crystals grow, flat spots develop on the surface of the sphere and facets begin to form. Growth velocities are obtained by measuring the diameter of the crystal in different directions.

Like other members of the alum family, potassium chrome alum ($KCr(SO_4)_2 \cdot 12H_2O$ is a cubic water-soluble crystal that is readily grown from saturated solution. Growth profiles of an alum sphere are illustrated in Fig. 32.2. Flat faces corresponding to the {111} octahedron, the {100} cube, and {110} dodecahedron are the first to develop. As growth proceeds, the slow-growing octahedral faces begin to dominate the external shape. Relative growth rates for different crystal faces are listed in Table 32.1. As indicated, the growth velocities depend markedly on the degree of supersaturation. High index faces grow very quickly and soon disappear.

Using the data from Table 32.1, the slowness surface can be visualized using the reciprocal growth velocities. Fig. 32.3 shows the profile of this surface in the (110) plane. Sharp peaks are observed along the slow growing ⟨111⟩, ⟨110⟩, and ⟨100⟩ directions with deep minima for the high index faces in between. The slowness surface applies to crystal dissolution as well as crystal growth. The cavities etched into silicon devices are bounded by the slow-growth surfaces.

These ideas go back to the work of Gibbs and Curie. Each face is assigned a *surface free energy* F_i and the crystals adopt a morphology which minimizes the total surface energy. The free energies F_i are determined by measuring the perpendicular distance h_i from the point of origin of the crystal to each face, such that

$$\frac{h_1}{F_1} = \frac{h_2}{F_2} = \frac{h_3}{F_3} = \cdots = \frac{h_i}{F_i}$$

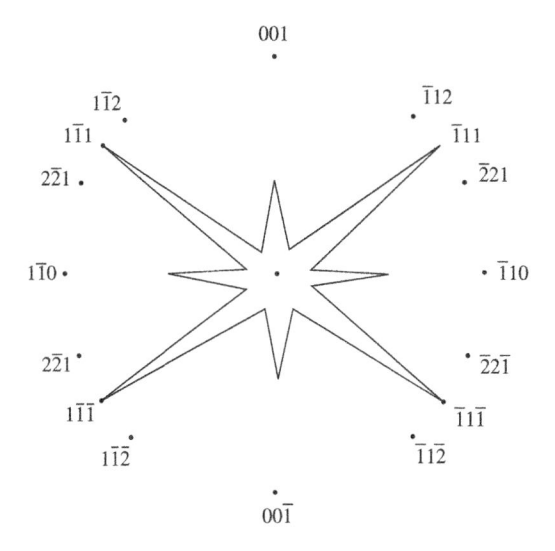

Fig. 32.2 Growth velocity surface of potassium chrome alum viewed along [110]. Fast growing faces disappear leaving the {111} faces to form an octahedron.

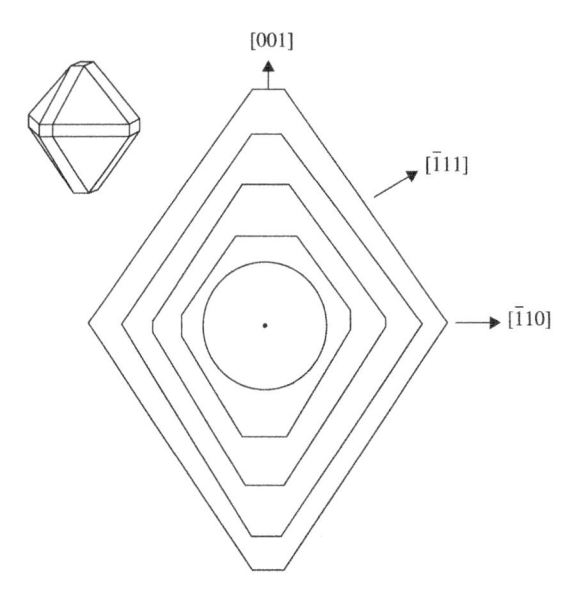

Table 32.1 Relative growth velocities of various faces formed on potassium chrome alum crystals (Buckley)

Supersaturation	(111)	(110)	(100)	(112)	(012)	(122)
Weak	1.0	1.9	2.1	6.6	8.3	9.5
Strong	1.0	4.3	9.2	17.9	34.7	

Fig. 32.3 The slowness surface for potassium chrome alum projected on the (110) plane.

is a constant. The relative areas of each face, therefore, depend on their surface free energies.

Wulff represented this same principle geometrically, and identified the individual surface free energies with the growth velocities. *Wulff plots* are often used in predicting equilibrium crystal shapes.

32.3 Crystal growth and crystal structure

Perhaps the simplest notion regarding the relationship between crystal morphology and crystal structure is the concept of "dangling bonds." Consider the coordination of the Na^+ and Cl^- atoms in rocksalt. Inside the crystal each ion is bonded to six neighbors forming an octahedron. But on the outer surface of a crystal the situation is different. On the (100) face of the crystal each sodium atom is bonded to four surface chlorines and to a fifth Cl neighbor inside the crystal. This leaves one "dangling" bond protruding from the surface. The unsatisfied bond becomes a point of attachment for an anion in the surrounding medium.

Other faces on the rocksalt crystal will have different numbers of dangling bonds per unit area, and therefore different growth rates. For the {110} dodecahedral faces each surface atom has four neighbors and two dangling bonds, and for {111} octahedral faces the surface atoms have three neighbors and three dangling bonds.

Because of the large number of dangling bonds, the atoms on {110} and {111} surfaces are not as well bonded as those on the cubic {100} faces. It also means that outside atoms or molecules will attach to the {110} and {111} faces more easily than to {100} faces. Therefore the ⟨111⟩ and ⟨110⟩ directions are fast-growing directions and ⟨100⟩ axes are slow-growing. During crystal growth, Na^+ and Cl^- ions are rapidly attached to the dangling bonds causing fast growth. As growth proceeds, fast-growing faces tend to disappear leaving the slow-growing faces behind. This explains why rocksalt generally grows as cubes, but does not explain why high-index faces generally grow faster than low-index faces.

Surface roughness also plays a role. The presence of steps or crevices on a surface increases the growth rate by raising the sticking coefficient of approaching atoms or molecules. As an example, compare the atomic smoothness of the (100), (120), and (150) planes of sodium chloride. All three planes have charge-neutral outer layers but the surface coverage is quite different. For (100), there is good coverage with 0.126 atoms/A^2, compared to 0.056 for (120) and 0.025 for (150). Rough high-order faces like (150) tend to grow quickly and disappear. Stated another way, it costs energy to put steps on a flat surface. For cubic crystals, the flattest surfaces are (100), (110), and (111). These are the peaks on the slowness surface of alum (Fig. 32.3).

Chemical anisotropy is also strongly influenced by surface layers of foreign atoms or molecules that attach preferentially to certain faces. This can have the effect of changing a fast-growing face into a slow-growing face, and altering the morphology of a crystal. In the crystal growth literature, these shape-changing additives are known as "poisons" or "habit modifiers." Lead chloride poisons the fast-growing (111) faces of KCl and changes the morphology from a cube to an octahedron. Similar changes in morphology take place in KBr and KI. An extensive listing of habit modifiers is given in the classic "Crystal Growth" by H.E. Buckley.

Another way of altering crystal shape is to make use of intermediate pseudomorph phases. A pseudomorph is a crystal that resembles a chemically and structurally different crystal. Pseudomorphs are common in nature where one mineral is converted to another without changing shape. This is often an atom-by-atom replacement process in which for example, cubic iron sulfide (pyrite, FeS_2) is converted to orthorhombic goethite (FeOOH) without losing the cubic morphology. A similar process has been used in the reverse direction to make magnetic tape. To obtain the desired needle-like morphology, the iron oxide crystallites are first prepared as orthorhombic goethite and then oxidized to cubic γ-Fe_2O_3 while retaining the fibrous morphology of goethite. Shape anisotropy is important in fixing the magnetization direction (Section 16.4).

Chemical anisotropy extends to the catalytic processes of interest to the petroleum and chemical industries. The activity of certain chemical reactions is observed to change dramatically when the experiments are carried out on different surfaces of well crystallized catalysts. The catalyzed conversion of linear hydrocarbons to aromatic ring compounds provides an important commercial application. Structure-sensitive conversion of n-heptane to toluene over platinum single crystals is illustrated in Fig. 32.4. The (111) surface is far more efficient than (100). Platinum crystals have the face-centered cubic structure with close-packed planes parallel to the (111) surfaces. It has been suggested that the hexagonal symmetry of the (111) surface promotes the formation of ring compounds more readily than the square network on (100) planes.

Reaction rates increase even further on the stepped and kinked surfaces corresponding to high-index planes. Maximum activity is achieved on stepped surfaces with hexagonal orientation. Similar results are obtained in the conversion of n-hexane to benzene. Again the hexagonal aromatic ring of benzene prefers the hexagonal arrangement of Pt atoms on the (111) surface.

Alkane isomerization reactions behave quite differently. The square atomic arrangement of Pt (100) gives a much higher reaction rate than Pt (111) for the isobutene to n-butane process. Steps and kinks do not provide much additional improvement.

The synthesis of ammonia over iron is also highly directional. Studies with iron single crystals show that the Fe (111) and (211) surfaces possess much higher reaction rates than (100), (210), and (110). One of the important attributes of transition metal catalysts like Fe and Pt is their ability to atomize diatomic molecules like H_2, N_2, and O_2. The Fe–H and Pt–H bonds on the surface of the catalyst are especially important in hydrocarbon conversion reactions.

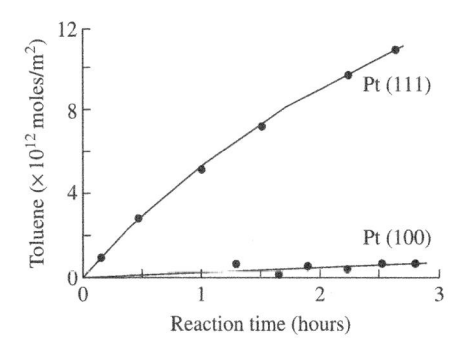

Fig. 32.4 Catalytic conversion rates of n-heptane to toluene at 573 K over Pt (111) and Pt (100) surfaces.

32.4 Surface structures and surface transformations

The arrangement of atoms on a surface usually differs from a simple planar termination of the bulk crystal structure. The "ideal" termination is the exception rather than the rule. For ionic insulators like NaCl, the {100} cleavage faces are probably close to ideal with a small puckering of the smaller Na$^+$ ions being shielded by Cl$^-$. These are neutral nonpolar surfaces that cleave easily indicating that the attractive forces to ions adjacent to the surface are relatively weak. Polar surfaces can be quite different.

The reconstruction of the polar surfaces found in noncentrosymmetric crystals is especially interesting because it can involve both chemical and crystallographic changes. The {111} surfaces of the semiconductor InSb consist of alternating indium and antimony layers. Opposing (111) and ($\bar{1}\bar{1}\bar{1}$) faces are positively and negatively charged in the ideal zincblende structure. Grazing incidence X-ray experiments show that a massive reconstruction takes place to achieve charge neutralization. On the indium surface (Fig. 32.5) one in four In atoms are ejected, and the remaining surface atoms withdraw into the layer beneath. The effective coordination of the In surface atoms is three rather than four, giving a neutral bilayer.

Solid surfaces undergo a wide variety of phase transformations as a function of temperature and chemical composition. Like phase transitions in the bulk, these surface structures are often metastable and highly dependent on the processing conditions. A cleaved (111) surface of silicon, for example, exhibits a metastable reconstruction that transforms to a very complex 7×7 superstructure when annealed at 380°C. Laser annealing at high temperatures returns the (111) surface to the ideal 1×1 unit cell. Orbital rehybridization is an important mechanism for surface reconstruction in covalent semiconductors. The surface buckles as the normal sp^3 hybrid changes to a deformed structure with fewer dangling bonds. Silicon atoms on (100) surfaces bend toward one another to form dimers (Fig. 32.6).

Surface reconstruction in metals is often more subtle with only small rearrangements of the mobile electrons and ion cores. At a surface, the electrons lower their kinetic energy by smoothing the electron density in a flat plane parallel to the surface. This smoothing process leaves the outer ion cores less well shielded. As a result, there are small ion movements away from the surface toward the interior of the crystal. Small motions can also take place parallel to the surface. The (100) surface of tungsten is an interesting example with a reversible transformation in the surface structure. At this low temperature transition the tungsten atoms make small shifts parallel to the [011] directions

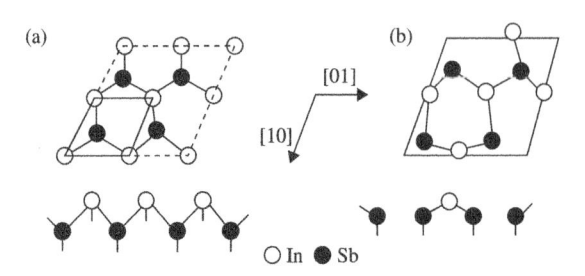

Fig. 32.5 Top and side views of the indium (111) surface. The ideal structure is shown in (a). The reconstructed surface in (b) involves removal of some In atoms and movements of others.

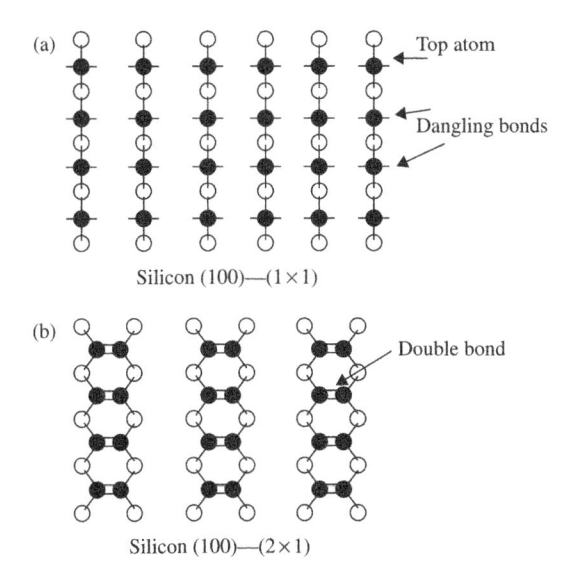

Fig. 32.6 (a) Ideally, the reactive (100) face of silicon has two dangling bonds per surface atom. (b) Electron diffraction experiments indicate, however, that the true surface is partially reconstructed with double bond formation. Locally the surface symmetry is 2*mm* rather than 4*mm*.

distorting the surface symmetry from 4*mm* to 2*mm*. Two-dimensional domain phenomena appear that are analogous to three-dimensional ferroic behavior.

A rich variety of surface structures are illustrated in the Atlas of Surface Structures, Vol. IA and IB. Beautiful examples of ideal and reconstructed surfaces, are shown, both with and without chemisorbed species.

32.5 Etch figures and symmetry relations

Cuprite (Cu_2O), the red oxide of copper, is a cubic mineral found in the weathered zones of copper deposits. Octahedral {111} morphologies are common, accompanied by smaller cubic {100} and dodecahedral {110} faces. When etched in dilute nitric acid, the etch pits on the {111} faces are equilateral triangles with three planes of symmetry. Square etch figures oriented at 45° angles to the crystal axes are observed on {100} faces. Simple canoe-shaped depressions are found on the dodecahedral {110} surfaces. The planar symmetry groups of these etch figures are 3*m* for {111}, 4*mm* for {100}, and *mm* for {110}. These planar groups correspond to the projection symmetries of cubic point group *m*3*m*. Similar results are achieved when cuprite crystals are etched in ammonium chloride and sulfuric acid. An expanded planar view of the etched surfaces is shown in Fig. 32.7

Note that it is not only the shape of the etch pits that is important, but their orientation with respect to the symmetry axes must conform as well. The striations on pyrite are a good example of orientation effects.

Pyrite (FeS_2) or "fool's gold" is a mineral that often occurs as beautiful golden cubes with a metallic luster. Striations are sometimes seen on the cube faces (Fig. 32.8(d)) with the striations on adjacent faces perpendicular to one another. Surface features such as these are characteristic of certain point groups.

Relationships between point groups and striations are illustrated in Fig. 32.8, where all five cubic point groups are represented. Cubes are observed in all five

362   *Chemical anisotropy*

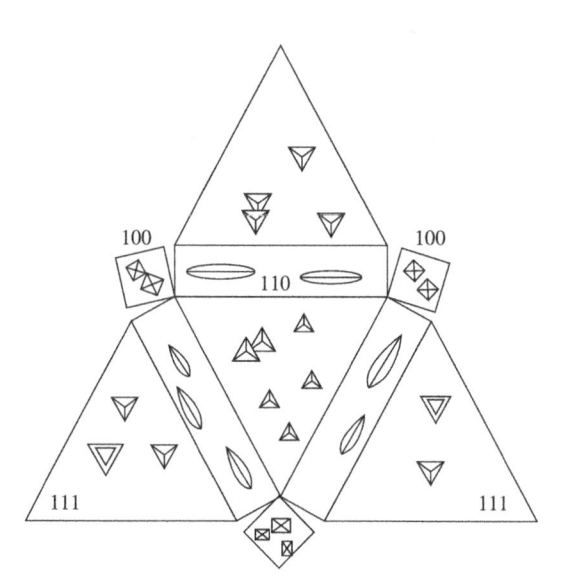

Fig. 32.7 Etch figure patterns on cuprite, Cu$_2$O. The shape and orientation of the etch pits are consistent with the most symmetric cubic point group, $m3m$ (Honess).

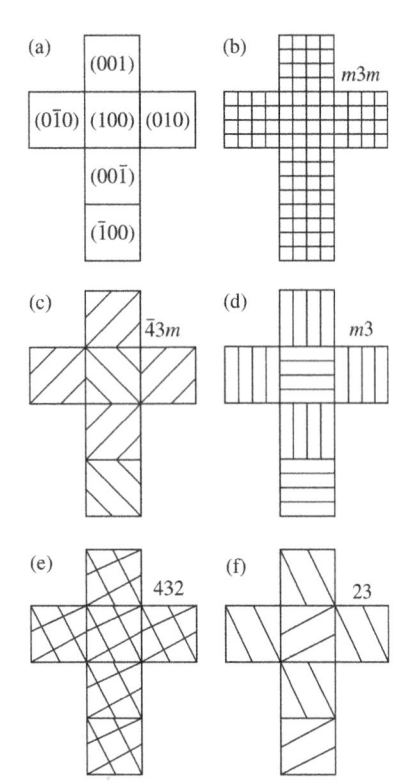

Fig. 32.8 Striation patterns representative of the five cubic point groups. Pyrite crystals belong to point group $m3$ with the perpendicular striations shown in (d).

groups but the striations take different orientations. The cross-hatched striations in Figs. 32.8(b) and (e), possess fourfold symmetry along the three ⟨100⟩ axes. These two patterns correspond to point groups $m3m$ and 432, respectively. The striations in Fig. 32.8(b) have mirror symmetry while those in Fig. 32.8(e) do not.

The remaining three patterns have twofold symmetry along the ⟨100⟩ axes. Fig. 32.8(c), (d), and (f) are representative of point groups $\bar{4}3m$, $m3$, and 23, respectively. Fig. 32.8(c) has mirror planes parallel to {110} faces, while those in Fig. 32.8(d) are parallel to {100} faces. Point group 23 patterns have threefold rotation axes parallel to the body diagonal ⟨111⟩ directions. This is the minimum symmetry common to all cubic crystals.

Etch figures are obtained by exposing crystal faces to a suitable reagent. Dissolution begins sporadically at dislocations and other defect sites. After a brief time a number of etch pits are formed with symmetries characteristic of the experiment. The experiment involves two symmetries: the symmetry of the crystal face and the symmetry of the reagent. The reagent is typically a liquid acid or base with randomly oriented molecules. If the molecules possess mirror or inversion symmetry, the symmetry of the reagent is $\infty\infty m$, the highest symmetry Curie group (Section 4.4). If the molecules possess handedness, like d- or l-lactic acid, the symmetry of the liquid is $\infty\infty$.

The symmetry of the crystal face corresponds to one of the ten two-dimensional point groups (Fig. 32.9). Ideally, the symmetry of a (100) cube face on rocksalt (point group $m3m$) is $4mm$. A (111) octahedral face has $3m$ symmetry, and a (110) dodecahedral face belongs to $2mm$. Each of these symmetries is a projection symmetry of $m3m$. Projection symmetries for the 32 crystal classes are listed in Table 32.2.

Polar crystals show different etch rates on opposite faces. LiNbO$_3$ (point group $3m$) is a polar crystal used as an electro-optic modulator. The [001] polar axis is terminated by (001) and (00$\bar{1}$) faces which etch differently in HF acid. The (00$\bar{1}$) face terminated by loosely bonded Li$^+$ ions etches faster than the more tightly bonded (001) face terminated by highly-charged niobium ions.

Similar symmetry relationships occur during crystal growth. Sodium chlorate (NaClO$_3$) belongs to point group 23 but often the crystals show higher symmetry. When grown rapidly from water solution only the {100} cube faces appear and the morphological symmetry is $m3m$. Only tetrahedral {111} faces are observed when sodium thiosulfate is present in the solution, and the apparent symmetry is $\bar{4}3m$. Both $m3m$ and $\bar{4}3m$ are supergroups of the true symmetry (23), in keeping with Neumann's Principle. Small additional faces showing the true symmetry appear when sodium chlorate is grown slowly from pure solution.

Symmetries lower than the true symmetry can occur if crystal growth takes place in chiral liquids. Lead chloride (PbCl$_2$) belongs to orthorhombic point group mmm, but when crystallized from solutions containing chiral dextrine molecules it forms bisphenoidal crystals with 222 symmetry. Mirror symmetry is lost because the left-handed dextrine molecules are preferentially adsorbed on half of the {111} faces, altering the crystal morphology. Stereographic projections can be used to verify these symmetries.

Curie's Principle determines the symmetry of the experiment. Suppose the liquid is water ($\infty\infty m$) and the surface is the (210) face of rocksalt. The projection symmetry is two-dimensional point group m. Curie's Principle says that the symmetry of the experiment is the symmetry common to that of the reagent and the crystal face. The symmetry common to $\infty\infty m$ and m is m. If the reagent were a chiral liquid with handed molecules (Curie group $\infty\infty$), the common symmetry with the (210) face would be point group 1.

Will the etch figure show these symmetries? Perhaps. Neumann's Principle says that the symmetry of the experiment must include the symmetry of the materials involved. If the etch rates were the same in all directions, no anisotropy will develop. Therefore, the symmetry of the etch figures may be higher than the predicted symmetry. If the etch rates change with directions, then the predicted symmetries appear. A number of these experiments were carried out by Honess using crystals of calcite, apatite, and other minerals with well-developed faces. Chiral reagents in right- and left-handed forms, as well as racemic mixtures, were used to demonstrate the Curie Principle.

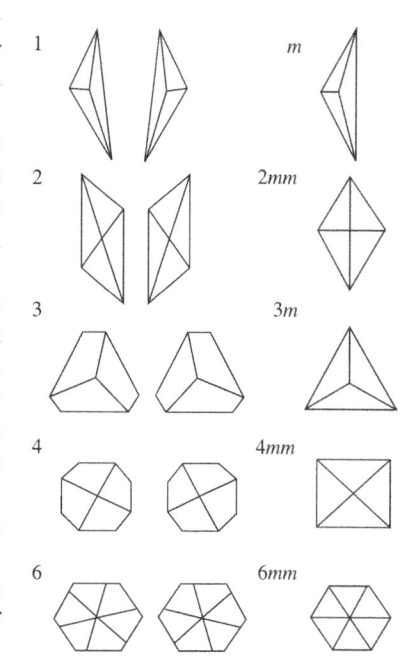

Fig. 32.9 Etch figures representative of the ten planar point groups. Five come in left- and right-handed pairs.

32.6 Micromachining of quartz and silicon

Etch figures are obvious manifestations of chemical anisotropy and crystallographic symmetry. Natural quartz crystals are generally twinned which ruins their piezoelectric performance. The atomic structures of Dauphine and Brazil twins were illustrated in Fig. 16.17. The presence of twinning can be detected by etching the crystals in hydrofluoric acid or ammonium bifluoride. The reaction of quartz (SiO$_2$) with hydrofluoric acid (HF) is given by

$$SiO_2 + 6HF \rightarrow SiF_4 + 2H_2O + 2HF \rightarrow 2H_2SiF_6 + 2H_2O.$$

Because of chemical anisotropy, the acid attacks quartz at different rates in different directions. When examined at glancing incidence, the twinned regions reflect light differently because of the etch pits. Synthetic quartz crystals are not twinned but are etched for a different reason.

The micromachining of synthetic quartz crystals has become important in the fabrication of tuning forks and other resonant structures for time and frequency

Table 32.2 Etch figure symmetries for various faces in the 32 crystal classes

Point group	{100}	{010}	{001}	{0kl}	{h0l}	{hk0}	{hkl}
1	1	1	1	1	1	1	1
$\bar{1}$	1	1	1	1	1	1	1
2	1	2	1	1	1	1	1
m	m	1	m	1	m	1	1
2/m	m	2	m	1	m	1	1
222	2	2	2	1	1	1	1
mm2	m	m	2mm	m	m	1	1
mmm	2mm	2mm	2mm	m	m	m	1

	{001}	{100}	{110}	{hk0}	{h0l}	{hkl}	{hkl}
4	4	1	1	1	1	1	1
$\bar{4}$	2	1	1	1	1	1	1
4/m	4	m	m	m	1	1	1
422	4	2	2	1	1	1	1
4mm	4mm	m	m	1	m	m	1
$\bar{4}$2m	2mm	2	2	1	1	m	1
4/mmm	4mm	2mm	2mm	m	m	m	1

	{0001}	{10$\bar{1}$0}	{11$\bar{2}$0}	{hki0}	{h0$\bar{h}$l}	{hh$\bar{2h}$l}	{hkil}
3	3	1	1	1	1	1	1
$\bar{3}$	3	1	1	1	1	1	1
32	3	1	2	1	1	1	1
3m	3m	m	1	1	m	1	1
$\bar{3}$m	3m	m	2	1	m	1	1
6	6	1	1	1	1	1	1
$\bar{6}$	3	m	m	m	1	1	1
6/m	6	m	m	m	1	1	1
622	6	2	2	1	1	1	1
6mm	6mm	m	m	1	m	m	1
$\bar{6}$m2	3m	2mm	m	m	1	1	1
6/mmm	6mm	2mm	2mm	m	m	m	1

	{100}	{111}	{110}	{hk0}	{hhl}	{hkl}	
23	2	3	1	1	1	1	
m3	2mm	3	m	m	1	1	
432	4	3	2	1	1	1	
$\bar{4}$3m	2mm	3m	m	1	m	1	
m3m	4mm	3m	2mm	m	m	1	

standards. The dissolution slowness surface for quartz is shown in Fig. 32.10(a). In reagents such as HF and NH_4HF_2, quartz dissolves much faster along Z_3 ([001]) than along Z_1 ([100]) or Z_2 ([120]). Large anisotropies of 30 to 100 times are observed. The slowness surface in Fig. 32.10(a) is projected along Z_1, the twofold symmetry axis of quartz. Note that the drawing has the required 180° rotational symmetry.

There are six major lobes on the slowness surface corresponding to the largest faces on quartz crystals. (01.0) and (0$\bar{1}$.0) are two of the six hexagonal prism faces (Fig. 32.10(b)). (01.1) and (0$\bar{1}$.1) are two of the six major rhombohedron faces, and (01.$\bar{1}$) and (0$\bar{1}$.1) are two of the six minor rhombohedron faces. In keeping with the general rule for crystal growth, fast-growing faces disappear, leaving the slow-growing faces to form the equilibrium shape. These slow-etching faces correspond to the lobes on the slowness surface.

Natural quartz is generally long and slender because of fast growth along Z_3, the c-crystallographic direction. The underlying connection to the crystal structure can be explained in terms of dangling bonds. Inside the crystal each silicon

32.6 Micromachining of quartz and silicon

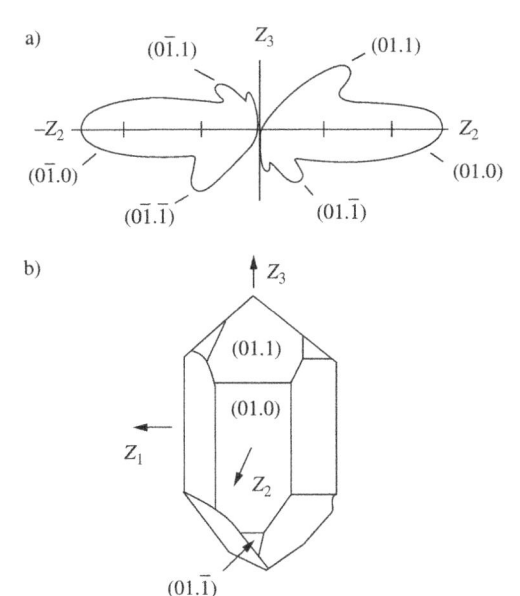

Fig. 32.10 (a) Dissolution slowness surface of quartz crystals. The slow-etching faces correspond to the major faces of natural quartz crystals as shown in (b).

atom is tetrahedrally bonded to four oxygen atoms. On the stable slow-growing prism and rhombodral faces, there is one dangling bond per surface silicon. However, on the fast-growing (00.1) face perpendicular to Z_3 there are two, leading to faster growth and faster dissolution.

Note that the commonly observed morphology shown in Fig. 32.10(b) has higher symmetry than the internal crystal structure. The morphological symmetry is $\bar{3}m$ while the true symmetry is 32. Point group 32 is a subgroup of $\bar{3}m$. The correct symmetry appears when the general form is present, as in Fig. 12.13. The small $\{6\bar{5}.1\}$ trapezohedral faces display the full symmetry but are seldom seen on mineral specimens. They are fast-growing faces that usually disappear during growth.

Similar phenomena occur during the micromachining of silicon. A wide variety of semiconductor sensors are made by the anisotropic etching of silicon crystals to form membranes, cantilevers, and many other shapes. Generally, the etch rate is slowest on the octahedral $\{111\}$ planes, and fastest on the $\{110\}$ and $\{100\}$ planes. The differences depend on the atom density and the number of dangling bonds per surface atom. The surface atoms on a (111) face of Si have one dangling bond while those on the more reactive (100) face have two (Fig. 32.6(a)). Surface reconstruction sometimes leads to changes in the chemical reactivity (Fig. 32.6(b)).

Several alkaline reagents display highly anisotropic etching characteristics that depend on temperature and chemical composition. S-EDP is a commonly used alkaline etchant made up of ethylene diamine ($NH_2(CH_2)_2NH_2$), pyrocathechol ($C_6H_4(OH)_2$), pyrazine ($C_4H_4N_2$), and water. Over the temperature range 50°C to 115°C, there is no solid residue with S-EDP. As with other chemical reactions, the etch rate R is exponentially dependent of temperature.

$$R = R_0 \exp\left(-\frac{E_a}{kT}\right),$$

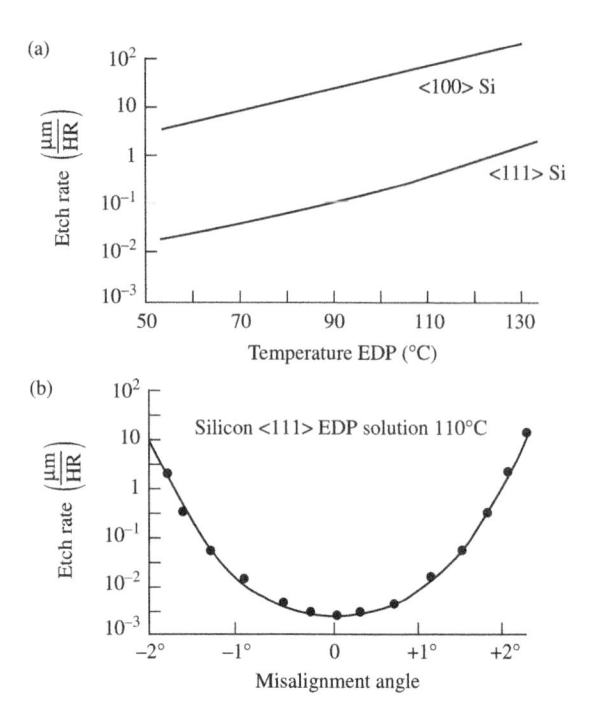

Fig. 32.11 Dissolution rates of silicon in EDP reagents. (a) (100) surfaces etch about a hundred times faster than the well bonded (111) faces. The rates increase exponentially with temperature. (b) The etching rates change rapidly with angle.

where the activation energy (E_a) and the pre-exponential factor (R_0) are determined experimentally. T is the absolute temperature and k is Boltzmann's Constant. The temperature dependence is illustrated in Fig. 32.11(a).

The etch-rate anisotropy for the three principal planes depends on temperature. At 115°C the ratio for (100):(110):(111) is 30:30:1, rising to 100:150:1 at 50°C. Even a small misalignment in orientation leads to large changes in the etching rate (Fig. 32.11(b)). Below 50°C, the anisotropy is even larger but the etching rates are too low for micromachining. S-EDP does not attack gold metallizations or silicon nitride coatings, and the ratio between Si(100) and SiO_2 exceeds 10,000. Many different structures with precise dimensional control can be etched into the (100) and (110) faces of silicon.

32.7 Tensor description

The surface of a crystal is generally a complex geometrical shape composed of a number of planar surface elements. The orientations of these planar surfaces are specified by Miller indices. When the growth (or dissolution) process is anisotropic, as it usually is, the various surface elements follow different trajectories as some faces grow in size while others recede and disappear. The movement of a surface element can be described as a velocity vector v whose magnitude and direction depend on the orientation of the element.

Tellier and coworkers have shown that the trajectory of the growth (dissolution) velocity vectors can be completely determined from the slowness vector $\vec{L}$ describing the so-called slowness surface. A straightforward way of describing the surface is to plot L, the magnitude of $\vec{L}$, as a function of direction. This can be done in terms of direction cosines N_1, N_2, and N_3 that specify the orientation of the slowness vector relative to the cartesian reference axes Z_1, Z_2, Z_3. Standard settings (Section 4.3) relate the reference axes to the crystal axes.

Since the slowness surface is a complex function of direction, L can be expressed as a polynomial series.

$$L = L_o + L_i N_i + L_{ij} N_i N_j + L_{ijk} N_i N_j N_k + L_{ijkl} N_i N_j N_k N_l + \cdots,$$

where L_o, L_i, L_{ij}, and L_{ijk} are tensor coefficients representing the growth and dissolution processes as a function of the direction cosines.

The next step is to simplify the tensor by symmetry arguments. In discussing the slowness surface we confine attention to the alum crystals described earlier in Section 32.2. They are typical of most cubic crystals that prefer simple shapes such as the cube and octahedron with {100} and {111} faces, respectively. For alum, the octahedron dominates with smaller cube faces.

Symmetry simplification proceeds as follows. All five cubic classes (23, $m3$, 432, $\bar{4}3m$, and $m3m$) possess twofold symmetry along the $\langle 100 \rangle$ cube edges and three fold symmetry along the $\langle 111 \rangle$ body diagonals. Applying these symmetry operations to the slowness surface gives

$$L = L_o + L_{11} + 6L_{123} N_1 N_2 N_3 + L_{1111}(N_1^4 + N_2^4 + N_3^4)$$
$$+ 6L_{1123}(N_1^2 N_2^2 + N_1^2 N_3^2 + N_2^2 N_3^2) + \cdots.$$

Alums belong to centric point group $m3$ for which all odd-rank tensor terms disappear. Hence $L_{123} = 0$. And noting that

$$N_1^4 + N_2^4 + N_3^4 = 1 - 2(N_1^2 N_2^2 + N_1^2 N_3^2 + N_2^2 N_3^2)$$

we find that

$$L = L_o + L_{11} + L_{1111} + 2(3L_{1122} - L_{1111})(N_1^2 N_2^2 + N_1^2 N_3^2 + N_2^2 N_3^2) + \cdots.$$

For a (100) face $N_1 = 1$, $N_2 = N_3 = 0$, and for (111), $N_1 = N_2 = N_3 = 1/\sqrt{3}$. These directions are the extreme values for this surface. If $3L_{1122} > L_{1111}$, (111) is the slowest growing face and the octahedron is the favored form. If $3L_{1122} < L_{1111}$, the cube is favored. Based on this equation, the slowness surface will be a fourth-order quartic function with maxima along the six $\langle 100 \rangle$ directions or the eight $\langle 111 \rangle$ directions, similar to the elastic constant surfaces in Chapter 13.

Now compare this result with the observed slowness surface for alum (Fig. 32.3). It is obvious this is not a quartic function, but has sharp peaks in the $\langle 111 \rangle$, $\langle 100 \rangle$, $\langle 110 \rangle$ directions, with deep depressions at intermediate directions. The tensor description with terms up to the fourth power gives poor agreement with the experimental results. A much better fit can be obtained by including higher order tensor coefficients. Etching patterns in silicon can be explained by including tensors up to rank 10 in the slowness surface. Higher power trigonometric functions are capable of modeling slowness surfaces like those of silicon, quartz, and alum.

It is not surprising that tensors of high rank are needed to describe chemical anisotropy. Every surface on a crystal is different. Each has its own chemistry, its own structure, and its own symmetry. Therefore the mathematical representation of a surface-related property such as etch rate is bound to be complicated.

It is complicated but it is also very important. Surface physics and surface chemistry are at the heart of modern materials technology. There is a rich variety of surface properties that remain to be investigated, and then engineered into useful systems. Nearly all the topics in this book need to be revisited—surface transport, surface waves, surface phase transitions—and then reformulated in terms of the appropriate surface tensors. New symmetry relations and new structure property relations must be developed. In common slang it might be said that "It's a whole new ball game", and that makes it a good place to end this book.

Further Reading

Auld, B.A. *Acoustic Fields and Waves in Solids (Two Volumes)*. Malabar, FL: Krieger Publishing Co. (1990).

Baldwin, J. *Introduction to Nonlinear Optics*. New York: Plenum Press (1974).

Beer, A.C. *Galvanomagnetic Effects in Semiconductors*. New York: Academic Press (1963).

Bhatia, A.B. *Ultrasonic Absorption*. New York: Dover Publications (1967).

Birss, R.R. *Symmetry and Magnetism*. Amsterdam: North Holland Publishing Co. (1964).

Blinov, L.M. *Electro-Optical and Magneto-Optical Properties of Liquid Crystals*. New York: John Wiley and Sons (1983).

Bond, W.L. The mathematics of the physical properties of crystals. *Bell Syst. Tech. J.* **XXII**, 1–72 (1943).

Bottcher, C.J.F. *Theory of Electric Polarisation*. Amsterdam: Elsevier Publishing Co. (1952).

Bottom, V. *Introduction to Quartz Crystal Unit Design*. New York: Van Nostrand Reinhold (1982).

Boyd, R.W. *Nonlinear Optics*. Amsterdam: Academic Press (2003).

Buckley, H.E. *Crystal Growth*. New York: John Wiley and Sons (1951).

Buerger, M.J. *Elementary Crystallography*. New York: John Wiley and Sons (1963).

Bundy, F.P. and H.M. Strong. Behavior of metals at high temperatures and pressures. *Solid State Physics* **13**, 81–147 (1962).

Burns, R.G. *Mineralogical Applications of Crystal Field Theory*. Cambridge: Cambridge University Press (1970).

Butcher, P.N. and D. Cotter. *The Elements of Nonlinear Optics*. Cambridge: Cambridge University Press (1990).

Cady, W.G. *Piezoelectricity*. New York: McGraw-Hill Book Co. (1946).

Carey, R. and E.D. Isaac. *Magnetic Domains and Techniques for Their Observation*. New York: Academic Press (1966).

Chikazumi, S. *Physics of Magnetism*. New York: John Wiley and Sons (1964).

Collings, P.J. *Liquid Crystals*. Princeton, NJ: Princeton University Press (1990).

Cook, W.R. and H. Jaffe. Electrooptic coefficients. *Landolt Bornstein Tables* **11**, 552–670 (1979).

Cook, W.R. and H. Jaffe. Piezoelectric, electrostrictive, and dielectric constants. *Landolt Bornstein Tables* **11**, 287–470 (1979).

de Launay, J. The theory of specific heats and lattice vibrations. *Solid State Physics* **2**, 220–290 (1956).

Drickamer, H.G., R.W. Lynch, R.L. Clendenen, and E.A. Perez-Albuerne. X-ray diffraction studies of the lattice parameters of solids under very high pressure. *Solid State Physics* **19**, 135–229 (1966).

Freeman, A.J. and H. Schmid. *Magnetoelectric Interaction Phenomena in Crystals.* London: Gordon and Breach (1975).

Girifalco, L.A. *Atomic Migration in Crystals.* New York: Blaisdell Publishing Co. (1964).

Goodenough, J.B. *Magnetism and the Chemical Bond.* New York: Interscience Publishers (1963).

Green, D.J. *An Introduction to the Mechanical Properties of Ceramics.* Cambridge: Cambridge University Press (1998).

Gschneider, K.A. Physical properties and interrelationships of metallic and semimetallic elements. *Solid State Physics* **16**, 276–426 (1964).

Hargittai, I. and M. Hargittai. *Symmetry Through the Eyes of a Chemist.* New York: Plenum Press (1995).

Harman, T.C. and J.M. Honig. *Thermoelectric and Thermomagnetic Effects and Applications.* New York: McGraw-Hill Book Co. (1967).

Hazen, R.M. and L.W. Finger. *Comparative Crystal Chemistry.* New York: John Wiley and Sons (1982).

Hearmon, R.F.S. *An Introduction to Applied Anisotropic Elasticity.* London: Oxford University Press (1961).

Hearmon, R.F.S. Elastic constants of crystals and other anisotropic materials. *Landolt Bornstein Tables* **11**, 1–242 (1979).

Hearmon, R.F.S. The third- and higher-order elastic constants. *Landolt Bornstein Tables* **11**, 245–86 (1979).

Heising, R.A. *Quartz Crystals for Electrical Circuits.* New York: Van Nostrand (1946).

Henry, N.F.M. and K. Lonsdale. *International Tables for X-Ray Crystallography. Vol. I Symmetry Groups.* Birmingham: Kynoch Press (1952).

Honess, A.P. *The Nature, Origin, and Interpretation of the Etch Figures on Crystals.* New York: John Wiley and Sons (1927).

Huntingdon, H.B. The elastic constants of crystals. *Solid State Physics* **7**, 214–351 (1958).

Ikeda, T. *Fundamentals of Piezoelectricity.* Oxford: Oxford University Press (1990).

Jaffe, H.W. *Crystal Chemistry and Refractivity.* New York: Dover Publications (1996).

Jan, J.P. Galvanomagnetic and thermomagnetic effects in metals. *Solid State Physics* **5**, 1–96 (1957).

Jessop, H.T. and F.C. Harris. *Photoelasticity: Principles and Methods.* New York: Dover Publications (1960).

Jona, F. and G. Shirane. *Ferroelectric Crystals.* New York: The MacMillan Co. (1962).

Juretschke, H.J. *Crystal Physics: Macroscopic Physics of Interaction Processes.* Reading: W.A. Benjamin (1974).

Kanzig, W. Ferroelectrics and antiferroelectrics. *Solid State Physics* **4**, 5–197 (1957).

Khoo, I.C. *Liquid Crystals*. New York: John Wiley and Sons (1995).

Klassen-Neklyudova. *Mechanical Twinning of Crystals*. New York: Consultants Bureau (1964).

Klemens, P.G. Thermal conductivity and lattice vibrational modes. *Solid State Physics* **7**, 1–99 (1958).

Klocek, P. *Handbook of Infrared Optical Materials*. New York: Marcel Dekker (1991).

Kocks, U.F., C.N. Tome, and H.R. Wenk. *Texture and Anisotropy*. Cambridge: Cambridge University Press (1998).

Kurtz, S.K., J. Jerphagnon, and M.M. Choy. Nonlinear dielectric susceptibilities. *Landolt Bornstein Tables* **11**, 671–743 (1979).

Lax, B. and K.J. Button. *Microwave Ferrites and Ferrimagnetics*. New York: McGraw-Hill Book Co. (1962).

Lazarus, D. Diffusion in metals. *Solid State Physics* **10**, 71–127 (1960).

Levine, S.M. *Selected Papers on New Techniques for Energy Conversion*. New York: Dover Publications (1961).

Liu, S.T. Pyroelectric coefficients. *Landolt Bornstein Tables* **11**, 471–494 (1979).

Lovett, D.R. *Tensor Properties of Crystals*. Bristol: Institute of Physics Publishing (1999).

Lowry, T.M. *Optical Rotatory Power*. New York: Dover Publications (1964).

Mason, W.P. *Crystal Physics of Interaction Processes*. New York: Academic Press (1966).

Mason, W.P. *Piezoelectric Crystals and Their Applications to Ultrasonics*. New York: D. Van Nostrand Co. (1950).

McCrum, N.G., B.E. Read, and G. Williams. *Anelastic and Dielectric Effects in Polymeric Solids*. New York: Dover Publications (1967).

Megaw, H.D. *Crystal Structures: A Working Approach*. Philadelphia: WB Saunders (1973).

Mendelssohn, K. and H.M. Rosenberg. The thermal conductivity of metals at low temperatures. *Solid State Physics* **12**, 223–275 (1961).

Monch, W. *Semiconductor Surfaces and Interfaces*. Berlin: Springer (1995).

Morrish, A.H. *The Physical Principles of Magnetism*. New York: John Wiley and Sons (1965).

Nassau, K. *The Physics and Chemistry of Color*. New York: John Wiley and Sons (1983).

Nelson, D.F. and R.F.S. Hearmon. Piezooptic and elastooptic constants of crystals. *Landolt Bornstein Tables* **11**, 495–551 (1979).

Newnham, R.E. *Structure–Property Relations*. Berlin: Springer-Verlag (1975).

Nye, J.F. *Physical Properties of Crystals*. Oxford: Clarendon Press (1957).

Oles, A., F. Kajzar, M. Kucab, and W. Sikora. *Magnetic Structures Determined by Neutron Diffraction*. Warsaw: Panstwowe Wydawnictwo Naukowe (1976).

Olsen, J.L. *Electron Transport in Metals*. New York: Interscience Publishers (1962).

Perelomova, N.V. and M.M. Tagieva. *Problems in Crystal Physics with Solutions*. Moscow: Mir Publishers (1983).

Phillips, F.C. *An Introduction to Crystallography*. London: Longmans, Green and Company (1946).

Ristic, V.M. *Principles of Acoustic Devices*. New York: John Wiley and Sons (1983).

Salje, E.K.H. *Phase Transitions in Ferroelastic and Co-Elastic Crystals*. Cambridge: Cambridge University Press (1990).

Samara, G.A. High-pressure studies of ionic conductivity in solids. *Solid State Physics* **38**, 1–81 (1984).

Seymour, R.B. and C.E. Carraher. *Structure–Property Relationships in Polymers*. New York: Plenum Press (1984).

Shannon, R.D. Dielectric Polarizabilities of Ions in Oxides and Fluorides. *J. Applied Physics* **73**, 348–366 (1993).

Shewmon, P.G. *Diffusion in Solids*. New York: McGraw-Hill (1963).

Shubnikov, A.V. and V.A. Koptsik. *Symmetry in Science and Art*. New York: Plenum Press (1974).

Shuvalov, L.A. *Modern Crystallography IV: Physical Properties of Crystals*. Berlin: Springer-Verlag (1988).

Simmons, G. and H. Wang. *Single Crystal Elastic Constants and Calculated Aggregate Properties*. Cambridge, MA: The M.I.T. Press (1971).

Sirotin, Y.I. and M.P. Shaskolskaya. *Fundamentals of Crystal Physics*. Moscow: Mir Publishers (1982).

Slack, G.A. The thermal conductivity of nonmetallic crystals. *Solid State Physics* **34**, 1–73 (1979).

Smit, J. and H.P.J. Wijn. *Ferrites*. New York: John Wiley and Sons (1959).

Smith, A.C., J.F. Janak, and R.B. Adler. *Electronic Conduction in Solids*. New York: McGraw Hill Book Co. (1967).

Smith, C.S. Macroscopic symmetry and properties of crystals. *Solid State Physics* **6**, 175–249 (1958).

Sugano, S. and N. Kojima. *Magneto-Optics*. Berlin: Springer (2000).

Sze, S.M. *Semiconductor Sensors*. New York: John Wiley and Sons (1994).

Takahashi, A. Polymer Permittivities IEEE Trans. **15**, 418–420 (1992).

Toledano, P. and A.M. Figueiredo. *Phase Transitions in Complex-Fluids*. Singapore: World Scientific (1998).

Truell, R., C. Elbaum, and B.B. Chick. *Ultrasonic Methods in Solid State Physics*. New York: Academics Press (1969).

Uchino, K. *Piezoelectric Actuators and Ultrasonic Motors*. Boston: Kluwer Academic Publishers (1997).

Van Hook, A. *Crystallization: Theory and Practice*. New York: Reinbold Publishing Corp. (1961).

Von Hippel, A.R. *Dielectric Materials and Applications*. Cambridge, MA: The M.I.T. Press (1954).

Wahlstrom, E.E. *Optical Crystallography*. New York: John Wiley and Sons (1951).

Watson, P.R., M.A Van Hove, and K. Hermann. *Atlas of Surface Structures*. New York: American Institute of Physics (1994).

Winchell, A.N. *Optical Properties of Organic Compounds*. New York: Academic Press (1954).

Winchell, A.N. *Elements of Optical Mineralogy (Three Volumes)*. New York: John Wiley and Sons (1933).

Wood, E.A. *Crystals and Light*. New York: Dover Publications (1977).

Wooster, W.A. *A Textbook on Crystal Physics*. Cambridge: Cambridge University Press (1938).

Yariv, A. and P. Yeh. *Optical Waves in Crystals*. New York: John Wiley and Sons (1984).

Young, K.F. and H.P.R. Frederikse. Compilation of the static dielectric constants of inorganic solids. *J. Phys. and Chem. Reference Data* **2**, 313–410 (1973).

Zemansky, M.W. *Heat and Thermodynamics*. New York: McGraw-Hill (1951).

Zheludev, I.S. Ferroelectricity and symmetry. *Solid State Physics* **26**, 429–464 (1971).

Zheludev, I.S. *Physics of Crystalline Dielectrics (Two Volumes)*. New York: Plenum Press (1971).

Zheludev, I.S. Piezoelectricity in textured media. *Solid State Physics* **29**, 315–360 (1974).

Zvezdin, A.K. and V.A. Kotov. *Modern Magnetooptics and Manetooptical Materials*. Bristol, UK: Institute of Physics Publishing (1997).

Index

8-N rule 195
Acoustic attenuation 262–265
Acoustic divergence 259
Acoustic impedance 261, 265
Acoustic slowness surface 259
Acoustic waves 249, 252, 253, 256, 257, 259, 261, 266–268, 270–272, 274, 278, 279, 300, 301
  Acoustic impedance 261, 265
  Christoffel Equation 249, 251–257, 267–270
  Group velocity 258
  Hexagonal crystals 249, 252–255, 257
  Longitudinal waves 249, 253, 261, 264, 266, 269
  Matrix representation 255, 268
  Nonlinear Effects 271–273
  Piezoelectric media 267, 270
  Polarization directions 252–254
  Pure modes 257
  Reflection 261, 273
  Transverse waves 249, 267, 272
  Wave normals 252, 254, 259, 270
  Wave velocities 249, 252, 253, 258, 259, 261
Acousto-optics 298, 299
  anisotropic media 300
  Bragg Diffraction 299
  materials 301
Actuators 99, 100, 159, 160, 167
adiabatic Demagnetization 48, 49, 136
anomalous Rotatory Dispersion 338
Anisotropic conductors 193, 194
atomic vibrations 46, 47
Axial tensors 2, 31, 32, 122, 341, 350

Band gap 195–198, 200, 201, 239, 286–288, 290, 293, 314
Batteries 220
Binary diffusion 220, 221
Bruhat's Rule 339
  Thermogyration 340, 341
  Verdet Coefficients 342–344

Cotton Effect 338
Cotton-Mouton Effect 342, 350

Carnot efficiency 238, 239
Chemical anisotropy 354, 358, 359, 363, 367
  Growth velocity 356
  Morphology 354, 356
  Surface Free Energy 356
  Wulff Plots 357
Chemical potential 220, 221
Christoffel Equation 249–257, 269
Colossal magnetoresistance 231
Color 286, 288–291, 336–340
Compressibility 43, 44, 113, 114
Cross-coupled diffusion 220, 221
Crystal classes 1, 4, 15, 17–20, 23, 25, 49, 52, 63, 91, 97, 163, 322, 329, 331, 344, 362
Crystal growth 198, 354, 356, 358, 363, 364
Crystal morphology 64, 354, 355, 358, 363
Crystal optics 2, 274–285, 315
  Absorption 288–293
  Dichroism 288–290
  Dispersion 286–288
  Maxwell Equations 274, 275
  Optical indicatrix 276, 277, 278
  Poynting vector 259, 278, 279
  Ray directions 278, 279, 280
  Structure-Property Relations 280–285
  Thermo-Optic Effect 292, 293
  Wave velocity surface 275, 276
Crystallographic point groups 15, 109, 126, 245, 304
Curie groups 27, 63, 91, 97, 276, 304, 331, 344
Curie Principle 1, 363

Debye Temperature 44–47
Density 5–8
Diamagnetism 135
Dichroic Dyes 336, 337
Dichroism 288–290
Dielectric constant 5, 49, 58, 59, 61–71
  Effect of Symmetry 4, 62, 68
  Experimental methods 63–65
  Geometric Representation 67–69
  Origins 58, 59
  Structure-Pr 69–71
  Tensor 60–63, 68

Diffusion 211–222
  Experiments 212, 220, 221
  Measured values
  Solid solutions 213, 219
  Structure-property relations 212–217
  Temperature dependence 212, 213, 216
  Tensor formulation 211, 221, 222
Dispersion 286–293
Dulong and Petit Law 44, 45

Elastic compliance 30–34, 37, 38, 103, 109, 110, 152, 162–164, 179, 272, 296
Elastic stiffness 37, 54, 103, 110, 113, 114, 117, 119, 148–151, 266, 270, 278, 296
Elasticity 30, 35, 103–121
  Compressibility 113, 114
  Effect of Symmetry 107–109
  Engineering Coefficients 109, 110
  Measurements 109
  Polycrystalline averages 114, 116
  Stiffness-Compliance Relations 106, 107
  Structure-Property Relations 110–113
  Temperature Coefficients 116–118
  Tensors and Matrices 1, 103–105
  Transformations 105, 106
Electric field dependence 63, 70, 162, 306
Electric permittivity 58, 60, 122, 134, 178, 203, 269, 315, 343
Electrical conductivity 2, 3, 188, 195, 197, 218, 228, 231, 292
Electrical resistivity 188–202
  Bandgap and mobility 196–199
  Electrode metals 191–193
  Measurements 189, 190
  nonlinear behavior 199–202
  PTC and NTC thermistoro 200, 201
  Quasicrystals 202
  Semiconductors 194–196
  Tensor form 188, 189
Electrocaloric effect 38, 39, 49, 50, 133
Electrode metals 191–193
Electrogyration 340, 341, 350
Electrolysis 221, 222
Electromagnetic waves 267, 268, 274–276
Electron mobility 191, 199
Electron wind 221, 222

Index

Electro-optics 302–310
  Half-wave Voltages 307, 308
  KDP and ADP 304–309, 317
  Kerr Effect 309, 310
  Pockels Effect 303–310
Electrostriction 151–154, 161, 164, 270, 296, 310
Enantiomorphism 331, 332, 333
Energy flow direction 259, 275, 279, 314
Entropy 1, 37, 39, 41, 48–50, 133, 220, 221, 236
Etch figures 361–363
Ettingshausen Effect 241
Eulerian Rotations 12

Faraday Effect 273, 342–346, 348, 350, 351
Ferrobielectricity 177, 178
Ferrobimagnetism 177, 178
Ferrobielasticity 179–181
Ferroelastoelectricity 179–181
Ferroelasticity 165–167
Ferroelectricity 174–177, 186
Ferroic crystals 162–187
  Ferroelasticity 165–168
  Ferroelectricity 174–177
  Ferromagnetism 168, 169, 184
  Free Energy 162–185
  magnetic anisotropy 170–173
  Primary ferroics 163
  Secondary ferroics 163
  Shape memory alloys 167, 168, 178, 184
Ferromagnetism 30, 123, 124, 130, 168, 169, 184
Ferromagnetoelasticity 182, 183
Fick's First Law 211, 220
Fuel cells 211, 220
Functionally graded materials 239

Galvanomagnetic effects 223, 224–242
  basic physics 1
  Hall Effect 226–241
  Magnetic materials 124, 134, 141, 153, 154, 157, 169, 170, 230, 334, 346
  Magnetoresistance 226–241
  Measurement 223, 226, 227, 229, 231
  Tensor form 224
Geometric representations 33, 124

Heat capacity 43–46, 235
Heckmann Diagram 1, 2, 37, 103
Hooke's Law 32, 37, 103, 104, 148, 251, 262, 270
Hagen-Ruben's Law 292

Ionic conductivity 211–222
  Cross-coupled Diffusion 220, 221
  Nernst-Einstein Relation 217, 218
  Pressure Dependence 222
  Superionic Conductors 219, 220
  Temperature Dependence 212, 219
  Transport numbers 218
Improper Ferroics

Jahn-Teller Effect 231
Joule heating 234–236

Kelvin Relations 236, 237
Kerr Effect 309, 310, 347, 350
Kohler's Rule 228

Lattice vibrations 44, 46, 48, 198, 265
Linde's Law 192
Linear Systems 37
Liquid crystals 333–336
  Cholesteric 326, 334–336, 341
  Discotic 334, 335
  Nematic 334–337
  Smectic 334, 335
Luster 291, 292

Magnetic anisotropy 170–173
Magnetic circular dichroism 348, 349
Magnetic field sensors 228, 229
Magnetic Linear birefringence 350
Magnetic Linear dichroism 350, 351
Magnetic materials 229–232
  Galvanomagnetic effects 229, 231
Magnetic permeability 134, 135
Magnetic point groups 125–130
Magnetic structures 124, 125
Magnetic susceptibility 134–137
Magnetism 133
  axial vectors 130, 131
  Curie Law 136
  magnetic point groups 125–130
  magnetic susceptibility 134–137
  magnetocalorie effect 133
  magnetoelectricity 138–142
  Origins and units 122, 123
  Paramagnetism 136
  Piezomagnetism 142–146
  Pyromagnetism 2, 49, 131–133
  Time Reversal 124, 125
Magnetocaloric Effect 48, 49
Magnetoconcentration 228, 229
Magnetoelectricity 2, 32, 33, 68
Magnetoelectric Optical Effects 351–353

Magnetoelectronics 231
Magneto-optics 342–353
  Curie Point Writing 347
  Faraday Effect 342, 343
  Inverse Faraday Effect 344
  Kerr Rotation 342, 347, 348
  microwave magnetics 345
  nonlinear Effects 350, 351
  Recording media 346–348
  Relation to Optical Rotation 347
  Tensor nature 343, 344
Magnetoresistance 223, 224, 226–233, 240, 241
  Colossal 231
  Giant magnetoresistance 231
  Magnetic materials 230
Magneto-Seebeck Effect 241
Magnetostriction 153, 154
Magnetothermoelectricity 240
Matrix relations 188, 189
Matthiessen's Rule 191, 192, 210
Maxwell Equations 274, 275
Maxwell Relations 38, 39, 48
Measurements 13, 34, 35, 40, 43, 47, 52, 59, 64, 71, 80, 84, 90, 96, 134 141, 152, 157, 181, 183, 187, 222, 242, 244, 277, 337, 339, 341, 356
Mechanical stress 2, 72, 73
Mobility 196–199
Mooser-Pearson Relation 195, 196

Nernst Effect 241
Nernst-Lindermann Equation 44, 241
Neumann's Principle 5, 34, 35, 51, 62, 63, 91, 134, 204, 224, 363
Nonlinear acoustics 270–273
  acoustic activity 272
  acoustic Faraday Effect 273
  acousticelectricity 271
  magnetoelastic Effect 273
Nonlinear magneto-optics 350, 351
Nonlinear optics 313–324
  miller's Rule 314, 317
  Parametric Devices 315, 316
  Phase matching 318–321, 324
  Second Harmonic Generation 315–317
  Structure-Property Relations 313, 314
  Tensor Formulation 315, 316
  Third harmonic Generation 322, 323
Nonlinear permittivity 147, 152
nonlinear Phenomena 147–161
  actuators 159, 160
  Dielectric 147, 148, 151, 152
  Elastic 148–150
  Electromagnetostriction 160, 161
  Electrostriction 151–153
  magnetostriction 154–159

Nonohmic resistivity 199
Numerical values 11, 115, 247, 262, 323

Ohm's Law 31, 188, 200, 224, 243
Onsager Principle 239
Onsager's Theorem 204
Optical absorption 337, 346, 347
Optical activity 325–341
    ambidextrous Behavior 326
    Circular Dichroism 337–339
    Dispersion 337–339
    Effect of symmetry 329, 330
    Enantiomorphism 331–333
    Gyration surface 330
    Liquid Crystals 333–337
    molecular Origins 325, 326
    Tensor description 327, 328
Optical indicatrix 276, 277, 279, 283, 292, 295, 298, 303, 304, 309
Order Parameter 183–187
Orthogonality conditions 10, 11, 67

Paramagnetism 123, 136
Peltier Effect 234, 235, 237
Penrose Tiling 202
Phase matching 318–321, 324
Phonon drag 237, 238
Photoelasticity 179, 295, 296
    Basic Concepts 294
    Elasto-Optic Coefficients 295–298, 301
    matrix form 295
Piezocaloric Effect 49
Piezoelectricity 1, 87–102
    Effect of Symmetry 87, 91
    Experimental measurements 93
    Hydrostatic Effect 97–99
    Poled ceramics 94, 97, 100
    Quartz crystals 100–102
    Structure-Property Relations 94–96
    Tensors and matrices 87, 88
    Transformations 89–91
Piezogyration 340, 341
Piezomagnetism 2, 142–145
Piezoresistance 243–248
    Matrix formulation 244, 245
    Structure-property relations 247
    Tensor description 243, 244
Planar Hall Effect 226, 233
Plastic Crystals 333, 334
Pockels Effect 303, 304, 307–310
Point groups 17–19
    Populations 20, 21
Polar axes 52, 53
Polar glass-ceramics 57
Polar tensors 31, 33, 35, 105, 163, 210, 224, 225, 242, 243, 341

Polymer fibers 208
Primary ferroics 2, 163
Proper Ferroics 160, 161, 185
Proper and Improper Ferroics 185, 186
Pseudopiezoelectricity 160, 161
Pure mode directions 256, 257
Pyroelectricity 1, 32, 33, 50–57
    applications 57
    Geometric Representation 53, 54
    materials 55
    measurements 54
    Primary and Secondary 54, 55
    Symmetry limitations 50, 51
    Temperature Dependance 55–57
Pyromagnetism 49, 131–133

Quartz transducers 253
    AC-cuts 253
    AT-cuts 101, 118–120
    BT-cuts 101, 118–120
    X-cuts 253
Quasicrystals 202

Ray directions 270, 278, 279
Reflectivity 291, 292
Refractive indices 275–278, 280–282, 286, 287, 291, 293, 295, 300, 301, 304, 307 308, 317, 319, 320, 326, 327, 350, 351
Retger's Rule 8

Saturation magnetization 131–133, 156, 169, 241, 345, 346
Second harmonic generation 316–318, 351, 352
Secondary ferroics 2, 163, 177, 179, 181–183
Seebeck Effect 234–236, 242
    Absolute Seebeck Effect 238
Semiconductors 194–200, 204, 228, 229, 238, 239, 262, 264, 286
Solid solutions 191, 209, 219, 219
Soret experiment 221
Specific heat 1, 38, 43–48, 57
Spiral heat flow 204
Standard settings 26, 367
Stereographic projections 15, 52, 363
Strain gages 243, 245
Stress 32, 37, 72–74, 103
    Transformation 74
Strain 75–77, 103
    Transformation 77
Stress tensor 72, 73, 76, 87, 244
Suhl Effect 228
Superinsulators 206

Superionic conductors 219, 220
Surface structure 360, 361
    Catalysis 359
    Etch Figures 361–363
    Micromachined Crystals 363–366
    Surface Symmetry 361
    Tensor Description 366–368
Surface symmetry 276, 361
Surface waves 261, 265, 368
Symmetry 14–22

Tammann's Rules 213
Temperature dependence 44, 49, 55–57, 79, 85, 86, 132, 202, 207–210
Tensor nature 58, 203–206, 343–345
Tensor relations 88, 220, 239, 240, 315, 348
Thermal conductivity 2, 30–32, 188, 191, 203–211, 221, 223, 232, 233, 236–241, 264, 347
    Experiments 203, 204, 205
    Mechanisms 203
    Temperature dependence 207, 208–210
    Tensor nature 203–206
Thermal diffusion 2, 9, 221, 222
Thermal expansion 1, 33, 35, 38, 42, 44, 49, 54, 55, 68, 79–86, 119, 191, 293, 341
    Effect of Symmetry 79, 80
    measurements 81, 82
    Structure-property relations 82–85
    Temperature Dependence 85, 86
    Volume expansivity 82
Thermistors 199, 200–202
    $BaTiO_3$ 200, 202
    ceramic 200, 201
    NTC 200, 201
    PTC 200, 201
    temperature 200
Thermocouples 205, 234, 237, 238
    Cu-Fe 234
Thermodynamic relations 37–43, 82, 133, 186, 236
Thermoelectricity 234–242
    Devices
        Seebeck and Peltier 238
        thermoelectric 238, 239
    Efficiency 238
        Carnot 238, 239
        Greater 238
    Ferromagnetic crystals 241
    Magnetic field dependence 240–242
    Kelvin Relations 235–237
    Peltier Effect 235, 237
    Practical materials 238, 239
    Seebeck Effect 234, 238, 241
    Tensor relations 239, 240
    Thomson Effect 235, 236

Thermomagnetic effects 232–233
  Maggi-Righi-Leduc Effect 233
  Magnetothermal resistance 232, 233
  Righi-Leduc Effect 233
  Thermal Hall Effect 232, 233
  Thermal planar Hall Effect 232
Thermo-optic Effect 292, 293
Thermopower 236–242
Third harmonic generation 315, 322–324
Time Reversal 124, 125

Thomson Effect 234–236
Transformations 1, 9–13
  Axis 9
Transport numbers 218, 220
  Cross-coupled effects 2, 223, 241
  Nernst-Einstein Equation 217
  Pressure dependence 150, 237
  Superionic conductors 219, 220
  Temperature dependence 44, 49, 55–57, 79, 85, 86, 132, 202, 207, 208–210

Varistors 199–202
Vegard's Law 8
Voigt-Reuss-Hill average 114–116

Wave normals 252, 254, 259, 270, 274, 276, 278–280
Wave velocity surfaces 275, 279, 287, 301
Wiedemann-Franz Law 207

Printed and bound by CPI Group (UK) Ltd, Croydon, CR0 4YY